Color Atlas of Genetics

Color Atlas of Genetics

Eberhard Passarge

174 color plates by Jürgen Wirth

1995
Georg Thieme Verlag Stuttgart · New York
Thieme Medical Publishers, Inc. New York

Eberhard Passarge, M.D.
Director, Institute of Human Genetics
University of Essen

Color plates by Jürgen Wirth
Professor of Visual Communication
Fachhochschule Darmstadt

English version prepared by Mary Fetter Passarge, M.D. and Eberhard Passarge, M.D.

Library of Congress Cataloging-in-Publication Data

Passarge, Eberhard.
[Taschenatlas der Genetik. English]
Color atlas of genetics / Eberhard Passarge : 174 color plates by Jürgen Wirth : English version prepared by Mary F. Passarge and Eberhard Passarge.
p. cm.
Includes bibliographical references and index.
1. Genetics--Atlases. 2. Medical genetics--Atlases. I. Title.
[DNLM: 1. Genetics, Medical--atlases. QZ 17 P286t 1995a]
QH436.P3713 1995
575.1'022'2--dc20
DNLM/DLC
for Library of Congress 95-12781

This book is an authorized translation of the German edition with revisions, published and copyrighted 1994 by Georg Thieme Verlag, Stuttgart, Germany.
Title of the German edition: Taschenatlas der Genetik

Rüdigerstraße 14, D-70469 Stuttgart
Thieme Medical Publishers, Inc., 381 Park Avenue South, New York, NY 10016

Typesetting by Druckhaus Götz, D-71636 Ludwigsburg
(CCS-Textline [Linotronic 630])

Printed in Germany by Appl, D-86650 Wemding

ISBN 3-13-100361-8 (GTV, Stuttgart)
ISBN 0-86577-587-7 (TMP, New York)

Preface

Genetics is a rapidly developing science. Knowledge about the structure and function of genes is accumulating at an accelerating pace. New genes in man, mouse, drosophila, yeast, worms, bacteria, and plants are being discovered almost daily. The genome, the entirety of all genes and genetic material of an organism, is being studied in depth in several organisms, including man. It is becoming increasingly difficult to obtain an overview of the many aspects of the field, its fundaments and methods, and its goals. Genetics has become especially important in medicine. An increasing number of diseases can now be understood by genetic analysis. This contributes to improved precision in diagnosis and prognosis, helps to counsel patients and their relatives, and provides an outlook toward treatment. This Color Atlas of Genetics aims at showing the current status of genetics, in particular the interaction between genetics and medicine.

In contrast to a standard textbook, an atlas consists mainly of maps or illustrations. In 1594 Mercator first used the word "atlas" for a collection of maps. Today, the term "atlas" in its current context does not refer just to maps, but to illustrations in general. The visual presentation of a field has the advantage that complex knowledge can be made more readily comprehensible. The disadvantage is a tendency toward fragmentation and oversimplification. Thus, this atlas should be considered a supplement to a textbook, not a substitute.

This book is addressed to students of the biological, premedical, and medical sciences. Physicians and biologists, and other persons interested in genetics may find it useful for an orientation. It conveys recent advances, which contribute with increasing importance to an understanding of living organisms and disease processes.

The contents are presented in three parts, (1) fundamentals of genetics (100 plates), (2) functional areas with emphasis on disease processes (64 plates), and (3) genome organization (10 plates). The last section contains maps of selected human genetic diseases for which the cause has been localized on a particular chromosome. The introduction gives an overview of the development of genetics since the beginning of this century. A list of references provides access to additional information. When space allowed, the text to individual plates also contains references. A definition of genetic terms (glossary) at the end should help to understand genetic terms.

The color plates are the main vehicle to present the contents. They have been designed to present a particular set of concepts or facts, divided into related sections within each plate. Each plate is accompanied by a closely corresponding text, beginning with a brief introduction to the problem and then describing each section of the plate. A color code is used within a comparable context. Thus, a similar color is a signal of similarity of a structure shown in other plates. For example, coding DNA is dark blue, noncoding DNA is light blue, RNA turquoise green, a gene product usually brown, etc. This should help the reader to remain oriented.

In a field as wide as genetics, it becomes essential to select the material that is to be included. I have made choices according to my experience of nearly 20 years in teaching medical students at the beginner and the advanced level, as well as my own interest in the field. Personal contacts with many colleagues in the field of human genetics also have influenced my decisions. Particular emphasis is on the close connection between theoretical and methodological fundaments in genetics and their application in medical genetic diagnosis. The genetic features of groups of human disorders and selected diseases are included to illustrate how intertwined genetics and medicine are today. The reader will have to select topics of interest and bypass others. In fact, each plate and its text is a small chapter of its own.

The design of the color plates is the result of a close and pleasant cooperation with Jürgen Wirth, Professor of Visual Communication in the Faculty of Design at the Fachhochschule Darmstadt. Jürgen Wirth has created all color plates as computer graphics, based on sketches, hand drawings, and graphic and photographic material submitted by the author. Each plate was structured according to didactic and graphic principles. To my knowledge, this is the first visual presentation in genetics based entirely on computer graphics. This is a new approach and requires adaption to the possibilities of visual design. Precision and adjustment to narrow space are indispensable. For this highly qualified work and many constructive suggestions I am deeply indebted to Jürgen Wirth. I also thank Mrs. Vera Wirth for the work required to enter the English text into the computer graphics.

I especially thank my wife, Mary Fetter Passarge, M.D., for careful work in the preparation of the American edition and many helpful suggestions. A number of persons at the Georg Thieme Verlag Stuttgart have helped to produce this book. The German version was supported by Dr. D. Bremkamp in the beginning and subsequently by Mrs. Margarete Hieber. The English version was done under the guidance of Dr. C. Bergman and Mrs. Allyson Famous–Thierauf. Mr. Fleischmann and co-workers I would like to thank for the excellent technical work.

Many colleagues both within and outside of my department have given useful suggestions. In this regard, I particularly thank Drs. G. Obe (Essen), B. Horsthemke (Essen), D. Lohmann (Essen), G. Scherer (Freiburg), and the late W. Lenz (Münster).

I thank the following colleagues for providing material of their own work for use in some of the plates: Drs. D. Bootsma (Rotterdam), K. Buiting (Essen), T. Cremer (Heidelberg), W. Engel (Göttingen), W. Friedl (Bonn), G. Gillessen-Kaesbach (Essen), R. Gold (Würzburg), W. Höpping (Essen), B. Horsthemke (Essen), N. Konietzko (Essen), D. Lohmann (Essen), H.J. Lüdecke (Essen), V.A. McKusick (Baltimore), L. Manuelidis (New Haven), R.K. Moyzes (Los Alamos), C.R. Müller-Reible (Würzburg), U. Pfeiffer (Bonn), P. Propping (Bonn), H. Schuster (Berlin), P. Steinbach (Ulm), and E. Zrenner (Tübingen).

Essen, Spring 1995

E. Passarge

Table of Contents (Overview)

Table of Contents in Detail

Fundamentals

Genome Organization

Introduction

The cells of living organisms contain a program that regulates their many functions. This program controls cell division; embryonic development with organ and tissue differentiation; energy production from nutrients; muscle contraction; maintenance of bone and other supporting tissues; nerve, brain, and sensory organ functions; immune defense; the production and regulation of hormones and other intra- and extracellular signals; the synthesis, transport, and degradation of biomolecules; and many other processes.

This program is genetically determined, which means that at each cell division it is transferred to both of the newly formed cells. Malfunctions result when the maintenance or transfer does not occur precisely. Since no biological system functions without fault, mistakes do, in fact, accompany all life processes, infrequently but with regularity. However, most, but not all, are corrected by a complex repair system.

The genetic program consists of discrete units of information, the genes. Each gene controls a defined function. Genes are in the nucleus of every cell, located on the chromosomes. The genes of unicellular organisms control cell division and interaction with the environment. Since the cells of multicellular organisms differentiate into tissues with different functions, only part, and not all, of their genetic program is active in any one given cell. The number of genes in different organisms varies with the degree of complexity of the organism. Bacteria possess a few thousand genes; mammals are thought to have up to about 100 000 genes.

Genes are arranged linearly on the chromosomes. Each gene has a defined position and structure. The information contained in a gene consists of a series of nucleotide bases in specific sequences, similar to letters in a text. The text is coded, and it must be transcribed and translated to produce a molecule (the gene product) corresponding to the text information.

Many genes are surprisingly similar, even in very distantly related organisms. This is due to their origin from a common ancestor. The differences among organisms, which arose during evolution, are the result of adaption to different environmental conditions (Dawkins 1989; Dobzhansky 1970; Mayr 1982; Stebbins 1982). In particular, genes with an important function have been preserved with little change. For this reason, the study of genes is no longer limited to one particular species, but embraces all.

Genetics is the science concerned with the structure and function of all genes in different organisms (analysis of biological variation). New investigative methods and observations, especially during the last 5 to 10 years, have helped genetics to develop into a field in the mainstream of biology and medicine. Today, genetics plays a central, unifying role comparable to that of cellular pathology at the beginning of this century. Genetics is relevant to virtually all medical specialties. Knowledge of basic genetic principles and their application in diagnosis are becoming an essential part of medical education today.

Classical Genetics Between 1900 and 1953

(see chronological table on p. 12)

In 1906, the English biologist William Bateson (1861–1926) proposed the term genetics for the new biological field devoted to investigating the rules governing heredity and variation. Bateson referred to heredity and variation when comparing the similarities and differences, respectively, of genealogically related organisms, two aspects of the same phenomenon. Bateson clearly recognized the significance of the Mendelian laws, which had been rediscovered in 1900 by Correns, Tschermak, and DeVries.

The Mendelian laws are named for the Augustinian monk Gregor Mendel (1822–1884), who conducted crossbreeding experiments on garden peas in his monastery garden in Brünn (Brno, Czech Republic) well over a century ago. In 1866, Mendel wrote that heredity is based on individual, defined factors that are independent of each other (see Brink and Styles 1965;

Mayr 1982). Transmission of these factors to the next plant generation, i.e., the distribution of different traits among the offspring, occurred in predictable proportions. Each factor was responsible for a certain trait. The term gene for such a heritable factor was introduced in 1909 by the Danish biologist Wilhelm Johannsen (1857–1927).

Johann Gregor Mendel

Starting in 1902, Mendelian inheritance was systematically analyzed in animals, plants, and also in man. Some human diseases were recognized as having a hereditary cause. A form of brachydactyly (type A1, McKusick Nr. 112500) observed in a large Pennsylvania sibship by W. C. Farabee (PhD thesis, Harvard University, 1902) was the first condition in man to be described as being transmitted by autosomal dominant inheritance (D. V. Haws & V. A. McKusick, Bull. Johns Hopkins Hosp. 16: 20–30, 1963).

In 1909, Archibald Garrod (1857–1936), later Regius Professor of Medicine at Oxford University, demonstrated that four congenital metabolic diseases (albinism, alkaptonuria, cystinuria, and pentosuria) are transmitted by autosomal recessive inheritance (A. Garrod: Inborn Errors of Metabolism, 1909). Garrod was the first to recognize that there are biochemical differences among individuals that do not lead to illness but that also have a genetic basis. However, the relationship of genetic and biochemical findings revealed by this concept was ahead of its time: The far-reaching significance for the genetic individuality of man was not recognized (Bearn 1993). Certainly part of the reason was that the nature of genes and how they function was completely unclear. Early genetics was not based on chemistry or on cytology (Dunn 1965; Sturtevant 1965). Chromosomes in mitosis (Flemming 1879) and meiosis (Strasburger 1888) were observed; the term chromosome was coined by Waldeyer in 1888, but a functional relationship between genes and chromosomes was not considered. An exception was the prescient work of Theodor Boveri (1862–1915) about the genetic individuality of chromosomes (1902).

Genetics became an independent scientific field in 1910 with the study of the fruit fly (*Drosophila melanogaster*) at Columbia University in New York by Thomas H. Morgan. Subsequent systematic genetic studies on *Drosophila* over many years (Dunn 1965; Sturtevant 1965; Whitehouse 1973) showed that genes are arranged linearly on chromosomes. This led to the chromosome theory of inheritance (Morgan 1915). However, even into the 1920s, a few scientists asserted that chromosomes had nothing to do with heredity.

The English mathematician Hardy and the German physician Weinberg recognized that Mendelian inheritance accounts for certain regularities in the genetic structure of populations (1908). Their work contributed to the successful introduction of genetic concepts into plant and animal breeding. Although genetics was well established as a biological field by the end of the third decade of this century, knowledge of the physical and chemical nature

Thomas H. Morgan

of genes was sorely lacking. Structure and function remained unknown.

That genes can change and become altered was recognized by DeVries in 1901. He introduced the term mutation. In 1927, H. J. Muller determined the spontaneous mutation rate in drosophila and demonstrated that mutations can be induced by roentgen rays. C. Auerbach and J. M. Robson (1941) and, independently, F. Oehlkers (1943) observed that certain chemical substances also could induce mutations. However, it remained unclear what a mutation actually was since the physical basis for the transfer of genetic information was not known.

The complete lack of knowledge of the structure and function of genes has contributed to misconceptions in the 1920s and 30s about the possibility of eliminating "bad genes" from human populations (eugenics). However, modern genetics has shown that the poorly accepted eugenic approach to eliminating human genetic disease is also ineffective. For a review of the astrocious misuse of eugenics by the National Socialist Government in Germany 1933–45 see Becker (1988), Müller-Hill (1984), Vogel & Motulsky (1986).

Thus, incomplete genetic knowledge was applied to human individuals at a time when nothing was known about the structure of genes. Indeed, up to 1949 no essential genetic findings had been gained from studies in man. Quite the opposite holds true today.

Today, it is evident that genetically determined diseases generally cannot be eradicated. No one is free from a genetic burden. Every individual carries about five or six severe genetic defects that are inapparent, but that may show up in offspring.

With the demonstration in the fungus *Neurospora* that one gene is responsible for the formation of one enzyme ("one gene, one enzyme," Beadle and Tatum 1941), the close relationship of genetics and biochemistry became apparent, quite in agreement with Garrod's concept of inborn errors of metabolism. Systematic studies in microorganisms led to other important advances in the 1940s: Genetic recombination was demonstrated in bacteria (Lederberg and Tatum 1946) and viruses (Delbrück and Bailey 1947). Spontaneous mutations were observed in bacterial viruses (bacteriophages, Hershey 1947). The study of genetic phenomena in microorganisms turned out to be as significant for the further development of genetics as the analysis of *Drosophila* had been 35 years earlier (for review, see Cairns, Stent, and Watson, 1994). A very influential, small book entitled "What Is Life?" by the physicist E. Schrödinger defined genes in molecular terms (1944). At that time, elucidation of the molecular biology of the gene became a central theme in genetics.

Genetics and DNA

A major advance occurred in 1944 when Avery, McLeod, and McCarty at the Rockefeller Institute in New York demonstrated that a chemically relatively simple long-chained nucleic acid (deoxyribonucleic acid, DNA) carried genetic information in bacteria (for historical review, see Dubos 1976; McCarty 1985). Many years earlier,

F. Griffith (1928) had observed that permanent (genetic) changes can be induced in pneumococcal bacteria by a cell-free extract derived from other strains of pneumococci ("transforming principle"). Avery and his co-workers showed that DNA was this transforming principle. In 1952, Hershey and Chase proved that genetic information is transferred by DNA alone. With this knowledge, the question of its structure became paramount.

This was resolved most elegantly by James D. Watson, a 24-year-old American on a scholarship in Europe, and Francis H. Crick, a 36-year-old English physicist, at the Cavendish Laboratory of the University of Cambridge. Their findings appeared in a three-quarter page article on April 25, 1953 in *Nature* ("A structure for deoxyribonucleic acid," Nature. 1953; 171: 737). In this famous article, Watson and Crick proposed that the structure of DNA is a double helix. The double helix is formed by two complementary chains with oppositely oriented alternating sugar (deoxyribose) and monophosphate molecules. Inside this helical molecule lie paired nucleotide bases, each pair consisting of a purine and a pyrimidine. The crucial feature is that the base pairs lie inside the molecule, not outside. This insight came from construction of a model of DNA that took into account stereochemical considerations and the results of previous X-ray diffraction studies by M. Wilkins and R. Franklin. That the authors fully recognized the significance for genetics of the novel structure is apparent from the closing statement of their article, in which they state, "It has not escaped our notice that the specific pairing we have postulated immediately suggests a possible copying mechanism for the genetic material." Vivid, albeit different, accounts of their discovery have been given by the authors (Watson 1968; Crick 1988).

The elucidation of the structure of DNA is regarded as the beginning of a new era of molecular biology and genetics. The description of DNA as a double-helix structure led directly to an understanding of the possible structure of genetic information.

When F. Sanger determined the sequence of amino acids of insulin in 1955, he provided the first proof of the primary structure of a protein. This supported the notion that the sequence of amino acids in proteins could correspond to the sequential character of DNA. However, since DNA is located in the cell nucleus and protein

Oswald T. Avery

Watson (left) and Crick (right)

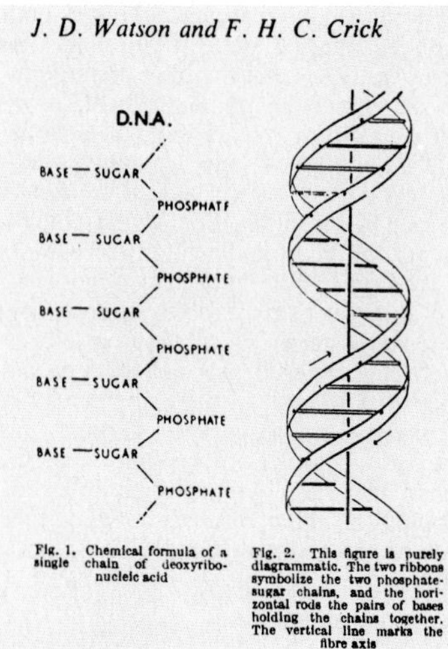

DNA-Struktur

synthesis occurs in the cytoplasm, DNA could not act directly. It turned out that DNA is first transcribed into a chemically similar messenger molecule (messenger ribonucleic acid, mRNA) (Crick, Barnett, Brenner, Watts-Tobin 1961) with a corresponding nucleotide sequence, which is transported into the cytoplasm. In the cytoplasm the mRNA then serves as a template for the amino acid sequence to be formed. The genetic code for the synthesis of proteins from DNA and messenger RNA was determined in the years 1963–1966 (Nirenberg, Mathaei, Ochoa, Benzer, Khorana, and others). Detailed accounts of these developments have been presented by Chargaff (1978), Judson (1979), Stent (1981), Watson & Tooze (1981), Crick (1988), Brown (1992), and others.

Important Methodological Advances in the Development of Genetics after About 1950

From the beginning, genetics has been a field strongly influenced by the development of new experimental methods. In the 1950s and 1960s, the ground work was laid for biochemical genetics and immunogenetics. Relatively simple but reliable procedures to separate complex molecules by different forms of electrophoresis, methods to synthesize DNA in vitro (Kornberg 1956), and other approaches were applied to questions in genetics. The development of cell culture methods was of particular importance for the genetic analysis of humans. Pontecorvo introduced the genetic analysis of cultured eukaryotic cells (somatic cell genetics) in 1958. The study of mammalian genetics, with increasing significance for studying human genes, was facilitated by methods to fuse cells in culture (cell hybridization, T. Puck, G. Barski, B. Ephrussi, 1961) and the development of a cell culture medium for selecting certain mutants in cultured cells (HAT medium, J. Littlefield 1964). The genetic approach that had been so successful in bacteria and viruses could now be applied in higher organisms, thus avoiding the obstacles of a long generation time and breeding experiments. A hereditary metabolic defect of man (galactosemia) was demonstrated for the first time in cultured human cells in 1961 (Krooth). The correct number of chromosomes in man was determined in 1956 (Tjio & Levan; Ford & Hamerton). Lymphocyte cultures were introduced for chromosomal analysis (Hungerford et al. 1960). The replication pattern of human chromosomes was described (J. German 1962). These developments further paved the way for expansion of the new field of human genetics.

Human Genetics

The medical aspects of human genetics (medical genetics) came to attention when it was recognized that sickle cell anemia is hereditary (Neel 1949) and caused by a defined alteration of normal hemoglobin (Pauling, Itano, Singer, and Wells 1949) and again when it was shown that an enzyme defect (glucose-6-phosphate deficiency, demonstrated in liver tissue by

Cori and Cori in 1952) was the cause of a hereditary metabolic disease in man (glycogen storage disease type I, or von Gierke disease). The American Society of Human Genetics and the first journal of human genetics (American Journal of Human Genetics) were established in 1949. In addition, the first textbook of human genetics appeared (Curt Stern, Principles of Human Genetics, 1st edn., 1949).

In 1959, chromosomal aberrations were discovered in some well-known human disorders (trisomy 21 in Down syndrome by J. Lejeune, M. Gautier, R. Turpin; 45,X0 in Turner syndrome by Ford et al.; 47,XXY in Klinefelter syndrome by Jacobs & Strong). Subsequently, other numerical chromosome aberrations were shown to cause recognizable diseases in man (trisomy 13 and trisomy 18, by Patau et al. and Edwards et al. 1960, respectively), and loss of small parts of chromosomes were shown to be associated with recognizable patterns of severe developmental defects (Lejeune et al. 1963, Wolf 1964, Hirschhorn 1964). The Philadelphia chromosome, a characteristic structural alteration of a chromosome in bone marrow cells of patients with adult type chronic myelogenous leukemia, was described by Nowell and Hungerford in 1962. The central role of the Y chromosome in establishing gender in mammals became apparent when it was realized that individuals without a Y chromosome are female and individuals with a Y chromosome are male, irrespective of the number of X chromosomes present. These observations further promoted interest in a new subspecialty, human cytogenetics.

Since early 1960, important knowledge about genetics in general has been obtained, often for the first time, by studies in man. Analysis of genetically determined diseases in man has yielded important insights into the normal function of genes in other organisms as well. Today, more is known about the general genetics of man than about that of any other species. Numerous subspecialties of human genetics have arisen, such as biochemical genetics, immunogenetics, somatic cell genetics, cytogenetics, clinical genetics, population genetics, teratology, mutational studies, and others. The development of the field has been well summarized by Vogel & Motulsky (1986) and McKusick (1992).

Genetics and Medicine

Most disease processes can be viewed as resulting from environmental influences interacting with the individual genetic makeup of the affected individual. A disease is genetically determined if it is mainly or exclusively caused by disorders in the genetic program of cells and tissues. More than 3000 defined human genetic diseases are known to be due to a mutation at a single gene locus (monogenic disease) and to follow a Mendelian mode of inheritance (McKusick 1994). They differ as much as the genetic information in the genes involved and may be manifest in essentially all age groups and organ systems. An important category of disease results from genetic predisposition interacting with precipitating environmental factors (multigenic or multifactorial diseases). This includes many relatively common chronic diseases (e.g., high blood pressure, hyperlipidemia, diabetes mellitus, gout, psychiatric disorders, certain congenital malformations). Further categories of genetically determined diseases are nonhereditary disorders in somatic cells (different forms of cancer) and chromosomal aberrations (see Weatherall 1991; Lindsten & Pettersson 1991; Scriver et al. 1995).

Due to new mutations, the laws of Mendelian inheritance, and small family size in developed countries, genetic disorders usually do not affect more than one member of a family. About 90% occur as isolated cases within a family. Thus, their genetic origin cannot be recognized by familial aggregation. Instead, they must be recognized by their clinical features. This may be difficult in view of the many different functions of genes in normal tissues and in disease. Since genetic disorders affect all organ systems and age groups and are frequently not recognized, their contribution to the causes of human diseases appears smaller than it actually is. Genetically determined diseases are not a marginal group, but make up a substantial proportion of diseases. More than one third of all pediatric hospital admissions are for diseases and developmental disorders that at least in part are caused by genetic factors (Weatherall 1991). The total estimated frequency of genetically determined diseases of different categories in the general population is about 3.5–5.0% (see Table 1).

The large number of individually rare genetically determined diseases and the overlap of diseases with similar clinical manifestations but different etiology (principle of genetic or etiological heterogeneity) cause additional diagnostic difficulties. This must be considered during diagnosis to avoid false conclusions about a genetic risk.

In 1966 Victor A. McKusick introduced a catalog of human phenotypes transmitted according to Mendelian inheritance (McKusick catalog, currently in its 11th edition; McKusick 1994). This catalog and the 1968–1973 Baltimore Conferences organized by McKusick (Clinical Delineation of Birth Defects) have contributed substantially to the systematization and subsequent development of medical genetics. The extent of medical genetics is reflected by the initiation since 1965 of several new scientific journals (Clinical Genetics, Journal of Medical Genetics, Human Genetics, Annales des Génétique, American Journal of Medical Genetics, Cytogenetics and Cell Genetics, European Journal of Human Genetics, Prenatal Diagnosis, Clinical Dysmorphology, and others).

In recent years, considerable, previously unexpected progress in clarifying the genetic etiology of human diseases, and thereby in furnishing insights into the structure and function of normal genes, has been achieved by molecular methods.

Molecular Genetics

The discovery in 1970 (independently by H. Temin and D. Baltimore) of reverse transcriptase, an unusual enzyme complex in RNA viruses (retroviruses), upset the dogma—valid up to that time—that the flow of genetic information went in one direction only, i.e., from DNA to RNA and from there to the gene product (a peptide). Not only is the existence of reverse transcriptase an important biological finding, but the enzyme provides a means of obtaining complementary DNA (cDNA) that corresponds to the coding regions of an active gene. Therefore, it is possible to analyze a gene directly without knowledge of its gene product, provided it is expressed in the tissue examined.

In addition, specific enzymes that cleave DNA at defined sites (restriction endonu-

Table **1** Frequency of genetically determined diseases

Type of genetic disease	Frequency per 1000 individuals
1. Monogenic diseases, total	4.5–15.0
Autosomal dominant	2–9.5
Autosomal recessive	2–3.5
X-chromosomal	0.5–2
2. Chromosomal aberrations	5–7
3. Multifactorial disorders*	7–10
4. Congenital malformations	19–22
Total	ca. 35–54

* Contribution of genetic factors variable (data based on Weatherall 1991)

cleases or, simply, restriction enzymes) were discovered in bacteria (W. Arber 1969; D. Nathans & H. O. Smith 1971). With appropriate restriction enzymes, DNA can be cut into pieces of reproducible and defined size, thus allowing easy recognition of a defined area to be studied. DNA fragments of different origin can be joined and their properties analyzed. Methods to produce multiple copies of DNA fragments and to sequence them (determine the sequence of their nucleotide bases) were developed between 1977 and 1985. These methods are collectively referred to as recombinant DNA technology (see chronology at the end of this introduction).

In 1977, recombinant DNA analysis led to a completely new and unexpected finding about the structure of genes in higher organisms, but also in yeast and drosophila: Genes are not continuous segments of coding DNA, but are usually interrupted by noncoding segments (see Watson & Tooze 1981; Watson et al., 1992). The size and sequence of coding DNA segments, or exons (a term introduced by Gilbert in 1978), and noncoding segments, or introns, is specific for each individual gene (exon/intron structure of eukaryotic genes).

Many genes have not arisen completely anew during evolution, but have been assembled from different parts of different genes. Most genes belong to structurally and functionally related groups (gene families), present in a wide spectrum of different organisms.

With the advent of molecular genetic DNA analysis, many different types of polymorphic DNA markers, i.e., individual heritable differences in the nucleotide sequence, have been mapped to specific sites on chromosomes (physical map). As a result, the chromosomal position of a gene of interest can now be determined (mapped) by analyzing the segregation of a disease locus in relation to the polymorphic DNA markers (linkage analysis). Once the chromosomal location of a gene is known, the latter can be isolated and its structure can be characterized (positional cloning, a term introduced by F. Collins). The advantage of such a direct analysis is that nothing must be known about the gene of interest aside from its approximate location. Prior knowledge of the gene product is not required.

Another, complementary approach is to identify a gene with possible functional relevance to a disorder (candidate gene), determine its chromosomal position, and then demonstrate mutations in the candidate gene in patients with the disorder. Positional cloning and identification of candidate genes have helped identify genes for many important diseases such as achondroplasia, degenerative retinal diseases, cystic fibrosis, Huntington chorea and other neurodegenerative diseases, Duchenne muscular dystrophy and other muscular diseases, mesenchymal diseases with collagen defects (osteogenesis imperfecta), Marfan syndrome (due to a defect of a previously unknown protein, fibrillin), immune defects, and numerous tumors.

The extensive homologies of genes that regulate embryological development in different organisms and the similarities of genome structures have contributed to leveling the boundaries in genetic analysis, which formerly existed for different organisms (e.g., *Drosophila* genetics, mammalian genetics, yeast genetics, bacterial genetics, etc.). Genetics has become a broad, unifying discipline in biology, medicine, and evolutionary research.

The Dynamic Genome

Between 1950 und 1953, remarkable papers appeared entitled "The origin and behavior of mutable loci in maize" (Proc Natl Acad Sci. 1950; 36: 344–355), "Chromosome organization and genic expression" (Cold Spring Harbor Symp Quant Biol. 1952; 16: 13–45), and "Introduction of instability at selected loci in maize" (Genetics. 1953; 38: 579–599). Here, the author Barbara McClintock of Cold Spring Harbor Laboratory described mutable loci in Indian corn plants (maize) and their effect on the phenotype of corn due to a gene that is not located at the site of the mutation. Surprisingly, this gene can exert a type of remote control. In addition, other genes can change their locations and cause mutations at distant sites.

In subsequent work, McClintock described the special properties of this group of genes, which she called controlling genetic elements (Brookhaven Symp Biol. 1955; 8: 58–74). Different controlling elements could be distinguished according to their effects on other genes and the mutations caused. However, her work received little interest (for review see Fox Keller 1983; Fedoroff & Botstein 1992).

Thirty years later, at her 1983 Nobel Prize lecture ("The significance of responses of the genome to challenge," Science. 1984; 226: 792–801), things had changed. Today we know that the genome is not rigid and static. Rather, it is flexible and dynamic because it contains parts that can move from one location to another (mobile genetic elements, the current designation). The precision of the genetic information depends on its stability, but complete stability would also mean static persistence. This would be detrimental to the development of new forms of life in response to environmental changes. Thus, the genome is subject to alterations, as life requires a balance between the old and the new.

References

Alberts B, Bray D, Lewis J, Raff M, Roberts K, Watson JD. Molecular Biology of the Cell. 3rd ed. New York: Garland; 1994.

Bearn AG. Archibald Garrod and the Individuality of Man. Oxford: Oxford University Press; 1993.

Becker PE. Zur Geschichte der Rassenhygiene. Wege ins Dritte Reich. Stuttgart: Thieme; 1988.

Brink RA, Styles ED, eds. Heritage from Mendel. Madison: University of Wisconsin Press; 1967.

Brown TA. Genetics: a molecular approach. 2nd ed. London: Chapman and Hall; 1992.

Cairns J, Stent GS, Watson JD. eds. Phage and the Origins of Molecular Biology. New York: Cold Spring Harbor Laboratory Press; 1978.

Chargaff E. Heraclitean Fire: Sketches from a life before nature. New York: Rockefeller University Press; 1978.

Crick F. What Mad Pursuit: A personal view of scientific discovery. New York: Basic Books; 1988.

Darnell J, Lodish H, Baltimore D. Molecular Cell Biology, 2nd ed. Scientific American Books. New York: Freeman; 1990.

Dawkins R. The Selfish Gene. 2nd ed. Oxford: Oxford University Press; 1989.

Dobzhansky T. Genetics of the Evolutionary Process. New York: Columbia University Press; 1970.

Dubos RJ. The Professor, the Institute, and DNA: O. T. Avery, his life and scientific achievements. New York: Rockefeller University Press; 1976.

Dunn LC. A Short History of Genetics. New York: McGraw-Hill; 1965.

Fedoroff N, Botstein D, eds. The Dynamic Genome: Barbara McClintock's ideas in the century of genetics. New York: Cold Spring Harbor Laboratory Press; 1992.

Fox Keller E. A Feeling for the Organism: the life and work of Barbara McClintock. New York: W. H. Freeman; 1983.

Hsu TC. Human and Mammalian Cytogenetics: a historical perspective. Springer Verlag, Heidelberg; 1979.

Judson HF. The Eighth Day of Creation. New York: Simon & Schuster; 1979.

Lindsten J, Pettersson U, eds. Etiology of Human Disease at the DNA Level. Nobel Symposium 80. New York: Raven Press; 1991.

Mayr E. The Growth of Biological Thought: diversity, evolution, and inheritance. Cambridge, Massachusetts: Harvard Univ. Press; 1982.

McCarty M. The Transforming Principle. New York: W. W. Norton; 1985.

McKusick VA. Presidential Address. Eighth International Congress of Human Genetics: The last 35 years, the present, and the future. Am J Hum Genet. 1992; 50: 663–670.

McKusick VA. Mendelian Inheritance in Man: A Catalog of Human Genes and Genetic Disorders. 11th ed. Baltimore: Johns Hopkins; 1994.

Müller-Hill B. Tödliche Wissenschaft: die Aussonderung von Juden, Zigeunern und Geisteskranken 1933–1945. Reinbek: Rowohlt; 1984.

Scriver CR, Beaudet AL, Sly WS, Valle D, eds. The Metabolic and Molecular Bases of Inherited Disease. 7th ed. New York: McGraw-Hill; 1995.

Schrödinger E. What Is Life? New York: Penguin Books; 1944.

Stebbins GL. Darwin to DNA: Molecules to Humanity. San Francisco: WH Freeman; 1982.

Stent GS. ed. James D. Watson: The Double Helix: a personal account of the discovery of the structure of DNA. A new critical edition including text, commentary, reviews, original papers. London: Weidenfeld & Nicolson; 1981.

Sturtevant AH. A History of Genetics. New York: Harper & Row; 1965.

Vogel F, Motulsky AG. Human Genetics: Problems and Approaches. 2nd ed. Heidelberg: Springer; 1986.

Watson JD. The Double Helix. London: Atheneum; 1968.

Watson JD, Tooze J. The DNA Story: a documentary history of gene cloning. San Francisco: WH Freeman; 1981.

Watson JD, Hopkins NH, Roberts JW, Steitz JA, Weiner AM. Molecular Biology of the Gene. 4th ed. Menlo Park, California: Benjamin/Cummings; 1987.

Watson JD, Gilman M, Witkowski J, Zoller M. Recombinant DNA. 2nd ed. Scientific American Books. New York: WH Freeman; 1992.

Weatherall DJ. The New Genetics and Clinical Practice, 3rd ed. Oxford: Oxford Univ. Press; 1991.

Whitehouse HLK. Towards an Understanding of the Mechanisms of Heredity. 3rd ed. London: Edward Arnold; 1973.

(additional references p. 370ff)

Chronology of Some Important Observations that Contributed to the Development of the Field of Genetics

(This list contains selected events and should not be considered complete, especially for the many important developments during the past several years.)

1839 Cells as the basis of living organisms (Schleiden, Schwann)

1859 Evolutionary concepts formulated (Charles Darwin)

1865 Mendelian laws (Gregor Mendel)

1879 Chromosomes in mitosis (Flemming)

1883 Quantitative aspects of heredity (F. Galton)

1897 Enzymes discovered (E. Büchner)

1900 AB0 blood group system

1902 Mendelian inheritance in man (Bateson, Garrod)
Individuality of chromosomes (Boveri)
Sex chromosomes (McClung)

1908 Population genetics (Hardy, Weinberg)

1909 Inborn errors of metabolism (Garrod)
Gene, genotype, phenotype (Johannsen)
Chiasma formation during meiosis (Janssens)

1910 Beginning of Drosophila genetics (Morgan)

1912 Crossing-over (Morgan & Cattell)
Genetic linkage (Morgan & Lynch)
Nondisjunction (Bridges)

1915 Genes located on chromosomes (chromosomal theory of inheritance) (Morgan, Sturtevant, Muller, Bridges)

1924 Blood group genetics (Bernstein)
Statistical analysis (Fisher)

1926 Enzymes are proteins (J. Sumner)

1927 Mutations induced by roentgen rays (Muller)
Genetic drift (Wright)

1928 Euchromatin/heterochromatin (Heitz)
Genetic transformation in pneumococci (Griffith)

1933 Pedigree analysis (Haldane, Hogben, Fisher, Lenz, Bernstein)
Polytene chromosomes (Heitz & Bauer, Painter)

1940 Genetic polymorphism (E. B. Ford)
Rhesus blood group (Landsteiner & Wiener)

1941 Evolution by gene duplication (E. B. Lewis)
Genetic control of biochemical reactions (Beadle & Tatum)
Mutation caused by mustard gas (Auerbach)

1944 DNA as the material basis of genetic information (Avery)

1946 Genetic recombination in bacteria (Lederberg & Tatum)

1947 Genetic recombination in viruses (Delbrück & Bailey, Hershey)

1949 Sickle cell anemia, a genetically determined molecular disease (Neel, Pauling)
X chromatin (Barr & Bertram)

1950 Nucleotide base relationship (Chargaff)

1952 Genes consist of DNA (Hershey & Chase)
Plasmids (Lederberg)
Transduction by phages (Zinder & Lederberg)
First enzyme defect in man (Cori & Cori)
First linkage group in man (Mohr)
Colchicine and hypotonic treatment in chromosomal analysis

(Hsu & Pomerat)
Exogenous factors as a cause of congenital malformations (Warkany)

1953 DNA structure (Watson & Crick, Franklin, Wilkins)
Non-Mendelian inheritance (Ephrussi)
Cell cycle (Howard & Pelc)
Dietary treatment of phenylketonuria (Bickel)

1954 DNA repair (Muller)
Leukocyte drumsticks (Davidson & Smith)
Turner syndrome is X-chromatin negative (Polani)

1955 Sequencing of insulin (F. Sanger)
Lysosomes (de Duve)
Buccal smear (Moore, Barr, Marberger)

1956 Chromosome number in man (Tijo & Levan, Ford & Hamerton)
DNA synthesis in vitro (Kornberg)
Genetic heterogeneity (Harris, Fraser)

1957 Hemoglobin molecule (Ingram)
Cistron (Benzer)
Genetic complementation (Fincham)
DNA replication is semiconservative (Taylor, Delbrück, Stent)
Genetic analysis of radiation effects in man (Neel & Schull)

1958 Somatic cell genetics (Pontecorvo)
Ribosomes (Roberts, Dintzis)
HLA antigens (Dausset)
Cloning of single cells (Sanford, Puck)
Synaptonemal complex (Moses)

1959 Chromosomal aberrations in man (Lejeune, Turpin, Jacobs, and others)
Isoenzymes (Vesell, Markert)
Pharmacogenetics (Motulsky, Vogel)

1960 Phytohemagglutinin-stimulated lymphocyte cultures (Nowell, Moorehead, Hungerford)

1961 Genetic code in triplets (Crick, Brenner, Barnett, Watts-Tobin)
Deciphering of the genetic code (Nirenberg, Mathaei, Ochoa)
Gene regulation, concept of operon (Jacob & Monod)
Galactosemia in cell culture (Krooth)
Cell hybridization (Barski, Ephrussi)

1962 X-chromosomal inactivation (Lyon, Beutler, Russell, Ohno)
Philadelphia chromosome (Nowell & Hungerford)
Blood group Xg (Mann, Race, Sanger)
PKU screening (Guthrie, Bickel)
Molecular characterization of immunoglobulins (Edelman, Franklin)
^{3}H-autoradiography of human chromosomes (German, Miller)
Replicon (Jacob & Brenner)
Thalidomide embryopathy (Lenz, McBride)

1963 Lysosomal storage diseases
Diseases due to chromosomal deletions

1964 Excision repair (Setlow)
MLC test (Bach & Hirschhorn, Bain & Lowenstein)
Microlymphotoxicity test (Terasaki & McClelland)
Selective cell culture medium HAT (Littlefield)
Spontaneous chromosomal instability (German, Schroeder)

1965 Limited life span of cells in culture (Hayflick & Moorehead)
Cell fusion with Sendai virus (Harris & Watkins)
Cell culture from amniotic fluid (Klinger)
Hereditary diseases in cell culture (Danes, Bearn, Krooth, Mellman)
Population cytogenetics (Court Brown)
Chromosomal aberrations in abortuses (Carr, Benirschke)

1966 Catalog of Mendelian phenotypes in man (McKusick)

1968 HLA-D the strongest histocompatability system (Ceppellini, Amos)
Repetitive DNA (Britten & Kohne)
Biochemical basis of the AB0 blood group substances (Watkins)
DNA excision repair defect in xeroderma pigmentosum (Cleaver)
Restriction endonucleases (Meselson & Yuan, Linn & Arber)
First assignment of an autosomal gene locus in man (Donahue, McKusick)

1970 Reverse transcriptase (Baltimore, Temin)
Synteny (Renwick)
Enzyme defects in lysosomal storage diseases (Neufeld, Dorfman)
Chromosome identification by stains producing banding patterns (Zech, Caspersson, Lubs, Drets & Shaw, Schnedl, Evans)
Y-chromatin (Pearson, Bobrow, Vosa)
Thymus transplantation for immune deficiency (van Bekkum)
Synthesis of a gene in vitro (Khorana)

1971 Two-hit theory of retinoblastoma (A. G. Knudson)

1972 High average heterozygosity (Harris & Hopkinson, Lewontin)
Association of HLA antigens and diseases

1973 Receptor defects in the etiology of genetic defects, genetic hyperlipidemia (Goldstein, Motulsky, Brown)
Demonstration of sister chromatid exchanges with BrdU (S. A. Latt)
Philadelphia chromosome as translocation (J. D. Rowley)

1974 Chromatin structure, nucleosome (Kornberg, Olins & Olins)
First cloning of a eukaryotic gene

1975 Asilomar conference
Monoclonal antibodies (Köhler & Milstein)

1976 Overlapping genes in phage φX174 (Barell, Air, Hutchison)
Loci for structural genes on each human autosome known (Baltimore Conference on Human Gene Mapping)

1977 Genes consist of coding and noncoding DNA segments (exon/intron structure)
First recombinant DNA molecule that contains mammalian DNA
Methods to sequence DNA (Sanger; Maxam & Gilbert)
X-ray diffraction analysis of nucleosomes (Finch et al.)

1978 β-Globin gene structure (Leder, Weissmann, Tilghman, and others)
Mechanisms of transposition in bacteria. Production of somatostatin with recombinant DNA

1979 First diagnosis with DNA technology (Y. H. Kan)
Total synthesis of a gene in vitro (Khorana)

1981 Sequencing of a mitochondrial genome (Anderson et al.)

1982 Tumor-suppressor genes (H. P. Klinger)

1983 Cellular oncogenes (H. E. Varmus and others)

1984 Localization of the gene for Huntington disease (Gusella)
Genetic imprinting in the mouse
Identification of T-cell receptor proteins

1985 Polymerase chain reaction (Saiki, Mullis et al.)
Characterization of the gene for clotting factor VIII (Gitschier)
Sequencing of the AIDS virus
Localization of the gene for cystic fibrosis
Hypervariable DNA segments

1986 Cloning of human genes
First structural analysis of a human gene based on its chromosomal location (Royer-Pokora et al.)
Genes for the embryonic development of drosophila
RNA as catalytic enzyme (T. Cech)

1987 Structure of an HLA molecule (Björkman, Strominger et al.)
Cloning of the gene for Duchenne muscular dystrophy (Kunkel)

1988 Start of the Human Genome Project
Successful gene therapy in vitro
Telomere structure of chromosomes

1989 Cloning of a defined region of a human chromosome obtained by microdissection
DNA microsatellites as frequent variants in the genome

1991 Cloning of the gene for cystic fibrosis
Increasing use of microsatellites as polymorphic DNA markers

1992 Trinucleotide expansion as a new class of pathogenic mutations
Complete map of chromosome 21 and the Y chromosome of man
X chromosome inactivation center identified
Complete sequence of a yeast chromosome

1993 Cloning of the gene for Huntington disease

1994 Physical map of the human genome in high resolution
Mutations in fibroblast growth factor receptor genes as cause of achondoplasia and other human diseases
Identification of genes for breast cancer

Fundamentals

The Cell and Its Components

Cells are the smallest structural and functional units of a multicellular organism. Although variously specialized, they are similar with regard to structure and function. Basically, there are two different types of cell structure: prokaryotic and eukaryotic. Prokaryotic cells have no nucleus. They include all bacteria and the blue-green algae. Plants, animals, true algae, fungi, and protozoa all are made up of eukaryotic cells with nuclei. Eukaryotic cells contain membranes that form functional compartments within the cell.

A. Eukaryotic cells

A eukaryotic cell consists of cytoplasm and a nucleus. It is enclosed by a plasma membrane. The cytoplasm contains a complex system of inner membranes that form cellular structures (organelles). The main organelles are the mitochondria (in which important energy-delivering chemical reactions take place), the endoplasmic reticulum (consisting of a series of membranes in which glycoproteins and lipids are formed), the Golgi apparatus (for certain transport functions), and peroxisomes (for the formation or degradation of certain substances). Eukaryotic cells contain lysosomes, in which numerous proteins, nucleic acids, and lipids are broken down. The cytoplasm contains a cytoskeleton of different fibrous proteins that lend the cell stability. Centrioles, small cylindrical particles made up of microtubules, play an essential role in cell division. Ribosomes are the sites of protein synthesis.

B. Cell nucleus

The eukaryotic cell nucleus contains the genetic information. It is enclosed by an inner and an outer membrane, which contain pores for the transport of substances between the nucleus and the cytoplasm. The nucleus contains a nucleolus and a fibrous matrix with different DNA–protein complexes. The nucleolus is the site of synthesis of most of the ribosomal RNA. The chromosomes, situated in the nucleus, are not visible individually, but only as a diffuse network (chromatin). They contain the genes.

C. Plasma membrane of the cell

The plasma membrane is composed of a double layer of phospholipids (lipid bilayer). It contains a series of channels made up of special proteins (protein channels) for the transport of substances into or out of the cell. The transport requires an active process (protein pump). The plasma membrane of a eukaryotic cell contains numerous transmembrane proteins that fulfill essential functions as cell surface receptors.

D. Comparison of animal and plant cells

Plant and animal cells have many similar characteristics. One fundamental difference is that plant cells contain chloroplasts for photosynthesis. In addition, plant cells are surrounded by a rigid wall of cellulose and other polymeric molecules and contain vacuoles for water, ions, sugar, nitrogen-containing compounds, or waste products. Vacuoles are enclosed in semipermeable membranes that are permeable to water, but not to the other substances enclosed in the vacuoles. When water enters, the vacuole expands and produces hydrostatic pressure (turgor) within the cell. Since animal cells have no vacuoles to take up additional water molecules, they swell in a hypotonic solution and shrink in a hypertonic solution.

References

Alberts B, et al. Molecular Biology of the Cell. 3rd ed. New York: Garland; 1994.

Darnell T, Lodish H, Baltimore D. Molecular Cell Biology. 2nd ed. New York: WH Freeman; 1990.

C. de Duve. Die Zelle. Spektrum der Wissenschaft, 1986.

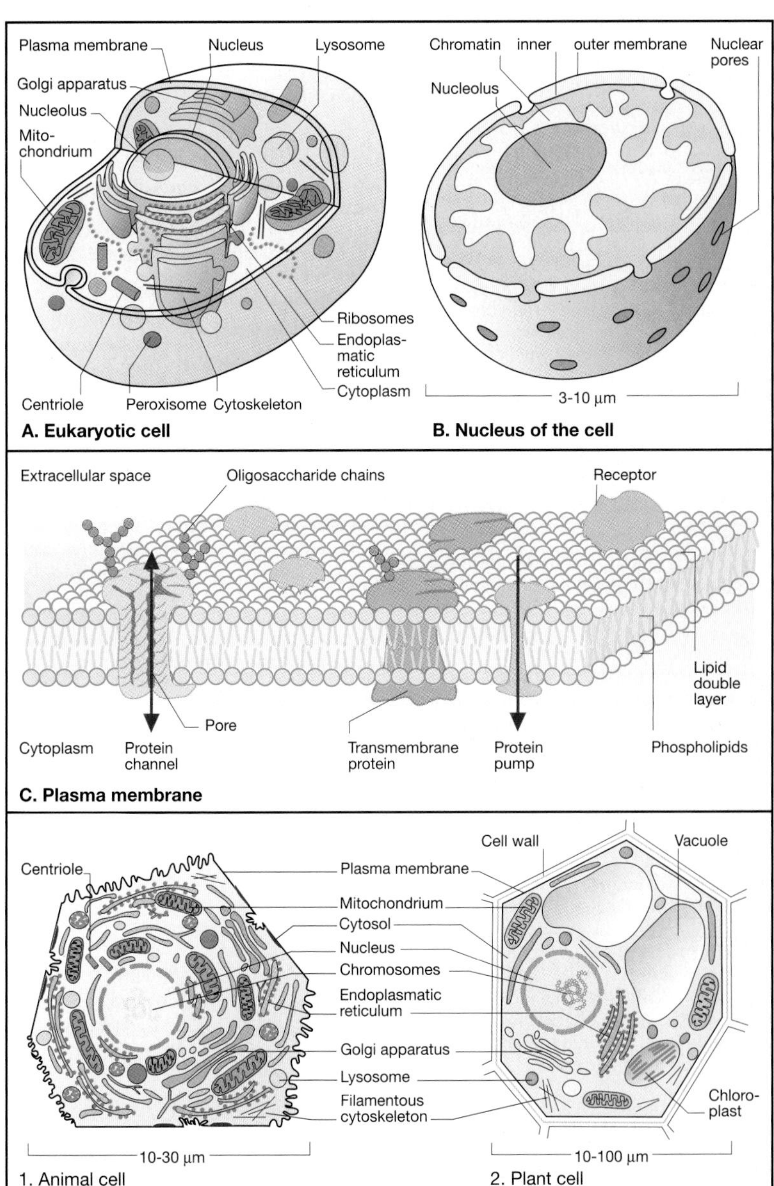

A. Eukaryotic cell

B. Nucleus of the cell

C. Plasma membrane

D. Comparison of the cell in animals and plants

Important Chemical Functional Groups and Molecular Compounds

Close to 99% of the weight of a living cell is composed of just four elements: carbon (C), hydrogen (H), nitrogen (N), and oxygen (O). Almost 50% of the atoms are hydrogen atoms; about 25% are carbon, and 25% oxygen. Apart from water (about 70% of the weight of the cell) almost all components are carbon compounds. Carbon, a small atom with four electrons in its outer shell, can form four strong covalent bonds with other atoms. But most importantly, carbon can combine with other carbon atoms to build chains and rings, and thus large complex molecules with specific biological properties.

A. Compounds of hydrogen (H), oxygen (O), and carbon (C)

Four simple combinations of these atoms occur frequently in biologically important molecules: hydroxyl (–OH; alcohols), methyl ($-CH_3$), carboxyl (–COOH), and carbonyl (C = O; aldehydes and ketones) groups. They impart the molecules with characteristic chemical properties, including possibilities to form compounds.

B. Acids and esters

Many biological substances contain a carbon–oxygen bond with weak acidic or basic (alkaline) properties. The degree of acidity is expressed by the pH value, which indicates the concentration of H^+ ions of a solution, namely from 10^{-1} (pH = 1, strongly acidic) to 10^{-14} (pH = 14, strongly alkaline). Pure water contains 10^{-7} moles H^+ per liter (pH = 7.0). An ester is formed when an acid reacts with an alcohol. Esters are frequently found in lipids and phosphate compounds.

C. Carbon–nitrogen bonds (C–N)

C–N bonds occur in many molecules: in amino groups, amines, and amides, especially in proteins. Of paramount significance are the amino acids (cf. p. 28), which are the subunits of proteins. All proteins have a specific role in the functioning of an organism.

D. Phosphate compounds

Ionized phosphate compounds play an essential biological role. A stable inorganic phosphate ion is HPO_4^{--}, from ionized phosphoric acid. A phosphate ion and a free hydroxyl group can form a phosphate esther. Phosphate compounds play an important role in energy-rich molecules and numerous macromolecules because they can store energy.

E. Sulfur compounds

Sulfur often serves to bind molecules together, especially when two sulfhydryl groups (–SH) react to form a disulfide bridge (–S–S–). Sulfur is a component of two amino acids (cysteine and methionine) and of some polysaccharides and sugars. Disulfide bridges play an important role in many complex molecules, serving to stabilize and maintain particular three-dimensional structures.

References

Alberts B, et al: Molecular Biology of the Cell. 3rd ed. New York: Garland; 1994.

Stryer L. Biochemistry. 4th ed. San Francisco: WH Freeman; 1995.

Koolman J, Röhm KH. Color Atlas of Biochemistry. Stuttgart: Thieme; 1995 [in press].

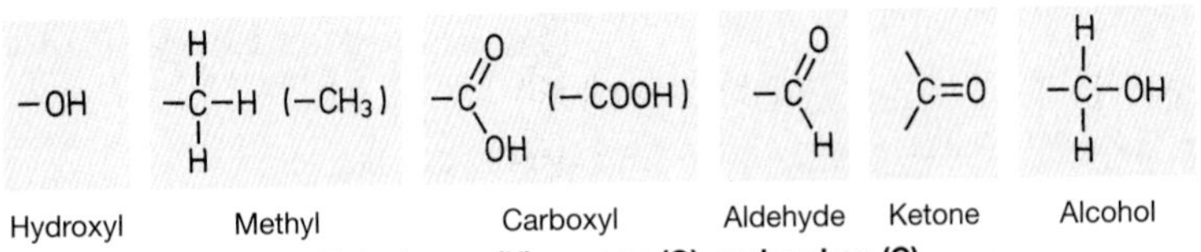

$-OH$ Hydroxyl

$-CH_3$ Methyl

$-COOH$ Carboxyl

Aldehyde

$C=O$ Ketone

$-C-OH$ Alcohol

A. Functional groups with hydrogen (H), oxygen (O), and carbon (C)

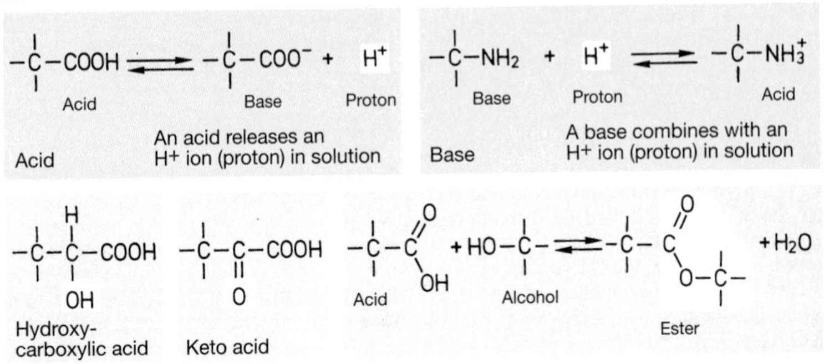

$-C-COOH \rightleftharpoons -C-COO^- + H^+$

Acid — Base — Proton

Acid

An acid releases an H^+ ion (proton) in solution

$-C-NH_2 + H^+ \rightleftharpoons -C-NH_3^+$

Base — Proton — Acid

Base

A base combines with an H^+ ion (proton) in solution

Hydroxy-carboxylic acid

Keto acid

Acid + $HO-C-$ Alcohol $\rightleftharpoons$ Ester $+ H_2O$

B. Acids and esters

$(-NH_2)$ Amino group

Amide

Amine

An amine accepts a proton in water

α-C Atom

Amino group

Side chain

Amino acid

$^+H_3N-C-COO^-$

Amino acids are ionized in aqueous solutions at pH 7

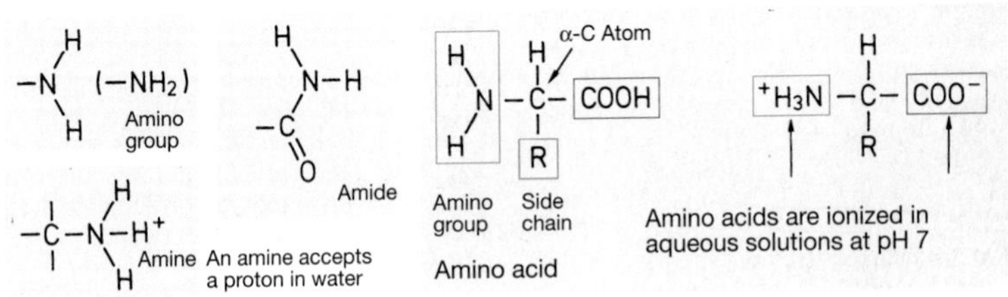

C. Carbon-nitrogen compounds (C-N)

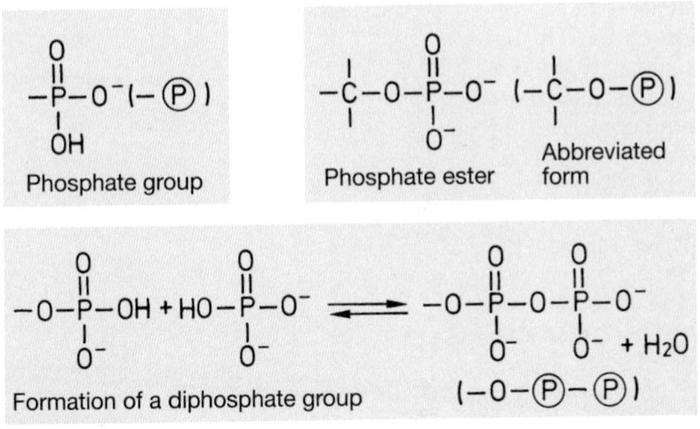

$-P-O^-$ (–Ⓟ)

Phosphate group

$-C-O-P-O^-$ (–C–O–Ⓟ)

Phosphate ester

Abbreviated form

$-O-P-OH + HO-P-O^- \rightleftharpoons -O-P-O-P-O^- + H_2O$

(–O–Ⓟ–Ⓟ)

Formation of a diphosphate group

D. Phosphate compounds

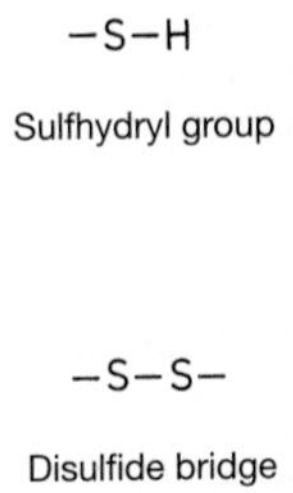

$-S-H$

Sulfhydryl group

$-S-S-$

Disulfide bridge

E. Sulfur groups

Carbohydrates

Carbohydrates in their various chemical forms and their derivatives are an important group of biomolecules for genetics. They provide the basic framework of DNA and RNA. Their flexibility makes them especially suitable for transferring genetic information from cell to cell.
Along with nucleic acids, lipids, and proteins, carbohydrates are one of the most important classes of biomolecules. Their main functions can be classified into three groups: (I) to deliver and store energy, (II) to help form DNA and RNA, the information-carrying molecules (see pp. 26 and 34), (III) to help form cell walls of bacteria and plants. Carbohydrates are often bound to proteins and lipids.
As polysaccharides, carbohydrates are important structural elements of the cell walls of animals, bacteria, and plants. They help form cell-surface structures (receptors) used in conducting signals from cell to cell. Since they can combine with numerous proteins and lipids, carbohydrates are important components of numerous cell structures. Finally, they function to transfer and store energy in intermediary metabolism.

A. Monosaccharides

Monosaccharides (simple sugars) are aldehydes (–C = 0,–H) or ketones (>C = O) with two or more hydroxy groups [general structural formula $(CH_2O)_n$]. The aldehyde or ketone group can react with one of the hydroxy groups to form a ring. This is the usual configuration of sugars that have five or six C atoms (pentoses and hexoses). The C atoms are numbered. The D- and the L-forms of sugars are mirror image isomers of the same molecule.
The naturally occurring forms are the D-(dextro) forms. These further include β- and α-forms as stereoisomers. In the cyclic forms the C atoms of sugars are not on a plane, but three-dimensionally take the shape of a chair or a boat. The β-D-glucopyranose configuration (glucose) is the intrinsic one, since all the axial positions are occupied by H atoms. The arrangement of the –OH groups can differ, so that stereoisomers such as, e.g., mannose or galactose are formed.

B. Disaccharides

These are compounds of two monosaccharides. The aldehyde or ketone group of one can bind to an α-hydroxy or a β-hydroxy group of the other. Sucrose and lactose are frequently occurring disaccharides.

C. Derivatives of sugars

When certain hydroxy groups are replaced by other groups, sugar derivatives are formed. These occur especially in polysaccharides. In a large group of genetically determined syndromes, complex polysaccharides can not be degraded due to reduced or absent enzyme function (mucopolysaccharidoses, mucolipidoses) (see p. 296).

D. Polysaccharides

Short (oligosaccharides) and long chains of sugars and sugar derivatives (polysaccharides) form essential structural elements of the cell. Complex oligosaccharides with bonds to proteins or lipids are part of cell surface structures, e.g., blood group antigens.

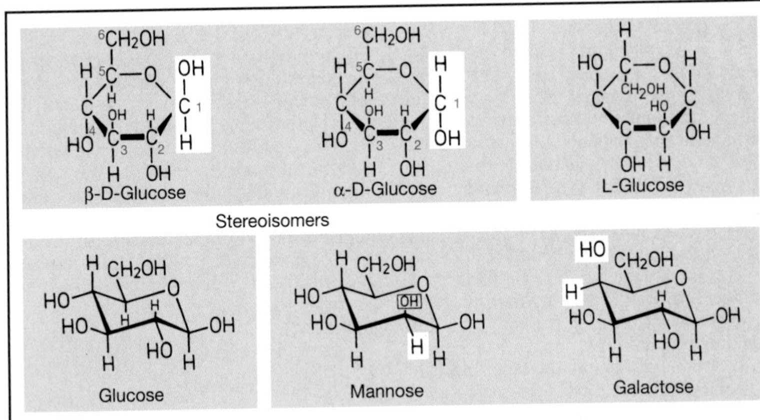

A. Monosaccharides

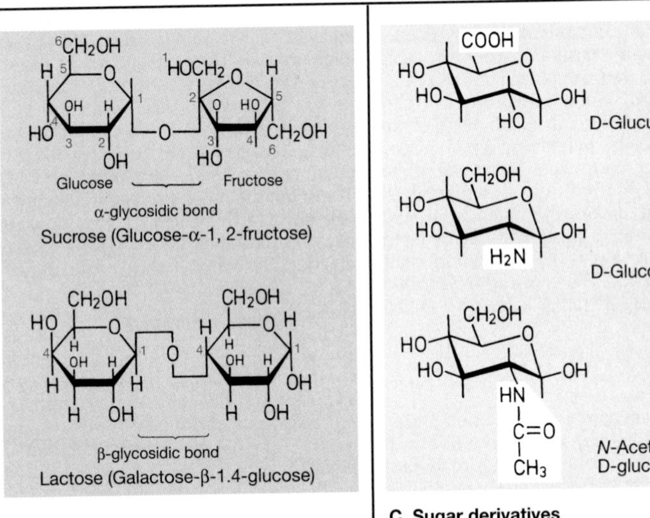

β-hydroxy

α-hydroxy

C1 Binding site of sugars

B. Disaccharides

N-Acetyl-D-glucosamine

C. Sugar derivatives

α-1.6 bonds at branching points

All others are α-1.4 bonds

D. Polysaccharides

Lipids (Fats)

Lipids usually occur as large molecules (macromolecules). They are essential components of membranes and precursors of other important biomolecules, such as steroids for the formation of hormones and other molecules for transmitting intercellular signals. In addition to fatty acids, compounds with carbohydrates (glycolipids), phosphate groups (phospholipids), and other molecules are especially important. A special characteristic is their pronounced polarity, with a hydrophilic (water-attracting) and a hydrophobic (water-repelling) region. This makes lipids especially suited for forming the outer limits of the cell (cell membrane).

A. Fatty acids

Fatty acids are composed of a hydrocarbon chain with a terminal carboxylic acid group. Thus, they are polar, with a hydrophilic (–COOH) and a hydrophobic end ($-CH_3$) and differ in the length of the chain and its degree of saturation. When one or more double bonds occur in the chain, the fatty acid is referred to as unsaturated. A double bond makes the chain relatively rigid and causes a kink. Fatty acids form the basic framework of many important macromolecules. The free carboxyl group (–COOH) of a fatty acid is ionized ($-COO^-$).

B. Lipids

Fatty acids can combine with other groups of molecules to form other types of lipids. As water insoluble (hydrophobic) molecules they are soluble only in organic solvents. The carboxyl group can enter into an ester or an amide bond. Triglycerides are compounds of fatty acids with glycerol.

Glycolipids (lipids with sugar residues) and phospholipids (lipids with a phosphate group and alcohol derivatives) are the structural basis of important macromolecules. Their intracellular degradation requires the presence of numerous enzymes, disorders of which have a genetic basis and lead to numerous genetically determined diseases.

Sphingolipids are an important group of molecules in biological membranes. Here, sphingosine, instead of glycerol, is the fatty acid-binding molecule. Sphingomyelin and gangliosides contain sphingosine. Gangliosides make up 6% of the central nervous system lipids. They are degraded by a series of enzymes. Genetically determined disorders of their catabolism lead to severe diseases, e.g., Tay–Sachs disease due to defective degradation of ganglioside G_{M2} (deficiency of β-N-acetylhexosaminidase).

C. Lipid aggregates

Due to their bipolar properties, fatty acids can form lipid aggregates in water. The hydrophilic ends form bonds with their aqueous surroundings; the hydrophobic ends protrude from the surface of the water and form a surface film. If completely under the surface, they form a micelle, compact and dry within. Phospholipids and glycolipids can form two-layered membranes (lipid membrane bilayer). These are the basic structural elements of cell membranes, which prevent molecules in the surrounding aqueous solution from invading the cell.

D. Other lipids: steroids

Steroids contain a multiring structure derived from simpler molecules, e.g., cholesterol. They are the structural basis of numerous hormones (steroid hormones). Hormones are signal-relaying molecules with high functional specificity.

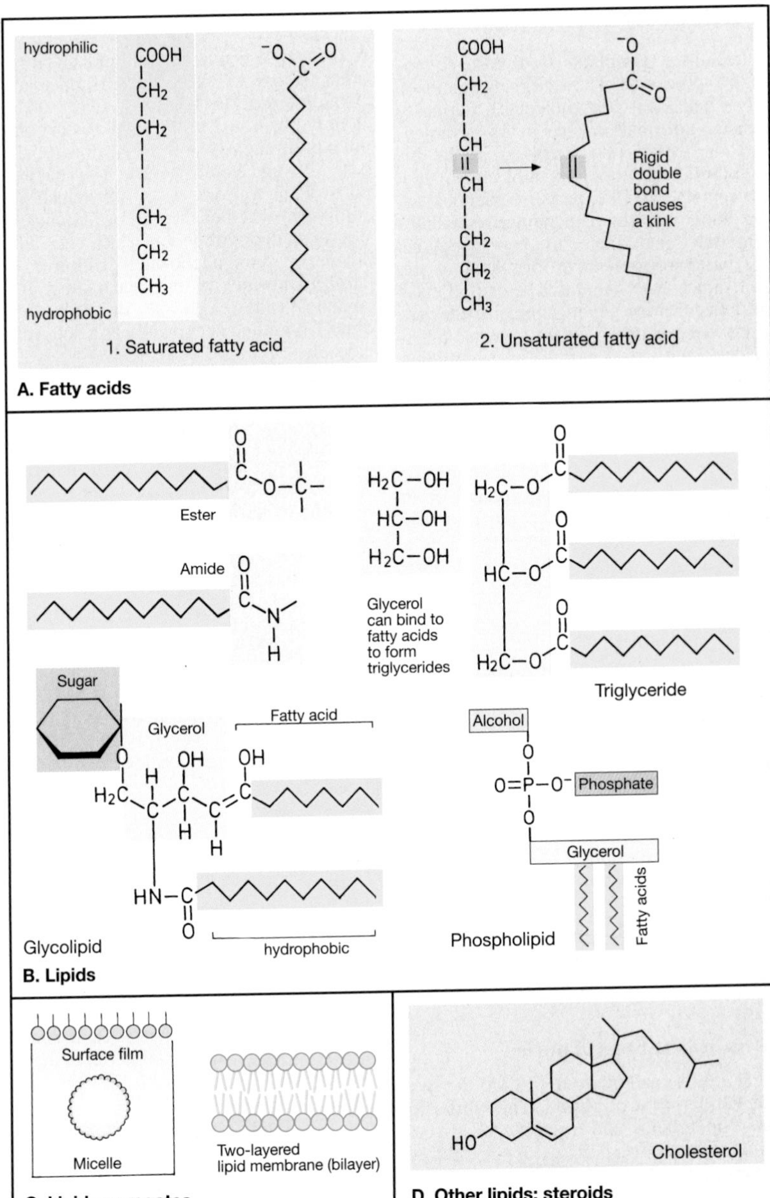
hydrophilic
COOH
CH_2
CH_2
CH_2
CH_2
CH_3
hydrophobic
1. Saturated fatty acid
COOH
CH_2
CH
CH
CH_2
CH_2
CH_3
Rigid double bond causes a kink
2. Unsaturated fatty acid
A. Fatty acids
Ester
Amide
H_2C-OH
$HC-OH$
H_2C-OH
Glycerol can bind to fatty acids to form triglycerides
Triglyceride
Sugar
Glycerol
Fatty acid
hydrophobic
Glycolipid
Alcohol
Phosphate
Glycerol
Fatty acids
Phospholipid
B. Lipids
Surface film
Micelle
Two-layered lipid membrane (bilayer)
C. Lipid aggregates
HO
Cholesterol
D. Other lipids: steroids

Nucleotides and Nucleic Acids

Nucleotides participate in almost all biological processes. They are the subunits of DNA and RNA, the molecules that carry genetic information (see p. 36). Nucleotide derivatives are involved in the biosynthesis of numerous molecules; they convey energy, are part of essential coenzymes, and regulate numerous metabolic functions. Since all these functions are based on genetic information of the cells, nucleotides represent a central class of molecules for genetics. Nucleotides are composed of three integral parts: phosphates, sugar, and purine or pyramidine bases.

A. Phosphate groups

Phosphate groups may occur alone (monophosphates), in twos (diphosphate) or in threes (triphosphate). They are normally bound to the hydroxy group of the C atom in position 5 of a five-C-atom sugar (pentose).

B. Sugar residues

The sugar residues in nucleotides are usually derived from either ribose (in ribonucleic acid, RNA) or deoxyribose (in deoxyribonucleic acid, DNA) (ribonucleoside or deoxyribonucleoside).

C. Nucleotide bases of pyrimidine

Cytosine (C), thymine (T), and uracil (U) are the three pyrimidine nucleotide bases. They differ from each other in their side chains ($-NH_2$ on C4 in cytosine, $-CH_3$ on C5 in thymine, O on C4 in uracil) and in the presence or absence of a double bond between N3 and C4 (present in cytosine).

D. Nucleotide bases of purine

Adenine (A) and guanine (G) are the two nucleotide bases of purine. They differ in their side chains and a double bond (between N1 and C6).

E. Nucleosides

A nucleoside is a compound of a sugar residue (ribose or deoxyribose) and a nucleotide base. The bond is between the C atom in position 1 of the sugar (as in compounds of sugars) and an N atom of the base (N–glycosidic bond). The nucleotides of the various bases are named according to whether they are a ribonucleoside or a deoxyribonucleoside, e.g., adenosine or deoxyadenosine, guanosine or deoxyguanosine, uridine (occurs only as a ribonucleoside), cytidine or deoxycytidine. Thymidine occurs only as a deoxynucleoside.

F. Nucleotides

A nucleotide is a compound of a five-C-atom sugar residue (ribose or deoxyribose) attached to a nucleotide base (pyrimidine or purine base) and a phosphate group. Nucleotides are the subunits of nucleic acids. The nucleotides of the individual bases are referred to as follows: adenylate (AMP, adenosine monophosphate), guanosine monophosphate (GMP), uridylate (UMP), and cytidylate (CMP) for the ribonucleotides (5'–monophosphates) and deoxyadenylate (dAMP), deoxyguanylate (dGMP), deoxythymidylate (dTMP), and deoxycytidylate (dCMP) for the deoxyribonucleotides.

G. Nucleic acids

Nucleic acids arise when nucleotides are joined to each other by means of phosphodiester bridges between the 3' C atom of one nucleotide and the 5' C atom of the next. The linear sequence of nucleotides is usually given in the 5' to 3' direction with the abbreviations of the respective nucleotide bases. For instance, ATCG would signify the sequence adenine (A), thymine (T), cytosine (C), and guanine (G) in the 5' to 3' direction.

Monophosphate

Diphosphate

Triphosphate

A. Phosphate groups

β-D-Ribose

β-D-Deoxyribose

B. Sugar residues (pentose)

Cytosine (C)

Thymine (T)

Uracil (U)

Pyrimidine

C. Nucleotide bases of pyrimidine

Guanine (G)

Adenine (A)

Purine

D. Nucleotide bases of purine

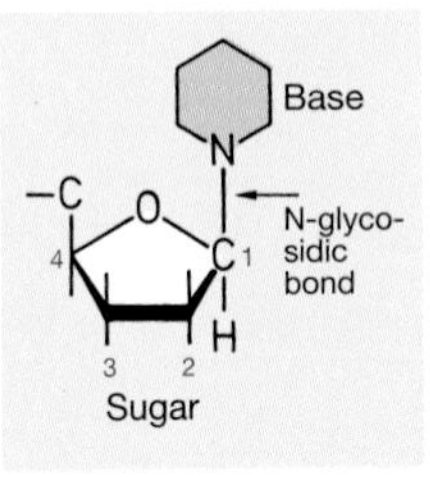

E. Nucleoside (base and sugar)

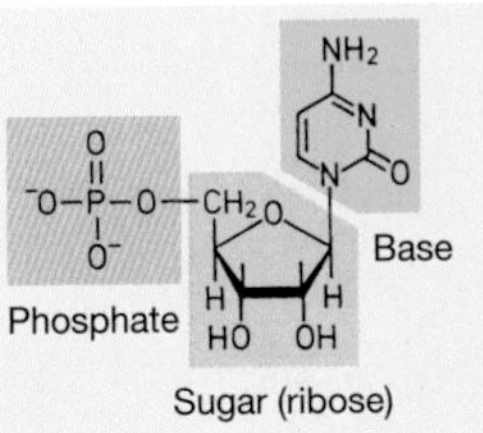

F. Nucleotide (base + sugar + phosphate)

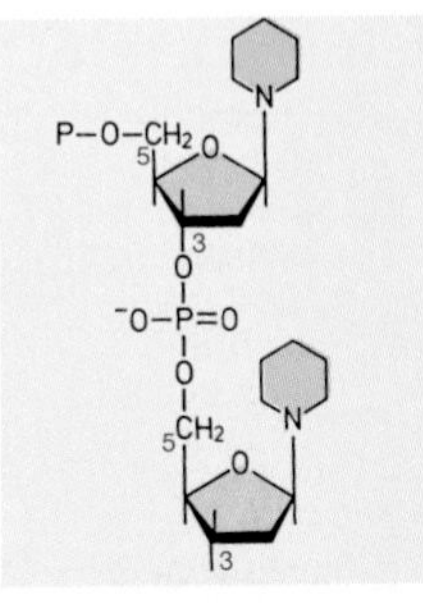

G. Nucleic acid

Amino Acids

Amino acids are the basic structural units of proteins. A defined linear sequence of the amino acids and a specific three-dimensional structure confer each protein quite specific physicochemical properties. An amino acid consists of a carbon with an amino group ($-NH_2$) on one end, a carboxyl group ($-COOH$) on the other, a hydrogen atom, and a variable side chain. Amino acids are ionized in neutral solutions, since the amino group takes on a proton ($-NH_3^+$) and the carboxyl group dissociates ($-COO^-$). The side chain determines the distinguishing characteristics of an amino acid, including the size, form, electrical charge or hydrogen-binding ability, and the total specific chemical reactivity. Amino acids can be differentiated according to whether they are neutral or not neutral (basic or acidic) and whether they have a polar or nonpolar side chain. Each amino acid has its own three-letter and one-letter abbreviation.

A. Neutral amino acids, nonpolar side chains

All neutral amino acids have a $-COO^-$ and an $-NH_3^+$ group. The simplest amino acids have a simple aliphatic side chain. For glycine this is merely a hydrogen atom (–H); for alanine, a methyl group ($-CH_3$). Larger side chains occur on valine, leucine, and isoleucine. These larger side chains are hydrophobic (water repellent) and make their respective amino acids less water soluble than do hydrophilic (water-attracting) chains. Proline has an aliphatic side chain, which unlike in other amino acids, is bound to both the central carbon and to the amino group, so that a ringlike structure is formed. Aromatic side chains occur in phenylalanine (a phenyl group bound via a methylene group [$-CH_2-$]) and tryptophan (an indol ring bound via a methylene group). These amino acids are very hydrophobic. Two amino acids contain sulfur (S) atoms. In cysteine this is in the form of a sulfhydryl group (–SH); in methionine, a thioether ($-S-CH_3$). Both are hydrophobic. The sulfhydryl group in cysteine is very reactive and participates in forming disulfide bonds (–S–S–). These play an important role in stabilizing the three-dimensional forms of proteins.

B. Neutral amino acids, polar side chains

Serine, threonine, and tyrosine contain hydroxyl groups (–OH). Thus, they are hydrolyzed forms of glycine, alanine, and phenylalanine. The hydroxyl groups make them hydrophilic and more reactive than the nonhydrolyzed forms. Asparagine and glutamine both contain an amino and an amide group. At physiological pH their side chains are negatively charged.

C. Nonneutral amino acids

These amino acids have either two ionized amino groups (basic) or two carboxyl groups (acidic). Basic amino acids (positively charged) are arginine, lysine, and histidine. Histidine has an imidazole ring and can be uncharged or positively charged, depending on its surroundings. It is frequently found in the reactive centers of proteins, where it takes part in alternating bonds (e.g., in the oxygen-binding region of hemoglobin). Aspartic acid and glutamic acid each have two carboxyl groups (–COOH) and are thus (as a rule) acidic.

Seven of the 20 amino acids have slightly ionizable side chains, making them highly reactive (Asn, Glu, His, Cys, Tyr, Lys, Arg).

COO^- $H_3\overset{+}{N}-C-H$ H Glycine Gly (G)	COO^- $H_3\overset{+}{N}-C-H$ CH_3 Alanine Ala (A)	COO^- $H_3\overset{+}{N}-C-H$ CH H_3C CH_3 Valine Val (V)	COO^- $H_3\overset{+}{N}-C-H$ CH_2 CH H_3C CH_3 Leucine Leu (L)	COO^- $H_3\overset{+}{N}-C-H$ CH H_3C CH_2 CH_3 Isoleucine Ile (I)
COO^- $H_2\overset{+}{N}$ (ring) Proline Pro (P)	COO^- $H_3\overset{+}{N}-C-H$ CH_2 (benzene ring) Phenylalanine Phe (F)	COO^- $H_3\overset{+}{N}-C-H$ CH_2 $C=CH$ NH (indole ring) Tryptophan Trp (W)	COO^- $H_3\overset{+}{N}-C-H$ CH_2 SH Cysteine Cys (C)	COO^- $H_3\overset{+}{N}-C-H$ CH_2 CH_2 S CH_3 Methionine Met (M)

A. Neutral amino acids, nonpolar side chains

COO^- $H_3\overset{+}{N}-C-H$ CH_2OH Serine Ser (S)	COO^- $H_3\overset{+}{N}-C-H$ $CHOH$ CH_3 Threonine Thr (T)	COO^- $H_3\overset{+}{N}-C-H$ CH_2 (benzene ring) OH Tyrosine Tyr (Y)	COO^- $H_3\overset{+}{N}-C-H$ CH_2 $H_2N-C=O$ Asparagine Asn (N)	COO^- $H_3\overset{+}{N}-C-H$ CH_2 CH_2 $H_2N-C=O$ Glutamine Gln (Q)

B. Neutral amino acids, polar side chains

1. basic (positively charged)			2. acid (negatively charged)	
COO^- $H_3\overset{+}{N}-C-H$ CH_2 CH_2 CH_2 NH $H_2N-C=NH_2^+$ Arginine Arg (R)	COO^- $H_3\overset{+}{N}-C-H$ CH_2 CH_2 CH_2 CH_2 $\overset{+}{N}H_3$ Lysine Lys (K)	COO^- $H_3\overset{+}{N}-C-H$ CH_2 (imidazole ring) N HN Histidine His (H)	COO^- $H_3\overset{+}{N}-C-H$ CH_2 COO^- Aspartic acid Asp (D)	COO^- $H_3\overset{+}{N}-C-H$ CH_2 CH_2 COO^- Glutamic acid Glu (E)

C. Charged amino acids

Proteins

Proteins are involved in practically all chemical processes in living organisms. Their universal significance is apparent in that as enzymes, they drive chemical reactions in living cells. Without enzymatic catalysis, the macromolecules involved would not react spontaneously. All enzymes are the products of one or more genes. Many genetically determined diseases are based on an inactive or absent enzyme resulting from a defective gene. Proteins also serve to transport small molecules, ions, or metals. For example, hemoglobin transports oxygen, and ferritin transports iron. Proteins have important functions in cell division during growth and in cell and tissue differentiation. They control the coordination of movements by regulating muscle cells and the production and transmission of impulses within and between nerve cells. They control blood homeostasis (blood clotting) and immune defense. They carry out mechanical functions in skin, bone, blood vessels, and other areas. Corresponding with the number of their functions, they occur in very different forms, sizes, and biological capacities. Nonetheless their basic pattern follows several relatively simple principles. Each protein consists of amino acids linked together in a specific order. The most important functions of proteins are (1) enzymatic catalysis of chemical reactions, (2) transport and storage functions, (3) immune defense, (4) muscle function, (5) mechanical functions (collagen molecules), (6) control of growth and differentiation, and (7) as receptor proteins.

A. Linking of amino acids (peptide bonds)

The basic units of proteins, amino acids, can be linked together very easily due to their dipolar ionization (zwitterions). The carboxyl group of one amino acid binds to the amino group of the next (peptide bond, sometimes also referred to as an amide bond). When a dipeptide is formed from two amino acids, a water molecule is released. When many amino acids are bound together by peptide bonds, they form a polypeptide chain. Each polypeptide chain has a defined direction, determined by the amino group ($-NH_2$) on one end and the carboxyl group ($-COOH$) on the other. By convention, the amino group represents the beginning, and the carboxyl group the end of a peptide chain.

B. Linking of peptide chains by disulfide bridges

The sulfhydryl group (–SH) of the amino acid cysteine can easily form a disulfide bridge (–S–S–) with the –SH of a neighboring cysteine. The resulting disulfide is called cystine. Disulfide bridges bind together various peptide chains of a protein, substantially contributing to the correct three-dimensional structure and the stability of the protein. Disulfide bridges mainly occur in extracellular proteins, seldom in intracellular. Some proteins, such as collagen, utilize lysine instead of cysteine to form cross-links to a side chain.

C. One-dimensional structure of a simple protein (insulin)

Insulin is an example of a relatively simple protein. It consists of two polypeptide chains, A and B, of defined lengths (21 amino acids in the A chain, 30 amino acids in the B) and defined sequences. The three-dimensional structure important for the specific function is stabilized by three disulfide bridges, one from amino acid residue 6 to 11 on the A chain and two from the A to the B chain (amino acids 7 to 7 and 20 to 19).

$$^{+}H_3N-CH(R^1)-COO^- + {}^{+}H_3N-CH(R^2)-COO^- \longrightarrow {}^{+}H_3N-CH(R^1)-CO-NH-CH(R^2)-COO^- \; (+H_2O)$$

Peptide bond

Amino acid 1 Amino acid 2 form a peptide

$H_2N-CH(R^1)-CO-NH-CH(R^2)-CO-NH-CH(R^3)-CO-NH-CH(R^4)-CO-NH-CH(R^5)-COOH$

Amino end (NH_2) Carboxy end (COOH)

Peptide composed of five amino acids (pentapeptide)

$-NH-CH(R^1)-CO-NH-CH(R^2)-CO-NH-CH(R^3)-CO-$

A. Joining of amino acids (peptide bond)

Cysteine CH_2 SH + SH CH_2 Cysteine $\longrightarrow$ $CH_2-S-S-CH_2$ Cystine

B. Joining of peptide chains by a disulfide bridge

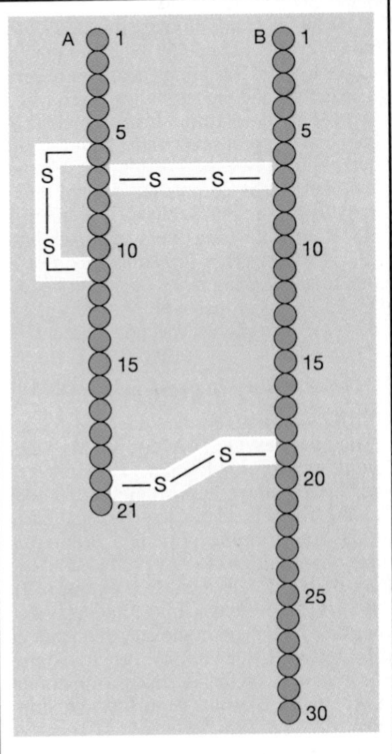

C. One-dimensional structure of a simple protein (insulin)

DNA as the Carrier of Genetic Information

Before the three-dimensional structure of DNA was recognized as a double helix, this chemically uniform molecule seemed poorly suited for coding manifold genetic information. An incidental, but precise observation (1928) and directed investigations (1944) indicated that DNA could be the carrier of genetic information.

A. The observation of Griffith

In 1928 the English microbiologist Fred Griffith made a remarkable observation. While investigating various strains of pneumococcus, he determined that mice injected with strain S (smooth) died (1). On the other hand, animals injected with strain R (rough) lived (2). When he inactivated the lethal S strain by heat, there were no sequelae, and the animal survived (3). Surprisingly, a mixture of the nonlethal R strain and the heat-inactivated S strain had a lethal effect like the S strain (4). And he found normal living pneumococci of the S strain in the animal's blood. Apparently, cells of the R strain were changed into cells of the S strain (transformed). For a time, this surprising result could not be explained and was met with skepticism. Its relevance for genetics was not apparent.

B. The transforming principle is DNA

Griffith's findings formed the basis for investigations by O. T. Avery, C. M. MacCleod, and M. J. McCarty (1944). Avery and co-workers at the Rockefeller Institute in New York elucidated the chemical basis of the transforming principle. From cultures of an S strain (1) they produced an extract of lysed cells (cell-free extract) (2). After all its proteins, lipids, and polysaccharides had been removed, the extract still retained the ability to transform pneumococci of the R strain to pneumococci of the S strain (transforming principle) (3).

With further studies, Avery and co-workers determined that this was attributed to the DNA alone. Thus, the DNA must contain the corresponding genetic information. This explained Griffith's observation. Heat inactivation had left the DNA of the bacterial chromosomes intact. The section of the chromosome with the gene responsible for capsule formation (S gene) could be released from the destroyed S cells and be taken up by some R cells in subsequent cultures. After the S gene was incorporated into its DNA, an R cell was transformed into an S cell (4). Page 85 shows that bacteria can take up foreign DNA so that some of their genetic attributes will be altered correspondingly.

C. Genetic information is transmitted by DNA alone

The final evidence that DNA, and no other molecule, transmits genetic information was provided by A. D. Hershey and M. Chase in 1952. They labeled the capsular protein of bacteriophages (see p. 82) with radioactive sulfur (^{35}S) and the DNA with radioactive phosphorus (^{32}P). When bacteria were infected with the labeled bacteriophage, only ^{32}P (DNA) entered the cells, and not the ^{35}S (capsular protein). The subsequent formation of new, complete phage particles in the cell proved that DNA was the exclusive carrier of the genetic information needed to form new phage particles, including capsular protein. Next, the structure and function of DNA needed to be clarified. The genes of all cells and some viruses consist of DNA, a long-chain threadlike molecule.

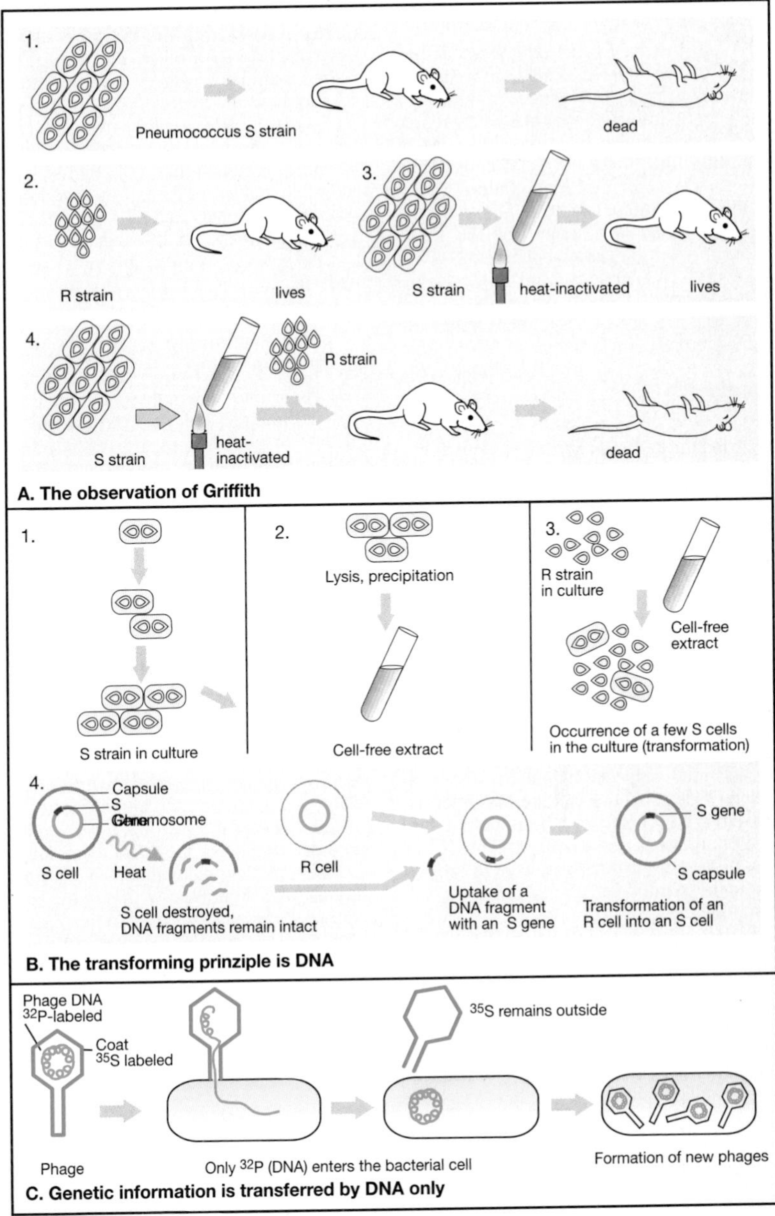

A. The observation of Griffith

B. The transforming prinziple is DNA

C. Genetic information is transferred by DNA only

DNA and Its Components

The information for the development and specific functions of cells and tissues is stored in the genes. A gene is a portion of the genetic information, definable according to structure and function. Genes lie on chromosomes in the nuclei of cells. They consist of a complex long-chained molecule, *d*eoxyribo*n*ucleic *a*cid (DNA). In the following, the constituents of the DNA molecule will be presented. DNA is a nucleic acid. Its chemical components are nucleotide bases, a sugar (deoxyribose), and phosphate groups. They determine the three-dimensional structure of DNA, from which it derives its functional consequence.

A. Nucleotide bases

DNA contains purines and pyrimidines as nucleotide bases. These include two types of purine, adenine (A) and guanine (G), and two types of pyrimidine, thymine (T) and cytosine (C). The nucleotide bases are part of a subunit of DNA, the nucleotide. It consists of one of the four nucleotide bases, a sugar (deoxyribose), and a phosphate group. The nitrogen atom in position 9 of a purine or in position 1 of a pyrimidine is bound to the carbon in position 1 of the sugar (N–glycosidic bond).
Ribonucleic acid (RNA) differs from DNA in two respects: it contains ribose instead of deoxyribose (unlike the latter, ribose has a hydroxyl group on the position 2 carbon atom) and uracil (U) instead of thymine. Uracil does not have a methyl group at position C5.

B. Nucleotide chain

DNA is a polymer of deoxyribonucleotide units. The nucleotide chain is formed by joining a hydroxyl group on the sugar of one nucleotide to the phosphate group attached to the sugar of the next nucleotide. The sugars linked together by the phosphate groups form the invariable part of the DNA. The variable part is in the sequence of the nucleotide bases A, T, C, and G.

A DNA nucleotide chain is polar. The polarity results from the way the sugars are attached to each other. The phosphate group at position C5 (the 5' carbon) of one sugar joins to the hydroxyl group at position C3 (the 3' carbon) of the next sugar by means of a phosphate diester bridge. Thus, one end of the chain has a 5' triphosphate group free and the other end, a 3' hydroxy group free (5' end and 3' end, respectively). By convention, the sequence of nucleotide bases is written in the 5' to 3' direction.

C. Spatial relationship

The chemical structure of the nucleotide bases determines a defined spatial relationship. Within the double helix, a purine (adenine or guanine) always lies opposite a pyrimidine (thymine or cytosine). Three hydrogen-bond bridges are formed between cytosine and guanine and two, between thymine and adenine. Therefore, only guanine and cytosine or adenine and thymine can lie opposite and pair with each other (complementary base pairs G–C and A–T). Other spatial relationships are not usually possible.

D. DNA double strand

DNA forms a double strand. As a result of the spatial relationships of the nucleotide bases within, a cytosine will always lie opposite to a guanine and a thymine to an adenine. The sequence of the nucleotide bases on a strand of DNA (in the 5' to 3' direction) is complementary to the nucleotide base sequence (or simply the base sequence) of the other strand in the 3' to 5' direction. The specificity of base pairing is the most important structural characteristic of DNA.

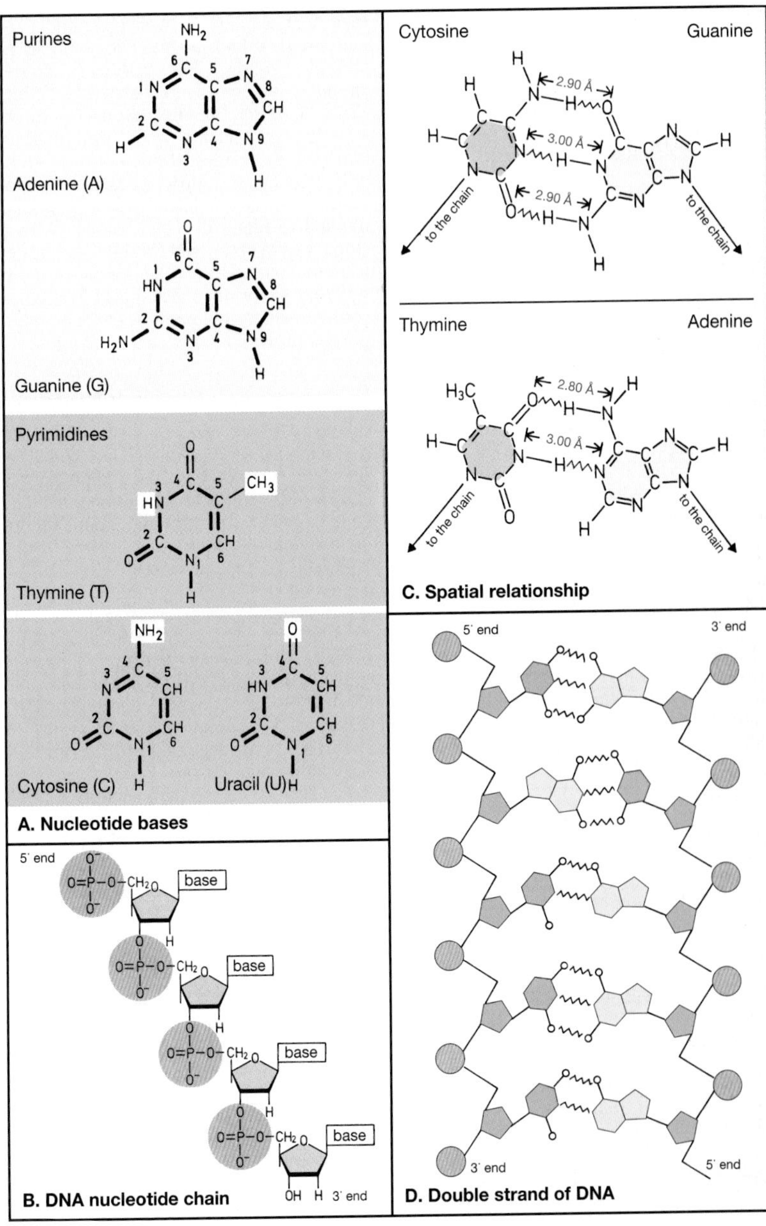
Purines
Adenine (A)
Guanine (G)
Pyrimidines
Thymine (T)
Cytosine (C)
Uracil (U)
A. Nucleotide bases
5' end
base
3' end
B. DNA nucleotide chain
Cytosine
Guanine
2.90 Å
3.00 Å
2.90 Å
to the chain
Thymine
Adenine
2.80 Å
3.00 Å
C. Spatial relationship
5' end
3' end
D. Double strand of DNA

DNA Structure

In 1953, James Watson and Francis Crick recognized that DNA must exist as a double helix. This structure explains both important functional aspects: replication and the transmission of genetic information.
The elucidation of the structure of DNA is considered as the beginning of the development of modern genetics. With it, gene structure and function can be understood at the molecular level.

A. DNA as a double helix

The double helix is the characteristic structural feature of DNA. The two helical polynucleotide chains are wound around each other along a common axis. The nucleotide base pairs (bp), either A–T or G–C, lie within. The diameter of the helix is 20 Å (10^{-7} mm). Neighboring bases lie 3.4 Å apart. The helical structure repeats itself at intervals of 34 Å, or every ten base pairs. Because of the fixed spatial relationship of the nucleotide bases within the double helix and opposite each other, the two chains of the double helix are exactly complementary. The form illustrated here is the so-called B form (B-DNA). Under certain conditions, DNA can also assume other forms (Z-DNA, A-DNA, see p. 216).

B. Replication

Since the nucleotide chains lying opposite each other within the double helix are strictly complementary, each can serve as a pattern (template) for the formation (replication) of a new chain when the helix is opened. DNA replication is semiconservative, i.e., one completely new strand will be formed, one strand retained.

C. Denaturation and renaturation

The noncovalent hydrogen bonds between the nucleotide base pairs are weak. Nevertheless, DNA is stable at physiologic temperatures because it is a very long molecule. The two complementary strands can be separated (denaturation) by means of a relatively weak chemical reagent (e.g., alkali, formamide, or urea) or careful heating. The resulting single-stranded molecules are relatively stable. With cooling, complementary single strands can reunite to form double-stranded molecules (renaturation). Noncomplementary single strands do not unite. This is the basis of an important method to identify nucleic acids: With a single strand of defined origin, it can be determined with which other single strand it will bind (hybridize). The hybridization of complementary segments of DNA is an important principle in the analysis of genes.

D. Transmission of genetic information

Genetic information lies in the sequence of nucleotide base pairs (A–T or G–C). A sequence of three base pairs represents a code word (codon) for an amino acid. The codon sequence determines a corresponding sequence of amino acids. These form a polypeptide (gene product).
The sequence of the nucleotide bases is first transferred (transcription) in the 3' to 5' direction from one DNA strand to a further information-bearing molecule (mRNA, messenger RNA). Then the nucleotide base sequence of the mRNA serves as a template for a sequence of amino acids corresponding to the order of the codons (translation).
A gene can be defined as a section of DNA responsible for the formation of a polypeptide (one gene, one polypeptide). One or more polypeptides form a protein. Thus, several genes may be involved in the formation of a protein.

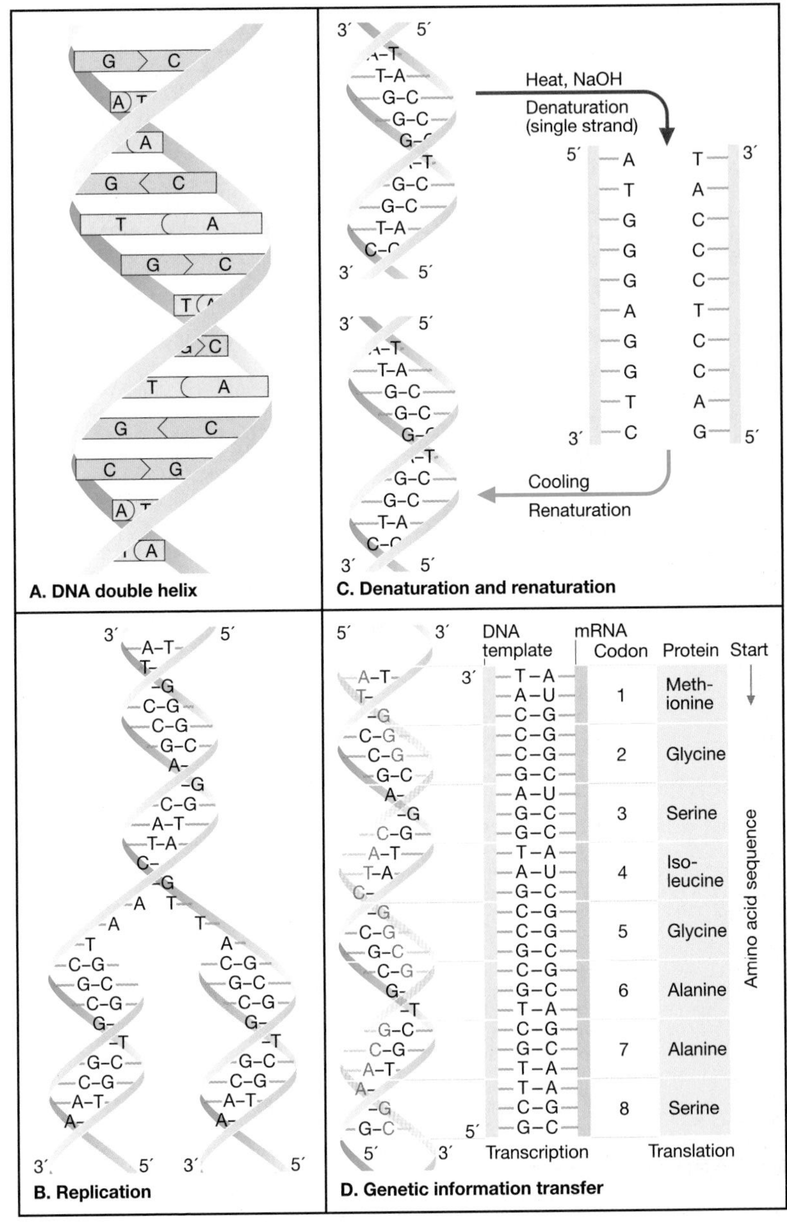

A. DNA double helix

B. Replication

C. Denaturation and renaturation

D. Genetic information transfer

DNA Replication

In their paper elucidating the structure of DNA, Watson and Crick (1953) noted in closing, "It has not escaped our attention that this structure immediately suggests a copying mechanism for the genetic material," at that time an unsolved problem. Although biochemically complex, DNA replication is genetically relatively simple. During replication, each strand of DNA serves as a template for the formation of a new strand (semiconservative replication).

A. Prokaryote replication begins at a single site

In prokaryote cells, replication begins at a defined point in the ring-shaped bacterial chromosome (1). From here, new DNA is formed at the same speed in both directions until the DNA has been completely duplicated and two chromosomes are formed. Replication can be visualized by autoradiography after the newly replicated DNA has incorporated tritium (^{3}H)-labeled thymidine (2).

B. Eukaryote replication begins at several sites

DNA synthesis occurs during a defined phase of the cell cycle (S phase). This would take a very long time if there were only one starting point. However, replication of eukaryotic DNA begins at numerous sites (replicons) (1). It proceeds in both directions from each replicon until neighboring replicons fuse (2) and all of the DNA is duplicated (3). The electron micrograph (4) shows replicons at four sites.

C. Scheme of replication

New DNA can be formed continuously in the 5' to 3' direction, but not in the 3' to 5' direction. A new nucleotide cannot be attached to the 5'-OH end of the new nucleotide chain. Only at the 3' end can nucleotides be attached continuously. New DNA at the 5' end will be replicated in small segments. This represents an obstacle at the 5' end of a chromosome (telomere, see p. 188).

D. Replication fork

At the replication fork, each of the two DNA strands serves as a template for the synthesis of new DNA. First, the double helix at the replication fork region is unwound by an enzyme system (topoisomerases). Since the parent strands are antiparallel, DNA replication can proceed continuously in only one DNA strand (5' to 3' direction) (leading strand). Along the 3' to 5' strand (lagging strand), the new DNA is formed in small segments of 1000–2000 bases (Okazaki fragments). In this strand a short piece of RNA is required as a primer to start replication. This is formed by an RNA polymerase (primase). The RNA primer is subsequently removed; DNA is inserted into the gap by polymerase I; and finally, the DNA fragments are linked by DNA ligase. The enzyme responsible for DNA synthesis (DNA polymerase III) is complex and comprises several subunits. There are different enzymes for the leading and lagging strands in eukaryotes. During replication, mistakes are eliminated by a complex proof-reading mechanism that removes any incorrectly incorporated bases and replaces them with the correct ones.

Reference

Watson JD, et al. Molecular Biology of the Gene. 3rd ed. 1987.

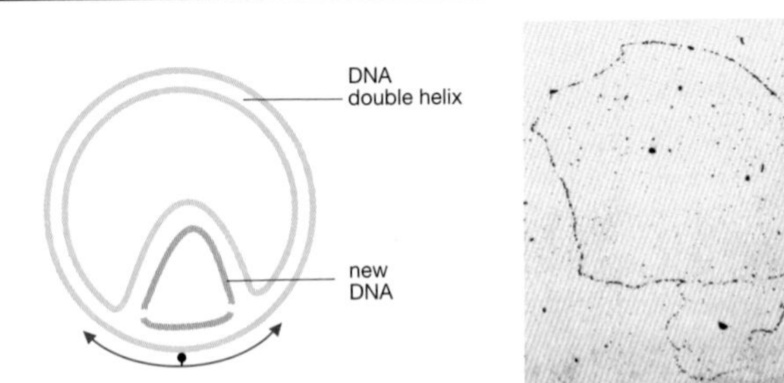

1. DNA replication in the bacterial chromosome

2. Prokaryotic replication in an autoradiogram in *E. coli* (J. Cairns)

A. Prokaryotic replication begins at one site

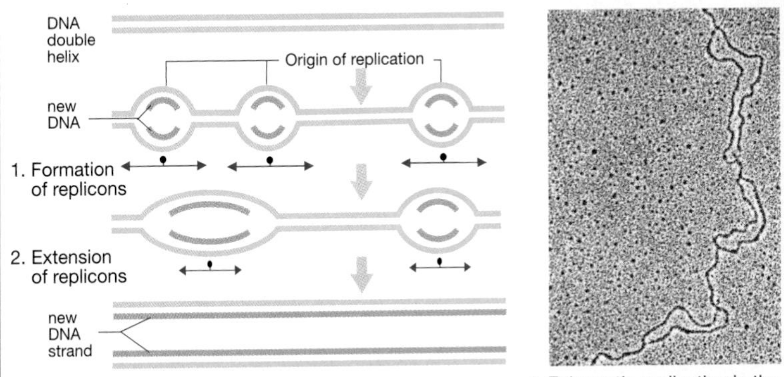

1. Formation of replicons

2. Extension of replicons

3. Replication completed

4. Eukaryotic replication in the EM (D. S. Hogness)

B. Eukaryotic replication begins at several sites

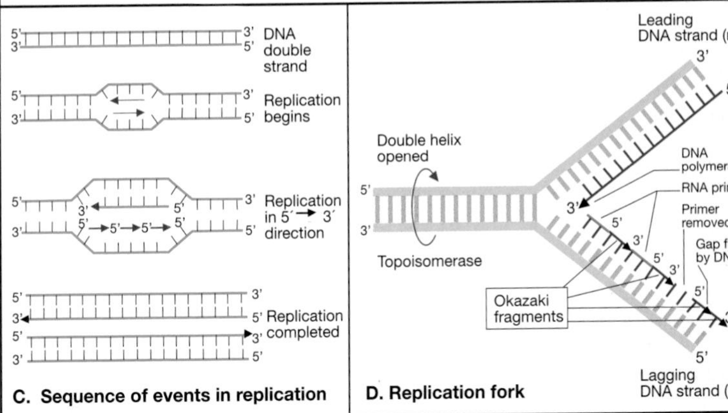

C. Sequence of events in replication

D. Replication fork

Repair of DNA Damage from UV Light and Reactive Chemicals

DNA can be altered by UV light, ionizing radiation, and a number of chemicals. However, the genetic information in DNA sequences must be reliably preserved for many cell divisions. This is essential for the survival of not only the individual but also of the species. The repair of DNA damage is a selective advantage; thus, a complex and efficient DNA repair system developed during evolution.

A. UV-induced DNA damage

A typical consequence of ultraviolet (UV) rays is the formation of a reactive double bond between neighboring thymines, neighboring cytosines, or a thymine and a cytosine. This disrupts the hydrogen bridges with the oppositely lying bases. These and other structural changes lead to a wrong base being incorporated into the new DNA strand during the next replication.

B. DNA repair by excision

An important form of repair is the excision of two thymines bound together (thymine dimer) on the same DNA strand. First, the affected DNA segment (strand 1) becomes distorted and is incised by an enzyme (endonuclease). Then, the synthesis of a new strand can begin. The opposite, unaffected strand (2) serves as a template. The damaged DNA section is excised by another enzyme (exonuclease). Finally, the newly formed strand is connected to the old strand by DNA ligase to complete the repair process. Prokaryotes and eukaryotes have numerous genes for repairing DNA damage (see p. 42).

C. Chemicals that react with DNA

Chemical substances that react with DNA can alter the structure or the sequence of the nucleotides (mutagens). Alkylating mutagens such as nitrosamine deliver a methyl or an ethyl group to a nucleotide base. This leads to alteration of the hydrogen bridges of the nucleotide base so that it pairs with the wrong base (base substitution). As a general principle, chemical substances that induce mutations can cause cancer. Although this has been well demonstrated under experimental conditions, it usually cannot be reliably demonstrated for individuals. As a precaution, threshold values have been established for many substances.

D. Alkylation of guanine

An especially sensitive site is the carbon in position 6 of guanine. It is very readily alkylated by accepting a methyl group. The resulting product, O^6-methylguanine, can form only two hydrogen bridges and thus pairs with thymine instead of cytosine. During the next replication, cytosine is replaced by a thymine in this position (transition).

E. GC to AT transition

If a guanine becomes alkylated (O^6), the usual GC pairing (three hydrogen bridges) is disrupted, and at the next replication, pairing is possible with thymine only. Thus, in the corresponding position of the newly formed daughter strand, a thymine will be inserted instead of a cytosine (C to T transition). After a further replication, a GC pair will have been replaced by an AT pair. A transition is the exchange of one purine or pyrimidine for another purine or pyrimidine base. Transversion is the replacement of a purine by a pyrimidine or vice versa. Both transitions and transversions occur in nature. A single-base substitution may be reversible (reversion).

Reference

Lewin B. Genes V. Oxford: University Press; 1994: 614–629.

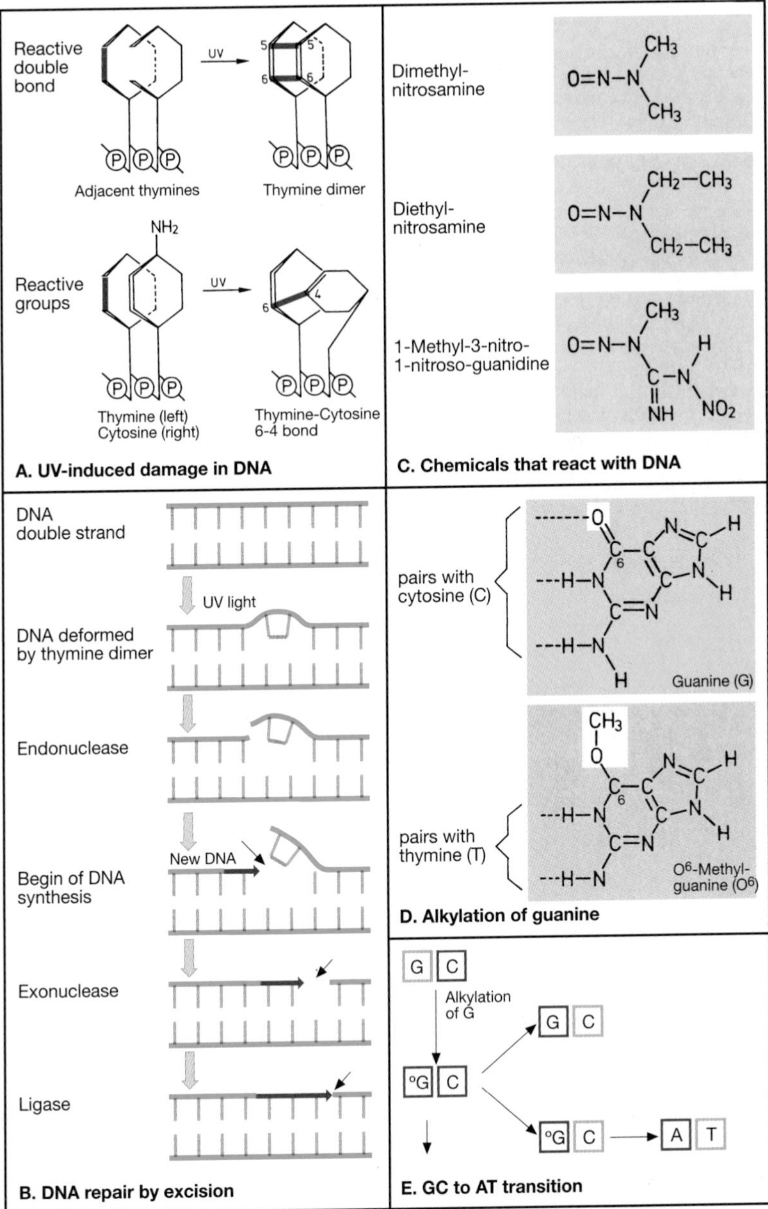
Reactive double bond
UV
Adjacent thymines
Thymine dimer
Reactive groups
Thymine (left) Cytosine (right)
Thymine-Cytosine 6-4 bond
A. UV-induced damage in DNA
Dimethyl-nitrosamine
Diethyl-nitrosamine
1-Methyl-3-nitro-1-nitroso-guanidine
C. Chemicals that react with DNA
DNA double strand
UV light
DNA deformed by thymine dimer
Endonuclease
New DNA
Begin of DNA synthesis
Exonuclease
Ligase
B. DNA repair by excision
pairs with cytosine (C)
Guanine (G)
pairs with thymine (T)
O6-Methyl-guanine (O6)
D. Alkylation of guanine
Alkylation of G
E. GC to AT transition

Xeroderma Pigmentosum

Xeroderma pigmentosum (XP) is a heterogeneous group of genetically determined skin disorders due to unusual sensitivity to ultraviolet light. They are manifested by dryness and pigmentation of the exposed regions of skin (xeroderma pigmentosum = "dry, pigmented skin"). The exposed areas of skin also show a tendency to develop tumors. The causes are different genetic defects of DNA repair.

Different types of DNA repair can be distinguished: (1) excision repair for damage control and (2) mismatch repair as copy editing for errors that have occurred during replication. Excision repair removes bulky UV-induced DNA lesions. Repair involves mechanisms similar to those involved in transcription and replication. The necessary enzymes are encoded by at least a dozen genes, which are highly conserved in bacteria, yeast, and mammals.

A. Clinical phenotype

The skin changes are limited to UV-exposed areas (1 and 2). Unexposed areas show no changes. Thus it is important to protect patients from UV light. An especially important feature is the tendency for multiple skin tumors to develop in the exposed areas (3). These may even occur in childhood or early adolescence. The types of tumors are the same as those occurring in healthy individuals after prolonged UV exposure.

B. Cellular phenotype

The UV sensitivity of cells can be demonstrated in vitro. When cultured fibroblasts from the skin of patients are exposed to UV light, the cells show a distinct dose-dependent decrease in survival rate compared with normal cells (1). Different degrees of UV sensitivity can be demonstrated. The short segment of new DNA normally formed during excision repair can be demonstrated by culturing cells in the presence of ^{3}H-thymidine and exposing them to UV light. The DNA synthesis induced for DNA repair can be made visible in autoradiographs. Since ^{3}H-thymidine is incorporated during DNA repair, these bases are visible as small dots caused by the isotope on the film (2). In contrast, xeroderma (XP) cells show markedly decreased or almost absent repair synthesis.

C. Genetic complementation in cell hybrids

If skin cells (fibroblasts) from normal persons and from patients (XP) are fused (cell hybrids) in culture and exposed to UV light, the cellular XP phenotype will be corrected (1). Normal DNA repair occurs. Also, hybrid cells from two different forms of XP show normal DNA synthesis (2) because cells with different repair defects correct each other (genetic complementation). However, if the mutant cells have the same defect (3), they are not be able to correct each other because they belong to the same complementation group. At present about ten complementation groups are known in xeroderma pigmentosum. They differ clinically in terms of severity and central nervous system involvement. Each complementation group is based on a mutation at a different gene locus. Several of these genes have been cloned and show homology with repair genes of other organisms, including yeast and bacteria.

References

Bootsma DA, Hoejmakers JHJ. The genetic basis of xeroderma pigmentosum. Ann Génét. 1991; 34: 143–150.

Hanawalt PC. Transcription-coupled repair and human diseases. Science. 1994; 266: 1957–1958.

Sancar A. Mechanisms of DNA excision repair. Science. 1994; 266: 1954–1956.

Wevrik R, Buchwald M. Mammalian DNA repair genes: Current Opinion Genet Develop. 1993; 3: 470–474.

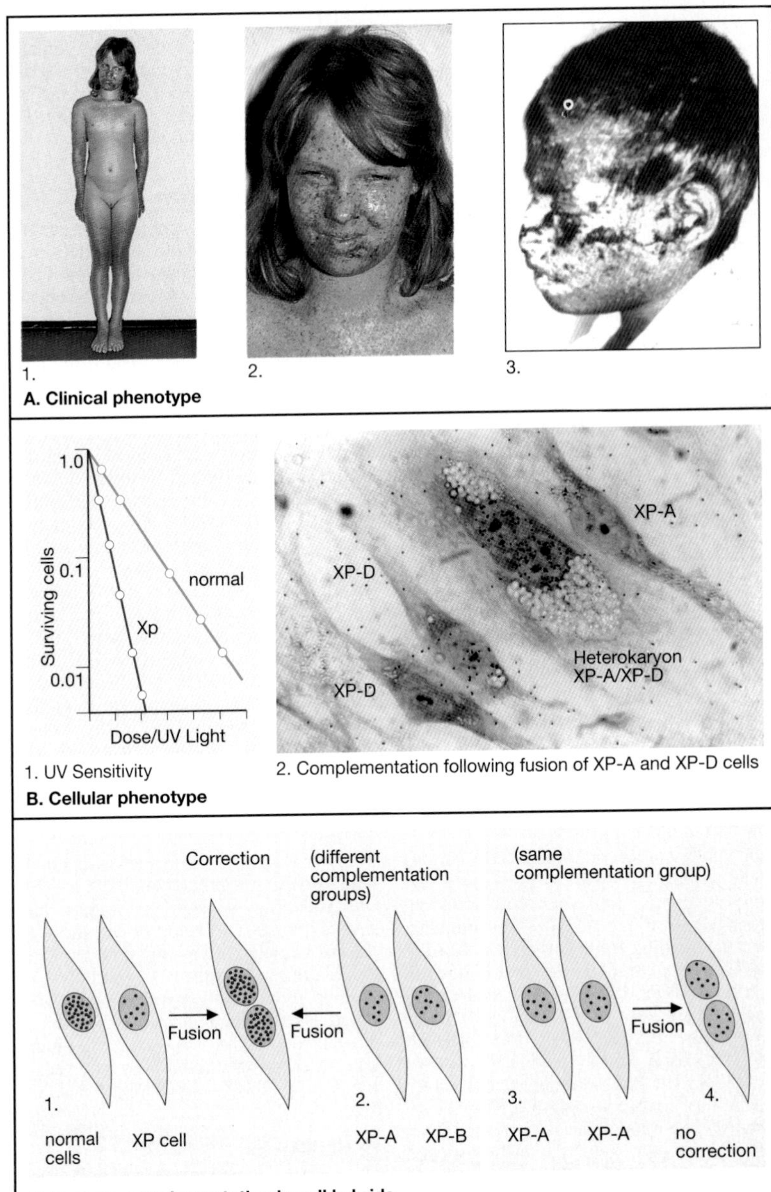

A. Clinical phenotype

1. UV Sensitivity

2. Complementation following fusion of XP-A and XP-D cells

B. Cellular phenotype

C. Genetic complementation in cell hybrids

Transcription and Translation

The genetic information stored in DNA must be converted into polypeptides with specific structures and functions (gene products). However, DNA does not directly serve as a template. The sequence of base pairs in DNA is first tranferred to information-transmitting messenger molecules of RNA (transcription). These subsequently serve as a template for the gene products (translation). The sequence of amino acids of the resulting polypeptides corresponds to the sequence of the codons in the DNA.

A. Transcription

Transcription is the process by which the sequence of nucleotide bases in DNA is converted into a corresponding sequence of nucleotide bases in RNA. First, the nucleotide sequence of one strand of DNA is transcribed into a complementary molecule of RNA (messenger RNA, mRNA). The DNA helix is opened and the DNA strand in the 3' to 5' direction serves as the template for the transcription into RNA. The synthesis of RNA proceeds in the 5' to 3' direction. The RNA strand used for translation (sense RNA) has the same orientation as the coding DNA strand. Transcription occurs through the action of a complex of enzymes (RNA polymerase, see pp. 196, 202).

B. Translation

During translation the sequence of codons made up of the nucleotide bases in mRNA is converted into a corresponding sequence of amino acids. Translation occurs in a reading frame which is defined at the start of translation (start codon). Amino acids are linked in the sequence determined by the mRNA nucleotide bases by a further class of RNA, transfer RNA (tRNA). Each amino acid has its own tRNA, which has a region that is complementary to its codon of the mRNA (anticodon). The anticodon CCG of the tRNA for glycine pairs with the mRNA codon GGC. Thus, in the example shown, only glycine can be attached at this site. The codons 1, 2, 3, and 4 of the mRNA are translated into the amino acid sequence methionine (met), glycine (gly), serine (ser), and isoleucine (ile), etc. Glycine and alanine will follow as 5 and 6. Codon 1 is always AUG (start codon).

C. Stages of translation

Translation (protein synthesis) in eukaryotes occurs outside of the cell nucleus in ribosomes in the cytoplasm. Ribosomes consist of subunits of numerous associated proteins and RNA molecules (ribosomal RNA, rRNA; p. 196). Translation begins with initiation (1): an initiation complex comprising mRNA, a ribosome, and tRNA is formed. This requires a number of initiation factors (IF1, IF2, IF3, etc.). Then elongation (2) follows: a further amino acid, determined by the next codon, is attached. A three-phase elongation cycle develops, with codon recognition, peptide binding to the next amino acid residue, and movement (translocation) of the ribosome three nucleotides further in the 3' direction of the mRNA. Translation ends with termination (3), when one of three mRNA stop codons (UAA, UGA, or UAG) is reached. The polypeptide chain formed leaves the ribosome, which dissociates into its subunits. The biochemical processes of the stages shown here have been greatly simplified.

D. Structure of transfer RNA (tRNA)

Transfer RNA has a characteristic, cloverleaflike structure, illustrated here by yeast phenylalanine tRNA (1). It has three single-stranded loop regions and four double-stranded "stem" regions. The three-dimensional structure (2) is complex, but various functional areas can be differentiated, such as the recognition site (anticodon) for the mRNA codon and the binding site for the respective amino acid (acceptor stem) on the 3' end (acceptor end).

Reference

Watson JD, et al. Molecular Biology of the Gene. 3rd ed. 1987.

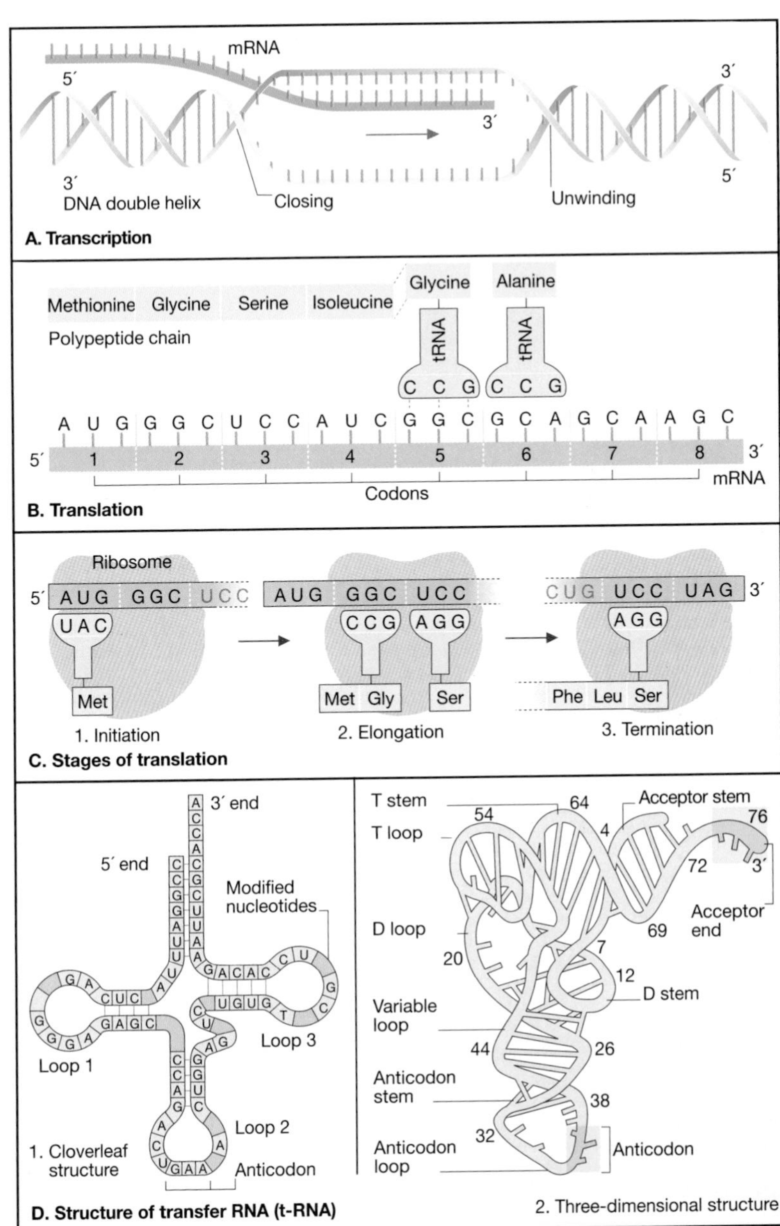

D. Structure of transfer RNA (t-RNA)

Genes and Mutation

The nucleotide sequence of DNA determines a corresponding sequence of amino acids through transcription and translation. If the DNA sequence becomes altered (mutation), the amino acid sequence may be altered at the corresponding site (mutation at the protein level). Thus, every mutation has a defined position. The decoding of mutations in bacteria contributed to the elucidation of the genetic code in 1966.

A. DNA and mutation

Coding DNA and its corresponding polypeptide are colinear. An alteration (mutation) of the DNA base sequence may lead to a different codon. The position of the resulting change in the sequence of amino acids corresponds to the postion of the mutation (1). **A** shows the gene for the protein tryptophan synthetase A of *E. coli* bacteria and mutations at four positions. At position 22, phenylalanine (phe) has been replaced by leucine (leu); at position 49, glutamic acid (glu) by glutamine (gln); at position 177, leu by arginine (arg). Every mutation has a defined position. Whether it leads to incorporation of another amino acid depends on whether the the corresponding codon has been altered. Different mutations at one position (one codon) are possible (2). Two different mutations have been observed at position 211: glycine (gly) to arginine (arg) and gly to glutamic acid (glu). Normally (in the wild type), codon 211 is GGA and codes for glycine (3). A mutation of GGA to AGA leads to a codon for arginine; a mutation to GAA, to a codon for glutamic acid (4).

B. Types of mutation

Basically, there are three different types of mutation involving single nucleotides (point mutation): 1. substitution (exchange), 2. deletion (loss), and 3. insertion (addition). With substitution, the consequences depend on whether a codon has been altered. Two types of substitution are distinguished: transition (exchange of one purine for another purine or of one pyrimidine for another) and transversion (exchange of a purine for a pyrimidine, or visa versa). A substitution may alter a codon so that a wrong amino acid is present at this site, but it has no effect on the reading frame (missense mutation), whereas a deletion or insertion causes a shift of the reading frame (frameshift mutation). Thus the following sequences no longer code for a functional gene product (nonsense mutation).

C. Transcription in prokaryotes and eukaryotes

Transcription differs between unicellular organisms without a nucleus such as bacteria (prokaryotes, 1) and multicellular organisms (eukaryotes, 2) with a cell nucleus. In prokaryotes, the mRNA serves directly as a template for translation. The sequences of DNA and mRNA correspond in a strict 1:1 relationship, i.e., they are colinear. Translation begins even before transcription has completely ended.

In contrast, a primary transcipt of RNA (precursor mRNA) is first formed in eukaryotic cells. This is a preliminary form of the mature mRNA. The latter is formed when the noncoding sections are removed from the primary transcript before it leaves the nucleus to be a template for forming a polypeptide (RNA processing).

The reason for these important differences is that functionally related genes generally lie together in prokaryotes and that noncoding segments (introns) are present in the genes of eukaryotes (see p. 50).

Reference

Watson JD, et al. Molecular Biology of the Gene. 3rd ed. 1987.

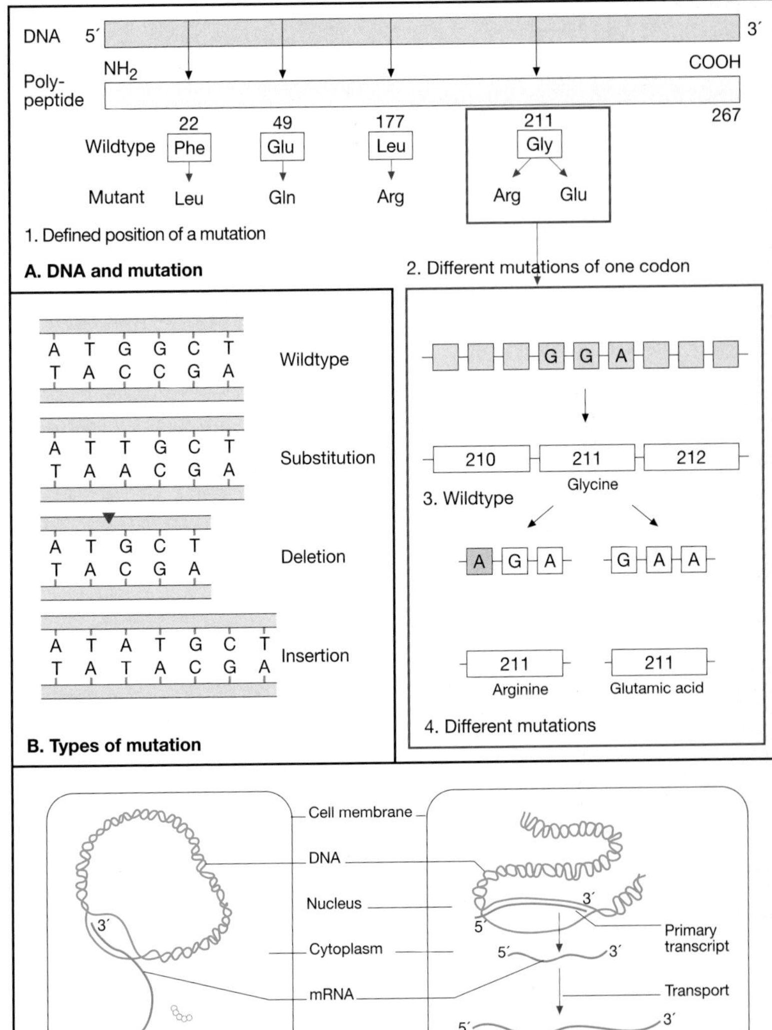
DNA 5′
3′
NH2
COOH
Poly-
peptide
22
49
177
211
267
Wildtype
Phe
Glu
Leu
Gly
Mutant
Leu
Gln
Arg
Arg
Glu
1. Defined position of a mutation
A. DNA and mutation
2. Different mutations of one codon
A T G G C T
T A C C G A
Wildtype
A T T G C T
T A A C G A
Substitution
A T G C T
T A C G A
Deletion
A T A T G C T
T A T A C G A
Insertion
B. Types of mutation
G G A
210
211
212
Glycine
3. Wildtype
A G A
G A A
211
Arginine
211
Glutamic acid
4. Different mutations
Cell membrane
DNA
Nucleus
Cytoplasm
mRNA
Ribosomes
Polypeptide
3′
5′
Primary
transcript
Transport
1. Prokaryote
2. Eukaryote
C. Transcription and translation in prokaryotes and eukaryotes

Genetic Code

The genetic code is the set of biological rules by which DNA nucleotide base pair sequences are translated into corresponding sequences of amino acids. Genes do not code for proteins directly, but through a messenger molecule (messenger RNA, mRNA). A code word (codon) for an amino acid consists of a sequence of three nucleotide base pairs (triplet codon). The genetic code also includes sequences for the beginning (start codon) and for the end (stop codon) of the coding region. The genetic code is universal; the same codons are used by different organisms.

A. Genetic code in mRNA for all amino acids

Each codon corresponds to one amino acid, but one amino acid may be coded for by different codons (redundancy of the code). For example, there are two possibilities to code for the amino acid phenylalanine: UUU and UUC, and there are six possibilities to code for the amino acid serine: UCU, UCC, UCA, UCG, AGU, and AGC. Many amino acids are determined by more than one codon. The greatest variation is in the third position (at the 3' end of the triplet). The genetic code was elucidated in 1966 by analyzing how triplets transmit information from the genes to proteins. mRNA added to bacteria could be directly converted into a corresponding protein. Synthetic RNA polymers such as polyuridylate (polyU), polyadenylate (polyA), and polycytidylate (polyC) could be directly translated into polyphenylalanine, polylysine, and polyproline in extracts of *E. coli* bacteria. This showed that UUU must code for phenylalanine, AAA for lysine, and CCC for proline. By further experiments with mixed polymers of different proportions of two or three nucleotides, the genetic code was determined for all amino acids and all nucleotide compositions.

B. Abbreviated code

Sequences of amino acids are designated with the single-letter abbreviations ("alphabetic code").
The start codon is AUG (methionine). Stop codons are UAA, UAG, and UGA. The only amino acids that are coded for by a single codon are methionine (AUG) and tryptophan (UGG).

C. Open reading frame

A segment of a nucleotide sequence can correspond to one of three reading frames (e.g., A, B, or C); however, only one is correct (open reading frame). In the example shown, the reading frames B and C are interrupted by a stop codon after three and five codons, respectively. Thus they cannot serve as reading frames for a coding sequence. On the other hand, A must be the correct reading frame: It begins with the start codon AUG and yields a sequence without stop codons (open reading frame).

D. Coding by similar nucleotide sequences

Since the genetic code is redundant, it is possible that different nucleotide sequences code for the same amino acid sequence. However, the differences are limited to one (or at most two) positions of a given triplet codon.

References

Rosenthal N. DNA and the genetic code. N Eng J Med. 1995; 331: 39–41.

Singer M, Berg P. Genes and Genomes: a changing perspective. Oxford–London: Blackwell Scientific Publications; 1991.

Nucleotide base					
First	Second				Third
	Uracil (U)	Cytosine (C)	Adenine (A)	Guanine (G)	
Uracil (U)	**F** Phenylalanine (Phe)	**S** Serine (Ser)	**Y** Tyrosine (Tyr)	**C** Cysteine (Cys)	U
	F Phenylalanine (Phe)	**S** Serine (Ser)	**Y** Tyrosine (Tyr)	**C** Cysteine (Cys)	C
	L Leucine (Leu)	**S** Serine (Ser)	Stop Codon	Stop Codon	A
	L Leucine (Leu)	**S** Serine (Ser)	Stop Codon	**W** Tryptophan (Trp)	G
Cytosine (C)	**L** Leucine (Leu)	**P** Proline (Pro)	**H** Histidine (His)	**R** Arginine (Arg)	U
	L Leucine (Leu)	**P** Proline (Pro)	**H** Histidine (His)	**R** Arginine (Arg)	C
	L Leucine (Leu)	**P** Proline (Pro)	**Q** Glutamine (Gln)	**R** Arginine (Arg)	A
	L Leucine (Leu)	**P** Proline (Pro)	**Q** Glutamine (Gln)	**R** Arginine (Arg)	G
Adenine (A)	**I** Isoleucine (Ile)	**T** Threonine (Thr)	**N** Asparagine (Asn)	**S** Serine (Ser)	U
	I Isoleucine (Ile)	**T** Threonine (Thr)	**N** Asparagine (Asn)	**S** Serine (Ser)	C
	I Isoleucine (Ile)	**T** Threonine (Thr)	**K** Lysine (Lys)	**R** Arginine (Arg)	A
	Start (Methionine)	**T** Threonine (Thr)	**K** Lysine (Lys)	**R** Arginine (Arg)	G
Guanine (G)	**V** Valine (Val)	**A** Alanine (Ala)	**D** Aspartic acid (Asp)	**G** Glycine (Gly)	U
	V Valine (Val)	**A** Alanine (Ala)	**D** Aspartic acid (Asp)	**G** Glycine (Gly)	C
	V Valine (Val)	**A** Alanine (Ala)	**E** Glutamic acid (Glu)	**G** Glycine (Gly)	A
	V Valine (Val)	**A** Alanine (Ala)	**E** Glutamic acid (Glu)	**G** Glycine (Gly)	G

A. Genetic code for all amino acids in mRNA

Start	AUG
Stop	UAA, UAG, UGA
A (Ala)	GCU, GCC, GCG, GCA
C (Cys)	UGU, UGC
D (Asp)	GAU, GAC
E (Glu)	GAG, GAA
F (Phe)	UUU, UUC
G (Gly)	GGU, GGC, GGG, GGA
H (His)	CAU, CAC
I (Ile)	AUU, AUC, AUA
K (Lys)	AAG, AAA
L (Leu)	CUU, CUC, CUG, CUA, UUG, UUA
M (Met)	AUG
N (Asn)	AAU, AAC
P (Pro)	CCU, CCC, CCG, CCA
Q (Gln)	CAG, CAA
R (Arg)	CGU, CGC, CGG, CAA, AGG, AGA
S (Ser)	UCU, UCC, UCG, UCA, AGU, AGC
T (Thr)	ACU, ACC, ACG, ACA
V (Val)	GUU, GUC, GUG, GUA
W (Trp)	UGG
Y (Tyr)	UAU, UAC
B (Asx)	Asn oder Asp
Z (Glx)	Gln oder Glu

B. Abbreviated code

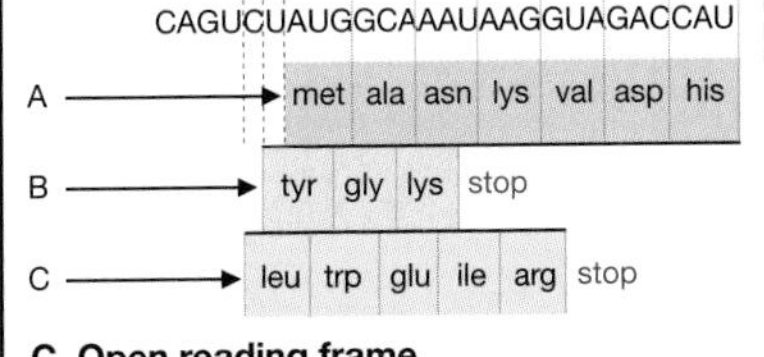

C. Open reading frame

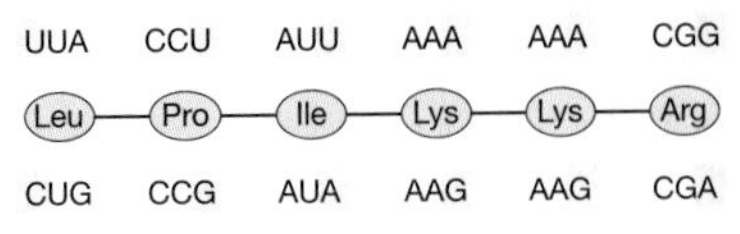

D. Coding by similar nucleotide sequences

The Structure of Eukaryotic Genes

In 1977, it was unexpectedly found that the DNA of a eukaryotic gene is longer than its corresponding mRNA. The reason is that certain sections of the initially formed primary RNA transcript are removed before translation occurs. The genes of eukaryotic organisms consist of sections of coding (exons) and noncoding DNA (introns).

A. Exons and introns

Electron micrographs show that DNA and its corresponding transcript (RNA) are of different lengths (1). When mRNA and its complementary single-stranded DNA are hybridized, loops of single-stranded DNA arise because mRNA hybridizes only with certain sections of the single-stranded DNA. In (2), seven loops (A to G) and eight hybridizing sections are shown (1 to 7 and the leading section L). Of the total 7700 DNA base pairs, only 1825 hybridize with mRNA. A hybridizing segment is called an exon. An initially transcribed DNA section that is subsequently removed from the primary transcript is an intron. The size and arrangement of exons and introns are characteristic for every eukaryotic gene (exon/intron structure) (Electronmicrograph from J. D. Watson et al., 1987).

B. Intervening DNA sequences

In prokaryotes, DNA is colinear with mRNA and contains no introns (1). In eukaryotes, mature mRNA is complementary to only certain sections of DNA because the latter contains introns (2).

C. mRNA splicing

Initially, a strand of DNA is completely transcribed to form the preliminary mRNA (primary transcript). Before the mRNA leaves the nucleus, defined sections of the primary transcript corresponding to the introns are removed, and the neighboring, remaining segments are joined together (mRNA splicing).

D. Splice junction

The sites of transition from exons to introns are referred to as splice junctions. The excision of an intron must be precise; otherwise, a shift in the reading frame will result. Therefore an exon always ends with AT (donor site) and an intron begins with GT (acceptor site). Since these sequences are identical in all higher organisms, they are referred to as consensus sequences. A splice junction may be altered by mutation so that an intron cannot be excised normally (splice mutant). Introns are removed enzymatically in a protein complex called the spliceosome.

E. Differential RNA splicing

A DNA segment can code for different forms of mRNA when different introns are removed from the primary transcript (alternative splicing). By means of alternative gene splicing, a gene can code for different, albeit similar gene products. This signifies a high degree of functional flexibility. Numerous examples for differential RNA splicing are known for mammalian genes. For example the primary transcript for the calcitonin gene is split so that two different forms of mature mRNA result. One is produced in the thyroid and codes for calcitonin (32 amino acids); the other codes for a calcitoninlike protein in the hypothalamus (calcitonin gene-related product, CGRP with 37 amino acids). In the thyroid, the CGRP sequences are removed; and in the hypothalamus, the calcitonin sequences are removed.

Reference

Watson JD, et al. 1987, and Stryer. 1995.

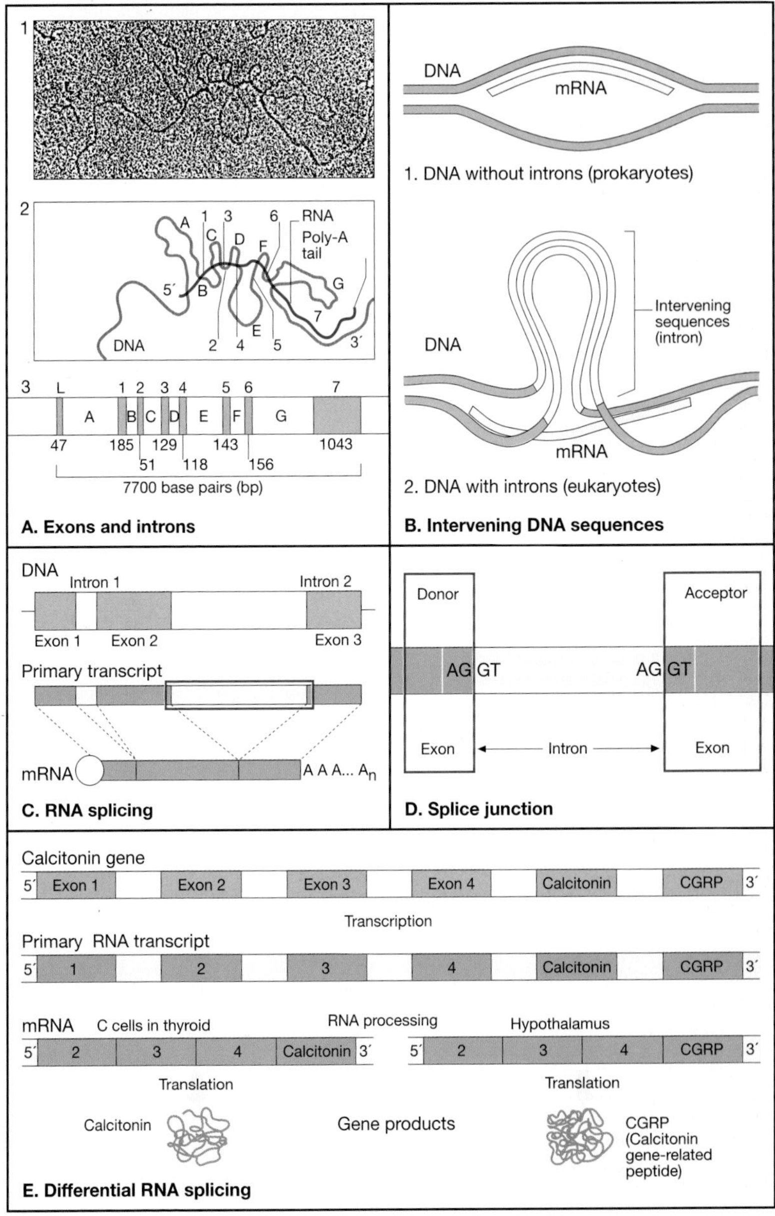

A. Exons and introns

B. Intervening DNA sequences

C. RNA splicing

D. Splice junction

E. Differential RNA splicing

DNA Sequencing by Chemical Methods (Maxam–Gilbert)

In the late 1970s, methods for determining the base sequences in DNA were developed. These methods are relatively simple and rapid, so that even long sections of DNA can be sequenced. Genetic information is thus directly accessible and does not require prior knowledge of the gene product. Two principal methods of DNA sequencing are available: chemical and enzymatic (p. 54).
Increasingly, nonradioactive procedures with specific fluorochrome staining of the different bases are being used. These are suitable for automation (see p. 356).

A. Principle

This method is based on base-specific cleavage of DNA. The DNA segment to be sequenced is analyzed for each of the four bases in reaction mixtures, where it is cleaved into fragments by chemical modification. The sizes of the fragments depend on the positions of the respective nucleotide bases. In the example shown (1, 2, and 3), guanine (G) lies in positions 3, 6, 9, and 14. When guanine is cleaved, particles are formed whose exact sizes depend on the positions of guanine. No other fragments have these sizes. By exact comparison of all the fragments formed in the four reaction mixtures, the positions of the individual nucleotide bases can be inferred and their sequence determined.

B. Preparation

First, the section of DNA (1) to be sequenced is radioactively labeled (2). Then single-stranded DNA (3), the starting material, is produced by denaturation. Cleavage of the DNA is attained by base-specific chemical modification: of guanine (G) by dimethyl sulphate, of guanine and adenine (G + A) by formic acid, of cytosine and thymine (C + T) by hydrazine, and of cytosine (C) by hydrazine with 5M NaCl. The modified base detaches from deoxyribose, and a break in the DNA strand is produced at this point (4). The reaction conditions are chosen so that on average, only one base per molecule is modified. This results in a distribution of fragments of different sizes such that all the sizes of the analyzed DNA segment that can end with the cleaved base will occur. All fragments are labeled at the 5' end.

C. Procedure and results

The test is done in four separate reaction mixtures (1) in which the chemically induced cleavage ensues for each of the four nucleotides. The four fragment mixtures are subsequently applied next to each other in four parallel lanes on a polyacrylamide gel where they are electrophoretically separated according to length (2). Since guanine (G) and cytosine (C) are simultaneously cleaved in the reaction mixtures for adenine (A) and thymine (T), respectively, fragments of the same size will also occur for A and T at the corresponding positions in the respective lanes for G and C. Thus, in these positions the gels contain two fragments of the same size. In contrast, the cleavage of A and T is specific, so that they are represented by one fragment. The sequencing gel can then be read from the smallest to the largest DNA fragments opposite the direction of fragment migration (3). In the diagram, the sequence of the starting fragment thus determined is (TAGTCGCAG-TACCGTA).

References

Rosenthal N. Fine structure of a gene – DNA sequencing. N Eng J Med. 1995; 332: 589–591.

Watson JD, et al. Molecular Biology of the Gene; 1987.

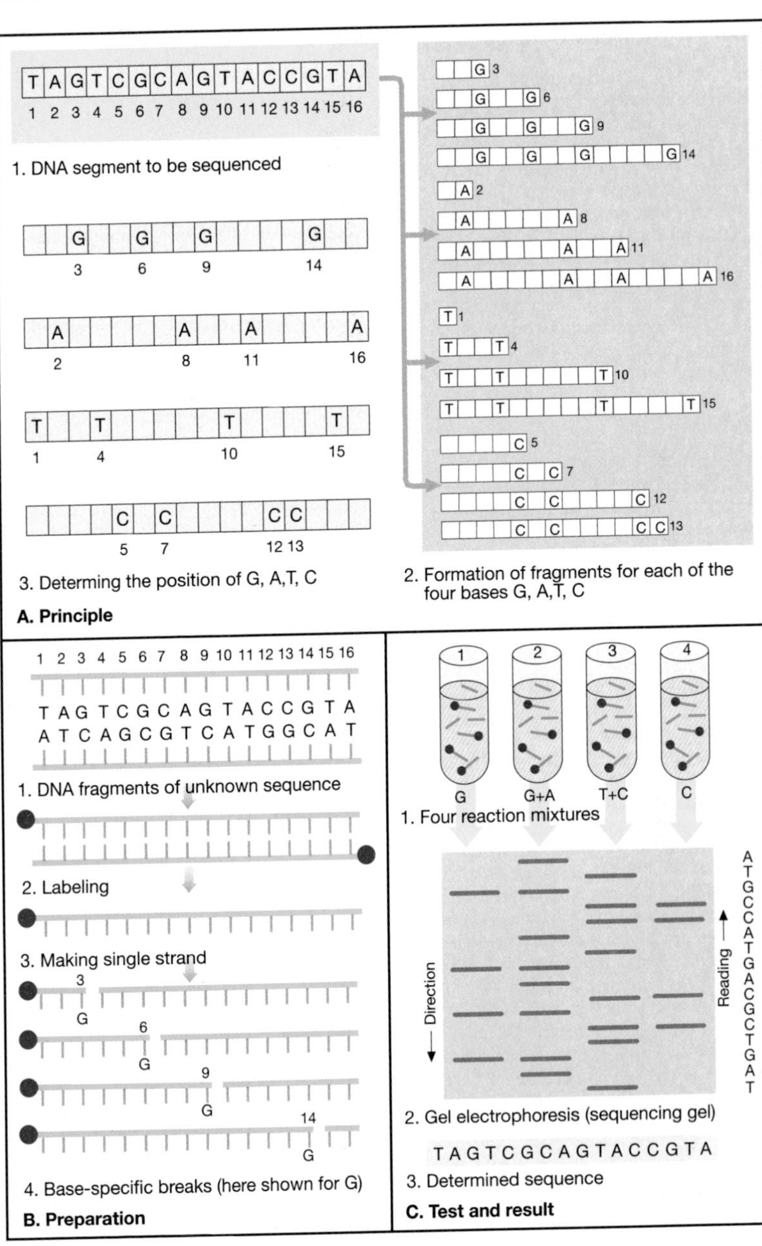
T A G T C G C A G T A C C G T A
1 2 3 4 5 6 7 8 9 10 11 12 13 14 15 16
1. DNA segment to be sequenced
G G G G
3 6 9 14
A A A A
2 8 11 16
T T T T
1 4 10 15
C C C C
5 7 12 13
3. Determing the position of G, A,T, C
2. Formation of fragments for each of the four bases G, A,T, C
A. Principle
1 2 3 4 5 6 7 8 9 10 11 12 13 14 15 16
T A G T C G C A G T A C C G T A
A T C A G C G T C A T G G C A T
1. DNA fragments of unknown sequence
2. Labeling
3. Making single strand
4. Base-specific breaks (here shown for G)
B. Preparation
1 2 3 4
G G+A T+C C
1. Four reaction mixtures
Direction
Reading
A T G C C A T G A C G C T G A T
2. Gel electrophoresis (sequencing gel)
T A G T C G C A G T A C C G T A
3. Determined sequence
C. Test and result

Enzymatic DNA Sequencing

The principle of the sequencing method according to F. Sanger lies in interrupting the synthesis of the 5' to 3' strand at specific bases of the DNA segment to be sequenced. As a consequence, DNA fragments of different sizes are formed. The sizes of the fragments reflect the positions of the respective bases. In this way, the base sequence of a given DNA segment can be determined. This method does not require the isolation of a labeled single strand, as with the chemical cleavage method of Maxam–Gilbert. Thus, it can also be applied to double-stranded DNA.

A. Principle

First, the DNA fragment to be sequenced (1) is cloned into plasmid DNA (2). DNA synthesis is initiated by a primer (e.g., an oligonucleotide with a sequence complementary to the plasmid DNA) (3). The incorporated DNA fragment is replicated with the plasmid DNA. DNA synthesis is interrupted at a given nucleotide by incorporating a nucleotide containing an H atom instead of the usual –OH group in the C3 position of the sugar component of that nucleotide (dideoxynucleotide, ddNTP) (4). When a dideoxynucleotide is incorporated, no bond between its position 3 and the next nucleotide is possible (interruption of the nucleotide chain).

The ratio of normal deoxynucleotide (dNTP) to dideoxynucleotide (ddNTP) is chosen so that in each reaction mixture, the nucleotide chains will be interrupted infrequently, on average each chain being interrupted at only one of the different positions of the respective nucleotide. For example, if thymine occurs three times in the DNA segment being sequenced, then three DNA fragments will be produced, the sizes of which reflect the locations of thymine in the sequence. Thus, a series of newly formed DNA fragments arises, all of which end with a dideoxynucleotide (5).

B. Procedure

Four reaction mixtures are set up for replicating the DNA segment to be sequenced (1). Each reaction vessel contains the normal building blocks of DNA (dATP, dTTP, dGTP, and dCTP) and DNA polymerase I and in addition, one of the four dideoxynucleotides (ddATP, ddTTP, ddGTP, or ddCTP). For subsequent demonstration of the DNA fragments either a radioactively or a fluorochrome-labeled primer is used. In the diagram, this is [α-^{32}P]-dCTP.

The fragments produced are separated according to size in the sequencing gel and demonstrated using autoradiography or a fluorochrome method (2). Finally, the nucleotide sequence is read from the gel in the direction opposite that of fragment migration (in the example shown, CTACGCATTAG).

The autoradiogram of a sequence of 105 nucleotide bases (3) begins below with CCACTCCTA... and ends above with ...CTAAATTTAAACTTCTT. Toward the top, the larger fragments are less easily distinguished than the smaller ones at the bottom. The difference in intensity of some bands is of technical origin and not of significance for sequencing.

The four lanes in the real sequence gel are arranged as in the schematic representation, i.e., GATC from left to right.

References

see p. 52

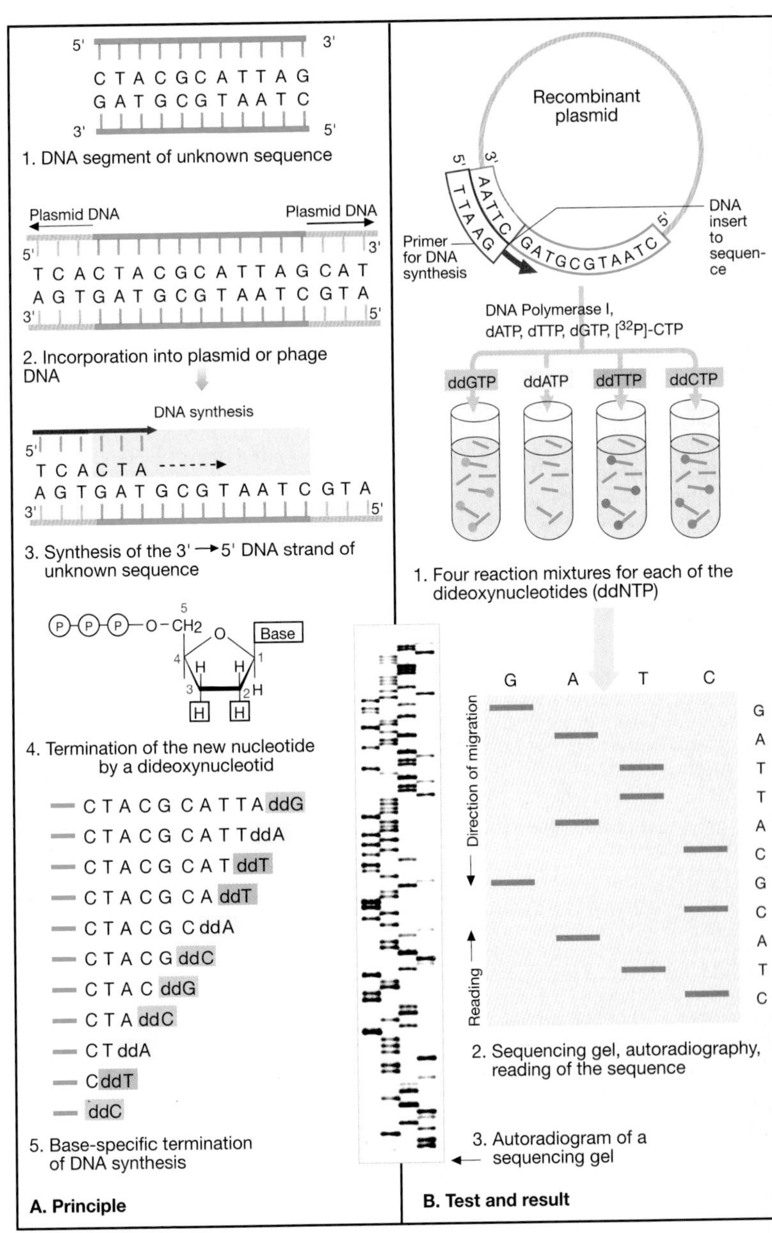

5' 3'
C T A C G C A T T A G
G A T G C G T A A T C
3' 5'
1. DNA segment of unknown sequence
Plasmid DNA
Plasmid DNA
5' 3'
T C A C T A C G C A T T A G C A T
A G T G A T G C G T A A T C G T A
3' 5'
2. Incorporation into plasmid or phage DNA
DNA synthesis
5'
T C A C T A
A G T G A T G C G T A A T C G T A
3' 5'
3. Synthesis of the 3' → 5' DNA strand of unknown sequence
P P P O CH2 O Base
5 4 3 2 1
H H H H H
4. Termination of the new nucleotide by a dideoxynucleotid
C T A C G C A T T A ddG
C T A C G C A T T ddA
C T A C G C A T ddT
C T A C G C A ddT
C T A C G C ddA
C T A C G ddC
C T A C ddG
C T A ddC
C T ddA
C ddT
ddC
5. Base-specific termination of DNA synthesis
A. Principle
Recombinant plasmid
5' 3'
T T A A G
A A T T C
G A T G C G T A A T C
5'
Primer for DNA synthesis
DNA insert to sequence
DNA Polymerase I,
dATP, dTTP, dGTP, [32P]-CTP
ddGTP
ddATP
ddTTP
ddCTP
1. Four reaction mixtures for each of the dideoxynucleotides (ddNTP)
G A T C
Direction of migration
Reading
G
A
T
T
A
C
G
C
A
T
C
2. Sequencing gel, autoradiography, reading of the sequence
3. Autoradiogram of a sequencing gel
B. Test and result

Cloning of a DNA Fragment

When a segment of DNA is to be investigated, for instance to characterize its structure and function or to demonstrate a pathogenic mutation, the starting material is not sufficient. Thus, multiple copies of the chosen DNA segment must be produced (cloning). Since bacteria grow very rapidly (a cell division every 20 to 40 minutes), the DNA segment in question can be reproduced manifold in a relatively short time in bacteria.

A. Cleavage of DNA at specific sites by restriction enzymes

To isolate a DNA fragment, restriction enzymes with cutting-site specificity are required. These enzymes, which are isolated from bacteria, cut DNA only at specific nucleotide sequences (recognition sites). For example, the restriction enzyme *Eco*RI cuts DNA wherever the sequence GAATTC occurs (1, 2). The resulting DNA fragments end on each side with a short, single-stranded segment of AATT– or –TTAA. Since the single–stranded ends are complementary, they can be joined together by DNA ligase (3). In this way, DNA fragments of different origin can be joined to each other as long as they have been cut by the same restriction enzyme.

B. Producing multiple copies (cloning) of a DNA fragment

The DNA fragment to be cloned must be introduced into bacteria by means of a vector. Thus it is first built into a plasmid. A plasmid is a small ring-shaped DNA molecule that is found in bacteria and can replicate independently of the bacterial chromosome. The plasmid and foreign DNA are cut with the same restriction enzyme (1). Some of the plasmids take up foreign DNA (hybrid plasmids of recombinant DNA) (2). Subsequently the bacteria containing the recombinant DNA must be identified and selectively cultured (3). Since plasmids contain genes for antibiotic resistance, they can be used to select for the desired bacteria. When the antibiotic resistance of a bacterium is altered by incorporating plasmids with recombinant DNA (loss of a resistance previously present or attainment of a resistance not previously present), the culture medium can be chosen so that only those bacteria with the desired plasmid DNA can grow (4). The DNA fragment in question can then be multiplied ad lib (5).

C. A plasmid vector (pBR322) for cloning

A previously frequently used plasma vector (pBR322) is presented. This plasmid contains cutting sites for the restriction enzymes *Pst*I, *Eco*RI, and *Sal*I in addition to genes for ampicillin and tetracycline resistance (1). If a foreign DNA fragment is incoporated into the plasmid after being cut with *Eco*RI, then tetracycline and ampicillin resistance are retained (2). If the enzyme *Pst*I is used, ampicillin resistance is lost (the bacterium becomes ampicillin sensitive), but tetracycline resistance is retained. If the enzyme *Sal*I is used, tetracycline resistance disappears (the bacterium becomes tetracycline sensitive), but ampicillin resistance is retained. This is the basis for selecting bacteria that have incorporated recombinant plasmids with a given DNA fragment.

Cloning in plasmids (bacteria) has become less important since artificial yeast chromosomes (YACs) have become available for cloning relatively large DNA fragments (see p. 104).

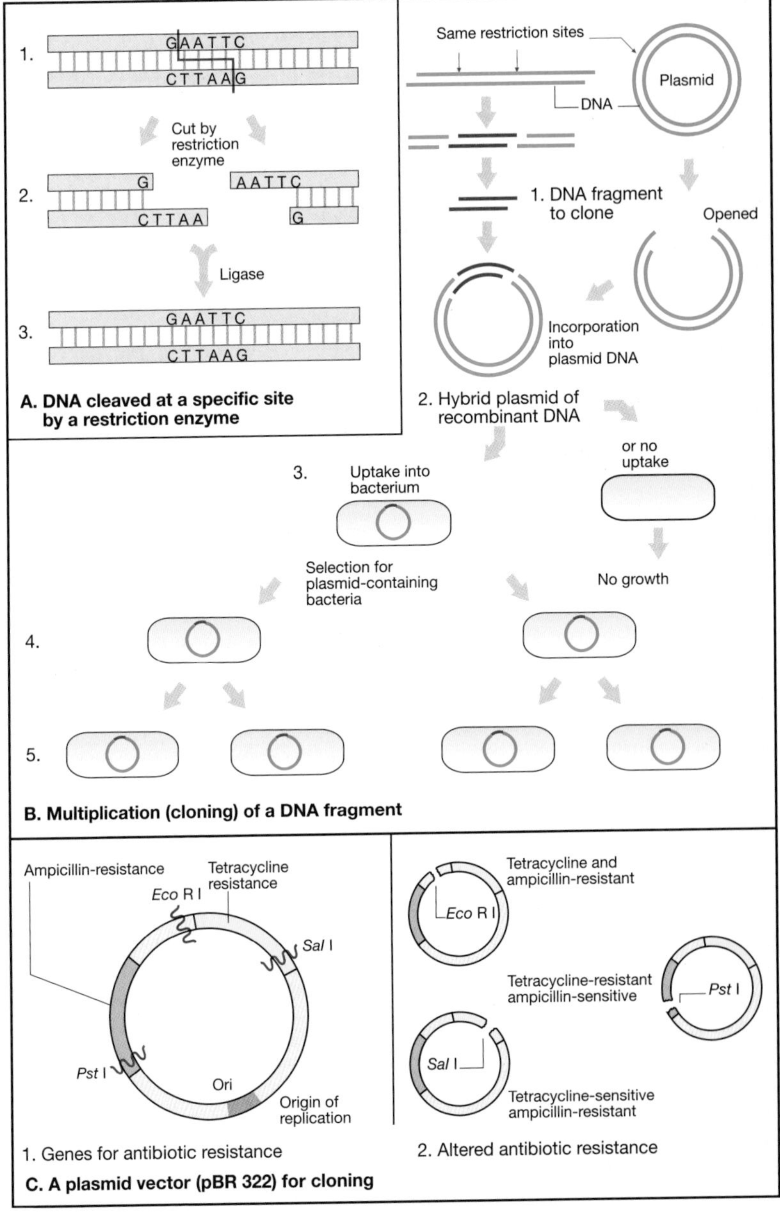

A. DNA cleaved at a specific site by a restriction enzyme

B. Multiplication (cloning) of a DNA fragment

C. A plasmid vector (pBR 322) for cloning

Libraries of Genomic DNA and cDNA

A DNA library is a collection of DNA fragments that in their entirety represent the genome, that is, a particular gene being sought and all remaining DNA. It is the starting point for cloning a gene of unknown chromosomal location. To produce a library, the total DNA is digested with a restriction enzyme, and the resulting fragments are incorporated into vectors and replicated in bacteria. A sufficient number of clones must be present so that every segment is represented at least once. This is a question of the size of the genome being investigated and the size of the fragments. Plasmids and phages are used as vectors. For larger DNA fragments, yeast cells may be employed. There are two different types of libraries: genomic DNA and cDNA.

A. Library of genomic DNA

Clones of genomic DNA are copies of DNA fragments (e.g., a–f) from all of the chromosomes (1). They contain coding and noncoding sequences. Restriction enzymes are used to split the genomic DNA into many fragments (2). The latter are incorporated into vectors, e.g., into phage DNA (3), and are replicated in bacteria (4) (p. 57). The complete collection of recombinant DNA molecules, containing all DNA sequences of a species or individual, is called a genomic library. To find a particular gene, a screening procedure is required (see p. 61).

B. Library of cDNA

cDNA is complementary DNA. It is formed from mRNA by the enzyme reverse transcriptase (see p. 62). Unlike a genomic library, which is complete and contains coding and noncoding DNA, a cDNA library consists only of coding DNA sequences. This specificity offers considerable advantages over genomic DNA. However, it requires that mRNA be available and does not yield information about the structure of the gene. mRNA can be obtained only from cells in which the respective gene is transcribed, i.e., in which mRNA is produced. In eukaryotes, the RNA formed during transcription (primary transcript) undergoes splicing to form mRNA (see p. 50). Complementary DNA (cDNA) is formed from mRNA by the enzyme reverse transcriptase. The cDNA can serve as a template for synthesis of a complementary DNA strand, so that complete double-stranded DNA can be formed (cDNA clone). Its sequence corresponds to the coding sequences of the gene exons. Thus it is well suited for use as a probe (cDNA probe). The subsequent steps, incorporation into a vector and replication in bacteria, correspond to those of the procedure to produce a genomic library. cDNA clones can only be won from coding regions of an active (mRNA-producing) gene; thus, the cDNA clones of different tissues differ according to genetic activity. Since cDNA clones correspond to the coding sequences of a gene (exons) and contain no noncoding sections (introns), cloned cDNA is the preferred starting material when further information about a gene product is sought by analyzing the gene. The sequence of amino acids in a protein can be determined from cloned and sequenced cDNA. Also, large amounts of a protein can be produced by having the cloned gene expressed in bacteria or yeast cells.

References

Rosenthal N. Stalking the gene – DNA libraries. N Eng J Med. 1994; 331: 599–600.

Watson JD, Gilman M, Witkowski J, Zoller M. Recombinant DNA. 2nd ed. New York: Scientific American Books; 1992.

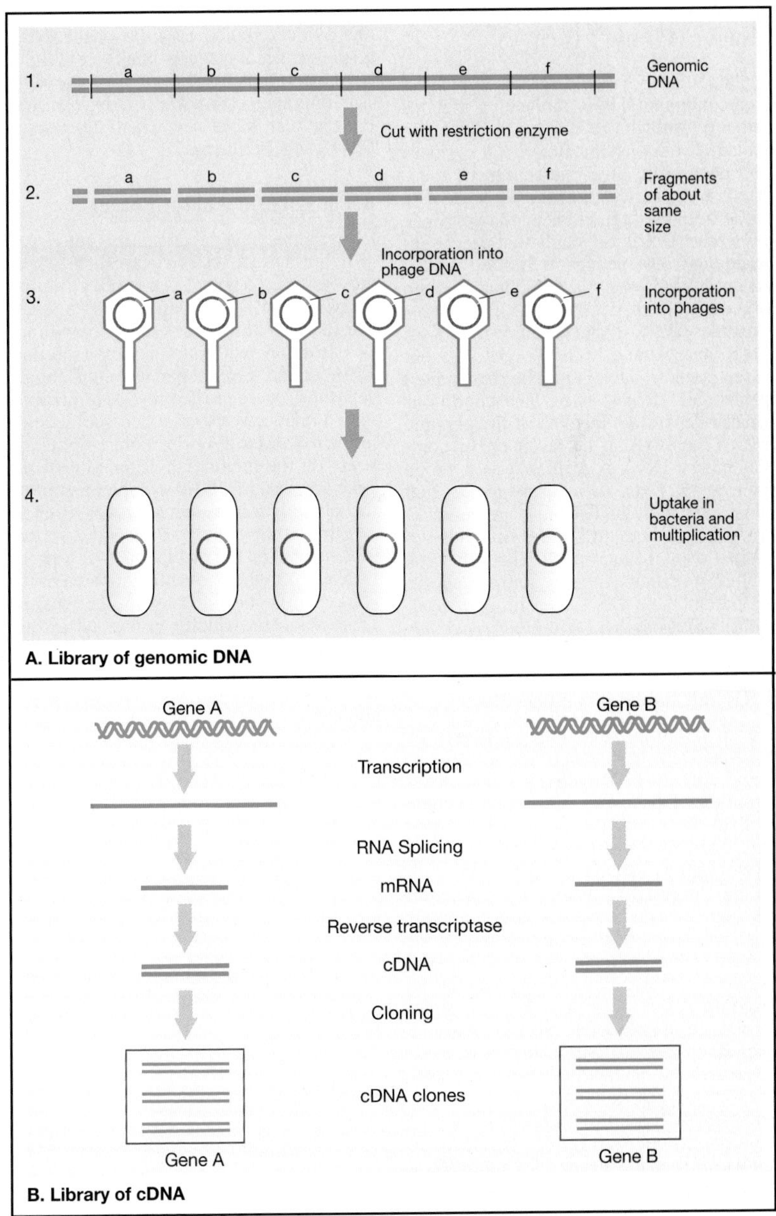

A. Library of genomic DNA

B. Library of cDNA

Storing of Eukaryotic DNA in Phages (Phage Library)

A phage library is a collection, in phages, of recombinant DNA molecules that together represent the genome of a higher organism. From this library, a DNA segment of a gene to be investigated can be identified and singled out (screening) if a probe from the segment is available. Since the size of DNA fragments that can be incorporated into phages is limited (up to about 20 kb), other carriers that take up larger fragments, e.g., yeast artificial chromosomes (YACs), are preferred. To isolate a gene from a library, every segment of the genomic DNA must be represented on average at least once. The conditions for this depend on the size of the genome and of the DNA to be taken up. If complementary DNA is available as a probe for a given gene, then (in principle) that gene can be isolated from a genomic library. In practice, difficulties occur due to limited ablity to incorporate the right size, limited specificity of selection procedures for recombinant DNA, and limited availability of probes.

A. Establishment of a phage library

The first step in constructing a library of genomic DNA (genomic library or gene library) consists in extraction (1) and cleaving of total DNA with a restriction enzyme so that the fragments are of suitable size (ca 15 kb) to be incorporated into phage DNA. This is usually accomplished with partial digestion using a restriction enzyme that cuts at not too large intervals (e.g., enzymes that recognize only four bases). Overlapping fragments (2) result. Fragments that are too large cannot be taken up by the vector, and those too small cannot be found later. Vectors may be phages, cosmids (plasmids that have a *cos* [*co*he*s*ive] site with complementary single strands), or yeast cells.

Since about 25 kb of the middle section of a phage genome (altogether ca. 48 kb in lambda) (3) is not needed for replication, this region can be removed and replaced with foreign DNA (4). The DNA fragments incorporated into the phage DNA form the library (gene bank) (5). Since only recombinant phages are of the right size, only they can multiply in bacteria (6). Phages that have not taken up foreign DNA cannot replicate.

B. Screening of a phage library

Bacteria that have taken up the vectors can grow on an agar-coated petri dish, where they form colonies (1). A replica imprint of the culture is taken on a membrane (2), and the DNA that sticks to the membrane is denatured with an alkaline solution. DNA of the gene segment being sought can then be identified by hybridization with a radioactively (or otherwise) labeled probe (3). After hybridization, a signal appears on the membrane at the site of the gene segment (4). DNA complementary to the labeled probe is located here; its exact position in the culture corresponds to that of the signal on the membrane (4). A probe is taken from the corresponding area of the culture (5). It will contain the desired DNA segment, which can now be further replicated (cloned) in bacteria. By this means, the desired segment can be enriched and is available for subsequent studies.

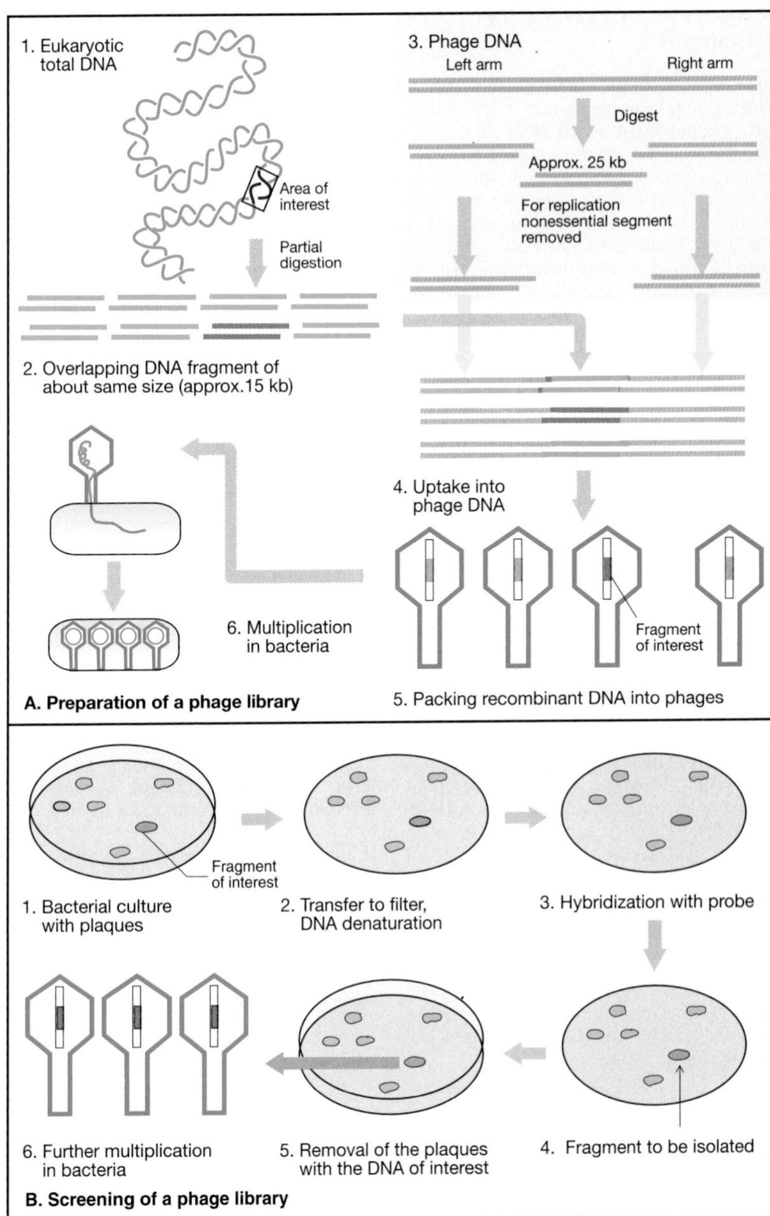

A. Preparation of a phage library

B. Screening of a phage library

Preparation of cDNA for Cloning

cDNA is a single-stranded segment of DNA that is complementary to the mRNA of a coding DNA segment or of a whole gene. It can be used as a probe (cDNA probe as opposed to a genomic probe) for the corresponding gene because it is complementary to coding sections (exons) of the gene. If the gene has been altered by structural rearrangement at a corresponding site, e.g., by deletion, the normal and mutated DNA can be differentiated. Thus, the preparation and cloning of cDNA is of great importance. From the cDNA sequence, essential inferences can be made about a gene and its gene product.

A. Preparation of cDNA

cDNA is prepared from mRNA. Therefore, a tissue is required in which the respective gene is transcribed and mRNA produced in sufficient quantities. First, mRNA is isolated. Then a primer is attached so that the enzyme reverse transcriptase can form complementary DNA (cDNA) from the mRNA. Since mRNA contains poly(A) at its 3' end, a primer of poly(T) can be attached. From here, the enzyme reverse transcriptase can start forming cDNA in the 5' to 3' direction. The RNA is then removed by means of alkali. The cDNA serves as a template for the formation of a new strand of DNA. This requires the enzyme DNA polymerase. The result is a double strand of DNA, one strand of which is complementary to the original mRNA. To this DNA, single sequences (linkers) are attached that are complementary to the single-stranded ends produced by the restriction enzyme to be used. The same enzyme is used to cut the vector, e.g., a plasmid, so that the cDNA can be incorporated for cloning.

B. Plasmids as vectors for cDNA cloning

Since only a small proportion of the plasmids take up foreign DNA, there must be selection for the bacteria that contain the recombinant plasmids. The plasmid DNA, which is ring shaped, is cut with a restriction enzyme and opened. In the process, short single-stranded DNA sequences arise at both ends, complementary to the previously added single-stranded sequences (linkers) of the DNA fragments to be cloned (e.g., AATT and TTAA). When such a fragment has been incorporated, the result is a recombinant plasmid. Bacteria that have taken in recombinant plasmids must be differentiated from those that have not. This is possible if the plasmids contain a gene for antibiotic resistance, e.g., ampicillin resistance (AmpR) (see p. 57). In this way, it is possible to select for bacteria that have taken up recombinant plasmids. This can be demonstrated by the changes of a gene (β-galactosidase) due to the insert.

C. cDNA cloning

Only those bacteria become ampicillin resistant that have incorporated a recombinant plasmid. Recombinant plasmids, which contain the gene for ampicillin resistance, transform ampicillin-sensitive bacteria into ampicillin-resistant bacteria. In an ampicillin-containing medium, only those bacteria grow that contain the recombinant plasmid with the desired DNA fragment. By further replication in these bacteria, the fragment can be cloned until there is enough material to be studied.

(Figures after J.D. Watson et al., 1987).

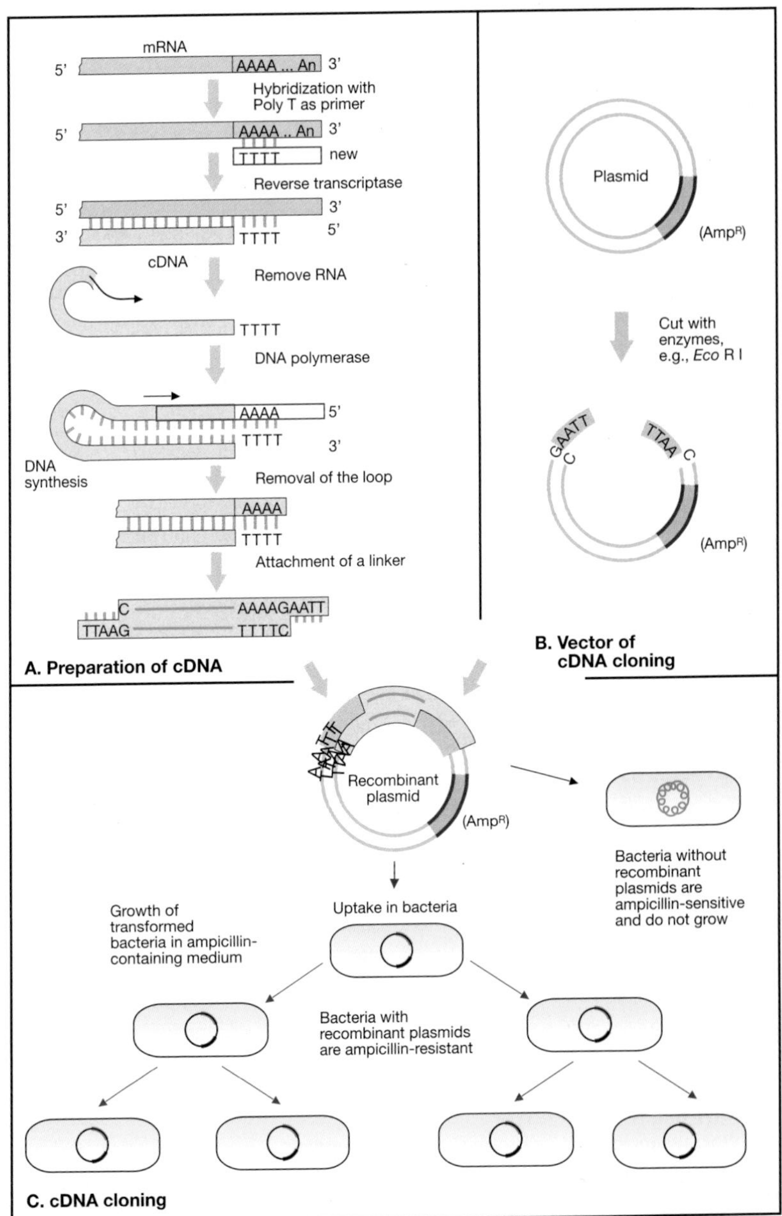
mRNA
5'
AAAA ... An
3'
Hybridization with
Poly T as primer
AAAA .. An
TTTT
new
Reverse transcriptase
TTTT
cDNA
Remove RNA
TTTT
DNA polymerase
AAAA
TTTT
DNA
synthesis
Removal of the loop
AAAA
TTTT
Attachment of a linker
C
AAAAGAATT
TTAAG
TTTTC
A. Preparation of cDNA
Plasmid
(AmpR)
Cut with
enzymes,
e.g., Eco R I
GAATT
C
TTAA
C
(AmpR)
B. Vector of
cDNA cloning
Recombinant
plasmid
(AmpR)
Bacteria without
recombinant
plasmids are
ampicillin-sensitive
and do not grow
Uptake in bacteria
Growth of
transformed
bacteria in ampicillin-
containing medium
Bacteria with
recombinant plasmids
are ampicillin-resistant
C. cDNA cloning

Restriction Analysis by the Southern Blot Procedure

Restriction analysis is the investigation of a DNA segment by means of restriction enzymes. It is an important method in the molecular genetic diagnosis of genetically determined diseases. It requires no cloning, sequencing, or other complex biochemical procedures.

A. Southern blot procedure for identifying a gene

The analysis starts with total DNA (1). The DNA is isolated and cut with restriction enzymes (2, 3). One of the not-yet-identified fragments contains the gene being sought or part of the gene. The fragments are sorted by size in a gel (usually agarose) in an electric field (electrophoresis) (4). The smaller the fragment, the faster it migrates; the larger, the slower. Next, the actual Southern blot (named for Dr. Edmund Southern of Edinburgh, who developed this method) is carried out: The fragments contained in the gel are transferred to a nitrocellulose or nylon membrane (5). There the DNA is denatured (made single-stranded) with alkali and fixed to the membrane by moderate heating (ca. 80° C) or UV cross-linkage. The sample is incubated with a probe of complementary single-stranded DNA (genomic DNA or cDNA) from the gene. The probe hybridizes solely with the complementary fragment being sought, and not with others. Since the probe is labeled with radioactive ^{32}P, the fragment being sought can be identified by placing an X-ray film on the membrane, where it appears as a black band on the film after development (autoradiogram) (6). The size, corresponding to position, is determined by running DNA fragments of known size in the electrophoresis.

B. Restriction fragment length polymorphism (RFLP)

In about every 100 base pairs of a DNA segment, the nucleotide sequence differs in some individuals (DNA polymorphism). As a result, the recognition sequence of a restriction enzyme may be present on one chromosome, but not the other. In this case the restriction fragment sizes differ at this site (restriction fragment length polymorphism, RFLP). An example is shown for two 5-kb (5000 base pair) DNA segments. In one, a restriction site in the middle is present (allele 1); in the other (allele 2) it is absent. With a Southern blot, it can be determined whether in this location an individual is homozygous 1–1 (two alleles 1, no 5-kb fragment), heterozygous 1–2 (one allele each, 1 and 2), or homozygous 2–2 (two alleles 2). If the mutation being sought lies on the chromosome carrying the 5-kb fragment, the presence of this fragment indicates presence of the mutation. The absence of this fragment would indicate that the mutation is absent.

C. Other DNA polymorphisms

Certain base sequences may occur repeatedly next to each other (tandem repeats). The number of repeats can vary as a polymorphism (variable number tandem repeat, VNTR). In such a segment, the number of repeats present determines the size of the fragments. Numerous DNA fragments of various sizes are formed. Such a DNA segment with four different fragments, of 2.0, 2.5, 3.0, and 3.5 kb, are presented. These can occur in a total of 10 different allelic combinations. Every individual has one of these combinations, and there is a high probability that two unrelated individuals differ in a given DNA segment. This can be used in genetic analysis (see Formal Genetics). Some DNA marker systems that show extreme individuality can be utilized for forensic purposes. More recently, polymorphic repeats of a few base pairs, e.g., CA repeats, are used to identify individual segments of DNA (minisatellites and microsatellites, see p. 353).

Reference

Housman D. Human DNA polymorphism. N Eng J Med. 1995; 332: 318–320.

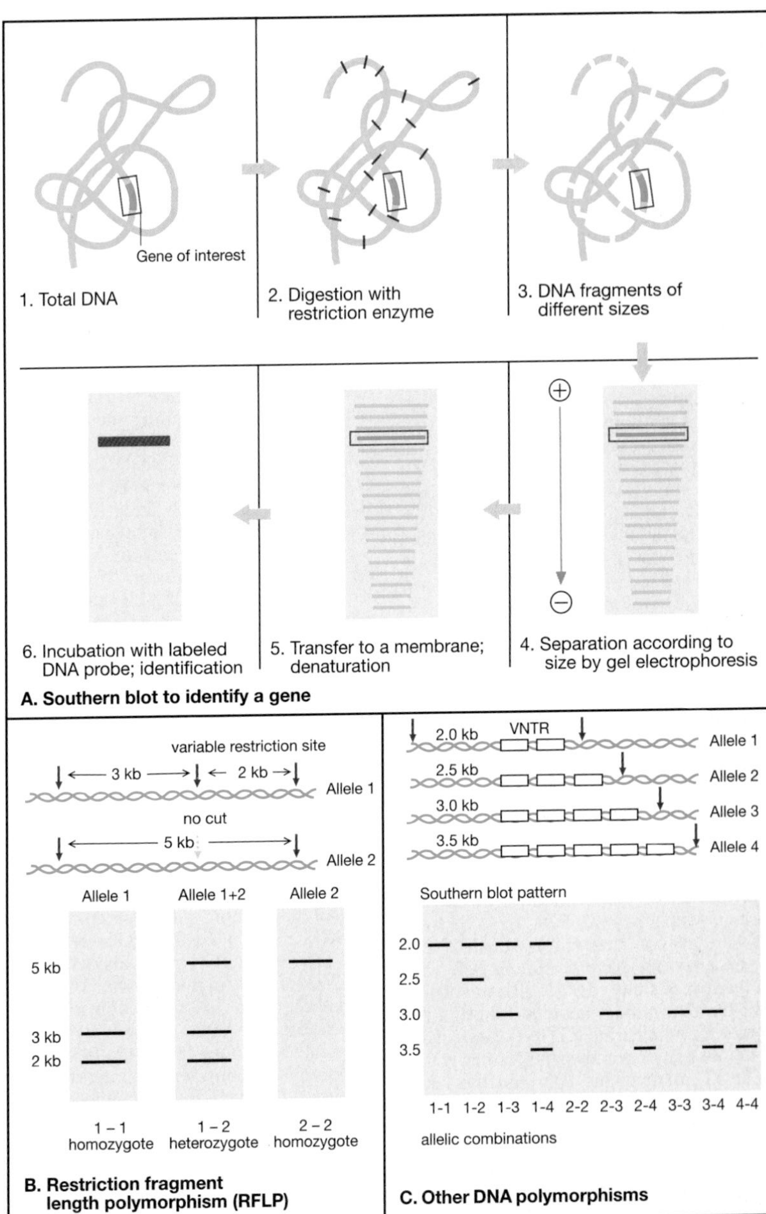

Gene of interest
1. Total DNA
2. Digestion with restriction enzyme
3. DNA fragments of different sizes
6. Incubation with labeled DNA probe; identification
5. Transfer to a membrane; denaturation
4. Separation according to size by gel electrophoresis
A. Southern blot to identify a gene
variable restriction site
3 kb
2 kb
Allele 1
no cut
5 kb
Allele 2
Allele 1
Allele 1+2
Allele 2
5 kb
3 kb
2 kb
1 – 1 homozygote
1 – 2 heterozygote
2 – 2 homozygote
B. Restriction fragment length polymorphism (RFLP)
2.0 kb
VNTR
Allele 1
2.5 kb
Allele 2
3.0 kb
Allele 3
3.5 kb
Allele 4
Southern blot pattern
2.0
2.5
3.0
3.5
1-1 1-2 1-3 1-4 2-2 2-3 2-4 3-3 3-4 4-4
allelic combinations
C. Other DNA polymorphisms

Restriction Mapping

Restriction endonucleases (restriction enzymes) are DNA-cutting enzymes. They are obtained from bacteria, which produce the enzymes as protection from foreign DNA. A given enzyme recognizes a specific sequence of 4–8 (usually 6) nucleotides (restriction site) where it cleaves the DNA. The sizes of the DNA fragments produced depend on the distribution of the restriction sites. More than 400 different types of restriction enzymes have been isolated.

A. Specific recognition sequences

The cutting patterns (recognition sequences) of three frequently used restriction enzymes, *Eco*RI, *Hin*dIII, and *Hpa*I, are presented. For *Eco*RI and *Hin*dIII the cut is "palindromic," i.e., the cut is asymmetric around an axis on which mirror-image complementary single-stranded DNA segments arise. Each corresponds to its opposite-lying strand in the reverse direction. Therefore, they can be joined to a DNA fragment whose ends contain complementary single-stranded sequences. *Hpa*I cuts both strands so that no single-stranded ends are formed. Frequently cutting and seldom cutting enzymes can be distinguished according to the frequency of occurrence of their recognition site.

B. Examples of restriction enzymes

The recognition sequences of some restriction enzymes are shown. The names of the enzymes are derived from those of the bacteria in which they occur, e.g., *Eco*RI from *E*scherichia *co*li *R*estriction enzyme *I*, etc. Some enzymes have a cutting site with limited specificity. In *Hin*dII it suffices that the two middle nucleotides are a pyrimidine and a purine (GTPyPuAC), and it does not matter whether the former is thymine (T) or cytosine (C), and the latter, adenine (A) or guanine (G). Such a recognition site occurs frequently and produces many relatively small fragments, whereas enzymes that cut very infrequently produce few and large DNA fragments.

C. Restriction fragments

In a given DNA segment, the recognition sequence of a restriction enzyme occurs irregularly. Thus, the distances between restriction sites differ. DNA fragments of various sizes (restriction fragments) result from digestion with a restriction enzyme. A given restriction enzyme will cleave a given segment of DNA into a series of DNA fragments of characteristic sizes. This leads to a pattern that can be employed for diagnostic purposes.

D. Determination of the positions of restriction sites

Since the fragment sizes reflect the relative positions of the cutting sites, they can be used to characterize a DNA segment (restriction map). If a 10-kb DNA segment cut by two enzymes, A and B, yields three fragments, of 2 kb, 3 kb, and 5 kb, then the relative location of the cutting sites can be determined by using enzymes A and B alone in further experiments. If enzyme A yields two fragments of 3 kb and 7 kb, and enzyme B two fragments of 2 kb and 8 kb, then the two cutting sites of enzymes A and B must lie 5 kb apart. To the left of the restriction site of enzyme A are 3 kb; to the right of the restriction site of enzyme B, 2 kb (1 kb = 1000 base pairs).

E. Restriction map

A given DNA segment can be characterized by the distribution pattern of restriction sites. In the example shown, a DNA segment is characterized by the distribution of the cutting sites for enzymes E (*Eco*RI) and H (*Hin*dIII). The individual sites are separated by intervals defined by the size of the fragments after digestion with the enzyme. A restriction map is a linear sequence of restriction sites at defined intervals along the DNA. Restriction mapping is of considerable importance in medical genetics and evolutionary research.

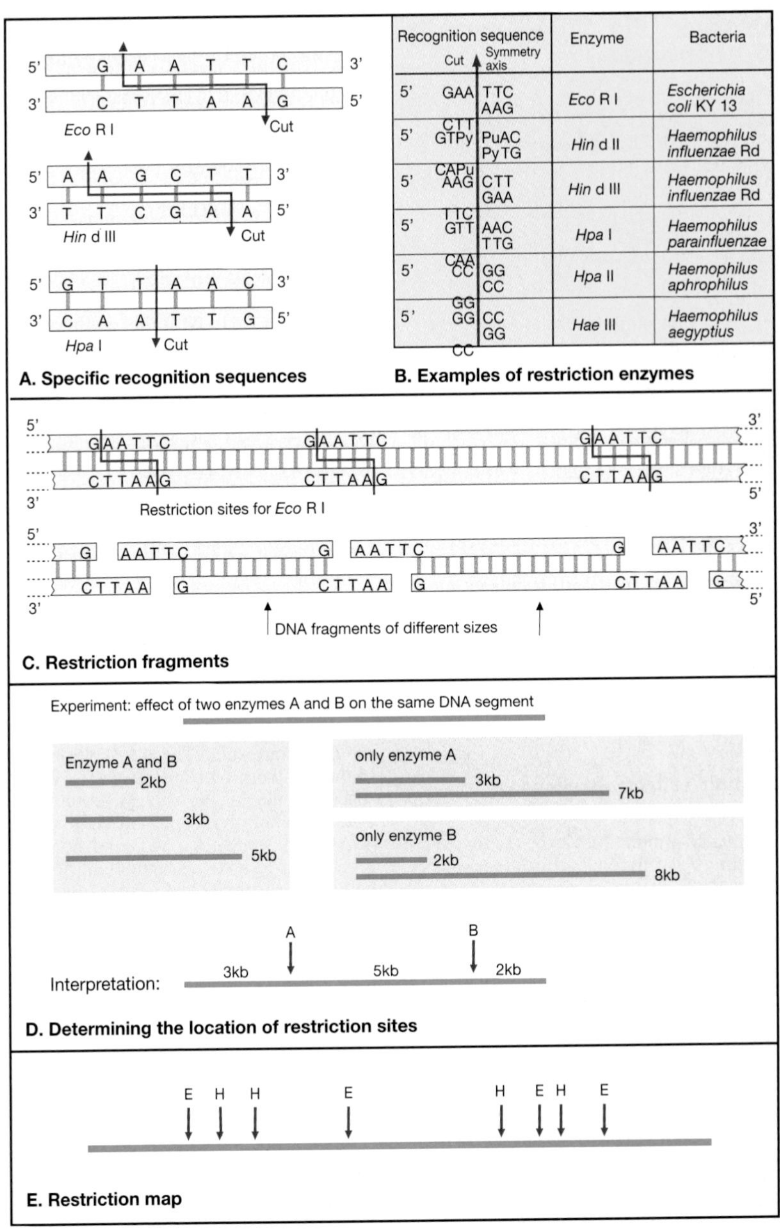

Recognition sequence (Cut, Symmetry axis)	Enzyme	Bacteria
5' GAA TTC / CTT AAG	*Eco* R I	*Escherichia coli* KY 13
5' GTPy PuAC / CAPu PyTG	*Hin* d II	*Haemophilus influenzae* Rd
5' AAG CTT / TTC GAA	*Hin* d III	*Haemophilus influenzae* Rd
5' GTT AAC / CAA TTG	*Hpa* I	*Haemophilus parainfluenzae*
5' CC GG / GG CC	*Hpa* II	*Haemophilus aphrophilus*
5' GG CC / CC GG	*Hae* III	*Haemophilus aegyptius*

A. Specific recognition sequences

B. Examples of restriction enzymes

C. Restriction fragments

D. Determining the location of restriction sites

E. Restriction map

Demonstration of Mutations by RFLP Analysis

Numerous procedures that do not require direct determination of the altered nucleotide base sequence can be used to demonstrate a mutation. An available probe of the gene or gene region being investigated and knowledge of the normal Southern blot pattern after restriction analysis of the gene are prerequisites (cf. p. 64).

A. Direct demonstration of a deletion

A partial deletion may cause an altered pattern of a Southern blot. Two genes (α2 and α1) of α-globin are presented (1). They are both located on a 14.5-kb restriction fragment. If partial deletion results in loss of a segment of, e.g., about 4.5 kb, that is part of both the α2 and the α1 gene, a fragment of 10.0 instead of 14.5 kb will result in this area (2). Three genotypes are possible (3): two normal genes, α2 and α1, represented by a 14.5-kb fragment; a normal DNA segment (14.5 kb) and one with a deletion (10.0 kb); or a deletion in both gene segments (only one fragment, of 10.0 kb). This can be demonstrated directly in the Southern blot pattern with a probe for the α1 gene (4).

B. Indirect demonstration of a mutation by RFLP analysis

Indirectly, a mutation can be demonstrated if there is an individual difference (polymorphism) in the base sequences of the mutant and the normal gene segments (restriction fragment length polymorphism, RFLP, see p. 64). For instance, if two of the same DNA segments differ in a polymorphism for the recognition sequence of a restriction enzyme, then DNA fragments of different sizes (here, either 7 kb and 6 kb, or 13 kb) result after cleavage with the enzyme (1). If the mutation has occurred within the 13-kb fragment, then this fragment indicates presence of the mutation. In the given DNA segment, there are three possibilities (genotypes): two fragments of 7 kb without mutation, one fragment of 7 kb (normal) and one fragment of 13 kb (which carries the mutation), and two mutation-carrying fragments of 13 kb (3). The Southern blot (4) shows whether the person being examined is homozygous normal (has no 13-kb fragment), is heterozygous (a 7-kb and a 13-kb fragment), or is homozygous for the mutation (two 13-kb fragments).

The prerequisite for this indirect analysis is previous knowledge of which of the DNA fragments contains the mutation. The observed difference is not the result of the mutation, as in **A**. If the Southern blot pattern of affected and unaffected does not differ, then this method will not be informative for the disorder.

C. Evidence of a point mutation from an altered restriction site

A restriction site may be altered by a mutation. For example, a sickle cell mutation in codon 6 of the β gene of hemoglobin (see p. 282) (1) causes loss of a restriction site for the enzyme *Mst*II (CCTN*A*GG instead of CCTN*T*GG) because the A (adenine) has been replaced by a T (thymine) (2). The normal allele (βA) in this area produces a 1.15-kb fragment after *Mst*II digestion, whereas the mutation eliminates the restriction site in the middle so that a 1.35-kb fragment results. The 1.35-kb fragment in the Southern blot indicates (3) the presence of the sickle cell mutation (ßS). Thus, homozygous normal individuals (AA), heterozygotes (AS), and homozygotes for the sickle cell mutation (SS) can be clearly distinguished; each of the three genotypes can be precisely diagnosed.

Increasingly, RFLP analysis is being replaced by the analysis of small polymorphic DNA repeats (microsatellites, see p. 353).

Reference

Housman D. Human DNA polymorphism. N Eng J Med. 1995; 332: 318–320.

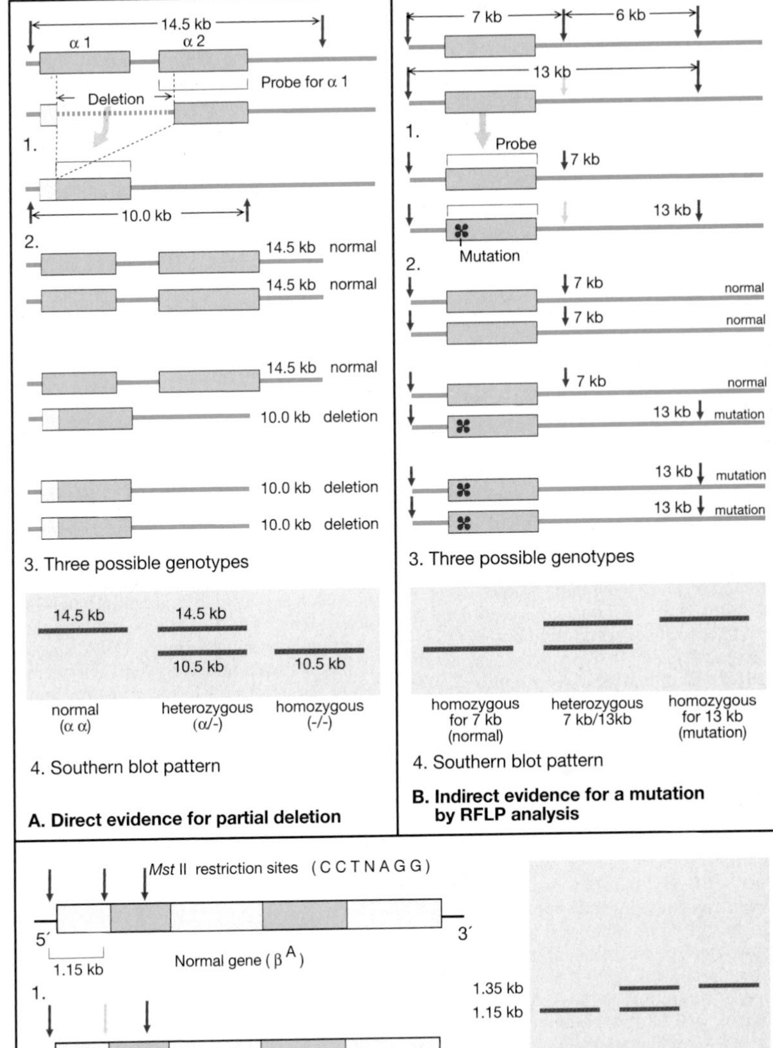

A. Direct evidence for partial deletion

B. Indirect evidence for a mutation by RFLP analysis

C. Demonstration of a point mutaton by an altered restriction site

Demonstration of Mutations without Sequencing

In addition to the detection of mutations by different DNA fragments in Southern blots (p. 64), there are methods based on differences in the hybridization of mutated and normal segments of DNA. Incomplete hybridization is determined by using short segments of single-stranded DNA (oligonucleotides) with a sequence complementary to the investigated region (see **A**). Other methods are based on demonstrating incomplete hybridization with mRNA (see **B**) or on the fact that a hybridized segment of normal and mutant DNA is less stable than normal DNA.

A. Demonstration of a point mutation by means of oligonucleotides

Short segments of DNA (oligonucleotides) are used to determine whether there is a mutation in a segment of DNA (1, normal DNA; 2, mutation from G to A). An oligonucleotide is a synthetically produced DNA segment about 20 nucleotides long; its sequence is complementary to a corresponding segment of the investigated gene. It hybridizes completely with its complementary segment (3). If a mutation, here from G to A (1), is located in this region, hybridization will not be perfect at this site (mismatch) (4). On the other hand, an oligonucleoide that is complementary to the DNA segment with the mutation will hybridize completely (allele-specific oligonucleotide, ASO) (5). This hybridizes incompletely with the normal DNA (6). By parallel use of both nucleotides, mutant and nonmutant DNA can be differentiated. The test results (7) show the hybridization of mutated DNA and of control DNA with the allele-specific oligonucleotides (ASO 1 for the control, ASO 2 for the mutation). Hybridization is indicated by a signal (dot-blot analysis).

B. Demonstration of a point mutation by ribonuclease A cleavage

The basis for this method is that a normal DNA strand hybridizes completely with mRNA from that region. Completely hybridized DNA and mRNA are protected from the effects of the RNA-splitting enzyme ribonuclease A (ribonuclease protection assay). Hybridization is incomplete in the area of a mutation. In this region, mRNA will be cleaved by ribonuclease A (RNase A). This can be demonstrated by Southern blot. There will be two fragments formed that together correspond to the size of the completely hybridized fragment (600 base pairs [bp], versus 400 and 200 bp).

C. Demonstration of a mutation by denaturation electrophoresis

This method exploits differences in the stability of DNA segments with and without mutation. While double-stranded DNA of a control person is completely complementary (homoduplex), a mutation leads to a mismatch at the site of mutation (heteroduplex). This DNA is less stable than completely complementary DNA strands (it has a lower melting point). If normal DNA (control) and DNA with the mutation are placed in a gel with an increasing concentration gradient of formamide (denaturation gradient gel), the mutant and normal DNA subsequently can be differentiated in a Southern blot. The normal DNA remains stable to higher concentrations of formamide and migrates further than mutant DNA, which dissociates earlier and therefore does not migrate as far.

References

Casky CT. Disease diagnosis by recombinant DNA methods. Science. 1987; 236: 1223–1229.

Dean M. Resolving DNA mutations. Nature Genet. 1995; 9: 103–104.

Mashal RD, Koontz J, Sklar J. Detection of mutations by cleavage of DNA heteroduplexes with bacteriophage resolvases. Nature Genet. 1995; 9: 177–183.

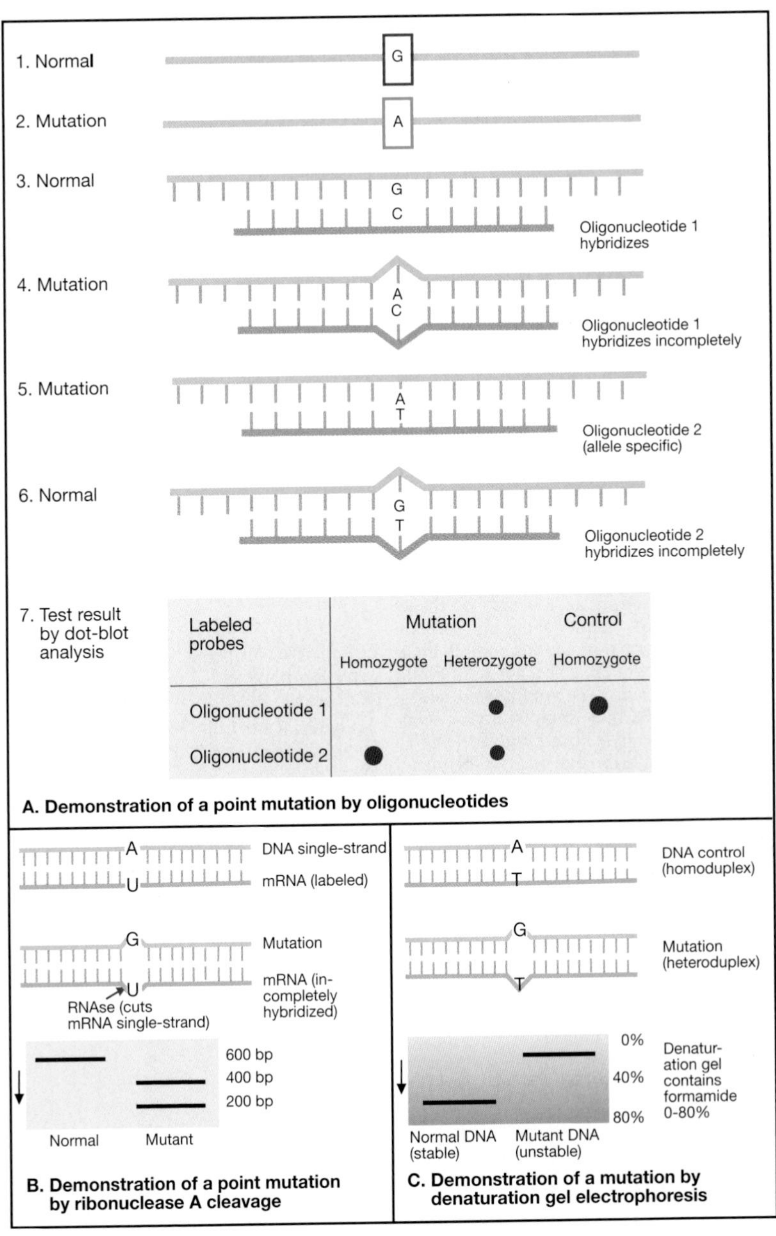

A. Demonstration of a point mutation by oligonucleotides

B. Demonstration of a point mutation by ribonuclease A cleavage

C. Demonstration of a mutation by denaturation gel electrophoresis

DNA Amplification (Polymerase Chain Reaction, PCR)

Multiple copies of a DNA segment can be produced by amplification using a system of consecutive DNA syntheses that does not require living cells such as bacteria. Since it involves a series of DNA polymerase reactions, this technique is called polymerase chain reaction, or PCR. With this method, introduced in 1985, a DNA segment can be reproduced a millionfold from a single original copy, so that sufficient amounts can be available for investigation, e.g., demonstration of a mutation.

A. Polymerase chain reaction (PCR)

First, the DNA segment to be amplified is separated into single strands (denatured). These serve as the template for synthesis of new DNA. Synthetic oligonucleotides are used as the primer for the DNA synthesis.

Each duplication cycle consists of three precisely timed (each depending on the other) consecutive reactions, for which different temperatures are needed. First, the DNA segment to be amplified is denatured (separated into single strands); then it is cooled and hybridized with the oligonucleotides (attachment of the primer); next, it is incubated with DNA polymerase and the four deoxyribonucleotide triphosphates so that new DNA can be formed (DNA synthesis). With each cycle, all the DNA present after the previous cycle serves as template. In this manner, the DNA is doubled during each cycle. The next cycle begins anew with heating to form single strands, then cooling to attach the primers, followed by synthesis of new DNA. With automation it is possible to complete up to 30 cycles in a few hours and produce an exponential number of copies. Since the PCR method makes analysis of very small amounts of material possible, it has become one of the most important procedures in medical genetics and in all of molecular genetics.

B. cDNA amplification

A partially known amino acid sequence (AS) of a protein can be used to amplify cDNA. From its sequence of amino acids, the sequence of the complementary DNA (cDNA) can be deduced. Synthetic oligonucleotides serve as primers for synthesis of the coding (sense) and for the noncoding (antisense) strands (1). cDNA can be formed from a series of different mRNA molecules by use of reverse transcriptase (2). After the mRNA has been removed by alkaline hydrolysis, the remaining cDNA can serve as a template for the formation of new DNA (see p. 62). The sense oligonucleotide serves as a primer for forming the nontranscribed (antisense) DNA strand, and the antisense oligonucleotide serves as the primer for the transcribing DNA strand (sense strand). The result is new DNA with a nucleotide sequence that corresponds to the original amino acid sequence. The new DNA can now be amplified by the PCR (3).

References

Erlich HA, Arnheim N. Genetic analysis with the polymerase chain reaction. Ann Rev Genet. 1992; 26: 479–506.

Volkenandt M, Löhr M, Dicker AP. Gen-Amplification durch Polymerase-Kettenreaktion. Dtsch Med Wschr. 1990; 17: 670–676.

White TJ, Arnheim N, Erlich HA. The polymerase chain reaction. Trends Genet. 1989; 5: 85–189.

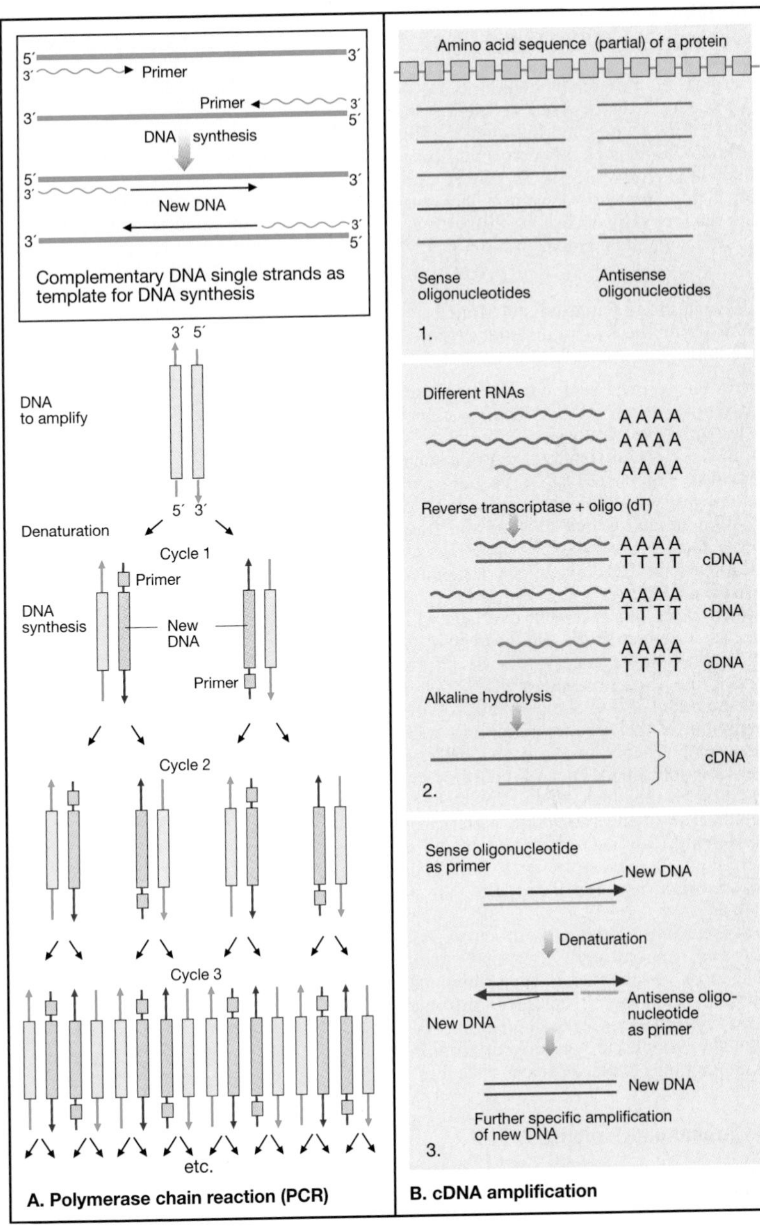

A. Polymerase chain reaction (PCR)

B. cDNA amplification

Locating a New Gene by Overlapping DNA Fragments

A region of previously identified DNA, e.g., a gene, can be used to initiate the search for a neighboring gene. This method requires a phage or cosmid library of genomic DNA (see pp. 58, 60). A number of important human genes, including the gene for cystic fibrosis (1989), were localized using this method.

A. Overlapping fragments of cloned DNA define the way to another region ("chromosome walking")

From the known segment A, one progresses by means of overlapping clones (20–40 kb), present in phages or cosmids, to the region of interest (chromosome walking). The cloned DNA ist only partially digested so that only some of the recognition sequences are cut and the fragments therefore overlap. A clone is isolated from the genomic library (fragment 1) with a probe (A) of genomic DNA or cDNA. Since this fragment overlaps with another (fragment 2), the latter can be identified with a subfragment (probe B) of fragment 1; the restriction pattern of the end of one fragment is the same as that of the beginning of the next (overlapping) fragment. Beforehand, one end (B) of fragment 1 is subcloned. Each of the following fragments is then identified in this manner, making use of the restriction patterns of overlapping fragments being identical at their ends. In this way, with numerous steps, several hundred base pairs can be bridged.
However, due to the size of eukaryotic genomes, chromosome walking using phages or cosmids is a very slow and laborious process. Therefore artificial yeast chromosomes (YACS) (see p. 104) or other strategies are employed (chromosome jumping) to reach greater distances.

B. Chromosome jumping

With this approach, DNA stretches of several hundred kb are traversed in one step. Restriction enzymes that cut infrequently are used (e.g., *Not*I: its recognition sequence occurs only about once every 1000 kb in the human genome) to produce large starting fragments (as opposed to about 20 kb with chromosome walking). The fragment is made into a ring ("circularized"). In this way, segments (here, A and E) that were far apart are approximated. In addition, a marker sequence for identifying recombinant clones is incorporated. Finally, the intermediate segments (B to D) are cut with the usual restriction enzymes and removed.
The fragments lying far apart in the genomic DNA but now together in the ring are cloned in phages and used to establish a library of overlapping fragments ("jumping" library). With the help of a probe for segment A, segment E can be isolated from the library. In this manner, the length of a jump can be extended to about several hundred kilobases.
A series of overlapping DNA fragments is called a contig (contiguous sequences) (see p. 271).

References

Davies KE, Reed AP. Molecular Basis of Inherited Disease. 2nd ed. Washington: IRL Press; 1992.

Watson JD, Gilman M, Witkowski J, Zoller M. Recombinant DNA. 2nd ed. Scientific American Books. New York: WH Freeman; 1992.

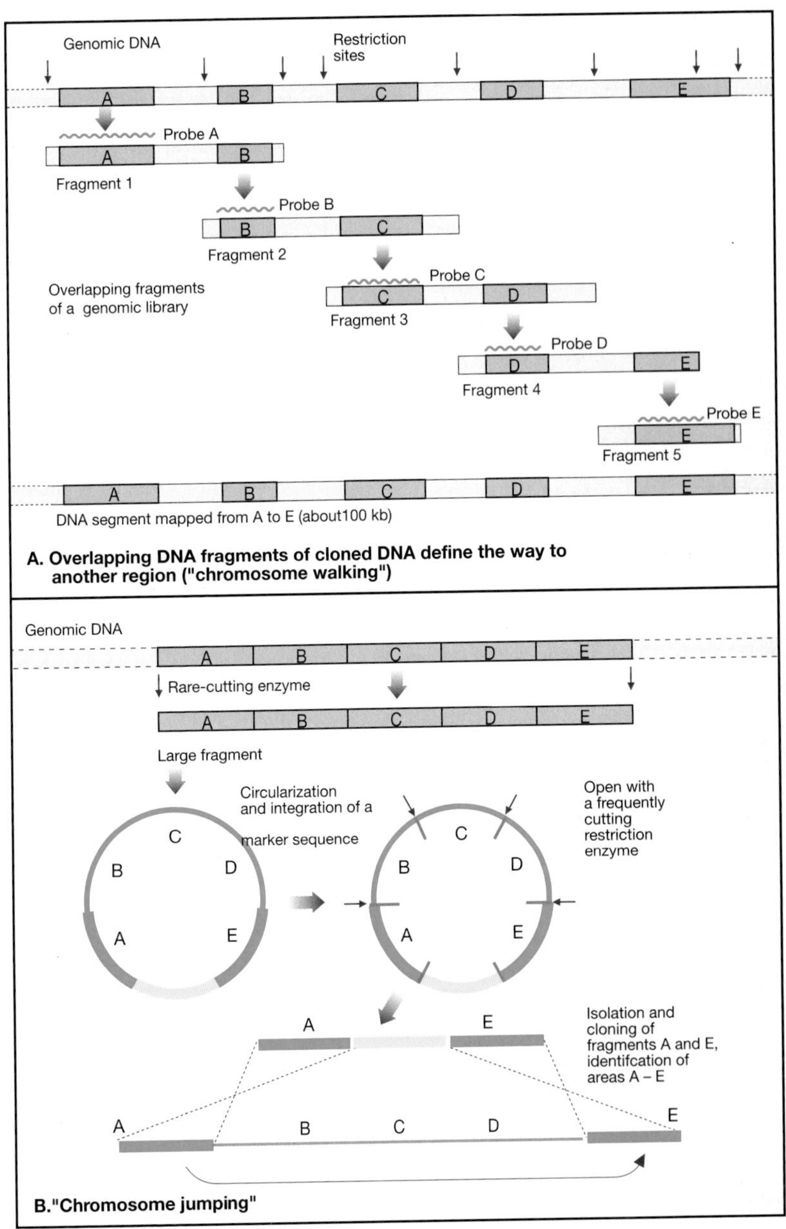

A. Overlapping DNA fragments of cloned DNA define the way to another region ("chromosome walking")

B. "Chromosome jumping"

Identification of a Coding DNA Segment

To identify a gene of interest, numerous methods are available that do not require large segments of DNA to be sequenced. Some examples are presented here.

A. Microdissection of metaphase chromosomes

If the chromosomal location of a gene of interest is known, this region can be cut out of a metaphase chromosome by means of microdissection (arrow). This has the advantage that all other chromosomal segments are eliminated. However this method, first applied by B. Horsthemke and co-workers (Lüdecke et al., 1989), is tedious (photo from K. Buiting et al., Genomics. 1990; 6:521–527).

B. Artificial yeast chromosomes (YACs)

Large DNA fragments (200–300 kb) can be replicated in yeast cells. They are inserted into artificial yeast chromosomes (*y*east *a*rtificial *c*hromosome, YAC, see p. 104) and replicated with them. A photograph of a transverse alternating field electrophoresis (TAFE) with nine lanes after ethidium bromide staining shows fragments of different sizes. These correspond to the naturally occurring yeast chromosomes. Six of the lanes (2–7) contain an additional band, which corresponds to an artificial yeast chromosome. They are marked with a yellow point: the lowest band in lane 2 (YAC9), the third band from the bottom in lanes 3, 4, and 6 (YAC41, YAC45, YAC51), and the lowest band in lane 7 (YAC52). In lane 5, YAC50 is masked by a yeast chromosome (3rd fragment from below). Lane 1 contains the standards for fragment size (preparation by K. Buiting and B. Horsthemke, Essen).

C. Exon trapping

In an unidentified segment of DNA, a gene can be recognized by the occurrence of coding segments (exons). To find and isolate an exon, the genomic fragment can be cloned in a vector that consists of a strong promoter gene and a reporter gene. An exon that is present is cut out by means of the donor and acceptor splice signals and expressed together with the genomic fragment (exon trapping). cDNA is produced from the mRNA and replicated by means of PCR. Finally, the trapped exon can be sequenced or characterized by other means (diagram after K. E. Davies and A. P. Read, 1992).

D. Single-strand conformation polymorphism (SSCP)

This procedure helps establish a difference in the nucleotide base sequence due to mutation or polymorphism. Whether DNA segments (single-stranded DNA) of common origin differ from each other is determined by their speed of migration in a polyacrylamide gel electrophoresis under different conditions, such as changes in temperature, in pH, etc. A base substitution may lead to a difference in spatial arrangement (conformation) and in mobility (lane 4, arrow, polyacrylamide gel electrophoresis and silver staining. D. Lohmann, Essen).

E. "Zoo blot"

The cross hybridization of DNA across species boundaries ("zoo blot") is an indication for coding sequences, since genes have similar structures in different organisms.

References

Davies KE, Read AP. Molecular Basis of Inherited Disease. 2nd ed. Oxford: IRL Press; 1992.

Lüdecke HJ, Senger G, Claussen U, Horsthemke B. Cloning defined regions of the human genome by microdissection of banded chromosomes and enzymatic amplification. Nature. 1989; 338: 348–350.

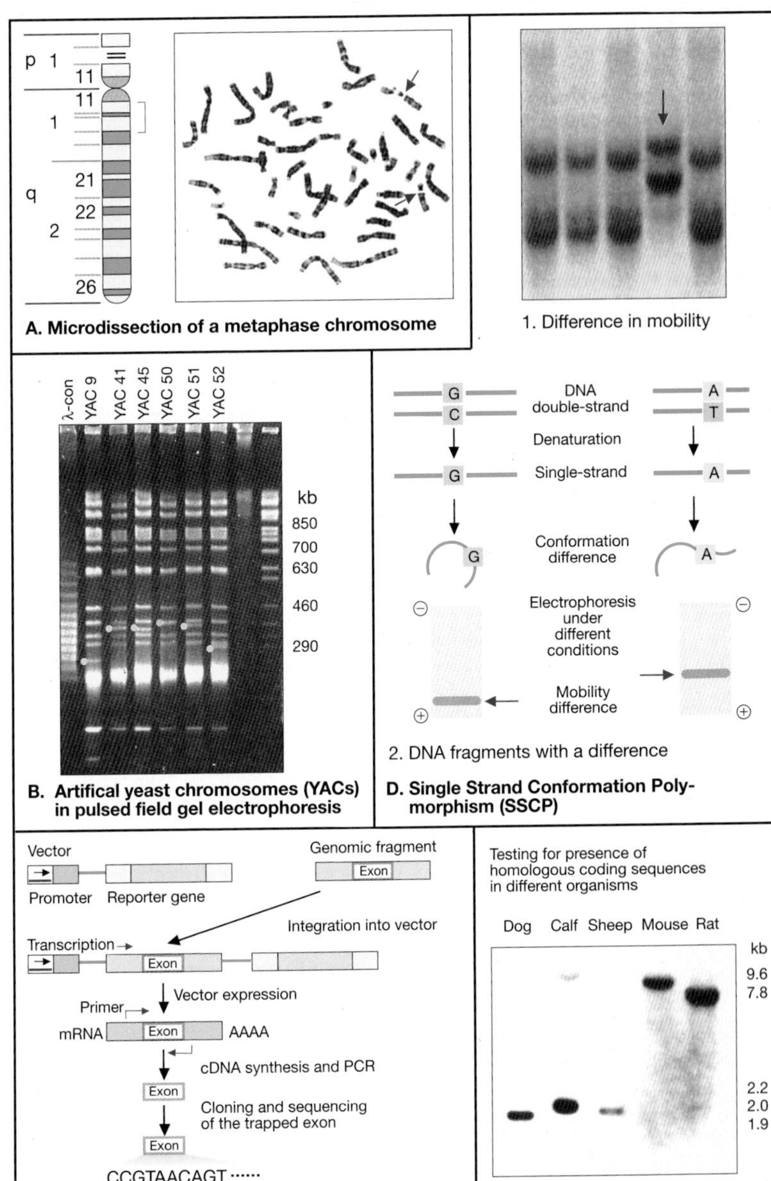

A. Microdissection of a metaphase chromosome

B. Artifical yeast chromosomes (YACs) in pulsed field gel electrophoresis

D. Single Strand Conformation Polymorphism (SSCP)

C. Exon trapping

E. "Zoo-blot"

Isolation of Mutant Bacteria

Important advances in genetics were made in the early 1950s through studies of bacteria. As prokaryotic organisms, bacteria have certain advantages over eukaryotic organisms because they are haploid and have an extremely short generation time. Mutant bacteria can be identified easily. The growth of some mutant bacteria depends on whether a certain substance is present in the medium (auxotrophism). Bacterial cultures are well suited for determining mutational events, since an almost unlimited number of cells can be tested in a short time. Without great difficulty, it is possible to detect one mutant in 10^7 colonies. Efficiency to this degree is not possible in the genetic analysis of eukaryotic organisms.

A. Replica plating to recognize mutants

In 1952, Joshua and Esther Lederberg developed replica plating of bacterial cultures. With this method, individual colonies on an agar plate can be taken up with a stamp and placed onto other culture dishes with media of different compositions. Some mutant bacteria differ from nonmutant in their ability to grow. Here several colonies are shown in the petri dish of the initial culture. Each of these colonies originated from a single cell. By means of replica plating, the colonies are transferred to two new cultures. One culture (right) contains an antibiotic in the culture medium; the other (left) does not. All colonies grow in normal medium, but only those colonies that are antibiotic resistant due to a mutation grow in the antibiotic-containing medium. In this manner, mutant colonies can be readily identified.

B. Mutant bacteria identified through an auxotrophic medium

Here it is shown how different mutants can be distinguished, e.g., after exposure to a mutagenic substance. After a colony has been treated with a mutagenic substance, it is first cultivated in normal nutrient medium. Mutants can then be identified by replica plating. The culture with the normal medium serves as the control. In one culture with minimal medium, from which a number of substances are absent, two colonies do not grow (auxotrophic mutants). Initially, it is not known for which of the substances the colonies are auxotrophic. If a different amino acid is added to each of two cultures with minimal medium, e.g., threonine (Thr) to one and arginine (Arg) to the other, it can be observed that one of the mutant colonies grows in the threonine-containing minimal medium, but not the other. The former colony is dependent on the presence of threonine (Thr^-), i.e., it is an auxotroph for threonine. The other culture with minimal medium had arginine added. Only here can the other of the two mutant colonies, an auxotroph for arginine (Arg^-), grow. After the mutant colonies requiring specific conditions for growth have been identified, they can be further characterized. This procedure is relatively simple and makes rapid identification of mutants possible. Many mutant bacteria have been defined by auxotrophism. The wild-type cells that do not have special additional growth requirements are called prototrophs.

Reference

Stent GS, Calendar R. Molecular Genetics. An Introductory Narrative. 2nd ed. San Francisco: WH Freeman; 1978.

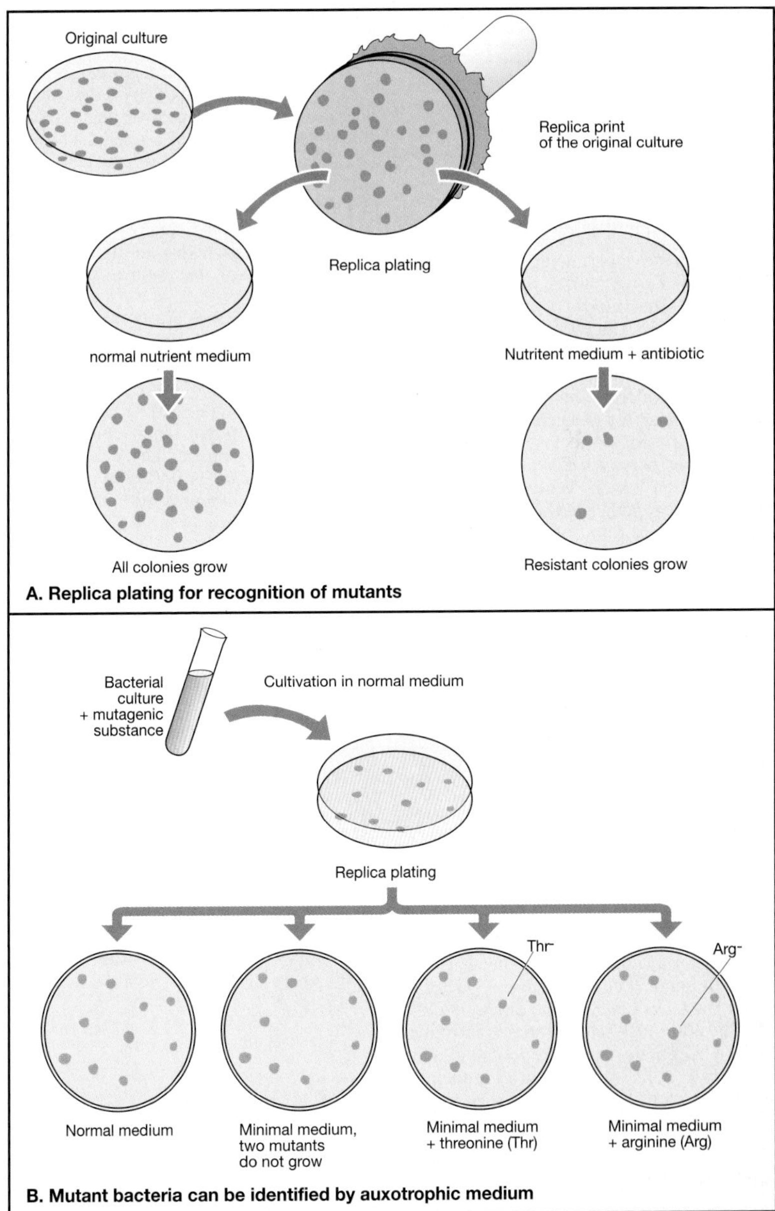

A. Replica plating for recognition of mutants

B. Mutant bacteria can be identified by auxotrophic medium

Recombination in Bacteria

In 1946, J. Lederberg and E. L. Tatum first demonstrated that genetic information can be exchanged between different mutant bacterial strains. This corresponds to a type of sexuality and leads to genetic recombination.

A. Genetic recombination in bacteria

In their classic experiment, Lederberg and Tatum used two different auxotrophic bacterial strains. One (A) was auxotrophic for methionine (met^-) and biotin (bio^-). This strain required methionine and biotin, but not threonine and leucine (thr^+, leu^+), to be added to the medium. The opposite was true for bacterial strain B, auxotrophic for threonine and leucine (thr^-, leu^-), but prototrophic for methionine and biotin (met^+, bio^+). When the cultures were mixed together without the addition of any of these four substances and then plated on an agar plate with minimal medium, a few single colonies (met^+, bio^+, thr^+, leu^+) unexpectedly appeared. Although this occurred rarely (about 1 in 10^7 plated cells), a few colonies with altered genetic properties usually appeared due to the large number of plated bacteria. The interpretation: genetic recombination between strain A and strain B. The genetic properties of the parent cells complemented each other (genetic complementation).

B. Conjugation in bacteria

Later, the genetic exchange between bacteria (conjugation) was demonstrated by light microscopy. Conjugation occurs with bacteria possessing a gene that enables frequent recombination. Bacterial DNA transfer occurs in one direction only. “Male” chromosomal material is introduced into a “female” cell. The so-called male and female cells of *E. coli* differ in the presence of a fertility factor (F). When F^+ and F^- cells are mixed together, conjugal pairs are formed with attachment of a male (F^+) sex pilus to the surface of an F^- cell (photo from Science. 1992; 257: 1037).

C. Integration of the F factor into an Hfr^- chromosome

The F factor can be integrated into the bacterial chromosome by means of specific crossing-over. After the factor is integrated, the original bacterial chromosome with the sections a, b, and c contains further genes, the F factor genes (e, d). Such a chromosome is called an Hfr chromosome (Hfr, high frequency of recombination) due to its high rate of recombination with genes of other cells owing to conjugation.

D. Transfer of F DNA from an F^+ to an F^- cell

Bacteria may contain the F factor (fertility) as an additional small chromosome, i.e., a small ring-shaped DNA molecule (F plasmid) of about 94 000 base pairs. This corresponds to about 1/40th of the total genetic information of a bacterial chromosome. It occurs once per cell and can be transferred to other bacterial cells. About a third of the F^+ DNA consists of transfer genes, including genes for the formation of sex pili. The transfer of the F factor begins after a strand of the DNA double helix is opened. One strand is transferred to the acceptor cell. There it is replicated, so that it becomes double stranded. The DNA strand remaining in the donor cell is likewise restored to a double strand by replication. Thus, DNA synthesis occurs in both the donor and the acceptor cell. When all is concluded, the acceptor cell is also an F^+ cell.

Reference

Watson JD, et al. Molecular Biology of the Gene. 3rd ed. 1987.

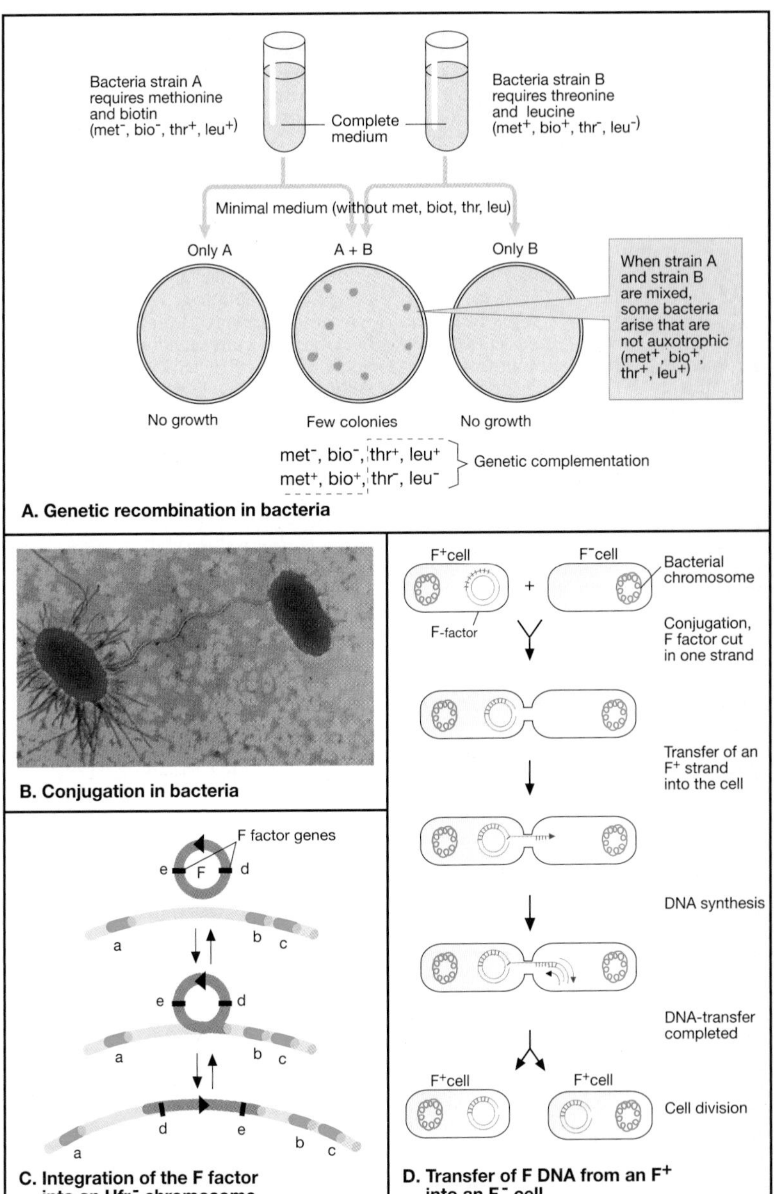

A. Genetic recombination in bacteria

B. Conjugation in bacteria

C. Integration of the F factor into an Hfr⁻ chromosome

D. Transfer of F DNA from an F⁺ into an F⁻ cell

Bacteriophages

The discovery of bacterial viruses (bacteriophages or phages) in 1941 opened a new era in the study of the genetics of prokaryotic organisms. Although they were disappointing in the original hope that they could be used to fight bacterial infections, phages served during the 1950s as vehicles for genetic analysis of bacteria. Unlike viruses that infect plant or animal cells, phages can be relatively easily analyzed in their host cells. Names associated with phage analysis are Max Delbrück, Salvador Luria, and Alfred D. Hershey (the "phage group," see: Cairns J, Stent GS, Watson JD. eds. Phage and the Origins of Molecular Biology. Cold Spring Harbor. 1966).

A. Attachment of a bacteriophage

Phages consist of DNA, a coat (coat protein) for protection, and a means of attachment (terminal filaments). Like other viruses, phages are basically nothing more than packaged DNA. One or more bacteriophages attach to a receptor on the surface of the outer cell membrane of a bacterium. The figure shows how an attached phage inserts its DNA into a bacterium. Numerous different phages are known, e.g., for *Escherichia coli* and *Salmonella* (phages T1, T2, P1, F1, lambda, T4, T7, and phi X174).

B. Lytic and lysogenic cycles of a bacteriophage

Phages do not reproduce by cell division as bacteria, but by intracellular formation and assembly of the different components. This begins with the attachment of a phage particle to a specific receptor on the surface of a sensitive bacterium. Different phages use different receptors, thus giving rise to specificity of interaction (restriction). The invading phage DNA contains the information for production of coat proteins for new phages and factors for DNA replication and transcription. Translation is provided for by cell enzymes. The phage DNA and phage protein manufactured in the cell are assembled into new phage particles. Finally, the cell disintegrates (lysis) and hundreds of phage particles are released. With attachment of a new phage to a new cell, the procedure is repeated (lytic cycle).

Phage reproduction does not always occur after cell invasion. Occasionally, phage DNA is integrated into the bacterial chromosome and replicated with it (lysogenic cycle). Phage DNA that has been integrated into the bacterial chromosome is designated a prophage. Bacteria containing prophages are designated lysogenic bacteria; the corresponding phages, lysogenic phages. The change from a lysogenic to a lytic cycle is rare. It requires induction by external influences and complex genetic mechanisms.

C. Insertion of a lambda phage into the bacterial chromosome by crossing over

A phage can be inserted into a bacterial chromosome by different mechanisms. With the lambda phage (λ), insertion results from crossing-over between the *E. coli* chromosome and the lambda chromosome. First, the lambda chromosome forms a ring. Then it attaches to a homologous section of the bacterial chromosome. Both the bacterial and the lambda chromosome are opened by a break and attach to each other. Since the homologies between the two chromosomes are limited to very small regions, phage DNA is seldom integrated. The phage is released (and the lytic cycle induced) by the reverse procedure.

(Figures adapted from J.D. Watson et al., Molecular Biology of the Gene, 1987).

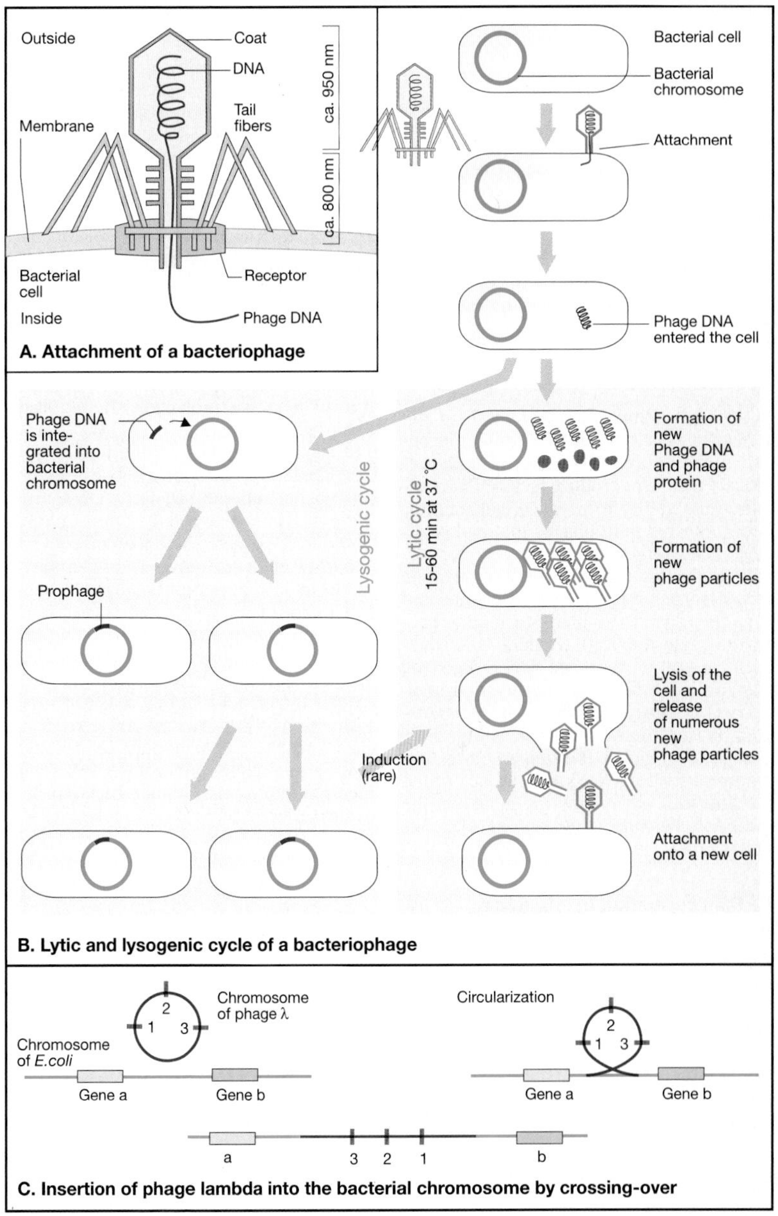
Outside
Coat
DNA
Tail fibers
Membrane
ca. 950 nm
ca. 800 nm
Bacterial cell
Receptor
Inside
Phage DNA
A. Attachment of a bacteriophage
Bacterial cell
Bacterial chromosome
Attachment
Phage DNA entered the cell
Phage DNA is integrated into bacterial chromosome
Lysogenic cycle
Lytic cycle
15-60 min at 37 °C
Formation of new Phage DNA and phage protein
Formation of new phage particles
Prophage
Lysis of the cell and release of numerous new phage particles
Induction (rare)
Attachment onto a new cell
B. Lytic and lysogenic cycle of a bacteriophage
Chromosome of phage λ
Chromosome of E.coli
Gene a
Gene b
Circularization
1 2 3
a
b
C. Insertion of phage lambda into the bacterial chromosome by crossing-over

DNA Transfer between Cells

Transfer of DNA occurs not only by fusion of gametes in sexual reproduction, but also between other cells of prokaryotic and eukaryotic organisms (conjugation of bacteria, transduction between bacteriophages and bacteria, transformation by plasmids in bacteria, transfection in cultures of eukaryotic cells). Cells altered genetically by taking up DNA are said to be transformed (see p. 32). The concept of transformation is used in different contexts and refers to the result, not the mechanism.

A. Transduction by viruses

In 1952, N. Zinder and J. Lederberg described a new type of recombination between two strains of bacteria. Bacteria previously unable to produce lactose (lac^-) acquired the ability to produce lactose after being infected with phages that had replicated in bacteria containing a gene for lactose production (lac^+). A small segment of DNA from a bacterial chromosome had been transferred by a phage to another bacterium (transduction). General transduction (insertion of phage DNA into the bacterial genome at any unspecified location) is distinguished from special transduction (insertion at a particular location). Genes regularly transduced together (cotransduction) were used to determine the positions of neighboring genes on the bacterial chromosome (mapping of genes in bacteria).

B. Transformation by plasmids

Plasmids are small, autonomously replicating, circular DNA molecules separate from the chromosome in a bacterial cell. Since they often contain genes for antibiotic resistance (e.g., ampicillin), their incorporation into a sensitive cell renders the cell resistant to the antibiotic (transformation). Only these bacteria can grow in culture medium containing the antibiotic (selective medium).

C. Multiplication of a DNA segment in transformed bacteria

Plasmids are well suited as vectors for the transfer of DNA. A selective medium is used so that only those bacteria that have incorporated a recombinant plasmid containing the DNA to be investigated can grow (see p. 56).

D. Transfection by DNA

The transfer of DNA between eukaryotic cells in culture (transfection) can be used to examine the transmission of certain genetic traits (transfection assay). Left, a DNA transfer experiment is shown in a culture of mouse fibroblasts; right, in a culture of human tumor cells (R. A. Weinberg, 1987).
The mouse fibroblast culture (see p. 114) is altered by the chemical carcinogen methylcholanthrene (D, left). DNA from these cells is precipitated with calcium phosphate, extracted, and then taken up by a normal culture (transfection). About 2 weeks later, cells appear that have lost contact inhibition (transformed cells). When these cells are injected into mice that lack a functional immune system (naked mice), tumors develop. DNA from cultured human tumor cells (right) also can transform normal cells after several transfer cycles. The DNA segment must be of limited size (e.g., a gene), since long DNA segments do not remain intact after repeated cycles of extraction and precipitation. Detailed studies of cancer-causing genes (oncogenes) in eukaryotic cells were first carried out using transfection (see p. 264).

References

Weinberg RA. Molekulare Grundlagen von Krebs. Heidelberg: Spektrum der Wissenschaft; 1987: pp. 66–68.

Weinberg RA. The action of oncogenes in the cytoplasm and nucleus. Science. 1985; 230: 770–776.

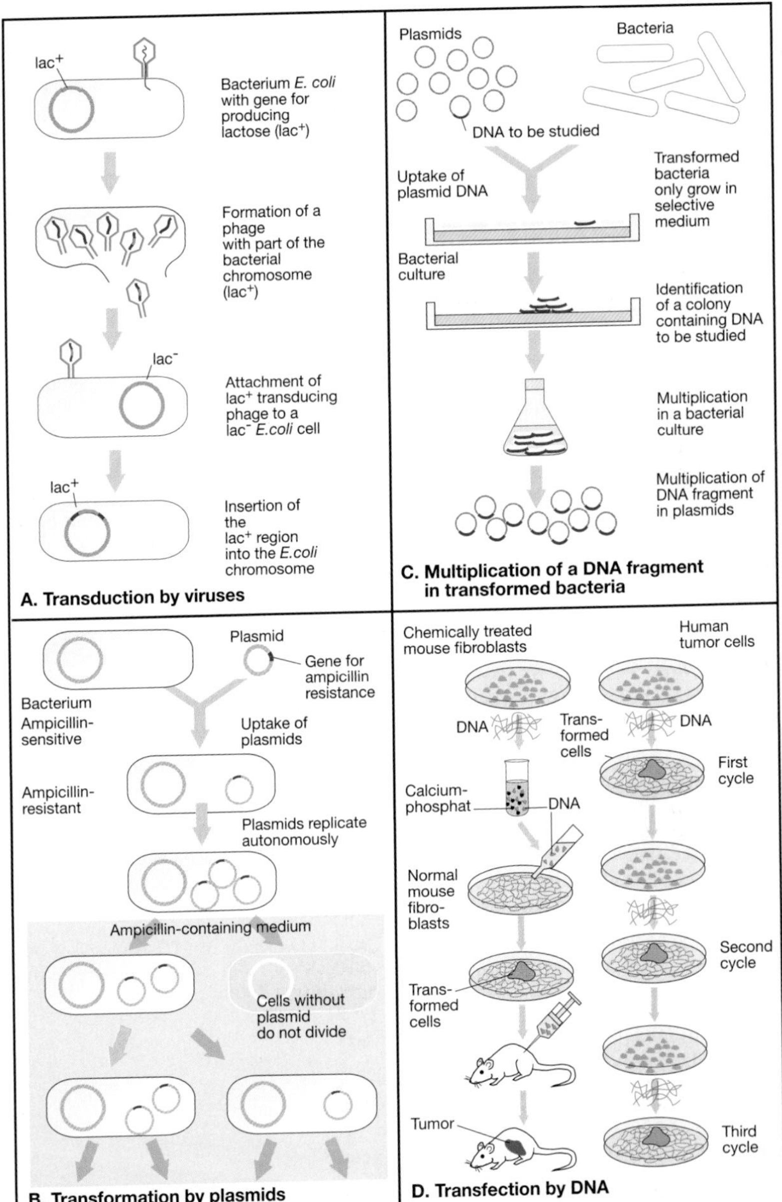
lac+
Bacterium E. coli with gene for producing lactose (lac+)
Formation of a phage with part of the bacterial chromosome (lac+)
lac-
Attachment of lac+ transducing phage to a lac- E.coli cell
lac+
Insertion of the lac+ region into the E.coli chromosome
A. Transduction by viruses
Plasmids
Bacteria
DNA to be studied
Uptake of plasmid DNA
Transformed bacteria only grow in selective medium
Bacterial culture
Identification of a colony containing DNA to be studied
Multiplication in a bacterial culture
Multiplication of DNA fragment in plasmids
C. Multiplication of a DNA fragment in transformed bacteria
Plasmid
Gene for ampicillin resistance
Bacterium
Ampicillin-sensitive
Uptake of plasmids
Ampicillin-resistant
Plasmids replicate autonomously
Ampicillin-containing medium
Cells without plasmid do not divide
B. Transformation by plasmids
Chemically treated mouse fibroblasts
Human tumor cells
DNA
Trans-formed cells
DNA
First cycle
Calcium-phosphat
DNA
Normal mouse fibro-blasts
Second cycle
Trans-formed cells
Tumor
Third cycle
D. Transfection by DNA

Viruses

Viruses are important pathogens in plants and animals, including man. The complete infectious viral particle is called a virion. Its genome carries a limited amount of genetic information, and it can replicate only in host cells. From analysis of the structure and expression of viral genes, fundamental biological processes such as DNA replication, transcription regulation, mRNA modification (RNA splicing, RNA capping, RNA polyadenylation), reverse transcription of RNA to DNA, viral genome integration into eukaryotic DNA, tumor induction by viruses, and cell surface proteins have been recognized and elucidated. The extracellular form of a virus particle includes a protein coat (capsid), which encloses the genome of DNA or RNA. The capsids contain multiple units of one or a few different protein molecules coded for by the virus. They usually have an almost spherical, icosahedral (20 plane surfaces), or occasionally a helical structure. Some viral capsids are surrounded by a lipid membrane envelope.

A. Classification of viruses

Viruses can be classified on the basis of the structure of their viral coat, their type of genome, and their organ or tissue specificity. The genome of a virus may be enclosed simply in a virus-coded protein coat (capsid) or in the capsid plus an additional phospholipid membrane, which is of cellular origin. The genome of a virus may consist of single-stranded DNA (e.g., parvovirus), double-stranded DNA (e.g., papovavirus, adenovirus, herpesvirus, and poxvirus), single-stranded RNA (e.g., picornavirus, togavirus, myxovirus, rhabdovirus), or double-stranded RNA (e.g., reovirus). Viruses with genomes of single-stranded RNA are classified according to whether their genome is a positive (plus RNA) or negative (minus RNA) RNA strand. Only an RNA plus strand can serve as a template for translation (5' to 3' orientation) (sense strand, see p. 44).

B. Replication and transcription of viruses

Since viral genomes differ, the mechanisms for replicating their genetic material also differ. Viruses must pack a lot of genetic information into a small genome; thus, one transcription unit (gene) of a viral genome is often used to produce several mRNAs by alternative splicing, and each mRNA codes for a different protein. In some RNA viruses, initially large precursor proteins are formed from the mRNA and subsequently split into several smaller functional proteins. An RNA plus strand can be used directly for protein synthesis. An RNA minus strand cannot be used directly; an RNA plus strand must be formed from the RNA minus strand by a transcriptase before translation is possible. RNA viruses contain a transcriptase to replicate their RNA genomes. RNA viruses in which DNA is formed as an intermediate step (retroviruses) contain a reverse transcriptase. This can form DNA from RNA. The DNA intermediate step in the replication of retroviruses becomes integrated into the host cell. Several RNA viruses have segmented genomes. They consist of individual pieces of RNA genome, each of which codes for one or more proteins (e.g., influenza virus). The exchange of individual pieces of RNA genome of different viral serotypes plays an important role in the formation of new viral strains (e.g., influenza strains).

(Figure after JD Watson et al. 1987).

Reference

Brock TD, Madigan MT. Biology of Microorganisms. 6th ed. Englewood Cliffs, New Jersey: Prentice Hall; 1991.

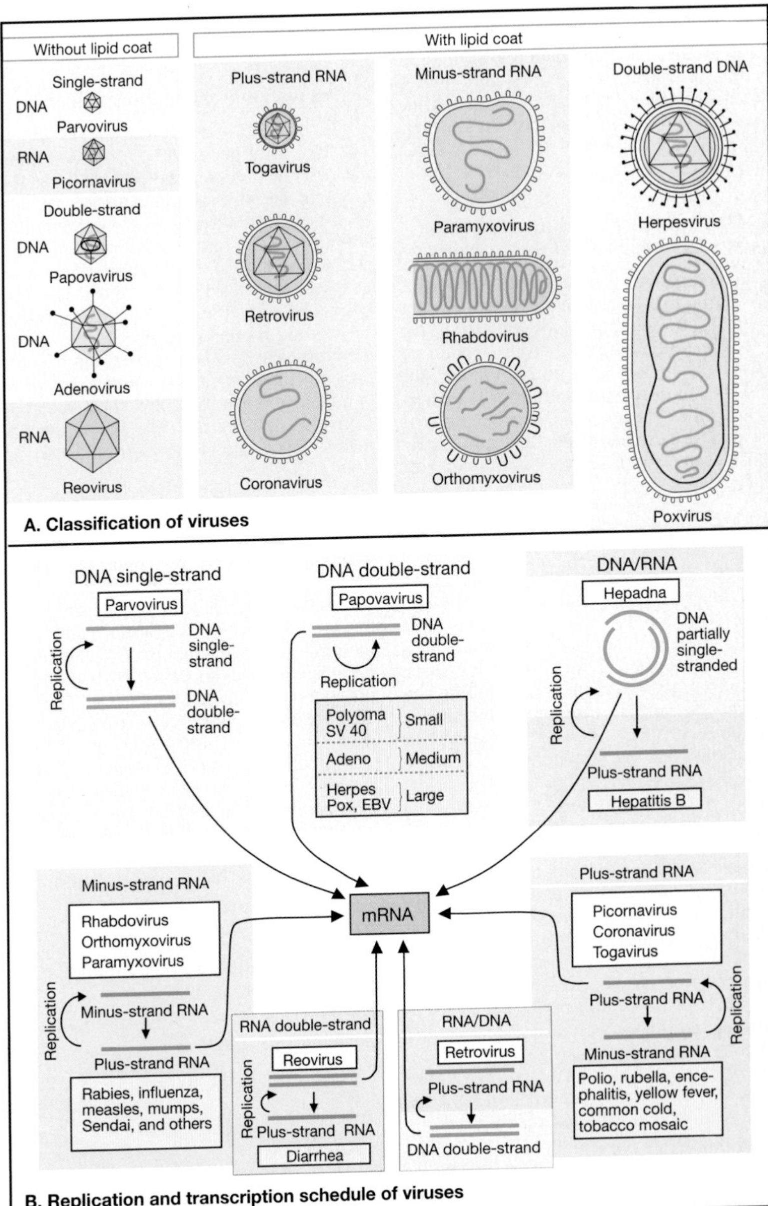

A. Classification of viruses

B. Replication and transcription schedule of viruses

Replication Cycle of Viruses

With all their different genomic structures, forms, and sizes, viruses basically have a relatively simple replication cycle. While only the genome of a bacteriophage enters a bacterium, the complete virus (genome and capsid) enters a eukaryotic cell.

A. General schedule of the replication cycle of a virus in a cell

The replication cycle of a virus consists of five principal consecutive steps: (1) entrance into the cell and release of the genome (uncoating), (2) transcription of the viral genes and translation of the mRNAs to form viral proteins, (3) replication of the viral genome, (4) assembly of new viral particles in the cell, and (5) release of the complete virions from the host cell.

B. Uptake of a virus by endocytosis

Besides fusion of the lipid membrane of membrane-enclosed viruses with the cell membrane of the host cell, the most frequent mechanism for a virion to enter a cell is by a special form of endocytosis. The virus attaches to the cell membrane by using cell surface structures (receptors), which serve other important functions for the cell, e.g., for the uptake of macromolecules (see p. 292). Like these, the virus is taken into the cytoplasm by a special mechanism, receptor-mediated endocytosis (coated pits, coated vesicles). Within the cell, the virus-containing vesicle fuses with other cellular vesicles (e.g., primary lysosomes). The viral coat is extensively dismantled in the endocytotic vesicle, and the viral core (genome, associated with virus-coded proteins) is released into the cytoplasm or nucleus, depending on the viral type. Replication and expression of the viral genome follow. Whether a cell can be infected by a virion depends on a specific interaction between the virus and a cellular receptor. Some viruses, such as the paramyxoviruses (e.g., mumps and Sendai virus), enter the cell by direct fusion of the viral and cellular membranes, mediated by a viral coat glycoprotein (F or fusion protein).

C. Transcription and replication of a virus

The first viral genes to be expressed after the virus has entered the cell are the early genes of the viral genome. Gene products of these early viral genes regulate transcription of the remaining viral genes and are involved in replicating the viral genome. Synthesis of the capsid proteins begins later (late genes), at the same time as genome replication, when new virions are formed from the genome and capsids (assembly). The virions (nucleocapsids = genome plus capsid) are then released from the cell by one of several mechanisms, depending on the type of virus.

D. Release of a virus by budding

The release of a virus coated by a lipid membrane occurs by budding. First, molecules of a viral-coded glycoprotein are built into the cell membrane, to which the virus capsid or virus core (containing the viral genome) attaches. Attachment of the genome leads to increased budding of that region of the cell membrane. Eventually, the entire virion is surrounded by a lipid membrane envelope of cellular origin containing viral proteins and is released. Virions can be expelled from the cell continuously and in great numbers without the virus-producing cell dying.

(Figures from JD Watson et al. 1987).

References

Alberts B, et al. Molecular Biology of the Cell. 4th ed. 1994.

Watson JD, et al. Molecular Biology of the Gene. 3rd ed. 1987

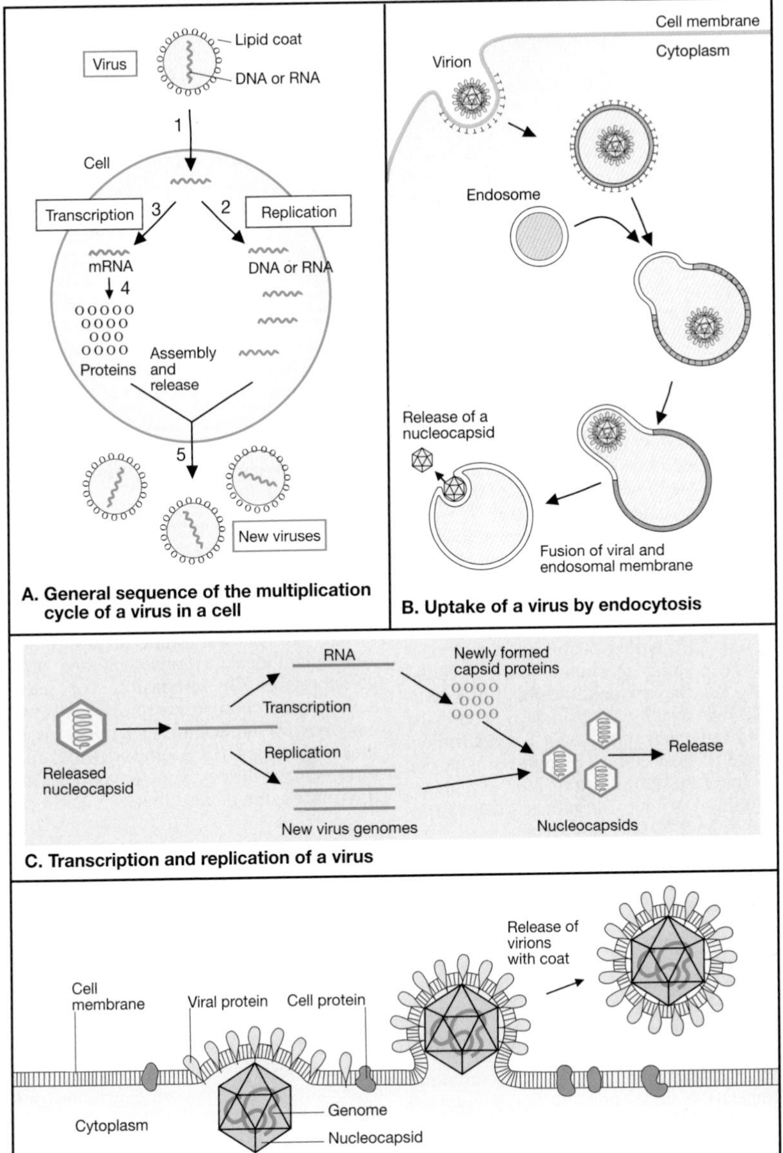
Virus
Lipid coat
DNA or RNA
1
Cell
Transcription
3
2
Replication
mRNA
4
DNA or RNA
Proteins
Assembly and release
5
New viruses
A. General sequence of the multiplication cycle of a virus in a cell
Cell membrane
Cytoplasm
Virion
Endosome
Release of a nucleocapsid
Fusion of viral and endosomal membrane
B. Uptake of a virus by endocytosis
RNA
Newly formed capsid proteins
Transcription
Replication
Released nucleocapsid
Release
New virus genomes
Nucleocapsids
C. Transcription and replication of a virus
Release of virions with coat
Cell membrane
Viral protein
Cell protein
Cytoplasm
Genome
Nucleocapsid
D. Release of a virus by budding

RNA Viruses: Genome, Replication, Translation

The genomes of RNA viruses consist of double- or single-stranded RNA. Single-stranded RNA can be of plus strands, which can be directly translated, or minus strands. The latter must first be transcribed into plus-strand RNA (sense RNA). Since eukaryotic cells do not have enzymes for copying RNA, RNA viruses contain enzymes necessary to transcribe RNA (transcriptase). In most of these RNA viruses, transcriptase molecules are included in the virus particle.

A. Genome of the poliovirus and its translation products

The genome of the poliovirus, a member of the enterovirus family, is one of the smallest genomes. It consists of about 7400 base pairs of plus-strand RNA of known nucleotide base sequence. It codes for a large precursor protein from which smaller, functional proteins are formed by proteolytic cleavage (VPO, 1, 2, 3, and 4), about 60 copies of each being present per virion. VP1 is responsible for attachment to the cellular receptor, which is found only on the epithelial cells, fibroblasts, and nerve cells of primates. Thus, the poliovirus can infect only a very limited range of host cells. The poliovirus is one of the fastest replicating animal viruses: Within 6–8 hours, an infected cell can release about 10 000 new virions.

B. Togavirus: replication and translation

Togaviruses (e.g., yellow fever, rubella, encephalitis viruses) have an RNA plus strand of about 12 000 base pairs as genome. It is enclosed in a capsid and a lipid membrane. Only the replicase protein can be translated from the genomic RNA, since start codons (AUG) for initiating translation lie 3' and are not recognized before replication. Thus, the capsid proteins are synthesized late after infection, i.e., after replication. This results in a differentiated regulation of viral gene expression.

C. Influenzavirus

The genome of the influenzavirus consists of eight segments of minus-strand RNA. Each codes for at least one, and some for more than one protein. The lipid membrane contains two virus-coded glycoproteins: hemagglutinin (HA), which recognizes and binds to cell surface receptors, and neuraminidase (NA), a receptor-destroying enzyme. In addition, the virions contain a matrix protein (M) and a nucleocapsid protein (N). If a cell has become infected with two different influenza virus strains, a new type of influenza virus can arise by exchange of genome segments. The recombinant virus is either not at all or only slowly recognized by the immune system.

D. Rhabdovirus (vesicular stomatitis virus, VSV)

The RNA minus-strand genome of rhabdoviruses is enclosed in a characteristically formed (bullet-shaped) outer membrane; it codes for five proteins, the nucleocapsid protein (N) within the virion, the matrix protein (M) between capsid and outer membrane, a transmembrane viral glycoprotein (G) responsible for interacting with cellular receptors, and two enzymes for replication and mRNA synthesis: proteins L (large) and NS (nonstructural). One of the best known types of rhabdoviruses is the rabies virus.

E. Transcription and translation of a minus-strand RNA virus

RNA viruses with minus-strand genomes (e.g., rhabdoviruses, myxoviruses) must first form an RNA plus strand by means of a virus-coded replicase contained in the viral particle. This serves as a template for the formation of a new genome (replication) and for mRNAs (transcription) for the synthesis of virus-coded proteins (translation). The new viral genomes are then packaged to form virions.

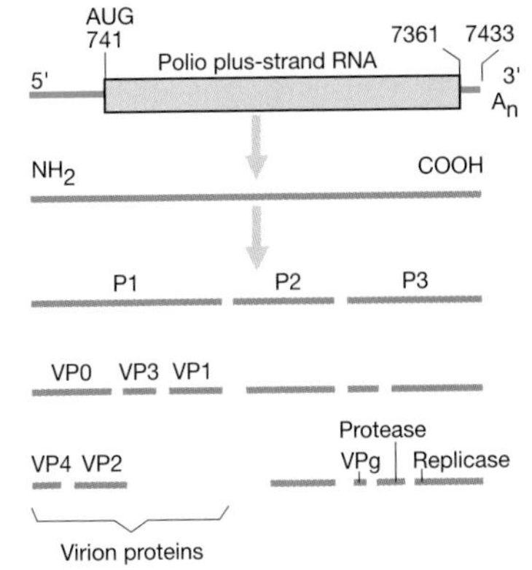

AUG
741
7361 7433
Polio plus-strand RNA
5'
3'
A_n
NH_2
COOH
P1 P2 P3
VP0 VP3 VP1
Protease
VP4 VP2
VPg Replicase
Virion proteins

A. Genome of poliovirus and its translation products

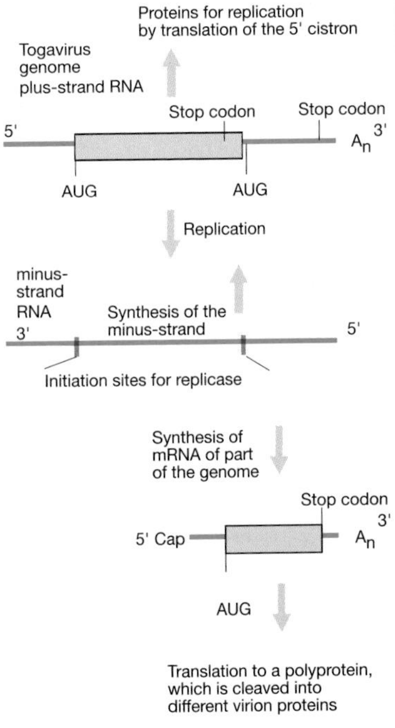

Proteins for replication by translation of the 5' cistron
Togavirus genome plus-strand RNA
Stop codon
Stop codon
5'
3'
A_n
AUG
AUG
Replication
minus-strand RNA
Synthesis of the minus-strand
3'
5'
Initiation sites for replicase
Synthesis of mRNA of part of the genome
Stop codon
3'
5' Cap
A_n
AUG
Translation to a polyprotein, which is cleaved into different virion proteins

B. Togavirus: replication and translation

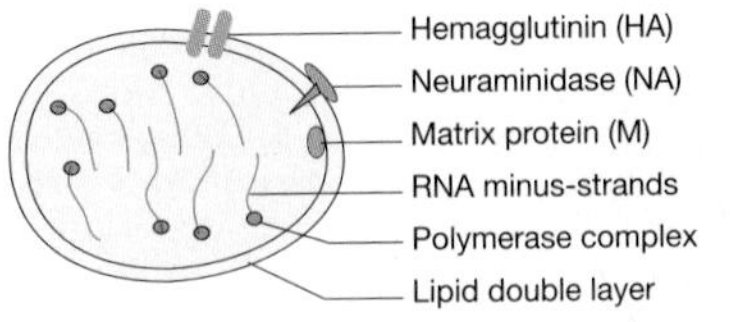

Hemagglutinin (HA)
Neuraminidase (NA)
Matrix protein (M)
RNA minus-strands
Polymerase complex
Lipid double layer

C. Influenzavirus

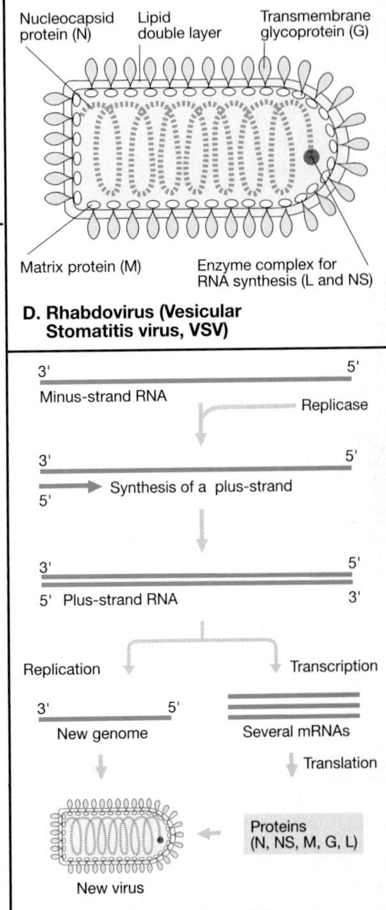

D. Rhabdovirus (Vesicular Stomatitis virus, VSV)

E. Transcription and translation of a minus-strand RNA virus

DNA Viruses

DNA viruses depend on host-cell DNA synthesis for their replication. Since the cellular proteins required for DNA synthesis can only be formed during the S phase of the cell cycle, most DNA viruses induce DNA synthesis in the host cell. A few contain genes for their own DNA polymerases and other proteins for DNA synthesis. Their genes are expressed in a well-defined chronological pattern. Important DNA viruses include herpesviruses and poxviruses.

A. SV40 virus

The SV40 (simian virus 40) virus has a double-stranded circular DNA genome (5243 base pairs) and belongs to the papovaviruses. The protein coat of the virion consists of three virus-coded proteins (VP1, VP2, VP3). An early and a late region of the viral genome can be distinguished according to their time of expression. Between them lies a regulative region. Two alternatively spliced mRNAs of the early region code for the large tumor antigen (T, approx. 90 kilodaltons [kDa]) and the small tumor antigen (t, ca. 20 kDa). They regulate further transcription, the initiation of replication, and the expression of specific cellular genes. The late region, coding for the capsid proteins, is transcribed at the onset of viral DNA synthesis in the opposing DNA strand and in the opposite direction. The late proteins are translated from differently spliced, overlapping mRNAs. A fourth, late gene codes for a small protein (agno protein) of unknown function. DNA replication of the viral genome begins at a defined point (OR, *o*rigin of *r*eplication).

B. Adenovirus

The adenovirus genome (1) consists of a linear double-stranded DNA of about 36 000 base pairs (36 kb). Both ends contain a repetitive nucleotide sequence (inverted terminal repeat), which is important for DNA replication. Early transcripts (E), which appear about 2–3 hours after infection and 6–8 hours before the onset of viral DNA replication, are transcribed at specific regions. The region E1a codes for proteins that initiate transcription of all other viral genes and that influence the expression of specific cellular genes. The E2 region codes for proteins directly involved in DNA replication, including a viral DNA polymerase (adenovirus-specific replication mechanism). A single promoter controls transcription of the late region (L), which is transcribed into a large primary RNA (2). From this, at least 20 different mRNAs are produced by alternative splicing. All mRNAs have the same 5' terminus. Unlike the early transcripts, the late transcripts are coded for by only one DNA strand. The late genes code mainly for viral coat proteins.

C. Parvovirus

Parvoviruses are small, single-stranded DNA viruses that can replicate only in proliferating cells. Replication of the single DNA strand is initiated (self-priming) by the formation of a specific genomic structure (hairpin structure).

D. Herpesvirus genome

Herpesviruses and poxviruses are large viruses with DNA genomes of 80–200 kb that contain repetitive sequences. They code for many (50–200) different proteins, including their own DNA polymerase. Other proteins interfere with the regulation of the cell's nucleotide metabolism. The group of herpesviruses includes the Epstein–Barr virus, the varicella-zoster virus, and the cytomegalovirus.

(Figures after Watson JD, et al. 1987).

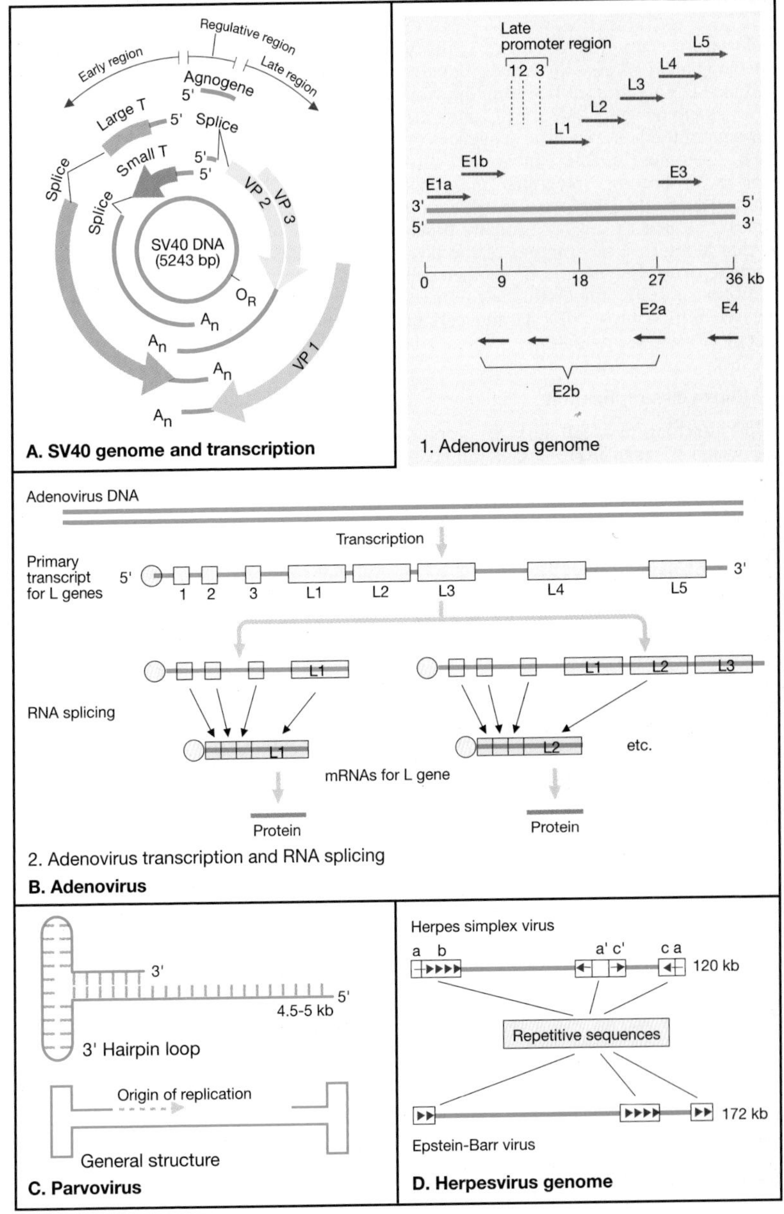

A. SV40 genome and transcription

1. Adenovirus genome

2. Adenovirus transcription and RNA splicing

B. Adenovirus

C. Parvovirus

D. Herpesvirus genome

Retroviruses

Retroviruses are RNA viruses with a developmental cycle in which double-stranded DNA is transcribed from the viral RNA genome and integrated into the genome of the host. Specific sequences of their genomes enable integrated proviruses to become integrated into new sites. When this happens, neighboring cellular sequences are occasionally transported along to a new region of the host genome or to the genome of another cell (retrotransposon). Important retroviruses are the AIDS virus (HIV I) and certain tumor viruses (see p. 264).

A. Retrovirus replication

After attaching to a cell surface receptor, the virion is taken into the cell. Immediately after entry into the cell, the viral RNA genome is transcribed into double–stranded DNA by an enzyme complex called reverse transcriptase, and the new DNA is integrated as a provirus into the DNA of the host cell. The RNAs transcribed from this DNA copy by cellular RNA polymerase II serve either as mRNA for the synthesis of virus proteins or as new virus genomes. Newly formed virions leave the cell by a specific process called exocytosis, without killing the cell.

B. Genomes of some retroviruses

Several retroviruses cause a tumor in mice (mouse leukemia) or chickens (Rous sarcoma). The only retrovirus identified to cause a tumor in humans is HTLV (type 1 and 2) (human T cell leukemia/lymphoma virus). HTLV viruses have genetic similarities with the AIDS virus, which causes acquired immune deficiency syndrome (p. 259).

The RNA genome of a typical retrovirus contains short repetitive (R) and singular (U) sequences at both ends (RU5 at the 5' end and U3R at the 3' end). As a rule, three protein-coding genes lie between them: *gag* (group-specific antigen), *pol* (reverse transcriptase), and *env* (a gene that codes for a glycoprotein that is built into the lipid membrane coat of the virion). The 5' end of the genome contains a nucleotide sequence that is complementary to the 3' end of a host cell tRNA. This nucleotide sequence binds to tRNA, which serves as a primer for the synthesis by reverse transcriptase of viral DNA from the virus genome RNA. At the 3' end of the genome, HTLV viruses carry several other genes (*px, lor, tat*, and others) that are involved in regulating the transcription of the viral genes. The AIDS virus (HIV I/II) contains further genes that do not occur in other retroviruses. These include *vif* (*vir*ion *i*nfectivity *f*actor, formerly designated *sor*), *rev* (*r*egulator of *e*xpression of *v*irion proteins), and *nef* (*ne*gative *f*actor, formerly *orf*). The general structure of the genome is more complex than shown here.

C. DNA synthesis of a retrovirus

Reverse transcriptase transcribes the RNA genome into DNA by RNA-dependent DNA polymerase activity and catalyzes subsequent steps by means of DNA-dependent DNA polymerase activity. Furthermore, the reverse transcriptase has RNAse activity for degrading the RNA of the newly formed RNA/DNA hybrid molecule. The first step in replicating the retrovirus genome is initiated by a primer of host cell tRNA, which is hybridized to the 5' end of the RNA genome (1). After synthesis of the first DNA strand and removal of the tRNA primer, synthesis of the second DNA strand (plus strand) begins at the RU5 region (2). Here, the previously formed minus DNA strand serves as the primer (3). As DNA synthesis is continued, the remaining RNA is degraded (4), and the DNA plus strand is synthesized to completion (5,6). The double-stranded DNA copy of the virus contains long terminal repeats (LTR) at both ends. These enable the viral DNA intermediary step to be integrated into the host cell DNA, and they contain the necessary regulatory sequences for transcribing the provirus DNA (see p. 96).

(Figures adapted from JD Watson et al., 1987).

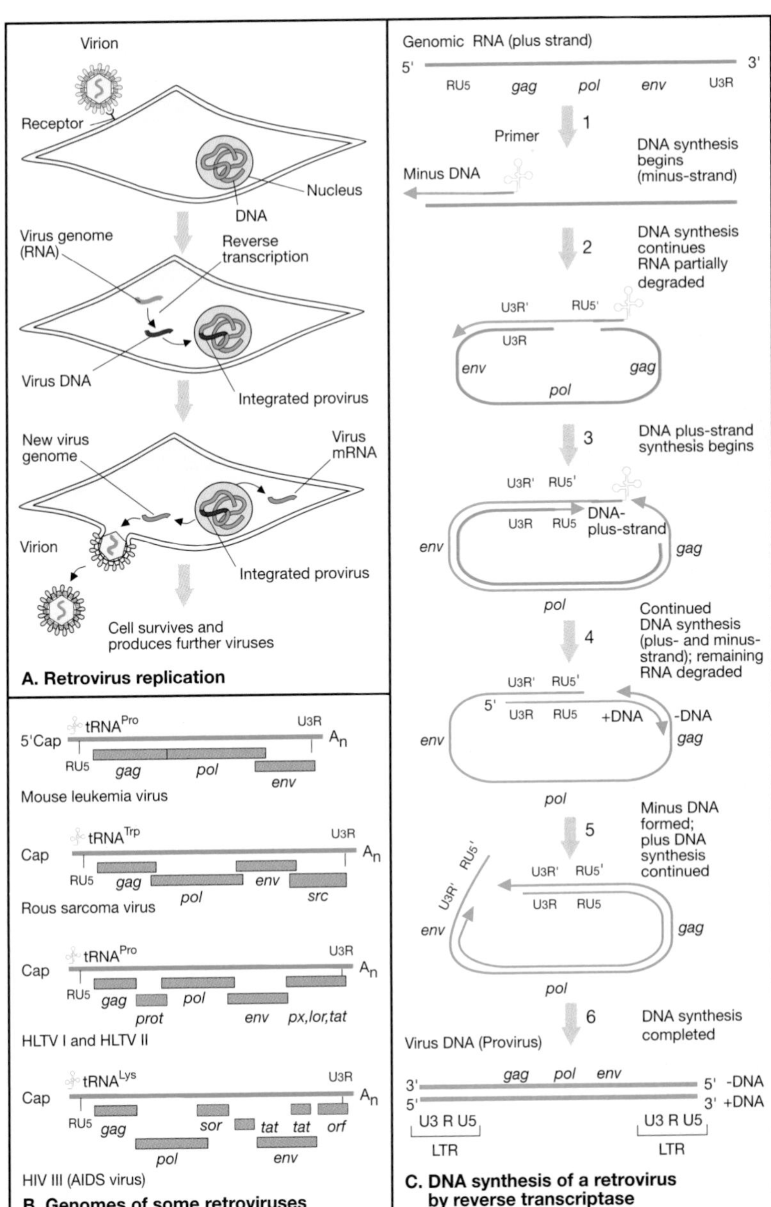

Virion
Receptor
Nucleus
DNA
Virus genome (RNA)
Reverse transcription
Virus DNA
Integrated provirus
New virus genome
Virus mRNA
Virion
Integrated provirus
Cell survives and produces further viruses
A. Retrovirus replication
5'Cap
tRNA Pro
U3R
An
RU5
gag
pol
env
Mouse leukemia virus
Cap
tRNA Trp
U3R
An
RU5
gag
pol
env
src
Rous sarcoma virus
Cap
tRNA Pro
U3R
An
RU5
gag
prot
pol
env
px,lor,tat
HLTV I and HLTV II
Cap
tRNA Lys
U3R
An
RU5
gag
sor
tat
tat
orf
pol
env
HIV III (AIDS virus)
B. Genomes of some retroviruses
Genomic RNA (plus strand)
5'
3'
RU5
gag
pol
env
U3R
1
Primer
DNA synthesis begins (minus-strand)
Minus DNA
2
DNA synthesis continues RNA partially degraded
U3R'
RU5'
U3R
env
gag
pol
3
DNA plus-strand synthesis begins
U3R'
RU5'
U3R
RU5
DNA-plus-strand
env
gag
pol
4
Continued DNA synthesis (plus- and minus-strand); remaining RNA degraded
U3R'
RU5'
5'
U3R
RU5
+DNA
-DNA
env
gag
pol
5
Minus DNA formed; plus DNA synthesis continued
RU5'
U3R'
U3R'
RU5'
U3R
RU5
env
gag
pol
6
DNA synthesis completed
Virus DNA (Provirus)
gag
pol
env
3'
5'
5' -DNA
3' +DNA
U3 R U5
U3 R U5
LTR
LTR
C. DNA synthesis of a retrovirus by reverse transcriptase

Retrovirus Integration and Transcription

Integration of the DNA copy of a retrovirus into the host cell DNA occurs at a random location. This may alter cellular genes (insertion mutation). The viral genes of provirus DNA are transcribed by cellular RNA polymerase II. The resulting mRNA serves either for translation or for the production of new RNA genomes, which are packaged into the virion. Some retrovirus genomes may contain an additional viral oncogene (*v*-onc). Viral oncogenes are parts of cellular genes (*c*-onc) previously taken up by the virus. If they enter a cell with the virus, they may change (transform) the host cell so that its cell cycle is altered and the cell becomes the origin of a tumor (tumor virus) (see p. 264).

A. Retrovirus integration into cellular DNA

In the nucleus of the host cell, the double-stranded DNA (1, virus DNA) produced from virus genome RNA first forms a ring-shaped structure (2) by joining LTRs (long terminal repeats). This is possible because the LTRs contain complementary nucleotide sequences. Recognition sequences in the LTRs and in the cellular DNA (3) allow the circular viral DNA to be opened at a specific site (4) and the viral DNA to be integrated into the host DNA. Viral genes can then be transcribed (6) from the integrated provirus (5). As a rule, the provirus remains in the genome of the host cell without disrupting the functions of the cell. One exception is the AIDS virus, which destroys a specific population of lymphocytes (helper T cells) (see p. 258). The genomes of vertebrates (including man) contain numerous DNA sequences that consist of endogenous proviruses. In mice, they may represent up to 0.5% of the total DNA. The genomes of higher organisms also contain LTR-like sequences that are very similar to those of an endogenous retrovirus. These sequences can change their location in the genome (mobile genetic elements or transposons). Since many of them have the fundamental structure of a retrovirus (LTR genes), they are designated retrotransposons (see pp. 94 and 354).

B. Regulation of retrovirus transcription

The LTRs are important not only for integration of the virus into cellular DNA, but also because they contain all regulatory signals necessary for efficient transcription of a viral gene. Typical transcription signals are the so-called "CCAAT" and "TATA" sequences of promoters, which are respectively located about 80 and 25 base pairs above the 5' end of the sequence to be transcribed. Further upstream (in the 5' direction) are nucleotide sequences that can increase the expression of viral genes (enhancer). Similar regulatory sequences are located at the 5' end of eukaryotic genes (see section on gene transcription). Newly synthesized viral RNA is structurally modified at the 5' end (formation of a cap). Furthermore, numerous adenine residues are added at the 3' end (polyadenylated, poly A, see p. 204).

C. Virus protein synthesis by posttranscriptional modification of RNA

The RNA transcripts synthesized from the provirus by cellular RNA polymerase II serve either for the translation into virus-coded proteins (*gag, pol*) or as the genome for the formation of new virions. Some of the RNA is spliced to form new mRNAs that code for coat proteins (*env*).

(Figures after Watson JD, et al. 1987).

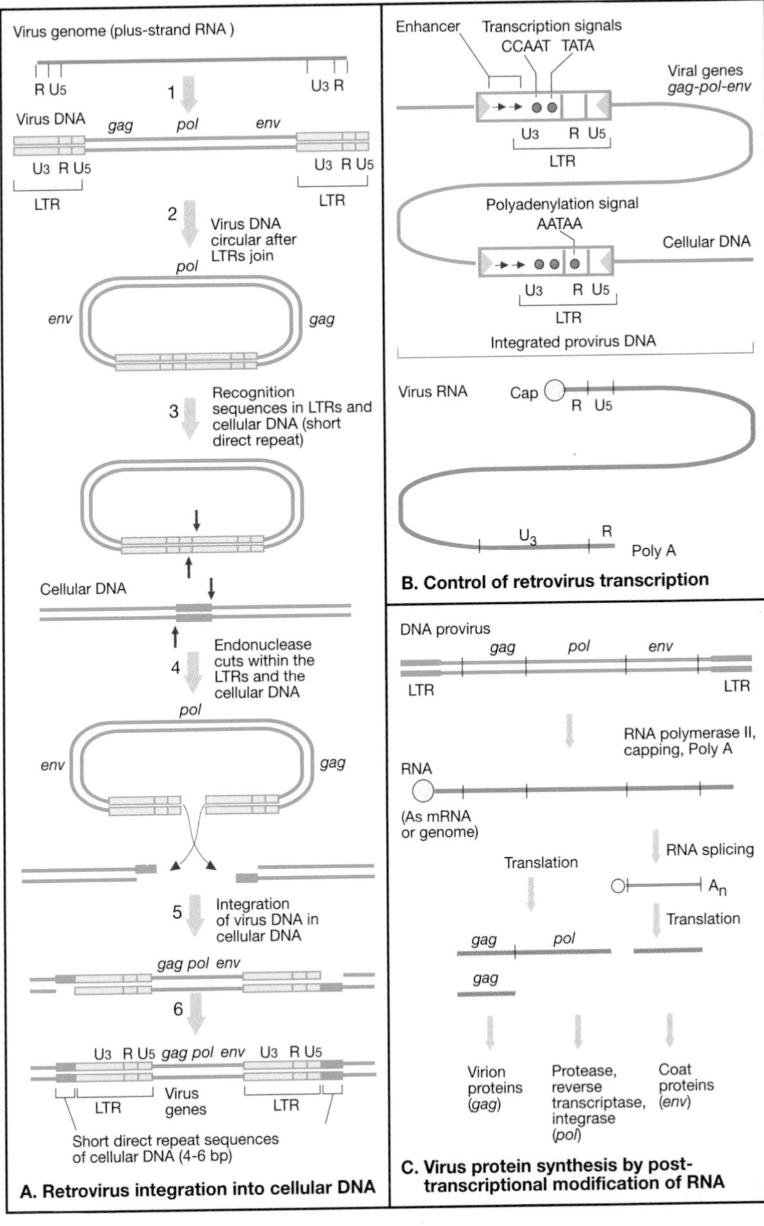
Virus genome (plus-strand RNA)
R U5
U3 R
1
Virus DNA
gag
pol
env
U3 R U5
U3 R U5
LTR
LTR
2
Virus DNA circular after LTRs join
pol
env
gag
3
Recognition sequences in LTRs and cellular DNA (short direct repeat)
Cellular DNA
4
Endonuclease cuts within the LTRs and the cellular DNA
pol
env
gag
5
Integration of virus DNA in cellular DNA
gag pol env
6
U3 R U5 gag pol env U3 R U5
LTR
Virus genes
LTR
Short direct repeat sequences of cellular DNA (4-6 bp)
A. Retrovirus integration into cellular DNA
Enhancer
Transcription signals
CCAAT TATA
Viral genes gag-pol-env
U3 R U5
LTR
Polyadenylation signal
AATAA
Cellular DNA
U3 R U5
LTR
Integrated provirus DNA
Virus RNA
Cap
R U5
U3
R
Poly A
B. Control of retrovirus transcription
DNA provirus
gag
pol
env
LTR
LTR
RNA polymerase II, capping, Poly A
RNA
(As mRNA or genome)
RNA splicing
Translation
An
Translation
gag
pol
gag
Virion proteins (gag)
Protease, reverse transcriptase, integrase (pol)
Coat proteins (env)
C. Virus protein synthesis by post-transcriptional modification of RNA

Yeast: Eukaryotic Cells with a Diploid and a Haploid Phase

Baker's yeast, *Saccharomyces cerevisiae*, is made up of cells of about 3 μm diameter, which can divide every 90 minutes under good nutritional conditions. The cells contain all compartments characteristc of eukaryotic cells, such as endoplasmic reticulum, Golgi apparatus, mitochondria, peroxisomes, and a vacuole analogous to a lysosome.

Yeasts are single-celled fungi. About 40 types are known. Yeast cells are eukaryotic; they divide with about the same frequency as microorganisms and pass through a haploid and a diploid phase. Thus, they are especially well suited for genetic studies of eukaryotic organisms.

The haploid genome of yeast contains about 1.4 x 10^7 base pairs in 16 chromosomes and is only 3 1/2 times larger than that of *Escherichia coli* (4 x 10^6 bp). For comparison, the genome of *Drosophila* (1.7 x 10^8 bp) contains 12 x the amount of DNA. Several chromosomes have been sequenced entirely. It is expected that by 1996, the entire yeast genome will have been sequenced. Many new genes with as yet unknown function have been discovered.

The yeast genome contains approximately 6000 nearly intronless genes. Many human genes have a significantly related yeast protein (Tugendreich, et al. 1994).

A. Life cycle of yeast *(S. cerevisae)*

Two haploid cells of opposite mating types, **a** and α, unite to form a diploid cell. When conditions are favorable for growth, the cells multiply by budding. Formally, this corresponds to a mitosis. The newly formed cell is at first smaller than the larger mother cell. Under favorable conditions the cells multiply by repeated mitotic division.

If conditions for growth become less favorable, especially due to nitrogen deficiency, meiotic division occurs. The recombination characteristic of meiosis (see p. 108) occurs in yeast about 10–20 times per chromosome, i.e., about 10 times more frequently than in man. Four haploid spores are formed, which are enclosed in a sacklike casing (ascus). When growing conditions again become favorable, each spore will be able to form a colony of haploid cells by vegetative reproduction (mitotic division). These can fuse to form diploid cells. Only cells of opposite mating types (**a** and α) will fuse.

The cell cycle of yeast cells is regulated by a set of genes (cell-cycle genes, *cdc* [*c*ell *d*ivision *c*ycle]). More than 50 *cdc* genes have been identified by mutation analysis. Genes of similar structure and function, for regulating cell cycle, also exist in vertebrates, including humans.

B. Determination of change of mating type

During its life cycle, a yeast cell may change its mating type. After an unequal division, the larger cell may change, e.g., from α to **a**. The smaller daughter cell cannot. This cell can change its mating type only when division occurs anew.

C. Genes for mating type

The mating type is determined by three genes on chromosome 3: locus MAT, locus α (HMLα), and locus **a** (HMRa) (1). The MAT (mating type) locus can be of either type α or type **a**. If it is α, the cell has the mating type α (2); if it is **a**, the cell is of mating type **a** (3). The change in mating type determination is regulated by the MAT locus. The loci HMLα and HMRa lie on opposite sides of the centromere (CEN). HMLα and HMRa contain additional silent copies of α and **a** sequences. The mating type determination in yeast cells is a model of gene regulation in eukaryotic cells (see p. 204).

(Figures after Darnell J, et al. 1990 and Watson JD, et al. 1987).

Reference

Tugendreich S, et al. Genes conserved in yeast and humans. Hum Mol Genet. 1994; 3: 1509–1517.

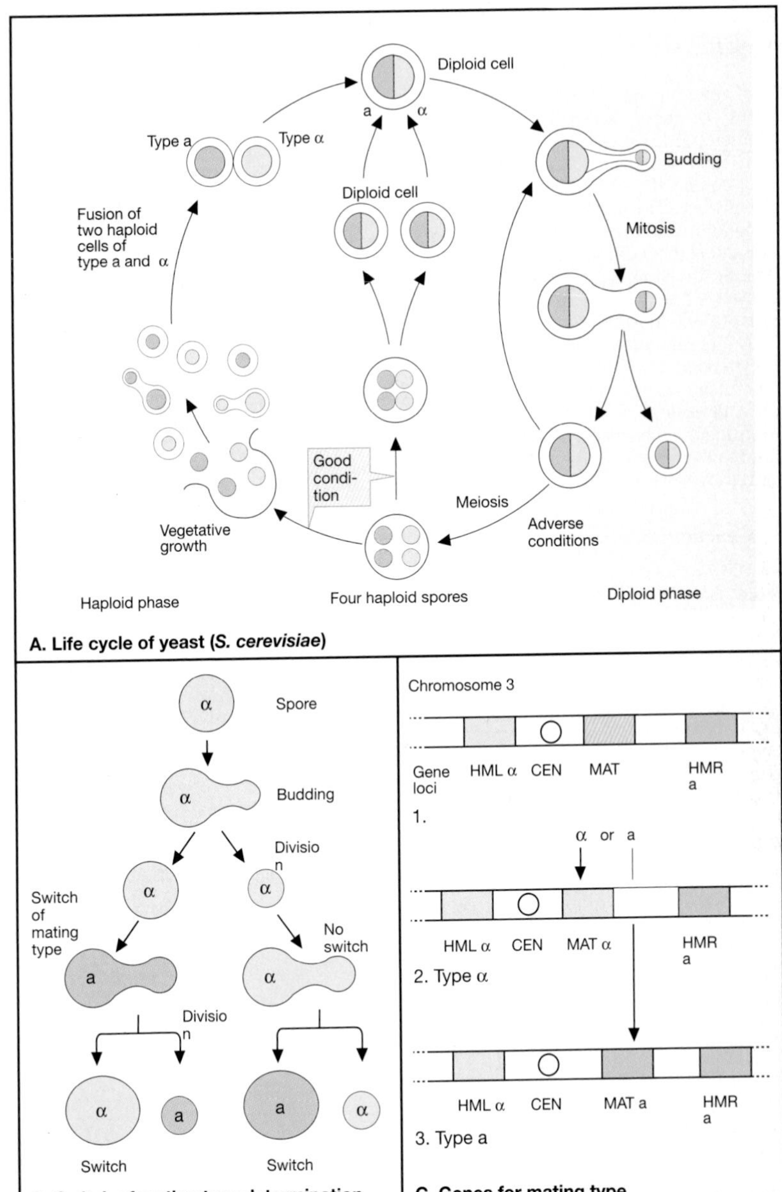

A. Life cycle of yeast (*S. cerevisiae*)

B. Switch of mating type determination

C. Genes for mating type

Mating-Type Determination in Yeast

The yeast genome contains about 6000 genes; of these, about 1900 have been characterized by a mutation, and more than 1200 sequenced. As a rule they are very compactly arranged and do not contain introns. The transition from mitotic to meiotic division occurs under various environmental conditions (see p. 98). After a meiotic division, yeast cells are haploid. Only two haploid cells of different mating types (α and **a**) fuse. Mating type is genetically determined. Alpha cells produce a polypeptide designated as α pheromone and an **a** receptor. **A** cells produce an **a** pheromone and an α receptor. Under poor nutritional conditions, the pheromones bind to their receptors, which stops further cell growth.

A. Genetic regulation of the MAT locus

The MATa locus codes for an a1 protein, which has no regulating effect on **a** cells. There is also an a2 protein, the function of which is not known. In the absence of a specific effect of the a1 protein, only **a**-specific genes are active, while the α-specific genes remain inactive. The MATα locus codes for two peptides, α2 and α1. Alpha 2 is a negative regulator of the **a**-specific genes, which become inactivated. As a positive regulator, α1 activates the α-specific genes. Haploid-specific genes are active in both **a** cells and α cells. In diploid cells, neither **a**-specific, α-specific, nor haploid-specific genes are expressed. Diploid cells produce an α2 protein, which forms a complex with the α1 protein and suppresses the haploid-specific genes. The α2/α1 complex also inhibits transcription of the α1 gene. Thus, altogether three proteins for regulating mating-type determination are formed (α1, α2, and a1 proteins). The cell type depends on which sequences are present on the MAT locus (see below).

B. Cassette model of mating-type determination

The mating type change in yeast cells is based on an exchange of DNA sequences at the MAT locus in one direction. MAT is the receiver; HML(α) and HMR(a) are donors. MATa is replaced by a copy of HML (α) (cassette model). The cassette model of mating-type determination means that MAT is provided with one active cassette (of α or **a** type), which in each case replaces an inactive cassette at the HML or HMR locus. Although all cassettes contain information for coding the mating type, only the active cassette at the MAT locus is expressed.

Mating change (mating switch) occurs when an active cassette at the MAT locus is replaced by another. The mating-type change is reversible. That is, an **a** type can become an α type. Silent α sequences at another location of the genome are copied and replace the **a** sequences at the MAT locus. The **a** sequences are removed and the MAT locus beomes occupied by the α sequences, thus effecting the change from mating type **a** to mating type α. With subsequent change from type α to type **a**, α sequences are removed from the MAT locus and replaced by a copy of a silent **a** gene. The removal of sequences at the MAT locus occurs by means of an endonuclease (HO endonuclease) regulated by an HO locus. A series of genes (SWI switch) code for proteins that are involved in the change of mating type.

(Data after Darnell et al. 1990).

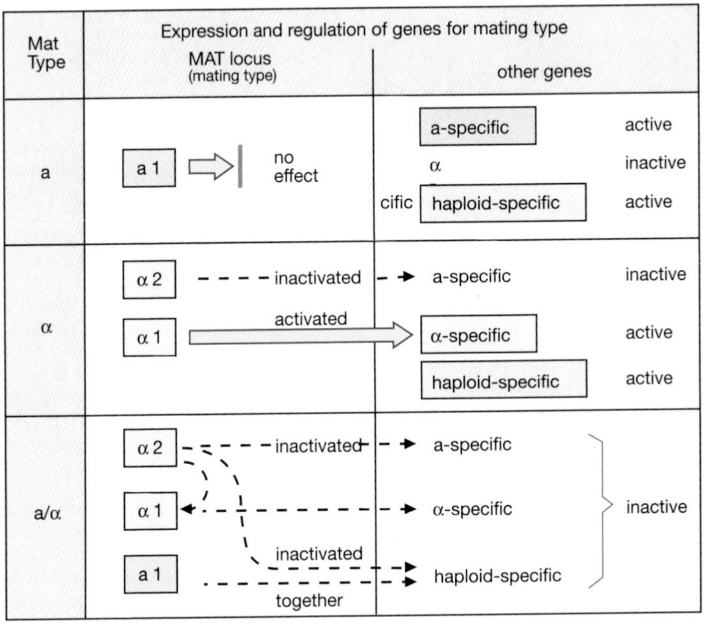

Mat Type	Expression and regulation of genes for mating type		
	MAT locus (mating type)	other genes	
a	a 1 → no effect	a-specific	active
		α	inactive
		cific haploid-specific	active
α	α 2 inactivated →	a-specific	inactive
	α 1 activated →	α-specific	active
		haploid-specific	active
a/α	α 2 inactivated →	a-specific	inactive
	α 1 →	α-specific	
	a 1 inactivated together →	haploid-specific	

A. Genetic regulation at the MAT locus

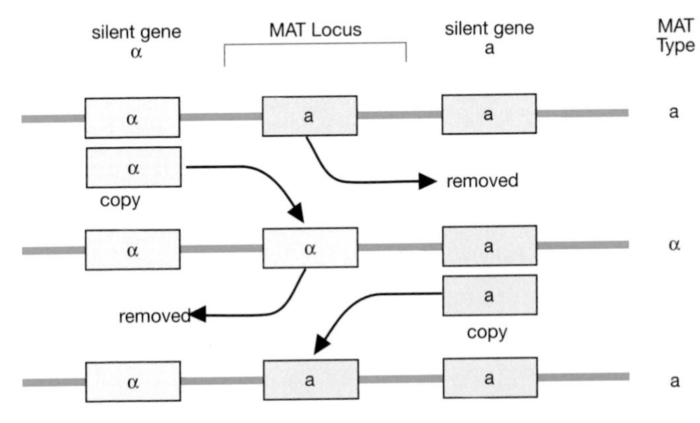

B. Cassette model of mating-type determination

Functional Elements in Yeast Chromosomes

The haploid genome of *S. cerevisiae* consists of 1.4×10^7 base pairs in 16 chromosomes. Several functional DNA sequences have been isolated and characterized from yeast chromosomes. They are necessary for replication and normal distribution of the chromosomes during mitosis (mitotic segregation). If they lie on the same chromosome, they are designated *cis* elements (as opposed to *trans* on the opposite-lying chromosome). They are genetically stable and can be joined in vitro to DNA of any origin to form stable artificial chromosomes in yeast cells (see p. 104). Three types of functional DNA sequences are known: autonomously replicating sequences (ARS), centromere sequences (CEN), and telomere sequences (TEL). All eukaryotic chromosomes contain these functional elements, but they can be especially well demonstrated in yeast cell chromosomes.

A. Autonomously replicating sequences (ARS)

Replication proceeds in two directions (bidirectional). The ARS represent the starting points of replication in chromosomes. Their functional significance can be recognized in a transformation experiment: A mutant yeast cell (e.g., with the inability to produce leucine, Leu^-) can be transformed into a leucine-producing (Leu^+) cell (1) by means of transfection with cloned plasmids containing the gene for leucine formation (Leu). However, the daughter cells remain leucine-dependent because the plasmid cannot replicate. If the plasmid contains autonomously replicating sequences (ARS) along with the Leu gene, a small fraction of the daughter cells will be Leu^+ because plasmid replication has occurred (2). But most of the daughter cells remain Leu^- because mitotic distribution (segregation) is defective.

B. Centromere sequences (CEN)

If the plasmid DNA (1) contains sequences from the centromere (CEN) of the yeast chromosome along with the gene for leucine (Leu) and the autonomously replicating sequences (ARS), then normal mitotic segregation takes place (2). This demonstrates that the CEN sequences are necessary for normal distribution of the chromosomes at mitosis (see p. 106).
The centromere sequences are similar but not identical in different yeast chromosomes. They contain three elements with a total of about 220 base pairs (bp), which occur in all chromosomes. Element I is a conserved sequence of 8 bp; element II is AT rich and has about 80 bp; element III has about 25 bp. This segment is nuclease-protected and very important for mitotic stability.

C. Telomere sequences (TEL)

If the plasmid is linear instead of circular as in B, transformation will occur, but the plasmid does not replicate (2). However, if telomere sequences (TEL) are attached to both ends of the plasmid (3) before it is incorporated (4), then transformation of the yeast cell after incorporating the plasmid is followed by normal replication and mitosis (5).
These observations prove that autonomously replicating sequences (starting point of replication), centromere sequences, and telomere sequences (see p. 188) are necessary for normal chromosome function (p. 164). These functional elements (ARS; CEN; TEL) can be utilized to produce artificial yeast chromosomes (YACs).

References

Darnell J, Lodisch H, Baltimore H. Molecular Cell Biology. 2nd ed. New York: Scientific American Books, W H Freeman; 1990: 327.

Traut W. Chromosomen. Klassische und molekulare Cytogenetik. Heidelberg: Springer; 1991: 148.

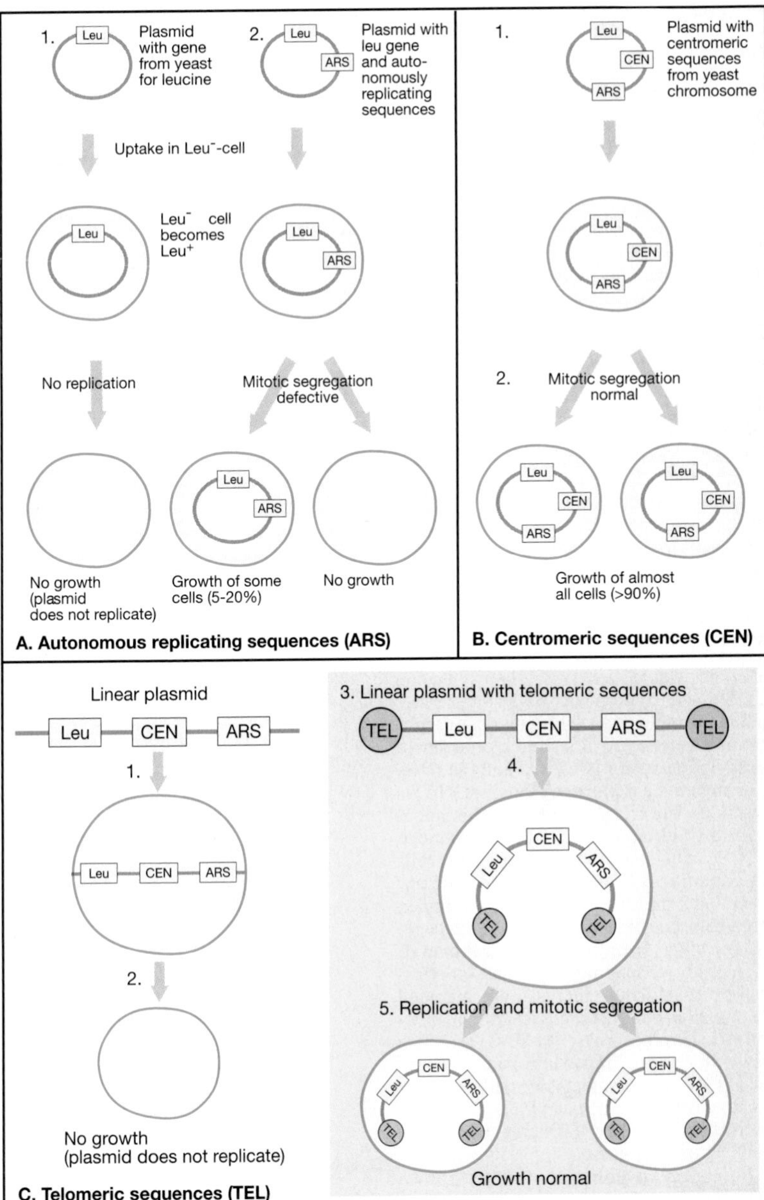
1.
Leu
Plasmid with gene from yeast for leucine
2.
Leu
ARS
Plasmid with leu gene and autonomously replicating sequences
Uptake in Leu⁻-cell
Leu
Leu⁻ cell becomes Leu⁺
Leu
ARS
No replication
Mitotic segregation defective
Leu
ARS
No growth (plasmid does not replicate)
Growth of some cells (5-20%)
No growth
A. Autonomous replicating sequences (ARS)
1.
Leu
CEN
ARS
Plasmid with centromeric sequences from yeast chromosome
Leu
CEN
ARS
2.
Mitotic segregation normal
Leu
CEN
ARS
Leu
CEN
ARS
Growth of almost all cells (>90%)
B. Centromeric sequences (CEN)
Linear plasmid
Leu
CEN
ARS
1.
Leu
CEN
ARS
2.
No growth (plasmid does not replicate)
C. Telomeric sequences (TEL)
3. Linear plasmid with telomeric sequences
TEL
Leu
CEN
ARS
TEL
4.
CEN
Leu
ARS
TEL
TEL
5. Replication and mitotic segregation
CEN
Leu
ARS
TEL
TEL
CEN
Leu
ARS
TEL
TEL
Growth normal

Artificial Chromosomes for Analyzing Complex Genomes

The mapping of large and complex genomes (for instance in humans with 3 billion base pairs) would not be possible if cloned fragments of only a few dozen to a few thousand base pairs were available (with cosmids, up to about 50 kb). Large distances could not be bridged because nonclonable DNA segments would lie between the mapped segments. Since physical mapping of these segments was not possible, attempts were made to develop vectors for cloning large DNA fragments. Artificial chromosomes can be constructed from yeast chromosomes (YACs, yeast artificial chromosomes). These can incorporate DNA fragments of about 100–1000 kb. Since they are stable during growth, they represent an important instrument for gene mapping.

A. Construction of artificial chromosomes (YACs) for cloning and mapping

The starting material is a linear vector. This consists of telomere sequences (TEL) and selectable markers, here leucine (LEU) and tryptophan (TRP), a starting point for replication (autonomously replicating sequences, ARS), and a centromere (CEN). The vector is incised and the DNA fragment to be replicated (foreign DNA) is inserted. The capacity of a YAC is about 750 kb. Subsequently, the incorporated DNA can be cloned in a YAC in yeast cells and amplified. In this way, a YAC library containing the entire genome of a complex organism can be produced in a manageable manner. From this, a physical map of large and overlapping fragments (contigs) can be constructed. Increasingly, artificial mammalian chromosomes (mammalian artificial chromosomes, MACs) are also being developed. These consist of one arm of a YAC with a selectable marker, a replication starting point, and a telomere attached to a fragment of mammalian DNA with its own telomere.

Artificial yeast chromosomes are shown on page 77, section B.

B. YAC vector modification

Different methods can be used to modify YACs. Sequences can be incorporated so that with their help, both ends of the YAC can be recovered as plasmids in *E. coli* (1). As a prerequisite, plasmid sequences with selectable markers (2), here for ampicillin resistance (Amp^R), neomycin resistance (Neo^R), and onset of plasmid replication (ORI), are included. Digestion with an appropriate restriction enzyme produces fragments of the incorporated foreign DNA and of both vector arms (3). The fragments are formed into a ring and incorporated into *E. coli.* The hybrid plasmids (4) can then be selected for (5) by means of the markers (Amp^R and Neo^R).

The use of YACs has considerably simplified the genetic analysis of large genomes. While a complete human-genome library requires about 500 000 clones of lambda phage vectors, a YAC library with 150-kb fragments reduces the number of clones to about 60 000.

References

Burke DT. Cloning of large segments of exogenous DNA into yeast by means of artificial chromosome vectors. Science. 1987; 236: 806–812.

Schlessinger D. Yeast artificial chromosomes: tools for mapping and analysis of complex genomes. Trends Genet. 1990; 6: 248–258.

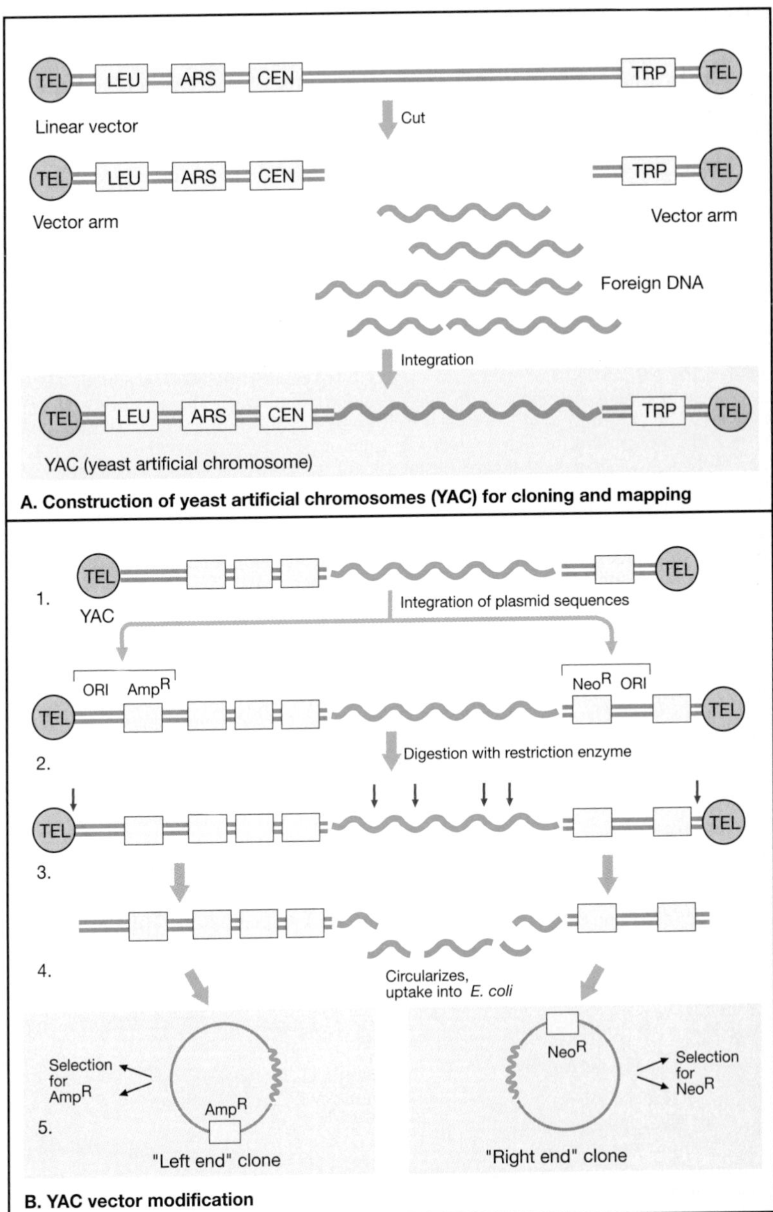

A. Construction of yeast artificial chromosomes (YAC) for cloning and mapping

B. YAC vector modification

Cell Division: Mitosis

Threadlike structures in dividing cells were first observed by Flemming in 1879. He introduced the term mitosis for cell division. Flemming also observed the longitudinal division of chromosomes during mitosis. In 1884, Strasburger coined the terms prophase, metaphase, and anaphase for the different stages of cell division. A mitosis results in two genetically identical daughter cells.

A. Mitosis

During the transition from interphase to mitosis, the chromosomes become visible as elongated threads (prophase). In early prophase, each chromosome is attached to a specific site on the nuclear membrane and appears as a double structure (sister chromatids), the result of the foregoing DNA synthesis. The chromosomes contract during late prophase to become thicker and shorter (chromosomal condensation). In late prophase, the nuclear membrane disappears and metaphase begins. At this point, the mitotic spindle becomes visible as thin threads. It begins at two polelike structures (centrioles). The chromosomes become arranged on the equitorial plate, but homologous chromosomes do not pair. In late metaphase during the transition into anaphase, the chromosomes divide, also at the centromere region. The two chromatids of each chromosome migrate to opposite poles, and telophase begins with the formation of a nuclear membrane. Finally the cytoplasm also divides (cytokinesis). In early interphase the chromosomal structures lose their discreteness. Interphase chromosomes are called chromatin (Flemming 1879), i.e, nuclear structures stainable by basic dyes.

B. Metaphase chromosomes

Waldeyer (1888) coined the term chromosome for the stainable threadlike structures visible during mitosis. A metaphase chromosome consists of two chromatids (sister chromatids) and the centromere, which holds them together. The centromere may divide each of the chromatids into two chromosome arms. The regions at both ends of the chromosome are the telomeres. The point of attachment to the mitotic spindle fibers is the kinetochore. During metaphase and prometaphase, chromosomes can be visualized under the light microscope as discrete elongated structures, 3–7 μm long (see p. 170).

C. Cell cycle (mitosis)

Eukaryotic cells go through cell division cycles (cell cycle). In eukaryotic cells, each cell division begins with a phase of DNA synthesis, which lasts about about 8 hours (S phase). This is followed by a phase of about 4 hours (G_2). During the G_2 phase (GAP-2), the whole genome is double. Mitosis (M) in eukaryotic cells lasts about an hour (see **A**). This is followed by a phase, G_1 (interphase), of extremely varied duration. It corresponds to the normal functional phase. Cells which no longer divide are in the G_0 phase. Every cell must have a "memory" to tell it whether it is in G_1 or in G_2, since division of the chromosomes before they have been doubled would be lethal. All phases of the cell cycle are regulated by specific proteins encoded by numerous cell-division cycle (cdc) genes. In particular, the transition from G_1 to S and from G_2 to M is regulated by specific cell-cycle proteins.

References

Whitehouse HLK. Towards the Understanding of the Mechanism of Heredity, 3rd ed. London: Edward Arnold; 1973.

Koshland D. Mitosis. Back to the basics. Cell. 1994; 77: 951–954.

Holm C. Coming undone. How to untangle a chromosome. Cell. 1994; 77: 956–957.

North G. Regulating the cell cycle. Nature. 1989; 339: 97–98.

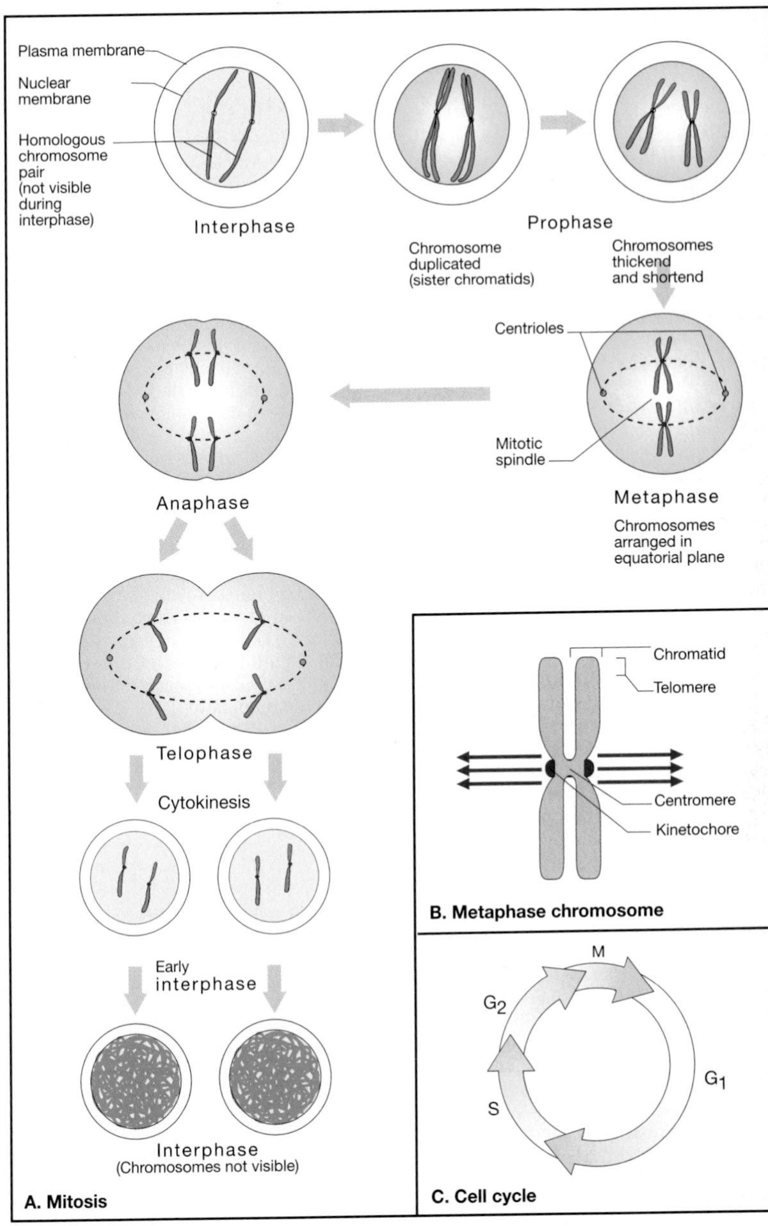
Plasma membrane
Nuclear membrane
Homologous chromosome pair (not visible during interphase)
Interphase
Prophase
Chromosome duplicated (sister chromatids)
Chromosomes thickend and shortend
Centrioles
Mitotic spindle
Metaphase
Chromosomes arranged in equatorial plane
Anaphase
Telophase
Cytokinesis
Early interphase
Interphase
(Chromosomes not visible)
A. Mitosis
Chromatid
Telomere
Centromere
Kinetochore
B. Metaphase chromosome
M
G_2
G_1
S
C. Cell cycle

Maturation Division (Meiosis)

Gametes are formed by a special type of cell division that differs from somatic cell division. Strasburger (1884) introduced the term meiosis (reduction division or maturation division) for this process.

Meiosis differs from mitosis in fundamental genetic and cytologic respects. Firstly, homologous chromosomes pair up. Secondly, exchanges between homologous chromosomes regularly occur (crossing-over). This results in chromosome segments with new constitutions (genetic recombination). Thirdly, the chromosome complement is halved during the first cell division (meiosis I). Thus, the daughter cells resulting from this division are haploid (reduction division).

Meiosis is a complex cellular and biochemical process. The cytologically observable course of meiosis and the genetic consequences do not correspond exactly in time. A genetic process occurring in one phase is usually not cytologically manifest until a subsequent phase.

A. Meiosis I

A complete meiosis consists of two cell divisions, meiosis I and meiosis II. The relevant genetic events, genetic recombination by means of crossing-over and reduction to the haploid chromosome complement, occur in meiosis I.

Meiosis begins with chromosome replication. Initially the chromosomes in late interphase are visible only as threadlike structures, as in mitosis. At the beginning of prophase I, the chromosomes are doubled, but this is not visible until a later period of prophase I (see p. 110). Subsequently, the pairing of homologous chromosomes can also be visualized. The pairing makes an exchange between homologous chromosomes (crossing-over) possible by juxtapositioning homologous chromatids (chiasma formation). The result of crossing-over is an exchange of material between two chromatids of homologous chromosomes (genetic recombination). This exchange has occurred by the time the cell has entered metaphase I. After the homologous chromosomes migrate to opposite poles, anaphase I begins.

B. Meiosis II

Meiosis II consists of longitudinal division of the doubled chromosomes (chromatids) and a further cell division. Each daughter cell contains one chromosome of a chromosome pair and is therefore haploid. Because of the recombination that occurred in prophase I, the chromosomes of the resulting haploid cell differ from those of the original cell. Thus, unlike in mitosis, the chromosomes of the daughter cells are not genetically identical with those of the original cell. On each chromosome, recombinant and nonrecombinant sections can be identified. The genetic events relevant to these changes have occurred in the prophase of meiosis I (see p. 110).

The distribution of chromosomes during meiosis explains the segregation (separation or splitting) of traits according to mendelian laws (1:1 segregation, cf. p. 126).

Recombination is the most striking event in meiosis. The molecular mechanisms of recombination are complex. Occasionally, recombination may occur during mitosis (mitotic recombination), e.g., during DNA repair. Gene conversion designates a unilateral, nonreciprocal exchange. With this, one allele is lost in favor of another.

References

Carpenter ATC. Chiasma function. Cell. 1994; 77: 959–962.

Moens PB, ed.: Meiosis. New York: Academic Press; 1987.

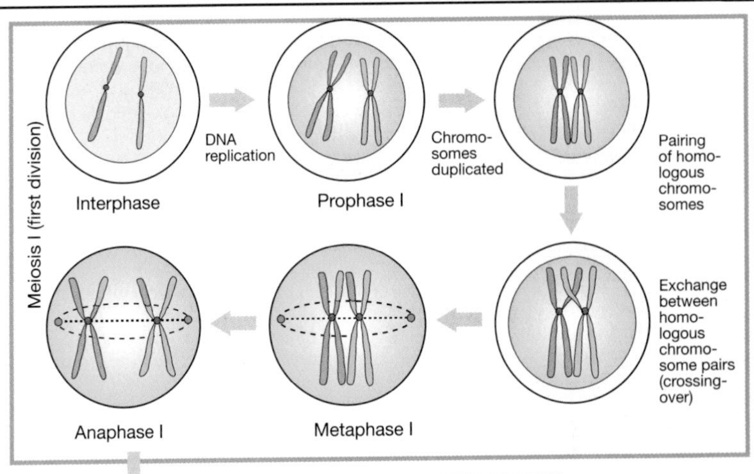

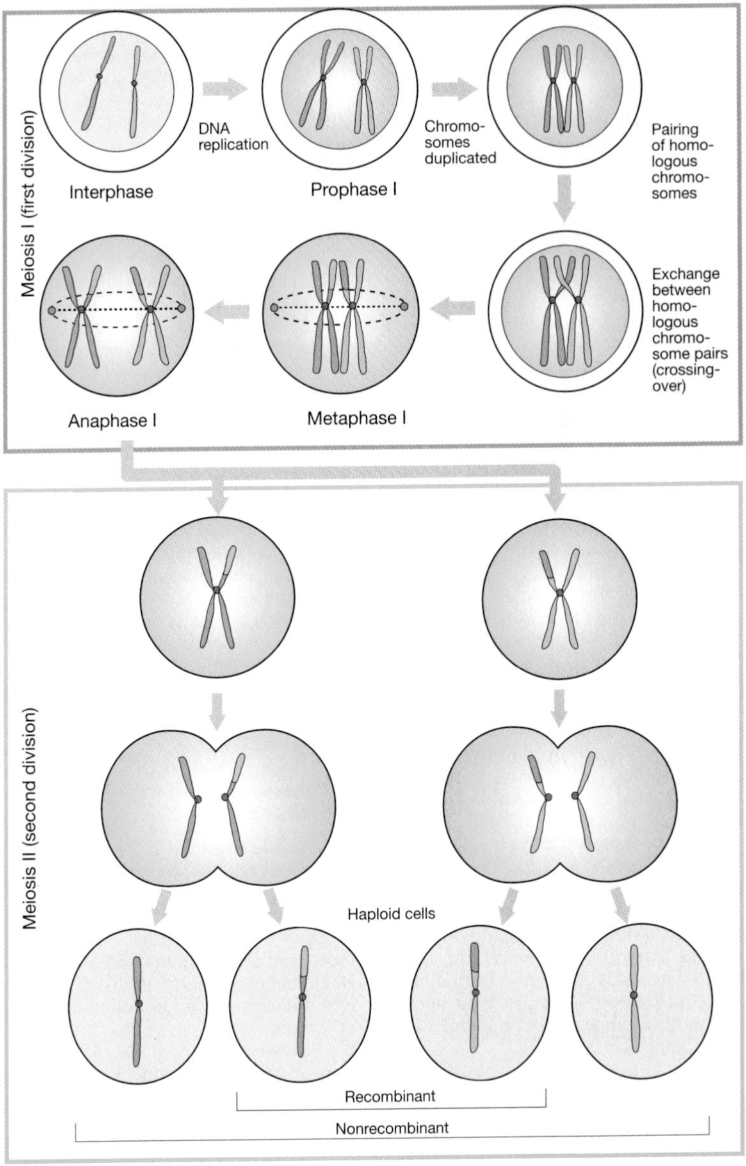

Meiosis II (second division)

Haploid cells

Recombinant

Nonrecombinant

A. Meiosis

Crossing-Over in Prophase I

Prophase of meiosis I is a complex period during which important cytological and genetic events occur (see p. 124). In this phase, exchanges between homologous chromosomes regularly occur by crossing-over. This results in new combinations of chromosome segments (genetic recombination).

A. Prophase of Meiosis I

The prophase of meiosis I goes through a number of stages that can be differentiated schematically, although they proceed continuously. In the leptotene stage, the chromosomes are first visible as fine threadlike structures (in **A** only one chromosome pair is shown schematically). In zygotene the chromosomes are visible as paired structures. By this time, every chromosome has been doubled and consists of two chromatids, which are held together at the centromere (each chromatid contains a DNA double helix). Two chromosomes that have paired are referred to as a bivalent. In the pachytene stage, the bivalents beome thicker and shorter. In diplotene the two homologous chromosomes separate for the most part, but still remain attached to each other at a few points (chiasmata). In subsequent stages of diplotene, each of the chromosome pairs separates even more extensively, especially at the centromere region, but not yet at one or more distally located points of attachment (chiasmata). Each chiasma corresponds to a region in which crossover has taken place. In the last stage of prophase I, diakinesis, the chromosomes are widely separated, although still attached at their distal ends. The chiasmata have shifted distally (terminalization). At the end of diakinesis, the nuclear membrane disappears and the cell enters metaphase I.

B. Synaptonemal complex

Shortly before the onset of the pachytene stage, homologous chromosomes move very close together and form a synaptonemal complex. This initiates chiasma formation and is the prerequisite for crossing-over and subsequent recombination. (Diagram after J.D. Watson et al. Molecular Biology of the Gene, 3rd ed. 1987)

C. Chiasma

When a chiasma is formed, either of the two chromatids of one chromosome pairs with one of the chromatids of the homologous chromosome (e.g., 1 and 3, 2 and 4, etc.). Chiasma formation is the cytological prerequisite for crossing-over and is important in the definitive separation (segregation) of the chromosomes. The centromere (Cen) plays an important role in chromosome pairing.

D. Genetic recombination through crossing-over

By crossing-over, new combinations of chromosome segments arise (recombination). As a result, recombinant and nonrecombinant chromosome segments can be differentiated. In the diagram, the areas A–E (shown in red) of one chromosome and the corresponding areas a–e (shown in blue) of the homologous chromosome become respectively a–b–C–D–E and A–B–c–d–e in the recombinant chromosomes.

E. Pachytene and diakinesis under the light microscope

In the micrograph, pachytene chromosomes are readily visualized as bivalents (a). In diakinesis (b), it can be seen that they have separated for the most part (photographs from Therman E. Human Chromosomes: structure and behavior. 2nd ed. Heidelberg: Springer; 1986). Today, electron micrographs are usually used for meiotic studies.

An unusual structure in pachytene is formed by the X and the Y chromosome. They appear to be joined end-to-end. Actually, short segments of the short arms in the regions with homologous sequences (pseudoautosomal region, see p. 324) have paired.

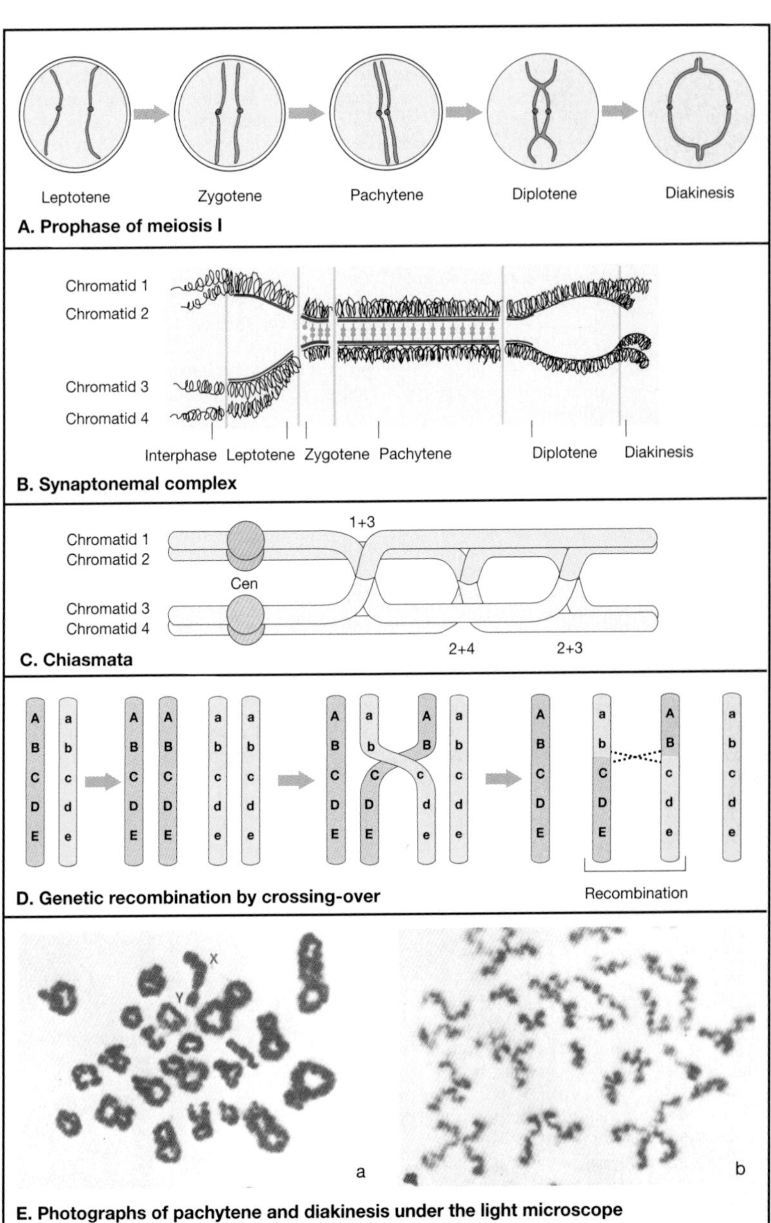

A. Prophase of meiosis I

B. Synaptonemal complex

C. Chiasmata

D. Genetic recombination by crossing-over

E. Photographs of pachytene and diakinesis under the light microscope

Formation of Gametes

Germ cells (gametes) are produced in the gonads. In females the process is called oogenesis (formation of oocytes) and in males, spermatogenesis (formation of spermatozoa). The primordial germ cells, which migrate to the gonads during early fetal development, increase in number by mitotic division. The actual formation of germ cells (gametogenesis) begins with meiosis. Meiosis differs in duration and results between males and females.

A. Spermatogenesis

Diploid spermatogonia are formed by repeated mitotic cell division. At the onset of puberty, some of the cells begin to differentiate into primary spermatocytes. The first meiotic cell division occurs in these cells. At the completion of meiosis I, a primary spermatocyte has given rise to two secondary spermatocytes, each of which has a haploid set of duplicated chromosomes (recombination is not illustrated here). Each chromosome consists of two sister chromatids, which become separated during meiosis II. In meiosis II, each secondary spermatocyte divides to form two spermatids. Thus, one primary spermatocyte forms four spermatids, each with a haploid chromosome complement. The spermatids differentiate into mature spermatozoa. Male spermatogenesis is a continuous process. In human males, the time lapse between differentiation into a primary spermatocyte at the onset of meiosis I and the formation of mature spermatocytes is about 6 weeks.

B. Oogenesis

Oogenesis (formation of oocytes) differs from spermatogenesis in timing and in the result. At first the germ cells, which have migrated to the ovary, multiply by repeated mitosis (formation of oogonia). In human females, meiosis I begins about 4 weeks before birth. Primary oocytes are formed. However, meiosis I is arrested in a stage of prophase designated dictyotene. The primary oocyte persists in this stage until ovulation. Only then is meiosis I continued (recombination is not shown here).

In females, the cytoplasm divides asymmetrically in both meiosis I and meiosis II. The result each time is two cells of unequal size: a larger cell that will eventually form the egg and a small cell, called a polar body. When the primary oocyte divides, the haploid secondary oocyte and polar body I are formed. When the secondary oocyte divides, again unequally, the result is a mature oocyte and another polar body (polar body II). The polar bodies do not develop further, but degenerate. On rare occasions when this does not occur, a polar body may become fertilized. This can give rise to an incompletely developed twin.

In the secondary oocyte, each chromosome still exists as two sister chromatids. These do not separate until the next cell division (meiosis II), when they enter into two different cells. In most vertebrates, maturation of the secondary oocyte is arrested in meiosis II. At ovulation the secondary oocyte is released from the ovary, and if fertilization occurs, meiosis is then completed. Faulty distribution of the chromosomes (nondisjunction) may occur in meiosis I as well as in meiosis II (see p. 342).

The maximal number of germ cells in the ovary of the human fetus at about the 5th month is 6.8×10^6. By birth this has been reduced to 2×10^6, and by puberty to about 200 000. Of these, about 400 are ovulated (Connor JM, Ferguson-Smith MA. Essential Medical Genetics. 4th ed. London: Blackwell Scientific; 1993).

The long period between meiosis I and ovulation is presumably a factor in the relatively frequent nondisjunction of homologous chromosomes in older mothers.

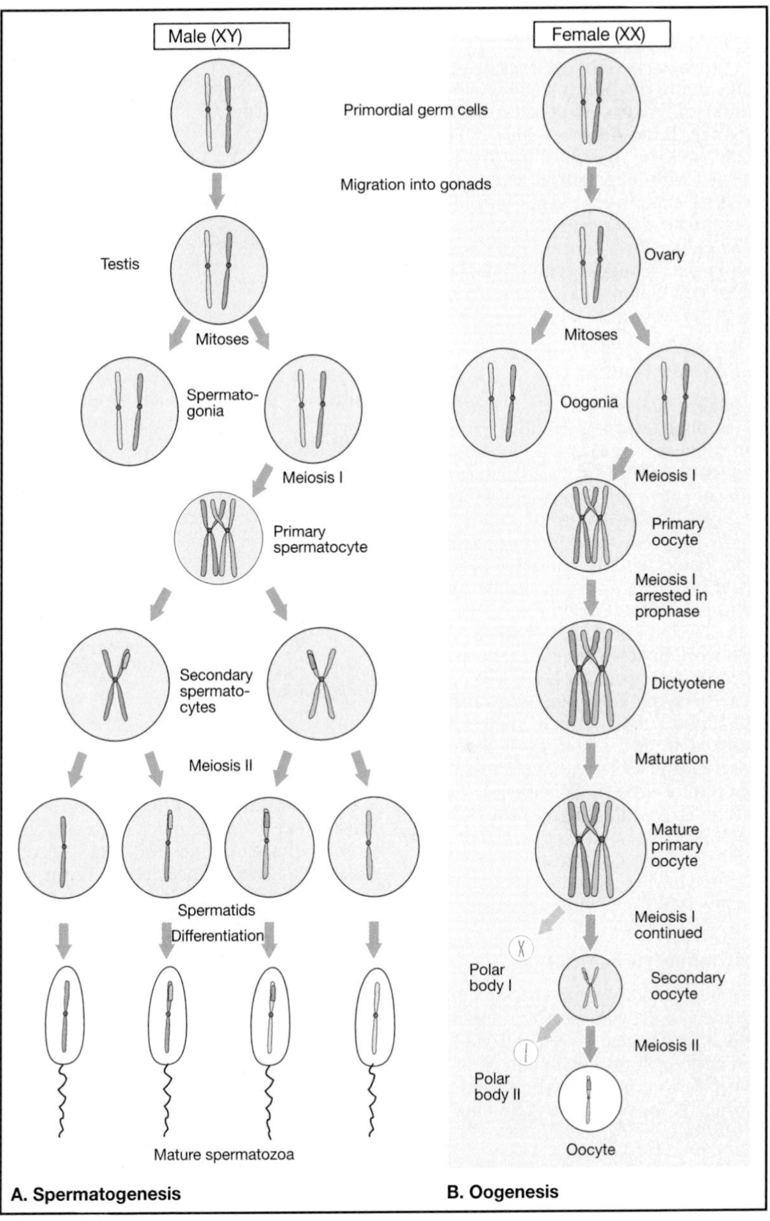
Male (XY)
Female (XX)
Primordial germ cells
Migration into gonads
Testis
Ovary
Mitoses
Mitoses
Spermato-
gonia
Oogonia
Meiosis I
Meiosis I
Primary
spermatocyte
Primary
oocyte
Meiosis I
arrested in
prophase
Secondary
spermato-
cytes
Dictyotene
Maturation
Meiosis II
Mature
primary
oocyte
Spermatids
Differentiation
Meiosis I
continued
Polar
body I
Secondary
oocyte
Meiosis II
Polar
body II
Oocyte
Mature spermatozoa
A. Spermatogenesis
B. Oogenesis

Cell Culture

The propagation of living cells in a culture (cell culture) is a fundamental requirement of modern genetics. With cell cultures, studies can be carried out that otherwise would not be possible for higher organisms with long generation times, or in breeding experiments, e.g., with humans. Appropriate experimentation using cultivated cells allows genetic analysis by detecting recombinants and mutants. The use of cell cultures has greatly facilitated the advances in gene mapping (p. 356).

A. Fibroblast cultures

A small piece of skin (2 x 4 mm) is obtained under sterile conditions and cut up into smaller pieces with smooth edges. The pieces are then placed into a vessel with suitable nutrient medium. Fibroblast-like cells begin to grow within about 10–20 days. The small pieces of skin must be in contact with the bottom of the culture vessel and may not swim in the medium (adhesion culture). Fibroblast cells begin to divide and within 2 to 4 weeks form a layer of juxtapositioned cells (monolayer). Cells in a monolayer will cease to divide (contact inhibition). However, they will divide again when the cells are detached from the bottom of the culture vessel and placed in a larger vessel or in two culture vessels (subculture). Cells in culture have a limited life span of an average of 20 to 30 cell divisions. Under appropriate conditions, viable cultured cells can be stored in vials submerged in liquid nitrogen at −196° C and reused at a later time.

B. Lymphocyte cultures

Peripheral blood lymphocytes placed in a culture vessel with suitable nutrient medium will replicate in suspension (suspension culture) for a maximium of 3–5 cell divisions if mitosis is stimulated or induced, e.g., by phytohemagglutinin.

C. Transformed cells in culture

Cells are said to be transformed when their behavior at cell division differs from the norm, e.g., by loss of contact inhibition. Transformed cells show some similarities to cancer cells in vivo.

D. Formation of hybrid cells

Cells can be fused (cell fusion) by adding Sendai virus or polyethylene glycol to a culture of mixed cells. If cells of different origins fuse, e.g., cell A and cell B, they form a common cell containing two different nuclei (heterokaryon). If the two nuclei fuse, a hybrid cell is formed that contains the chromosome complement of the two original cells in its nucleus. Hybrid cells are recognized by a selective medium that suppresses growth of the parental cells. A frequently used selective medium is HAT (containing hypoxanthine, aminopterin, and thymine). It selects for hybrid cells derived from mutant cells with HGPRT deficiency (hypoxanthine phosphoribosyltransferase) and thymidine kinase deficiency (TK^-). The division of TK^- cells is blocked by the enzyme deficiency; the division of the $HGPRT^-$ cells, by aminopterin.

E. Colonies of hybrid cells

Hybrid cells can be grown like normal cells by subculturing. With repeated cell division, one after the other of the chromosomes of the original cell types become lost at random. If the original parent cell types differed by certain traits, then hybrid cells can be identified by using selective medium, i.e., the medium is selected so that only hybrid cells divide, and not the original parent cells. Colonies of hybrid cells are then formed (cell clones) that differ in their genetic traits, since they contain different chromosomes. The chromosomal localization of genes for traits of interest can be deduced by comparison of the presence or absence of genetic traits with the presence or absence of certain chromosomes.

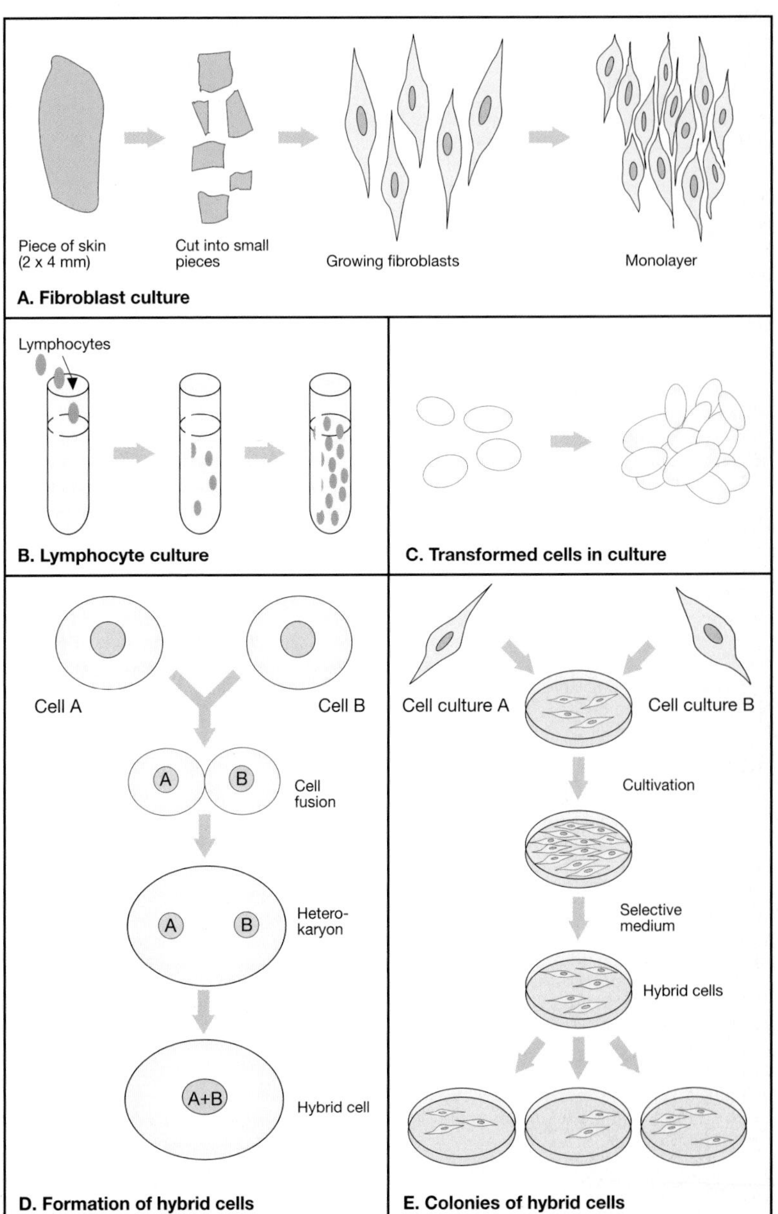
Piece of skin
(2 x 4 mm)
Cut into small
pieces
Growing fibroblasts
Monolayer
A. Fibroblast culture
Lymphocytes
B. Lymphocyte culture
C. Transformed cells in culture
Cell A
Cell B
A
B
Cell
fusion
Hetero-
karyon
A+B
Hybrid cell
D. Formation of hybrid cells
Cell culture A
Cell culture B
Cultivation
Selective
medium
Hybrid cells
E. Colonies of hybrid cells

Genetically Controlled Energy-Delivering Processes in Mitochondria

Eukaryotic organisms contain essential genetic information separate from the nuclear DNA, in extrachromosomal genomes. The mitochondria of all eukaryotes and the chloroplasts of green plants and algae contain cytoplasmic DNA molecules (mitochondrial DNA, mtDNA). As such, they are exclusively maternally inherited, and not in a mendelian manner. Mitochondria and chloroplasts are the sites of essential energy-delivering processes and photosynthesis. Unlike nuclear DNA, the genome in mitochondria and choroplasts is ring shaped. Sequence homologies indicate evolutionary relationships. In particular, evolutionary transfer of DNA segments from chloroplasts to mitochondria, and from chloroplasts to nuclear DNA of eukaryotic organisms, has been demonstrated. The genes contained in mitochondrial DNA code for proteins of the respiratory chain, subunits of the ATPase complex, subunits of the NADH-dehydrogenase complex (ND), and genes for transfer RNA (tRNA). A number of diseases due to mutations and deletions in mtDNA are known in humans.

A. Fundamental processes in a mitochondrion

The essential energy-conserving process in mitochondria is oxidative phosphorylation. Relatively simple energy carriers such as NADH and FADH2 (nicotinamide-adenine dinucleotide in the reduced form and flavin adenine dinucleotide in the reduced form) are produced from the degradation of carbohydrates, fats, and other foodstuffs by oxidation. The important energy carrier adenosine triphosphate (ATP) is formed by oxidative phosphorylation of adenosine diphosphate (ADP) through a series of biochemical reactions in the inner membrane of mitochondria (respiratory chain).

B. Oxidative phosphorylation (OXPHOS) in mitochondria

Adenosine triphosphate (ATP) plays a central role in the exchange of energy in biological systems. ATP is a nucleotide consisting of adenine, a ribose, and a triphosphate unit. It is energy-rich because the triphosphate unit contains two phospho-anhydride bonds. Energy (free energy) is released when ATP is hydrolyzed to form ADP. The energy contained in ATP and bound to phosphate is released, e.g., during muscle contraction.

C. Electron transfer in the inner mitochondrial membrane

The genomes of mitochondria and chloroplasts contain genes for the formation of the different components of the respiratory chain and oxidative phosphorylation. Three enzyme complexes regulate electron transfers: the NADH-dehydrogenase complex, the b–c1 complex, and the cytochrome oxidase complex. Intermediaries are quinone (Q) derivatives such as ubiquinone and cytochrome c. Electron transport leads to the formation of protons (H^+). These lead to the conversion of ADP and Pi (phosphorus) into ATP (oxidative phosphorylation). ATP represents a phosphate-bound reservoir of energy, which serves as an energy supplier for all biological systems. Thus it is understandable that genetic defects in mitochondria become manifest primarily as diseases with reduced muscle strength and other degenerative signs.

Reference

Alberts B, et al. Molecular Biology of the Cell. 3rd ed. 1995.

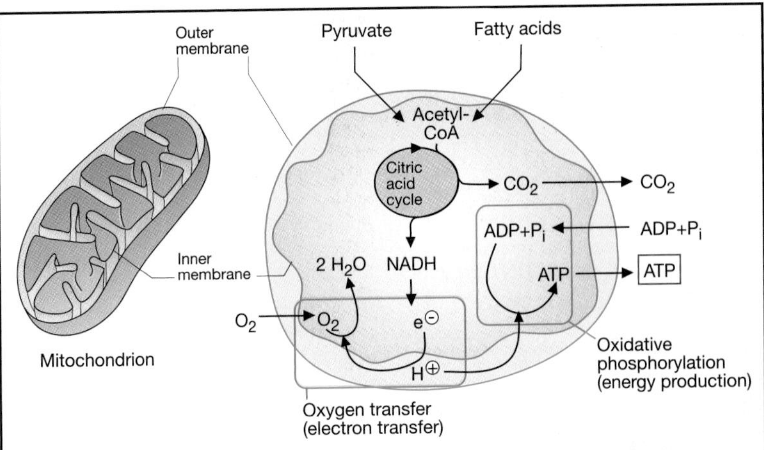

A. Principal events in mitochondria

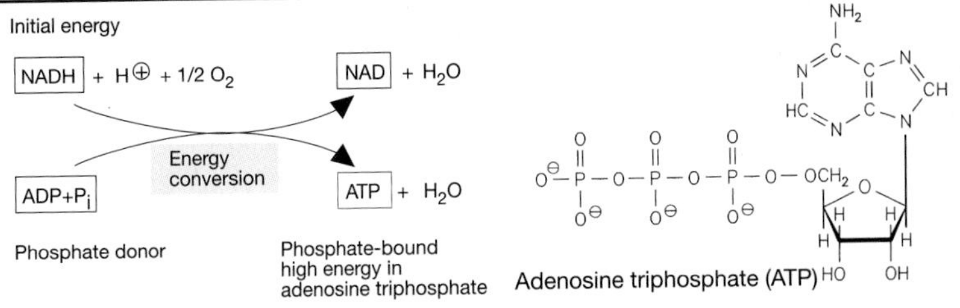

B. Oxidative phosphorylation (OXPHOS) in mitochondria

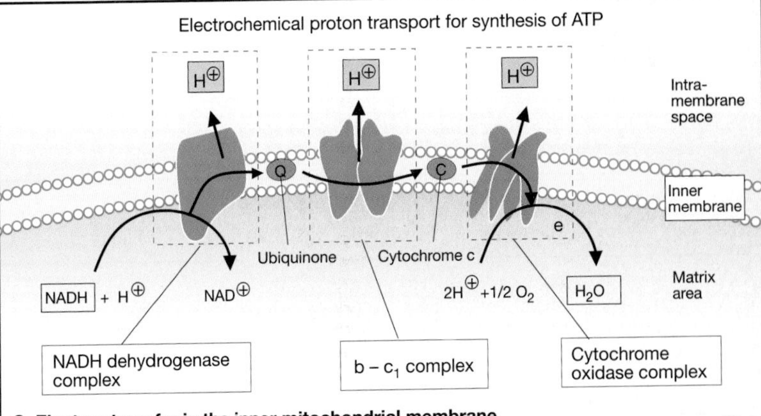

C. Electron transfer in the inner mitochondrial membrane

The Genome in Chloroplasts and Mitochondria

Chloroplasts of higher plants and mitochondria of eukaryotic cells contain genomes of circular DNA. About 12 000 base pairs (12 kb) of the genomes of chloroplasts and mitochondria are homologous. Furthermore, homologous regions are found in nuclear DNA. Thus it is assumed that the DNAs of chloroplasts, mitochondria, and nuclear DNA are evolutionarily related.

A. Genes in the chloroplasts of a moss *(Marchantia polymorpha)*

The genome in chloroplasts is large: 121 kb in the moss *M. polymorphia* and 155 kb in the tobacco plant. Yet the organization and number of their genes are similar. Protein synthesis shows certain similarities with that of bacteria. Many of the ribosomal proteins are homologous with those of *E coli*. Genes in the chloroplast genomes are interrupted and contain introns. The chloroplast genome codes for two copies each of four ribosomal RNAs (16S rRNA, 23S rRNA, 4.5S rRNA, and 5S rRNA). The genes for ribosomal RNA are located in two DNA segments with opposite orientation (inverted repeats), which are characteristic of chloroplast genomes. An 18- to 19-kb segment with short single gene copies lies between the two inverted repeats. The genome of chloroplasts contains genetic information for about 30 tRNAs and about 50 proteins. The proteins belong to photosystem I (2 genes), photosystem II (7 genes), the cytochrome system (3 genes), and the H^+-ATPase system (6 genes). The NADH dehydrogenase complex is coded for by 6 genes; ferredoxin, 3 genes; and ribulose, 1 gene. Twenty-nine genes have not been identified to date (data after B. Levine, Genes V. Oxford University Press, 1994).

B. Mitochondrial genes in yeast *(C. cerevisiae)*

The mitochondrial genome of yeast is large (ca. 84 kb) and contains introns. It contains genes for the tRNAs, the respiratory chain (cytochrome oxidase 1, 2, and 3; cytochrome b), for 15S and 21S rRNA, and for subunits 6, 8, and 9 of the ATPase system. The yeast mitochondrial genome is remarkable because its ribosomal RNA genes are separated. The gene for 21S rRNA contains an intron. About 25% of the mitochondrial genome of yeast contains AT-rich DNA without a coding function.

The genetic code of the mitochondrial genome differs from the universal code in nuclear DNA with respect to usage of some codons. The nuclear stop codon UGA codes for tryptophan in mitochondria, while the nuclear codons for arginine (AGA and AGG) function as stop codons in mammalian mitochondria.

Differences between the genetic code of the mitochondrial genome and the universal code of nuclear DNA

Codon	Nuclear DNA	Mitochondrial genome			
		Mammals	Drosophila	Yeast	Plants
UGA	Stop	Trp	Trp	Trp	Stop
AUA	Ile	Met	Met	Met	Ile
CUA	Leu	Leu	Leu	Thr	Leu
AGA } AGG }	Arg	Stop	Ser	Arg	Arg

(from B. Alberts et al., Molecular Biology of the Cell)

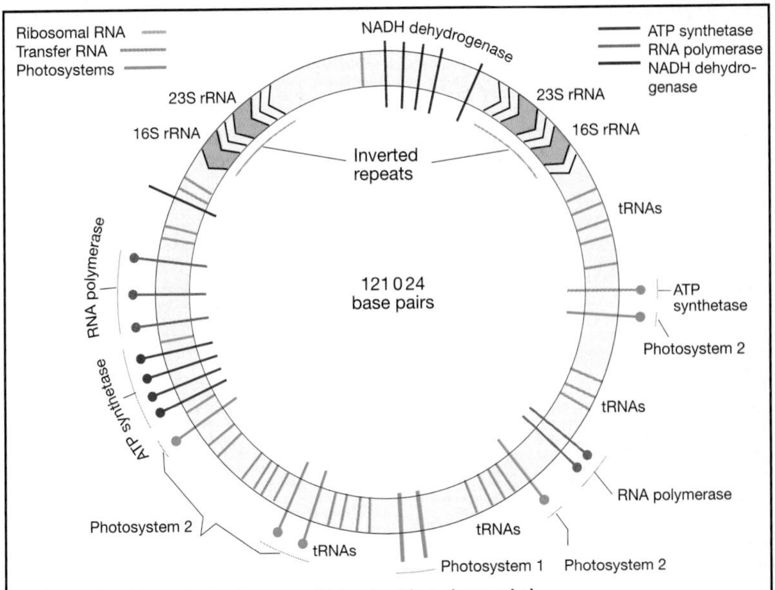

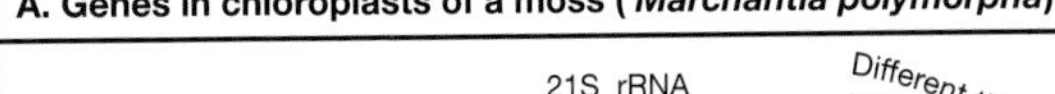

A. Genes in chloroplasts of a moss (*Marchantia polymorpha*)

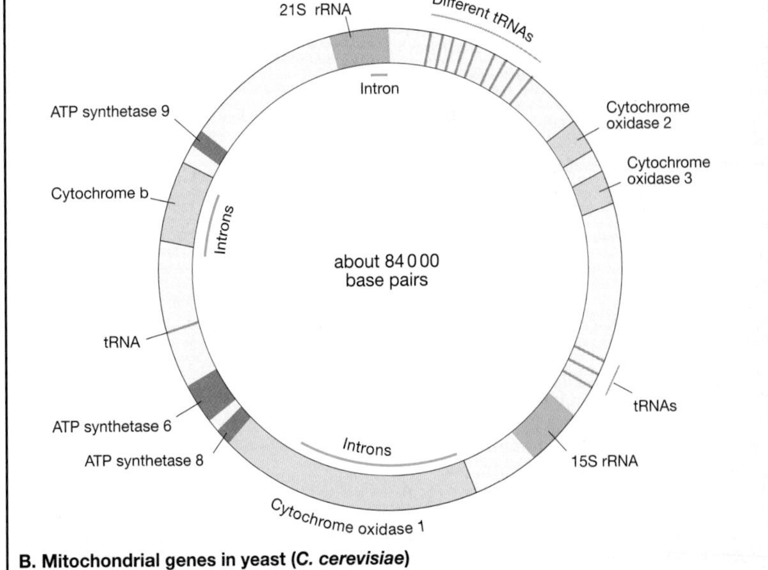

B. Mitochondrial genes in yeast (*C. cerevisiae*)

The Mitochondrial Genome of Man

The mitochondrial genome in mammals is small and compact. It contains no introns, and in some regions the genes overlap, so that practically every base pair belongs to a gene. The mitochondrial genomes of humans and mice have been sequenced and contain extensive homologies. Each consists of about 16.5 kb, i.e., is considerably smaller than a yeast mitochondrial genome.

A. Mitochondrial genes in humans

The mitochondrial genome of humans contains 13 protein-coding regions: genes for the cytochrome c oxidase complex (subunits 1, 2, and 3), for cytochrome b, and for subunits 6 and 8 of the ATPase complex. Unlike that of yeast, mammalian mitochondrial DNA contains seven subunits for NADH dehydrogenase (ND1, ND2, ND3, ND4L, ND4, ND5, and ND6). Of the mitochondrial coding capacity, 60% is taken by the seven subunits of NADH reductase (ND). A heavy (H) and a light (L) single strand can be differentiated by a density gradient. Most genes are found on the H strand. The L strand codes for a protein (ND subunit 6) and 14 tRNAs. From the H strand, two RNAs are transcribed, a short one for the rRNAs and a long one for mRNA and most of the tRNAs. A single transcript is made from the L strand. A 7S RNA is transcribed in a counterclockwise manner close to the origin of replication (ORI) (not shown).

B. Cooperation between the mitochondrial and nuclear genomes

Some mitochondrial proteins are aggregates of gene products of nuclear and mitochondrial genes. These gene products are transported into the mitochondria after nuclear transcription and cytoplasmic translation. In the mitochondria, they form functional proteins from subunits of mitochondrial and nuclear gene products. This explains why a number of mitochondrial genetic disorders show Mendelian inheritance, while purely mitochondrially determined disorders show exclusively maternal inheritance.

C. Evolutionary relationships of mitochondrial genomes

Homologous segments of the DNA sequences in chloroplasts, mitochondria, and nuclear DNA can be viewed as the result of evolutionary relationships. It has been shown that DNA sequences have been transferred from chloroplasts to mitochondria, from chloroplasts to nuclear DNA, and from mitochondria to nuclear DNA. Mitochondrial DNA evolves much more rapidly than nuclear DNA. Mutations accumulate about 10 times faster. Marked intraspecies polymorphism has been the result.

A report in 1987 that all mitochondrial DNA in humans could be traced back to one woman ("Eve"), who would have lived about 200 000 years ago, aroused considerable interest. The conclusion is controversial. However, some information about certain evolutionary or ancestral relationships can be gained from mtDNA. (Vigilant L, et al. Science. 1991; 253:1503–1507. Templeton AR, Hedges SB, et al. Science. 1991; 255: 737–739).

References

Singer M, Berg P. Genes and Genomes. Oxford: Blackwell Scientific; 1991.

Wallace DC. Mitochondrial diseases: genotype versus phenotype. Trends Genet. 1993; 9: 128–133.

Wallace DC. Review. Mitochondrial DNA sequence variation in human evolution and disease. Proc Nat Acad Sci. 1994; 91: 8739–8746.

Suomalainen A, et al. An autosomal locus predisposing to deletions of mitochondrial DNA. Nature Genet. 1995; 9: 146–151.

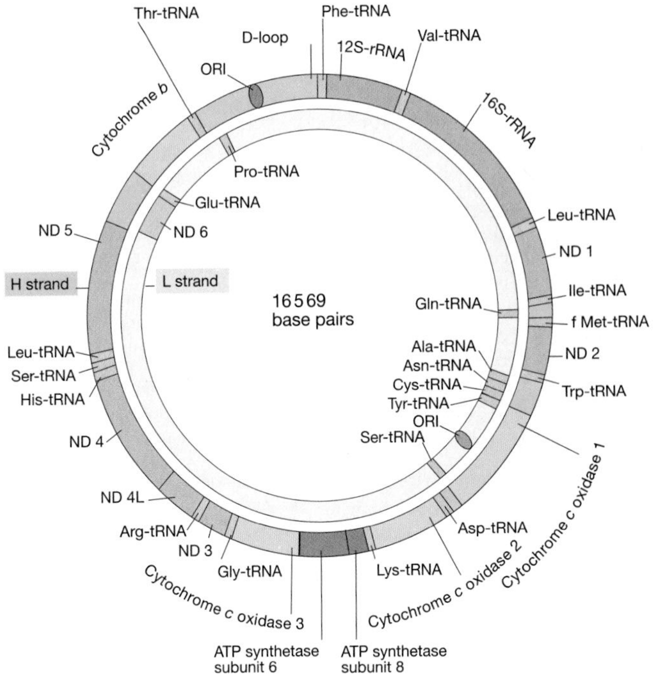

Thr-tRNA
Phe-tRNA
D-loop
12S-rRNA
Val-tRNA
ORI
Cytochrome b
16S-rRNA
Pro-tRNA
Glu-tRNA
Leu-tRNA
ND 5
ND 6
ND 1
H strand
L strand
Ile-tRNA
16569
base pairs
Gln-tRNA
f Met-tRNA
Leu-tRNA
Ala-tRNA
ND 2
Ser-tRNA
Asn-tRNA
His-tRNA
Cys-tRNA
Trp-tRNA
Tyr-tRNA
ORI
ND 4
Ser-tRNA
Cytochrome c oxidase 1
ND 4L
Arg-tRNA
Asp-tRNA
ND 3
Gly-tRNA
Lys-tRNA
Cytochrome c oxidase 2
Cytochrome c oxidase 3
Cytochrome c oxidase 2
ATP synthetase subunit 6
ATP synthetase subunit 8

A. Mitochondrial genes in man

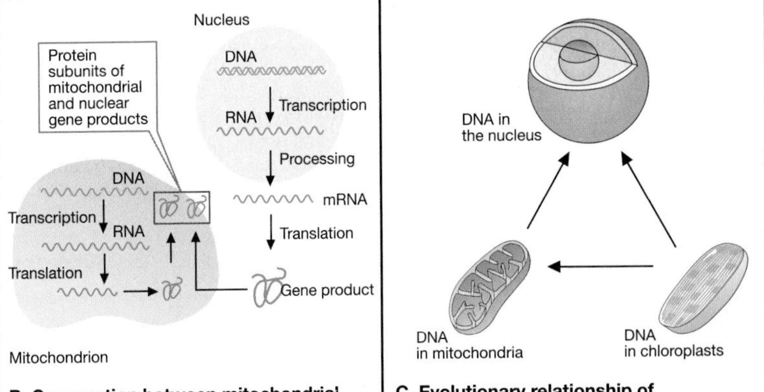

B. Cooperation between mitochondrial and nuclear genome

C. Evolutionary relationship of mitochondrial genomes

Mitochondrial Diseases in Humans

Numerous diseases resulting from changes in mitochondrial DNA are known in humans. Since oxidative phosphorylation is the primary source of energy in brain, muscle, heart, liver, kidney, and the insulin-forming islets of the pancreas, mitochondrial disorders are manifest predominantly in these organs. The clinical picture and the severity result from the type of mtDNA mutation, the proportion of mutant DNA, and the tissue distribution.

A. Mutations and deletions in mitochondrial DNA in man

Point mutations are known, e.g., for mitochondrial encephalomyopathy (MELAS; at position 3243, the number of the affected nucleotide base pair), for Leber's hereditary optic neuropathy (LHON), and for other disorders (see Table). A number of deletions have also been observed. One of the largest known deletions includes 10.4 kb, from about position 1400 to about 4500 (Wallace DC, Science. 1992; 256: 628).

B. Maternal inheritance of a mitochondrial cytopathy

Hereditary mitochondrial diseases are transmitted only through the maternal line since spermatozoa contain hardly any mitochondria. Thus the disease will not be transmitted from an affected man to his children, but will be transmitted from an affected mother to all her children.

C. Heteroplasmia for mitochondrial mutations

Mutations or deletions in mitochondria are more frequently limited to a single tissue (mitochondrial cytopathy) than germline mutations. In such cases, the cells contain different proportions of affected mitochondria (heteroplasmia). The proportion of defective mitochondria varies after repeated cell divisions. This contributes to the considerable variability of mitochondrial diseases.

Examples of diseases due to mutations and deletions in mitochondrial DNA

Abbreviation	MIM No.	Designation
LHON	535000	Leber's hereditary optical neuropathy
MELAS	540000	Mitochondrial encephalomyopathy
	540050	Lactic acidosis with stroke-like signs
MERRF	545030	Myoclonic epilepsy and ragged red fibers
MMC	590050	Maternally inherited myopathy and cardiomyopathy
NARP	551500	Neurogenic muscular weakness with ataxia and retinitis pigmentosa
CEOP	258470	Progressive external ophthalmoplegia
KSS	530000	Kearns-Sayre syndrome (ophthalmoplegia, pigmental degeneration of the retina, and cardiomyopathy)
PEAR	557000	Pearson syndrome (bone marrow and pancreatic failure)
ADMIMY	157640	Autosomal dominant inherited mitochondrial myopathy with mitochondrial deletion in the D loop (type Zeviani)

(MIM: Mendelian Inheritance in Man, McKusick's Catalog of Human Genes and Genetic Disorders. 11th ed., Baltimore: Johns Hopkins University Press; 1994.)

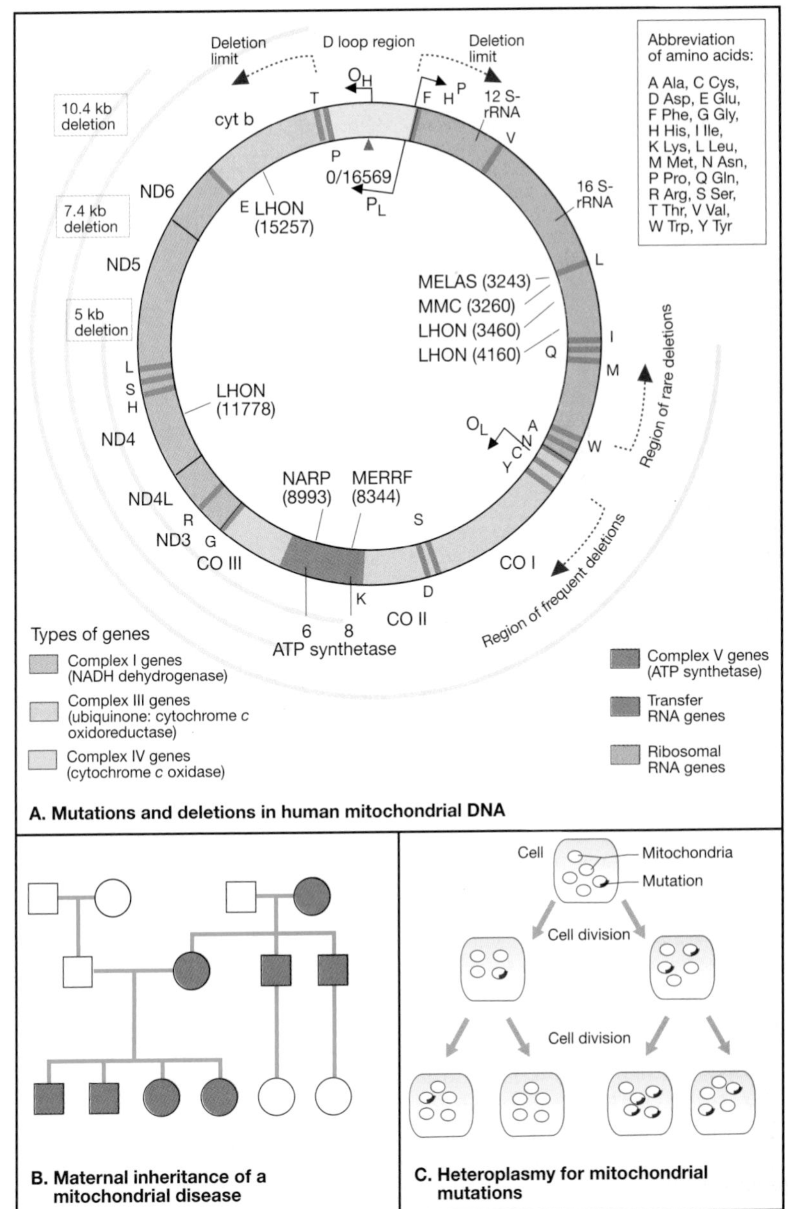

A. Mutations and deletions in human mitochondrial DNA

B. Maternal inheritance of a mitochondrial disease

C. Heteroplasmy for mitochondrial mutations

The Mendelian Traits

In 1865, the Augustinian monk Gregor Mendel published some remarkable observations in the *Berichte der Naturgeschichtlichen Vereinigung von Brünn* (Proceedings of the Natural History Society of Brno), which received little attention at the time. In this work, entitled "Experiments on Hybrid Plants," Mendel observed that certain traits in garden peas (*Pisum sativum*) are inherited independently of one another. Moreover, Mendel described certain regularities in the pattern of occurrence of individual traits in consecutive generations. His experimental system, the plants, and the observed traits will be presented here.

A. The pea plant (*Pisum sativum*)

The plant consists of a stem, leaves, blossoms, and seedpods. In the blossom, the (female) pistil (comprising the stigma, style, and ovary) and the (male) stamen comprising the anther and filament can be differentiated. The garden pea normally reproduces by self-fertilization. Pollen from the anther falls onto the stigma of the same blossom. However, it is relatively easy to carry out cross-fertilization. To do this, Mendel opened a blossom and removed the anther before pollen could escape and used pollen from another blossom instead. The resulting seedpod contains the seeds from which new plants develop.

B. The observed traits (phenotypes)

Mendel observed a total of seven characteristic traits: (1) height of the plants, (2) location of the blossoms on the stem of the plant, (3) the color of the pods, (4) the form of the pods, (5) the form of the seeds, (6) the color of the seeds, and (7) the color of the blossoms. Mendel observed that each pair of traits was inherited independently from all other pairs of traits. Mendel's main observation was that indepedent traits are inherited in certain predictable patterns. This was a fundamental new insight into the process of heredity. Since it distinctly deviated from the prevailing concepts about heredity at that time, its significance was not immediately recognized.

Today it is known that genetically determined traits are independently inherited (segregation) only when they are located on different chromosomes or far enough apart on the same chromosome to be separated each time by recombination (i.e., when there is no genetic linkage; see pp. 136, 138). This is true of the genes investigated by Mendel. In recent years, some of these genes have been cloned and their molecular structures have been characterized.

Deviation from the Mendelian pattern of inheritance

Mendelian traits do not always occur in the expected proportions (see p. 126). Due to the phenomenon of meiotic drive, one trait may occur much more frequently than others. Examples exist in the t complex of the mouse (about 99% instead of 50% of offspring of heterozygous t/+ male mice are also heterozygous) and Segregation Distorter (SD) in Drosophila. Recently, a mouse population in Siberia was described in which 85% and 65% of offspring were heterozygous for an inversion. Homozygosity for the inversion leads to reduced fitness and is a selective disadvantage. The shift was not due to early embryonal death. Possibly, deviations from Mendelian laws are more frequent than previously assumed. Further deviations occur with genomic imprinting (see p. 214) and germline mosaicism.

References

Corcos AF, Monaghan FV. Gregor Mendel's Experiments on Plant Hybrids. New Brunswick: Rutgers Univ. Press; 1993.

Pomiankowski A, Hurst DL. Siberian mice upset Mendel. Nature. 1993; 363: 396–397.

Weiling F. Johann Gregor Mendel: Der Mensch und Forscher. II Teil. Der Ablauf der Pisum Versuche nach der Darstellung. Med Genetik. 1993; 2: 208–222.

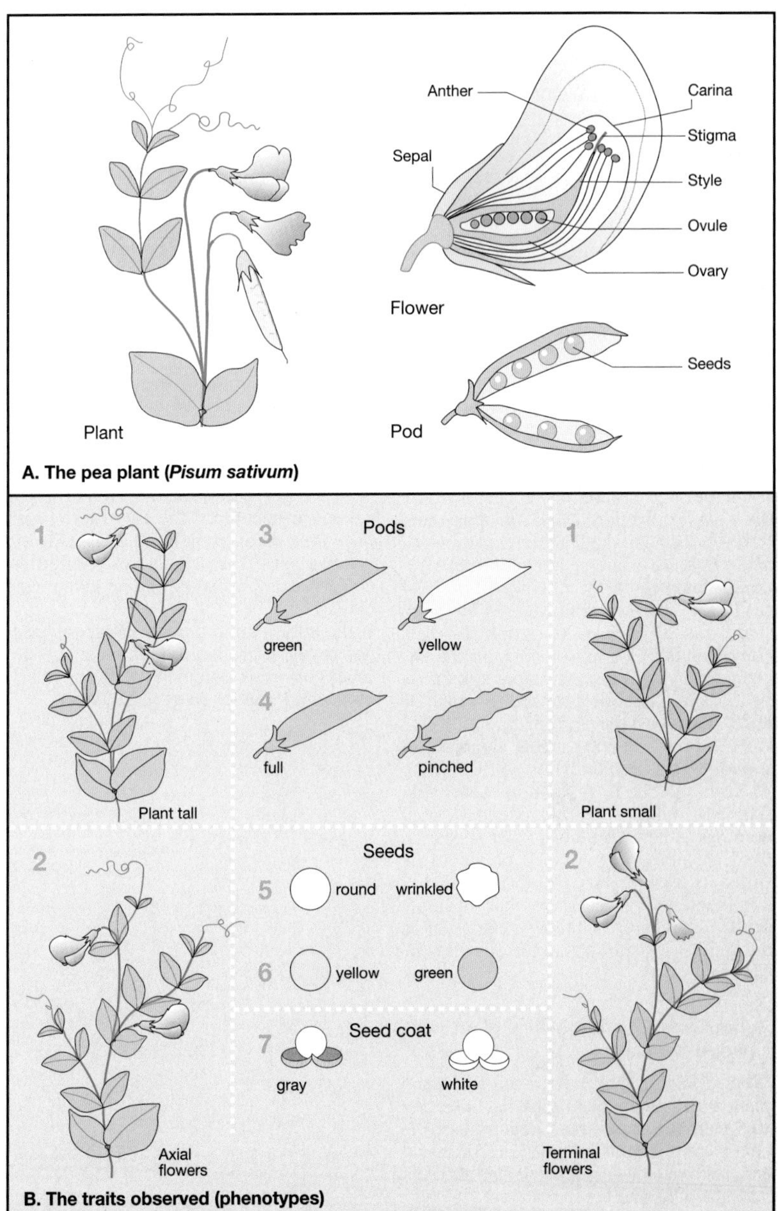

A. The pea plant (*Pisum sativum*)

B. The traits observed (phenotypes)

Distribution (Segregation) of Mendelian Traits

Mendel observed different characteristics of the pea plant and followed their occurrence in consecutive generations. From these observations, certain regularities in their patterns of occurrence (Mendelian laws) became apparent.

A. Segregation of dominant and recessive traits

In two different experiments, Mendel observed the shape (smooth or wrinkled) and the color (yellow or green) of the seeds. When he crossed the plants of the parental generation P, i.e, smooth x wrinkled or yellow x green, he observed that in the first filial (daughter) generation, F_1, all seeds were respectively smooth and yellow.
In the next generation, F_2, which arose by the self-fertilization usual for peas, the traits observed in the P generation (smooth and wrinkled, or green and yellow, respectively) reappeared. Among a total of 7324 seeds of one experiment, 5474 were smooth and 1850 were wrinkled. This corresponded to a ratio of 3:1. In the experiment with different colors (green vs yellow), Mendel observed that in a total of 8023 seeds of the F_2 generation, 6022 were yellow and 2001 green, again corresponding to a ratio of 3:1.
The trait that the F_1 generation exclusively showed (round or yellow), Mendel called dominant; the trait that did not appear in the F_1 generation (wrinkled or green) he called recessive. His observation that a dominant and a recessive pair of traits occur (segregate) in the F_2 generation in the ratio 3:1 is known as the first law of Mendel.

B. Backcross of an F_1 hybrid plant with a parent plant

When Mendel backcrossed the F_1 hybrid plant with a parent plant showing the recessive trait (1), both traits occurred in the next generation in a ratio of 1:1 (106 round and 102 wrinkled). This is called the second law of Mendel.

The interpretation of this experiment (2), the backcross of an F_1 hybrid plant with a parent plant, is that different germ cells (gametes) are formed. The F_1 hybrid plant (round) contains two traits, one for round (R, dominant over wrinkled, r) and one for wrinkled (r, recessive to round, R). This plant is a hybrid (heterozygote) and therefore can form two types of gametes (R and r).
In contrast, the other plant is homozygous for wrinkled (r). It can form only one type of gamete (r, wrinkled). Half of the offspring of the heterozygous plant receive the dominant trait (R, round), the other half the recessive trait (r, wrinkled). The resulting distribution of the observed traits is a ratio of 1:1, or 50% each.
The observed trait is called the phenotype (the observed appearance of a particular characteristic). The composition of the two factors (genes) R and r , (Rr) or (rr), is called the genotype. The alternative forms of a trait (here, round and wrinkled) are called alleles. They are the result of different genetic information at one given gene locus.
If the alleles are different, the genotype is heterozygous; if they are the same, it is homozygous (this statement is always in reference to a single, given gene locus).

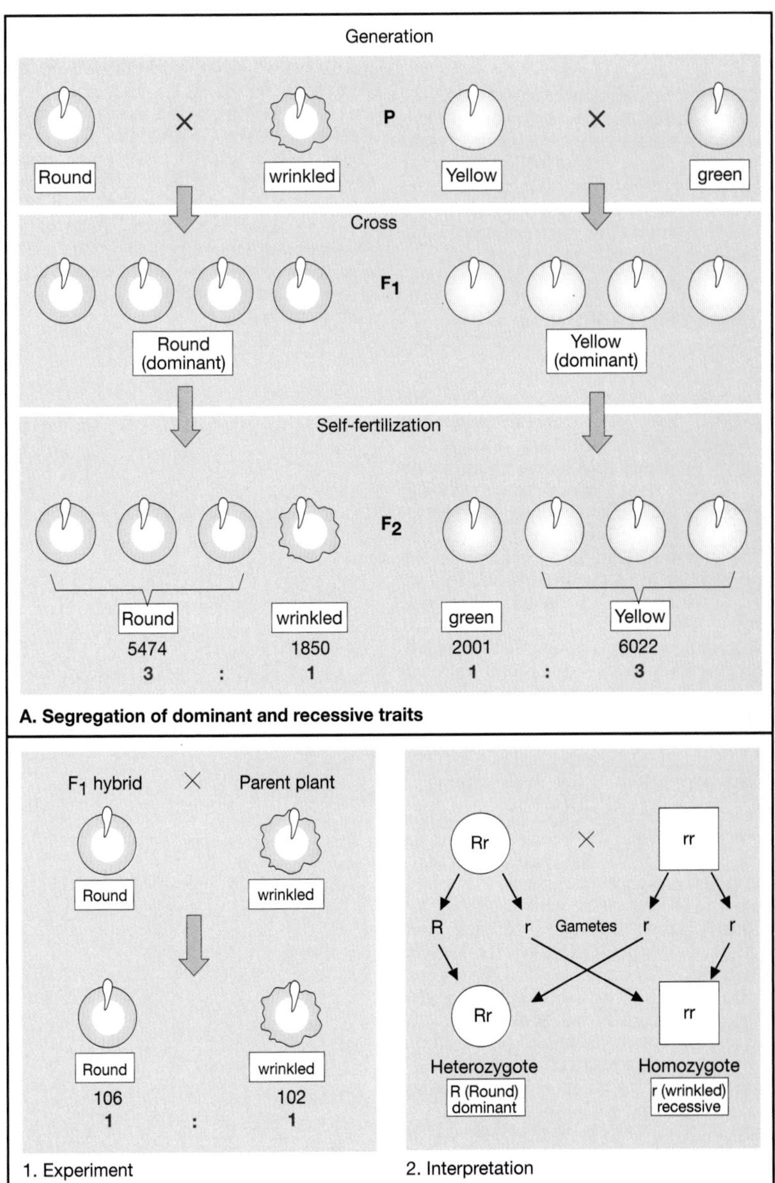

A. Segregation of dominant and recessive traits

1. Experiment

2. Interpretation

B. Back cross of an F_1-hybrid with a parent plant

Independent Distribution of Two Different Traits

In a further experiment, Mendel observed that two different traits are inherited independently of each other. Each pair of traits shows the same 3:1 distribution of the dominant over the recessive trait in the F_2 generation as he had previously observed. The segregation of two pairs of traits again followed certain patterns.

A. Independent distribution of two traits

In one experiment, Mendel investigated the crossing of the trait pairs round/wrinkled and yellow/green. When he crossed plants with round and yellow seeds with plants with wrinkled and green seeds, only round and yellow seeds occurred in the F_1 generation. This corresponded with the original experiments, as shown on p. 127. In the F_2 generation, the two pairs of traits occurred in the following distribution: 315 seeds yellow and round, 108 yellow and wrinkled, 101 green and round, 32 green and wrinkled, corresponding to a segregation ratio of 9:3:3:1. This is referred to as the third Mendelian law.

B. Interpretation of the observations

If we assign the capital letter **G** to the dominant gene *yellow*, a lowercase **g** to the recessive gene *green,* the capital letter **R** to the dominant gene *round*, and the lowercase **r** to the recessive gene *wrinkled*, the following nine genotypes of these two traits can occur: **GGRR, GGRr, GgRR, GgRr** (all *yellow* and *round*); **GGrr, Ggrr** (*yellow* and *wrinkled*); **ggRR, ggRr** (*green* and *round*); and **ggrr** (*green* and *wrinkled*). The distribution of the traits shown in A is the result of the formation of gametes of different types, i.e., depending on which of the genes they contain. The ratio of the dominant trait yellow (**G**) to the recessive trait green (**g**) is 12:4, or 3:1. Also, the ratio of dominant round (**R**) to wrinkled (**r**) seeds was 12:4, i.e., 3:1.

The square (Punnett square, named after an early geneticist) shows the different genotypes which can be formed in the zygote after fertilization. Altogether there are 9/16 yellow round seeds (**GRGR, GRGr, GrGR, GRgR, gRGR, GRgr, GrgR, gRGr, grGR**), 3/16 green round (**gRgR, gRgr, grgR**), 3/16 yellow wrinkled (**GrGr, Grgr, grGr**), and 1/16 green wrinkled seeds (**grgr**). Each of the two traits (dominant yellow versus recessive green or dominant round versus recessive wrinkled) occurs in a 3:1 ratio (dominant vs recessive).

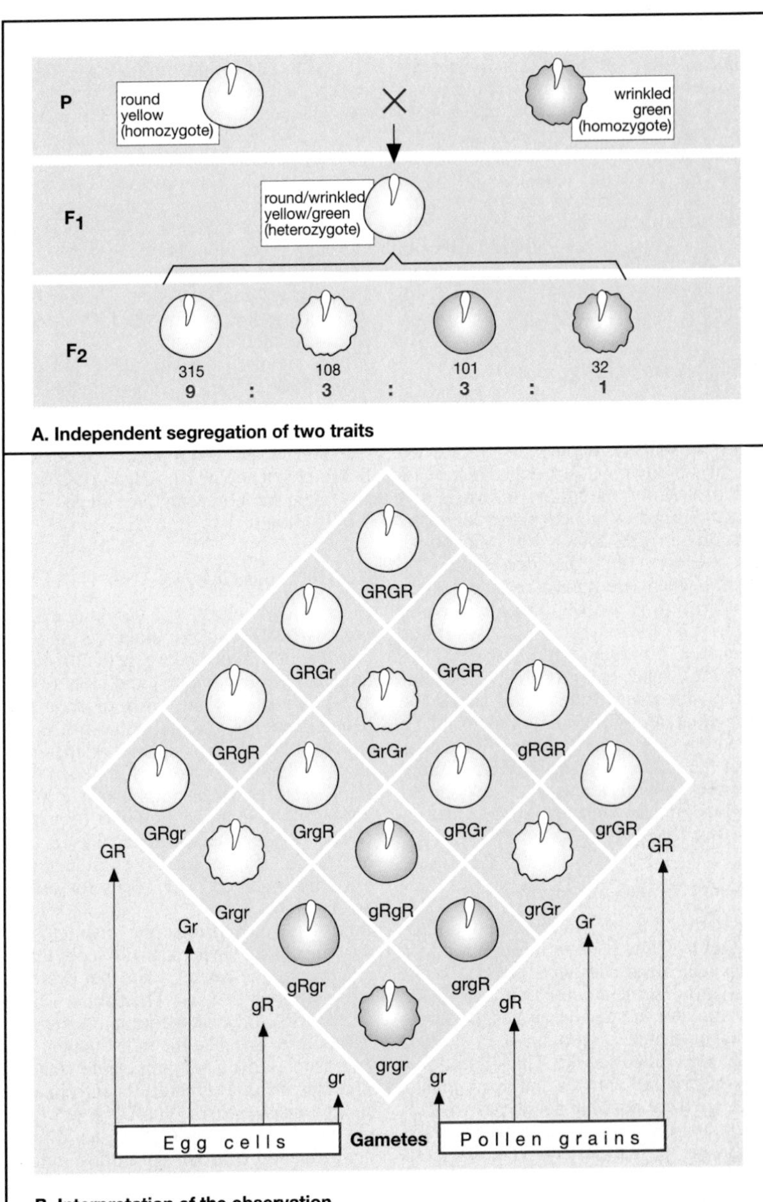

A. Independent segregation of two traits

B. Interpretation of the observation

Phenotype and Genotype

Formal genetic analysis examines the genetic relationship of individuals based on their kinship. These relationships are presented in a pedigree (pedigree analysis). An observed trait is called the phenotype. This could be a disease, a blood group, a protein variant, or any other attribute determined by observation. The phenotype depends to a great degree on the method and accuracy of observation. The term genotype refers to the genetic information on which the phenotype is based.

A. Symbols in a pedigree drawing

The symbols shown here represent a common way of drawing a pedigree. Males are shown as squares, females as circles. Individuals of unknown sex (e.g., because of inadequate information) are shown as diamonds. In medical genetics, the degree of reliability in determining the phenotype, e.g., presence of a disorder, should be stated. In each case it must be stated which phenotype (e.g. which disease) is being dealt with. Established diagnoses (data complete), possible diagnoses (data incomplete), and questionable diagnoses (statements or data doubtable) should be differentiated. False assignment of a phenotype can lead to false conclusions about the mode of inheritance. A number of further symbols are used, e.g., for heterozygous females with X-chromosomal inheritance (see p. 134).

B. Genotype and phenotype

The definitions of genotype and phenotype refer to the genetic information at a given gene locus. The gene locus is the site on a chromosome at which the genetic information for the given trait lies. Different forms of genetic information at a gene locus are called alleles. In diploid organisms, i.e., all animals and many plants, there are three possible genotypes with respect to two alleles at any one locus: (1) homozygous for one allele, (2) heterozygous for the two different alleles, and (3) homozygous for the other allele.

Alleles can be differentiated according to whether they can be recognized in the heterozygous state or only in the homozygous state. If they can be recognized in the heterozygous state, they are called dominant. If they can be recognized in the homozygous state only, they are recessive. The concepts dominant and recessive are an attribute of the accuracy in observation and do not apply at the molecular level. If the two alleles can both be recognized in the heterozygous state, they are designated codominant (e.g., the alleles A and B of the blood group system ABO; O is recessive to A and B).

If there are more than two alleles at a gene locus, there will be correspondingly more genotypes. With three alleles there are six genotypes, e.g., in the ABO blood group system: AA, AO (both phenotype A), BB, BO (both phenotype B), AB, and OO (actually, there are more than three alleles in the ABO system).

Genetic counseling

Genetic counseling is a communication process relating to the diagnosis and the possible occurrence of a genetically determined disease in a family and in more distant relatives. On the basis of an established diagnosis, the individual risk is determined for the consultand or that person's children. The goal of genetic counseling is to provide comprehensive information, including all possible decisions, course of the disease, medical care, and treatment. Professional confidentiality must be observed. The counselor makes no decisions.

In genetic counseling, one distinguishes the consultand, the person who seeks information, and the patient, since they are very often different persons. The patient whose disease first directed attention to a particular pedigree is called the index patient (or: proposita [female] or propositus [male]). The increasing availability of information about a disease based on a DNA test (predictive DNA testing) prior to disease manifestation calls for the utmost care in establishing whether it is in the interest of a given individual to have a test carried out.

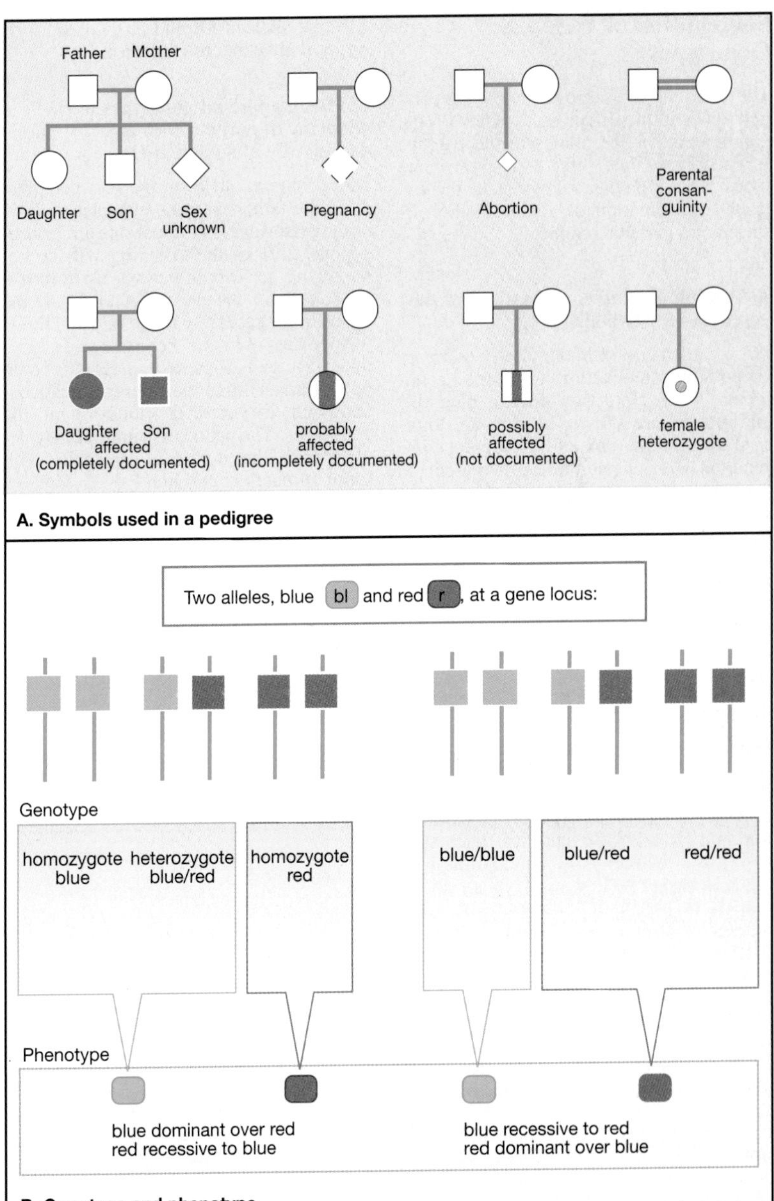

A. Symbols used in a pedigree

B. Genotype and phenotype

Segregation of Parental Genotypes

The distribution (segregation) of parental genotypes in the offspring depends on the combination of the alleles in the parents. In each case, they refer to a given gene locus. The Mendelian laws state the expected combination of alleles in the offspring of a parental couple.

A. Possible mating combinations of the genotypes for two alleles

For a gene locus with two alleles there are six possible combinations of parental genotypes (1–6). Here two alleles, blue (bl) and red (r), are shown, blue being dominant over red. In three of the parental combinations (1, 3, 4) neither of the parents is homozygous for the recessive allele (red). In three parental combinations (2, 5, 6), one or both parents manifest the recessive allele because they are homozygous. The distribution patterns of genotypes and phenotypes in the offspring of the parents are shown in B.

B. Distribution pattern in the offspring of parents with two alleles A and a

With three of the parental mating types for the two alleles **A** (dominant over **a**) and **a** (recessive to **A**), there are three combinations that lead to segregation (separation during meiosis) of allelic genes. These correspond to the parental combinations 1, 2, and 3 shown in A. In mating types 1 and 2, one of the parents is a heterozygote (**Aa**) and the other parent is a homozygote (**aa**). The distribution of observed genotypes expected in the offspring is 1:1; i.e., 50% (0.50) are **Aa** heterozygotes and 50% (0.50), **aa** homozygotes.
If both parents are heterozygous **Aa** (mating type 3 in A), the proportions of expected genotypes of the offspring (**AA, Aa, aa**) occur in a ratio of 1:2:1. In each case, 25% (0.25) of the offspring will be homozygous **AA**, 50% (0.50) heterozygous **Aa**, and 25% (0.25) homozygous **aa**. If the two parents are homozygotes for different alleles (**AA** and **aa**), all their offspring will be heterozygotes.

C. Phenotypes and genotypes in the offspring of parents with one dominantly inherited allele

One dominant allele (in the first pedigree, **A**, in the father) is to be expected in 50% of the offspring. If both parents are heterozygous, 25% of the offspring will be homozygous **aa**. If both parents are homozygous, one for the dominant allele **A**, the other for the recessive allele **a**, then all offspring are obligate heterozygotes (i.e., they have to be heterozygotes). It should be emphasized that the figures are percentages for expected distributions of the genotypes. The actual distribution may deviate from the expected, especially with small numbers of children.

Expected distribution of genotypes for parents with different combinations of genotypes for a dominant allele A *and a recessive allele* a

Parents	Offspring	Distribution
AA × AA	AA	1
AA × Aa	AA, Aa	1 : 1
Aa × Aa	AA, Aa, aa*	1 : 2 : 1
AA × aa	Aa	1
Aa × aa	Aa, aa	1 : 1
aa × aa	aa	1

* Dominant phenotype to recessive phenotype 3 : 1

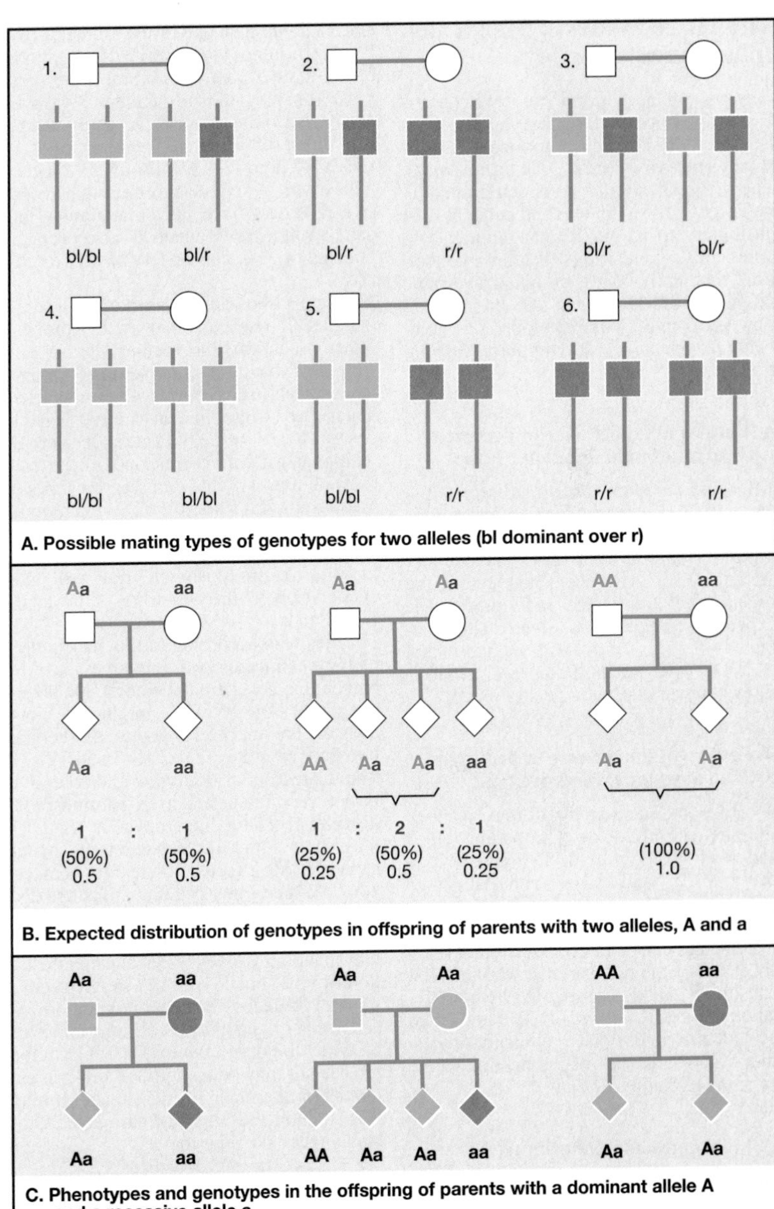

A. Possible mating types of genotypes for two alleles (bl dominant over r)

B. Expected distribution of genotypes in offspring of parents with two alleles, A and a

C. Phenotypes and genotypes in the offspring of parents with a dominant allele A and a recessive allele a

The Three Modes of Mendelian Inheritance

A pair of alleles **A** (dominant) and **a** (recessive) can show one of three modes of inheritance: autosomal dominant, autosomal recessive, and X-chromosomal. The mode of inheritance is determined by pedigree analysis. In medical genetics, it is customary to assign Roman numerals to consecutive generations. Within a generation, each individual is assigned an Arabic numeral. Individuals can also be assigned nonoverlapping combinations of numbers, for instance for computer calculations.

A. Pattern of inheritance in pedigrees with an autosomal dominant trait

Affected individuals are directly related in one or more successive generations unless a new mutation has occurred for the first time in a family. Both males and females are affected. The trait does not occur in the offspring of unaffected individuals. The expected segregation of affected and unaffected offspring of an affected individual is 50% (0.50) each. Pedigrees 2 and 3 show a new mutation in generation II.

B. Pattern of inheritance in pedigrees with an autosomal recessive trait

An autosomal recessive trait may occur in siblings of both sexes. The parents (II–3 and II–4 of pedigree 1, I–1 and I–2 of pedigree 2, and III–1 and III–2 of pedigree 3) are obligate heterozygotes. In pedigree 3, the homozygosity of the affected child can be traced back to common ancestors of the two parents, who are first cousins. Parental consanguinity (blood relationship) of III–1 and III–2 is indicated by a double line. For an autosomal recessive disease, the risk of the occurrence of an affected child is 25%.

C. X-chromosomal inheritance

Since a male has only one X chromosome, his daughter(s) will always inherit this chromosome plus one of the mother's two X chromosomes (1). A son will inherit one of his mother's two X chromosomes, but none from his father. A daughter always receives her father's only X chromosome (2). The proportion of new mutations is relatively high (3). A third of all affected males with severe X-chromosomal recessive disorders have new mutations (Haldane's rule). New mutations also occur in females (4), but can be less clearly recognized.

Pedigree 1 shows the distribution of three parental X chromosomes (one from the father, two from the mother) in the offspring. Typical X-chromosomal inheritance (5) is easy to recognize. Affected males may occur in consecutive generations, always through a female lineage (pedigree patterns correspond to the pattern of the knight's moves in chess). Father–son transmission of an X-chromosomal trait is not possible, since the son inherits his father's Y, and not his X. Thus, a familial X-chromosomal trait will not occur again in the offspring of an unaffected male.

Females who according to the distribution of the phenotypes and genotypes have to be heterozygotes are called obligate heterozygotes. Those who might be, but possibly are not heterozygotes are facultative heterozygotes (e.g., III–5 and IV–2). The distinction of dominant and recessive is not very important in X-chromosomal diseases because heterozygous females may show some manifestation and males carrying the affected allele are hemizygous and therefore always manifest the disease.

X-chromosome inactivation (see p. 190) may have an influence on the manifestation of an X-linked trait in a heterozygous female. There may be either nonrandom inactivation of one of the X chromosomes or a wide random fluctuation. Germline mosaicism may be a source of error in establishing the mode of inheritance. It may lead to affected sibs and simulate autosomal recessive inheritance.

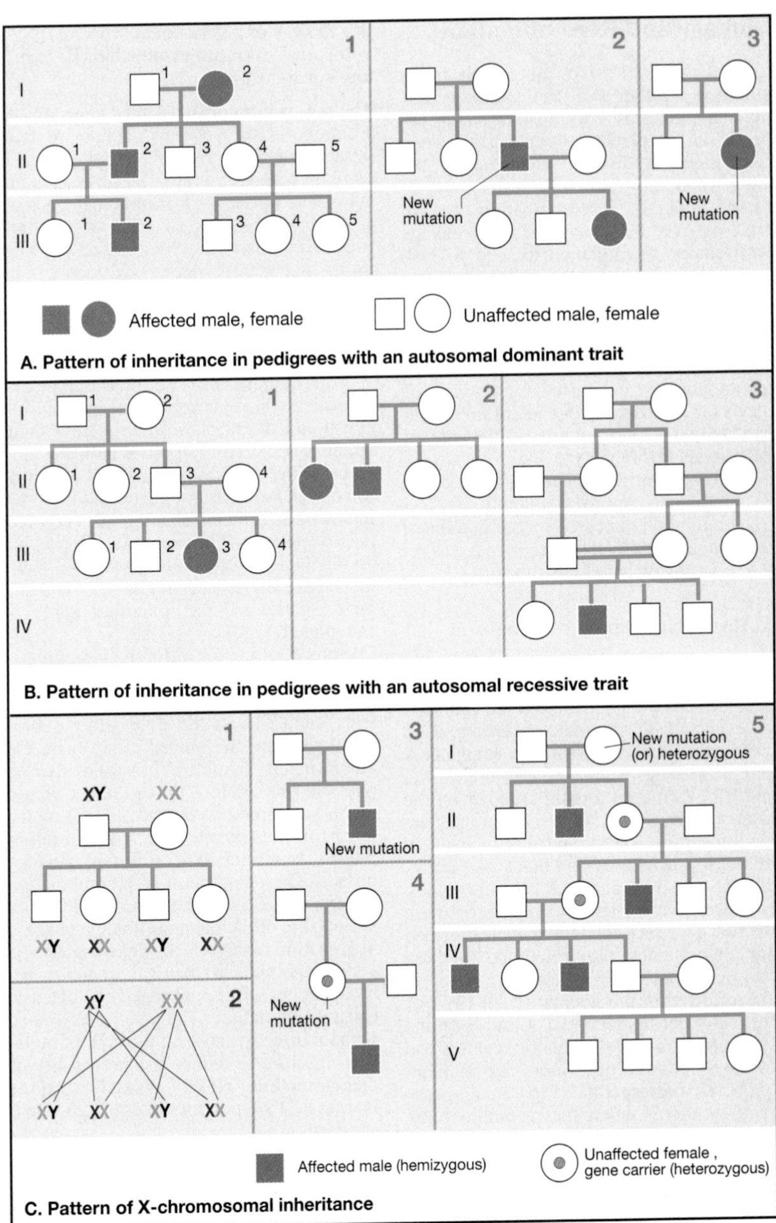
1
2
3
I
II
III
New mutation
New mutation
Affected male, female
Unaffected male, female
A. Pattern of inheritance in pedigrees with an autosomal dominant trait
IV
B. Pattern of inheritance in pedigrees with an autosomal recessive trait
XY
XX
New mutation
(or) heterozygous
4
5
V
Affected male (hemizygous)
Unaffected female, gene carrier (heterozygous)
C. Pattern of X-chromosomal inheritance

Linkage and Recombination

Linkage refers to two or more genes being inherited together as a result of their location on the same chromosome. This depends on the distance between their loci. The closer they lie next to each other, the more frequently they will be inherited together (linked). Recombination due to crossing-over between the loci (breakage and reunion during meiosis, see p. 110) leads to the formation of a new combination of linked genes. When the loci are very close together, recombination is rare; when they lie further apart, recombination is more frequent. In fact, the frequency of recombination can be used as a measure of the distance between gene loci. Linkage relates to gene loci, not to specific alleles. Alleles at different gene loci that are inherited together are called an association. If this occurs more frequently or less frequently than expected by the individual frequencies of the alleles involved, it is referred to as linkage disequilibrium.

A. Recombination by crossing-over

Whether neighboring genes on the same parental chromosome remain together or become separated depends on the cytological events (1) during meiosis. If there is no crossing-over between the two gene loci A and B, having the respective alleles A,a and B,b, then they remain together on the same chromosome (linked). The gamete chromosomes formed during meiosis in this case are not recombinant and correspond to the parental chromosomes. However, if crossing-over occurs between the two gene loci, then the gametes formed are recombinant with reference to these two gene loci. The cytological events (1) are reflected in the genetic result (2). For two neighboring gene loci A and B on the same chromosome the genetic result is one of two possibilities: not recombinant (gametes correspond to parental genotype) or recombinant (new combination). The two possibilities can be differentiated only when the parental genotype is heterozygous for both gene loci (Aa and Bb).

B. Linkage of a gene locus with an autosomal dominant mutation (B) to a marker locus (A)

The segregation of two linked gene loci in a family is shown here. There are two possibilities: 1, no recombination and 2, recombination. One locus (B) represents an autosomal dominant mutation that leads to a certain phenotype, e.g., that of an autosomal dominant inherited disorder. The father and three children (red symbols in the pedigree) are affected. The other locus (A) is a neighboring marker locus. All three affected children have inherited the mutant allele B as well as the marker allele A from their father. The three unaffected individuals have inherited the normal allele b and the marker allele a from their father. The paternal allele a indicates absence of the mutation (i.e., B not present). Recombination has not occurred (1).

In situation 2, recombination has occurred in two (indicated) persons: An affected individual has inherited alleles a and B from the father, instead of A and B. An unaffected individual has inherited allele A and allele b.

The precondition for differentiating the paternal genotypes is heterozygosity at the father's loci. In the case presented, the alleles A and B lie on one of the father's chromosomes, and the alleles a and b on the other (in cis position). It would also be possible that allele A in the father would lie on one chromosome and allele B on the other (in trans position). These two possibilities represent two different linkage phases. The recognition of recombination as opposed to nonrecombination assumes knowledge of the parental linkage phase.

Segregation analysis of linked genes is very important in medical genetics because the presence or absence of a disease-causing mutation can be determined without directly knowing the type of mutation. In order to reduce the probability of recombination, closely linked, flanking markers (DNA polymorphisms, see p. 64) are used.

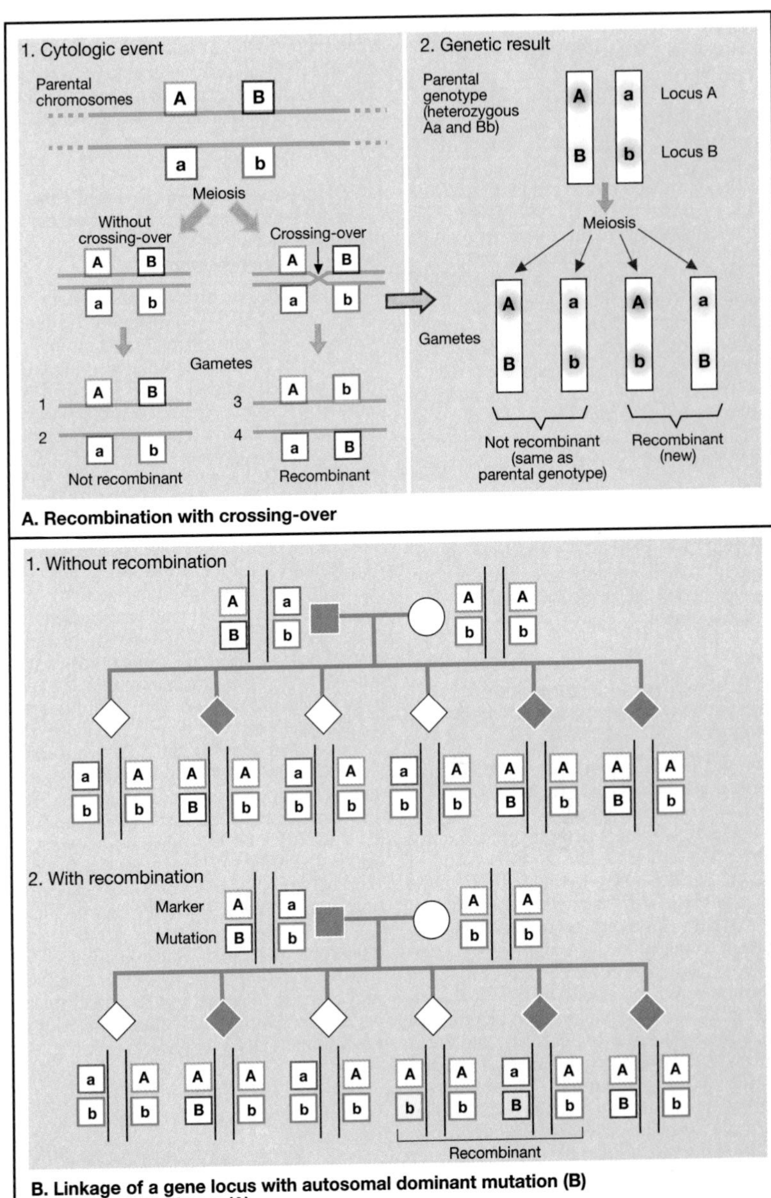

A. Recombination with crossing-over

B. Linkage of a gene locus with autosomal dominant mutation (B) with a marker locus (A)

Genetic Distance between Two Gene Loci and Recombination Frequency

The closer together two gene loci are located, the more frequently they are inherited together (genetic linkage); the further apart, the more frequently they become separated by recombination. The highest possible frequency of recombination is 50% (0.50), because this corresponds to the frequency of segregation of genes on different chromosomes. Thus, the frequency of recombination reflects the distance between two loci (genetic distance). This distance can be expressed as the frequency of genetic recombination (as opposed to the physical distance, which is given as the number of DNA base pairs lying between the two loci, see p. 356).

Synteny (HJ Renwick; 1971) refers to gene loci being located on the same chromosome, whether or not they are linked. Thus the term synteny also includes unlinked, widely separated loci on the same chromosome.

A. Recombination frequency as consequence of the distance between two loci

Two neighboring gene loci A and B in the parents may either become recombinant or remain nonrecombinant (see p. 136). If one of the parents is heterozygous for two alleles Aa and Bb, then homozygosity at only locus A (1) or only locus B (2) in the offspring will be the result of recombination. The observed recombination frequency between locus A and locus B (3%) results from the distance between them. These two loci are said to be 0.03 recombination units (3 cM) apart. One recombination unit is a centimorgan (cM), and 1 cM corresponds to a recombination frequency of 1% (0.01). In mammals, recombination occurs more often in female meiosis than in male meiosis, so that the genetic distance in females is about 1.5 times greater than in males (see p. 357). The term Morgan is derived from the name of the American geneticist who in 1911 first described recombination in Drosophila. At that time, the observation of linkage and recombination was an important argument for genes being linearly arranged along the chromosomes.

B. Determination of the order of three gene loci and their relative distances from each other by measuring recombination frequency

Not only the relative distances between gene loci, but also their order can be determined by comparing recombination frequencies. In the example presented, the order of three gene loci A, B, and C of unknown distance from each other is to be determined (1). In plants and animals, the distance between any two of the loci (locus A from locus B, locus B from locus C, locus A from locus C) can be established by observing the recombination frequency among the offspring of parents who are homozygous for either of the loci.

In traditional experimental genetics, such hybridization experiments were used for this purpose. For a test cross, parental genotypes of different homozygosity can be used. One homozygous parental genotype is crossed with another. The observed recombination frequencies (2) indicate the relative distances of the loci from each other. In the example presented, the distance from locus A to locus C is 0.08 (8%); the distance between locus B and locus C, 0.23 (23%); and the distance between A and B, 0.31 (31%). Thus, with the distance 0.08 between A and C and 0.23 between C and B, they are located in the order A – C – B (3).

This type of indirect determination of the relative location of gene loci in their correct order in classical experimental genetics has been replaced by direct methods of recombinant DNA technology.

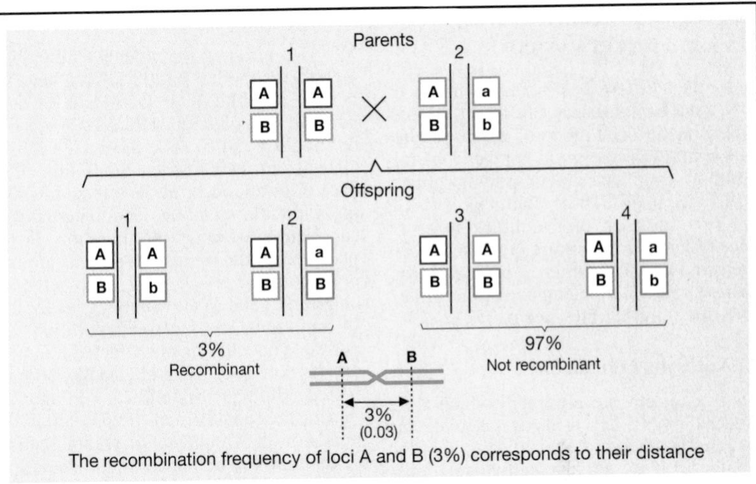

A. Recombination frequency as a consequence of the distance of two loci

1. Gene loci A, B, C of unknown distance

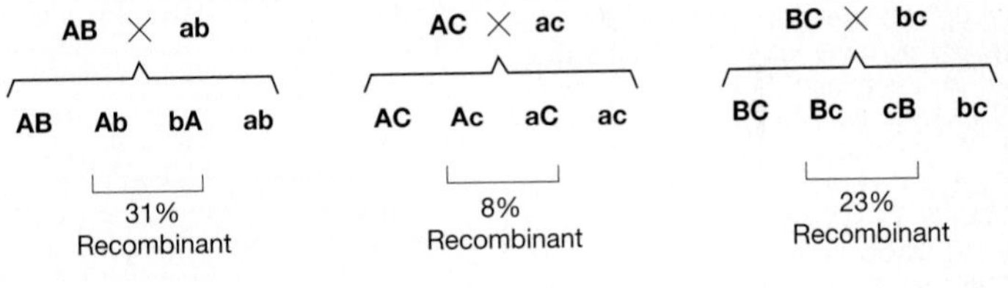

3. Relative distance

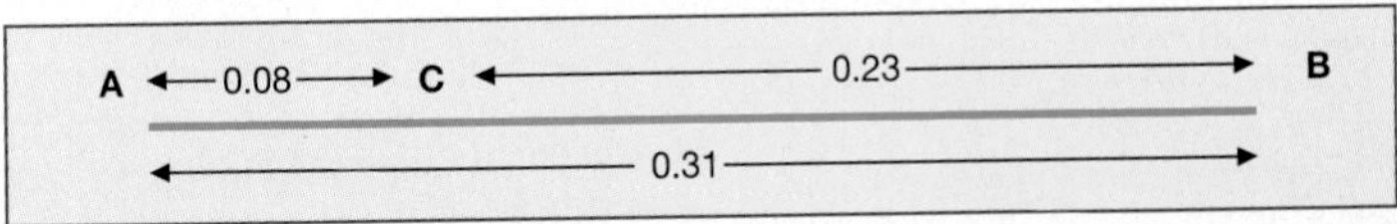

B. Determination of the relative distance and sequence of three gene loci by measuring the frequency of recombination

Segregation Analysis with RFLP Markers

Individual differences in the sequence of DNA nucleotide bases can lead to differences in the occurrence of cleavage sites of a restriction enzyme along the DNA (restriction fragment length polymorphism, RFLP, see p. 64). Three examples of RFLP analysis for each of the three modes of Mendelian inheritance are given below. Increasingly, other markers with high information content are being used (microsatellites and minisatellites, see p. 352).

A. Autosomal dominant

Two pedigrees are presented which were studied by RFLP analysis. Above, the pedigree is shown; below, diagrams of the Southern blots of the individuals analyzed. The DNA probe hybridizes to ("recognizes") two DNA fragments of different size, 2.1 and 1.6 kb. A given individual will be either homozygous (2.1, 2.1), heterozygous (2.1, 1.6), or homozygous (1.6, 1.6) at this locus (marker locus). In the pedigree without recombination (1), the mother (individual 2) and two children (individuals 4 and 5) are affected with an autosomal dominant disorder. The mother is heterozygous for the two DNA fragments, 2.1 and 1.6 kb. The father is homozygous for the 1.6-kb allele. All affected individuals in this pedigree carry the 2.1-kb fragment, all unaffected do not. Thus, the allele carrying the mutation is represented by the 2.1-kb fragment. In family 2 recombination has occurred. The affected father (individual 1) is heterozygous (alleles of 2.1 and 1.6 kb), as is the affected child (individual 4). The 1.6-kb fragment represents the mutation-carrying allele. However, an unaffected child (individual 5) also has this allele. Thus, recombination must have occurred. The frequency of recombination determines the frequency of a false prediction. For this reason, very closely linked markers with very low recombination frequencies are used. Ideally one employs markers that flank the disease locus on both sides to recognize whether recombination has occurred.

B. Autosomal recessive

The two affected individuals (individuals 4 and 7) in the left-hand pedigree have inherited a 4.0-kb allele from each of their parents. Two of the unaffected siblings (individuals 5 and 6) have received a 2.6-kb allele from their mother. Since this allele does does not occur in the affected, it cannot carry the mutation. The maternal 4.0-kb allele must carry the mutation. In the pedigree on the right, both parents are heterozygous (2.6/4.0). The two affected children (3 and 5) are homozygous for 2.6. In individual 6, recombination has occurred. This child is not affected although it is homozygous for allele 2.6 like both affected siblings (individuals 3 and 5). Again, the reliability of the prediction depends on the frequency of recombination. (The minus sign indicates the mutation-carrying allele as deduced from the observed phenotype and the genotype at the marker locus.)

C. X chromosomal

Here, the segregation of an X-linked disease occurring in individuals 5, 8, 10, and 13 is shown in relation to the segregation of a linked marker locus on DNA fragments of 6.2 and 4.1 kb. Three affected male individuals (5, 8, 10) are hemizygous for the 6.2-kb allele. Thus, this allele represents the mutation. All unaffected male individuals are hemizygous for the 4.1-kb allele. The female individuals 2 and 6 must be heterozygous (obligate heterozygotes) for the mutation. This is confirmed by demonstration of a 6.2-kb and a 4.1-kb DNA fragment. However, recombination must have occurred in individual 13. Individual 9 is homozygous for the 4.1-kb allele. Thus, she is not heterozygous for the mutation. Individual 12 is heterozygous and probably a carrier of the mutation.
The examples shown here are typical for the type of information obtained by indirect DNA diagnosis of genetically determined diseases.

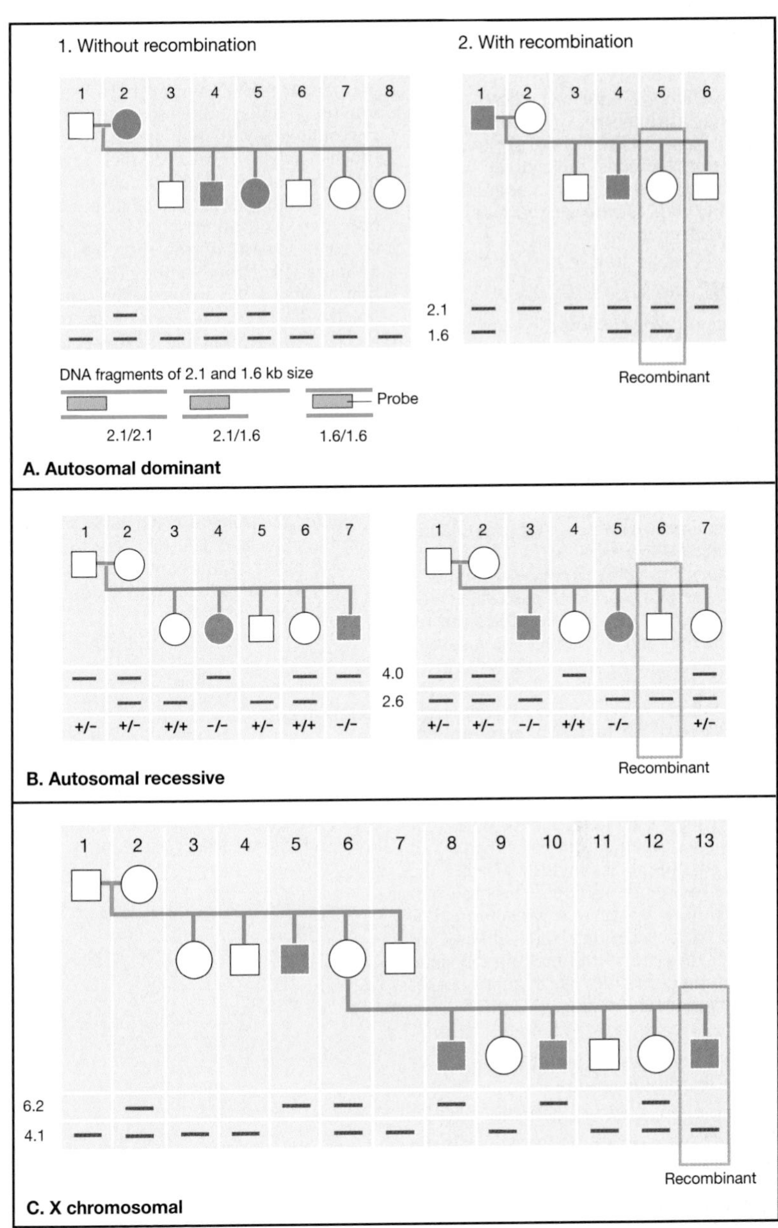
1. Without recombination
1 2 3 4 5 6 7 8
2.1
1.6
DNA fragments of 2.1 and 1.6 kb size
Probe
2.1/2.1
2.1/1.6
1.6/1.6
2. With recombination
1 2 3 4 5 6
Recombinant
A. Autosomal dominant
1 2 3 4 5 6 7
4.0
2.6
+/- +/- +/+ -/- +/- +/+ -/-
1 2 3 4 5 6 7
+/- +/- -/- +/+ -/- +/-
Recombinant
B. Autosomal recessive
1 2 3 4 5 6 7 8 9 10 11 12 13
6.2
4.1
Recombinant
C. X chromosomal

Linkage and LOD Score

For the probability of genetic linkage of two neighboring gene loci, there are two opposite assumptions, *for* linkage and *against* linkage. The ratio of the two is referred to as the odds. Linkage, i.e., a close relationship of two gene loci, is assumed when the odds for linkage reach or exceed a minimal value.

A. LOD scores

Linkage of two gene loci is assumed when the probability for linkage as opposed to the probability against linkage is equal to or greater than the ratio of 1000:1 (10^3:1). The logarithm of this ratio (odds) is called the LOD score (logarithm of the odds). A LOD score of 3 corresponds to an odds ratio of 1000:1. The frequency of linkage as opposed to nonlinkage for different recombination fractions (recombination frequencies), expressed as LOD scores, is determined by observations in families. The closer two loci lie to each other, the less frequently they are separated by recombination, and the higher the resulting LOD score. The table shows (in a simplified manner) the LOD score for close linkage with a recombination fraction under 0.05 (a), high probability of linkage with a recombination fraction of 0.15 (b), weak linkage (c), and no linkage (d). If the LOD score is less than 0, linkage is excluded (not shown in the table). With a recombination frequency of more than 0.5, it is not possible to distinguish whether the loci are located on different chromosomes and therefore segregate separately in 50% of the cases (i.e., are not linked), or whether they are so far apart on the same chromosome that they are regularly separated by recombination.

B. LOD score curves for different recombination fractions

The diagram shows a simplified form of the LOD score curves for the values in the table in A. (Figure after: Emery AEH. Methodology in Medical Genetics. 2nd ed. Edinburgh: Churchill Livingstone)

C. Multilocus analysis

Because of the great number of available markers at the DNA level (RFLPs, minisatellites), linkage analysis is usually carried out today with multiple markers (multilocus analysis). However, their sequence must be known. With the chromosomal position of the marker taken into consideration, the localization score is determined as the logarithm of the probability quotient (likelihood ratio). The localization score of the locus being sought is noted with respect to each of the marker loci (A, B, C, D). Each of the four peaks expresses linkage. The highest peak marks the probable location of the gene being sought. If there is no peak, linkage is not present, and the locus being studied does not map to the region tested (exclusion mapping).

For the linkage calculations, a number of computer programs are available to deal with the extensive data resulting from the study of many families and gene loci (e.g., program LIPED [Ott J. Am J Hum Genet. 1974; 26: 588.], MLINK [G L Lathrop et al., Proc Natl Acad Sci. 1984; 81: 3443], and others).

References

Emery AEH. Methodology in Medical Genetics. 2nd ed. Edinburgh: Churchill Livingstone; 1986.

Ott J. Analysis of Human Genetic Linkage. Baltimore: Johns Hopkins University; 1991.

Recombination fraction

	<0.05	0.05	0.10	0.15	0.20	0.30	0.35	0.40	0.45	0.50
a	3	0.7	0.3	0.2	0.01	0				
b		0	0.1	3	0.2	0				
c	0	0.2	0.7	**1.6**	1.0	0				
d		0	0.1	0.2	**0.3**	0.2	0.1	0.1	0	

A. LOD scores

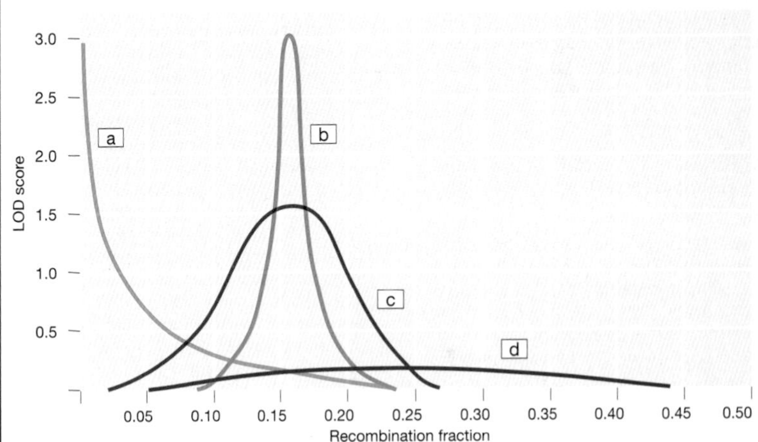

B. Lod score curves for close (a), probable (b), and no linkage (c,d)

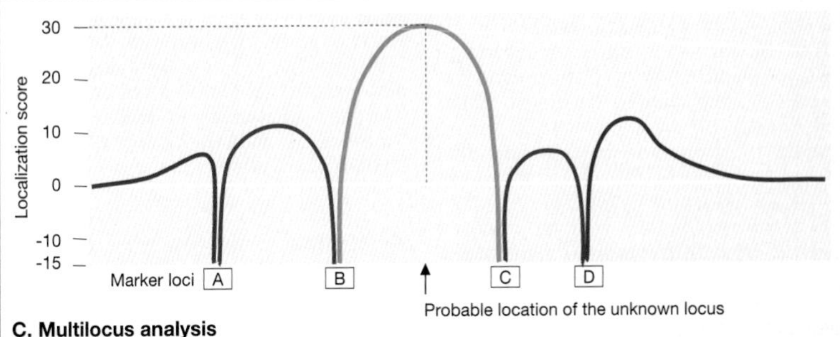

C. Multilocus analysis

Quantitative Differences of Genetic Traits

The phenotypes that can be observed to follow Mendelian laws of inheritance differ in a qualitative manner, i.e., presence or absence of a trait. However, some phenotypes can only be distinguished by quantitative differences. As such, they do not show Mendelian inheritance, although they may at least in part be genetically determined. The genetic factors cannot be demonstrated individually, and the phenotypic trait is called polygenic. A trait is called multifactorial if multiple genes are assumed to interact with environmental factors. Since neither the environmental factors nor the genetic factors can be identified individually, the differentiation of polygenic and multifactorial is somewhat arbitrary. Therefore, the two terms are often used interchangeably in practice.

A. Height of the corolla in *Nicotiana longiflora* after different parent plants are crossed

The length of the corolla stem of the tobacco plant *Nicotiana longiflora* is determined by quantitative differences of genetic traits. Parent generation (P) plants with respective average heights of 40 cm and 90 cm are shown. One parent type consists of a population of short (35–45 cm height), the other parent type of tall plants (85–95 cm height). The average height of first filial generation (F_1) plants, 65 cm, deviates from the parent plant heights. All plants of the F_1 generation are taller than the short-parent plants but shorter than the tall-parent plants. In the next generation (F_2), the distribution of plant heights increases at both ends of the range. Taller and shorter plants arise than in the F_1 generation. If one now breeds plants that are relatively short (50 cm), of average height (60 cm), or of greater height (about 80 cm), F_3 generations will be produced with different distributions of plant height. From these observations, it can be concluded that the height of a plant depends on genes that under the same environmental conditions, predispose to shortness or tallness of the plants. Parent plants of average height usually have offspring of average height. Similar observations have been made for numerous quantitatively defined differences in plants, animals, and humans.

In humans, these traits include height, intellectual development, and certain congenital malformations (e.g., neural tube defect, cleft lip and palate, some congenital heart defects, and others). (Figure after F.J. Ayala and J.A. Kiger 1984).

B. Frequency distribution in the F_2 generation with different numbers of gene loci

The observations shown in diagram A can be better understood if they are reduced to a limited number of different gene loci. In the parent generation (P), one gene (with the alleles a and A), two genes (with the alleles a and A, and b and B), three genes, or any number of genes is taken as a basis. When the parent generation is homozygous for different alleles at each of the gene loci, then all individuals of the F_1 generation must be heterozygotes. Therefore, for a quantitatively observed trait all individuals of the F_1 generation would correspond to an average value of the two parental traits. If there were no influence from environmental variation, then in a further generation (F_2 generation) a segregation according to Mendel's laws would result. However, this can only be observed with one gene locus. Even with only two gene loci, a quantitative distribution results. The more gene loci that have an influence on the expression of the trait, the less possible it is to distinguish the individual traits or factors from one another. Through additional influences and environmental variation, an increasingly smooth distribution curve results.

References

King R, Rotter J, Motulsky AG, eds. The Genetic Basis of Common Diseases. Oxford: University Press; 1992.

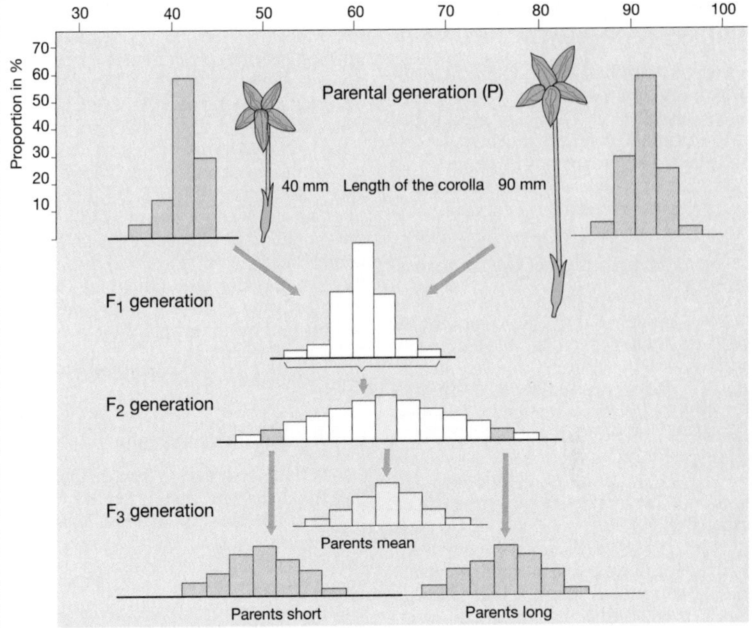

30 40 50 60 70 80 90 100

Proportion in %

70 60 50 40 30 20 10

Parental generation (P)

40 mm Length of the corolla 90 mm

F_1 generation

F_2 generation

F_3 generation

Parents mean

Parents short

Parents long

A. Length of the flower of *Nicotiana longiflora* after a cross between different types of parental plants

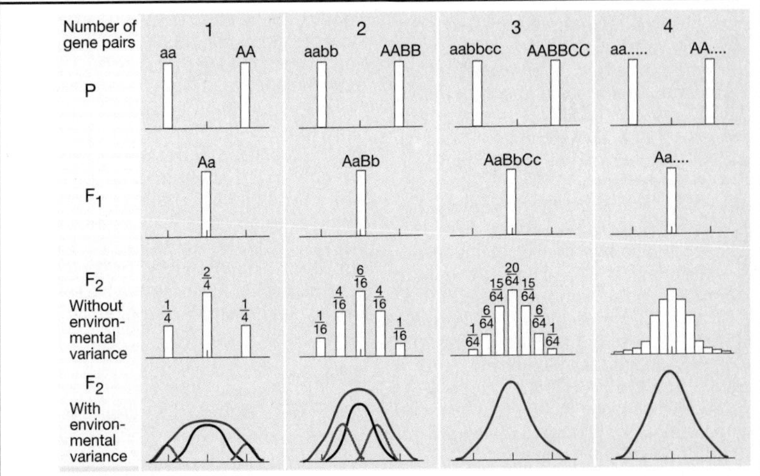

B. Distribution of frequency in the F_2 generation with a different number of gene loci

Normal Distribution and Multifactorial Threshold Model

Phenotypic variation of a population comprises individual data that vary in a continuous distribution. Different phenotypes and their genetic determinants must be distinguished. The proportion of genetic variation as opposed to environmentally caused variation must be established.

A. Normal distribution of the frequency of a trait

The measurements of a sample can be presented by dividing the data into two equal halves (a and b) separated by the average value ($\overline{x}$) (1). If the curve is divided at points that lie one standard deviation below (–1s) and above (+1s) $\overline{x}$ (2), then the segments c and d lie respectively below (–1s) or above (+1s) one standard deviation. The area below the curve can be further divided, e.g., –2s and +2s or –3s and +3s. If the areas are of equivalent sizes, the resulting curve is called a normal or Gaussian distribution. The average value $\overline{x}$ is determined by the sum of the individual single measurements (Σx) and the number n of the individuals in the sample (Formula 1). Many single measurements yield the total number or frequency fx (2). The variance σ^2 (3) is an expression of the variability. Here x represents any individual measurement in the population, μ the average value, and N the number of individuals. Since the population average and total number of individuals can usually not be established directly, the variance (σ^2) is determined from a sample. For this purpose a correction factor, n /n–1, is introduced since the number of independent measurements is n–1. This results in an estimate of the sample variance of S^2 (4, 5) and standard deviation S (Formula 6). Since the actual distribution of the investigated population is inferred from a sample, knowledge of the method of measurement and of the sample size and an appropriate statistical analysis are required for correct interpretation.

B. Multifactorial threshold model

For a trait that does not follow Mendelian laws of distribution, a genetic predisposition (or liability) for the occurrence of a disorder is assumed. The predisposition follows a normal Gaussian distribution. If a trait such as a disorder or a congenital malformation shows a qualitative difference: affected or not affected, one speaks of a threshold value. The threshold divides the population into affected and unaffected individuals (1). First-degree relatives of an affected individual differ in their liability for the disorder (genetic predisposition) by about half of the average of the general population (1/2 X) (2); second degree relatives, by about 1/4 X (3).

C. Different threshold values

For certain disorders a sex difference in the frequency can be observed, e.g., a higher proportion of affected males than females. It is assumed that affected females have a higher threshold than affected males (1). As a result, the first-degree relatives of a male proband (2) are less frequently affected than relatives of a female proband (3). This is explained by differences in the liability to be affected. The sex that is more frequently affected in the general population has a lower proportion of genetic determinants than the one less frequently affected. Conversely, a higher proportion of genetic factors are required before the threshold is reached in the less frequently affected sex. Although the concept of multifactorial inheritance with a threshold of liability to disease is of practical importance in establishing empirical risk figures for diseases, it has great limitations. Increasingly, a few major genes are being identified in disorders previously assumed to be of polygenic origin.

Reference

Burns GW, Bottino PJ. The Science of Genetics. 6th ed. New York: Macmillan; 1989.

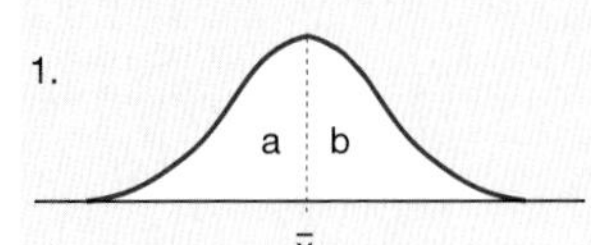

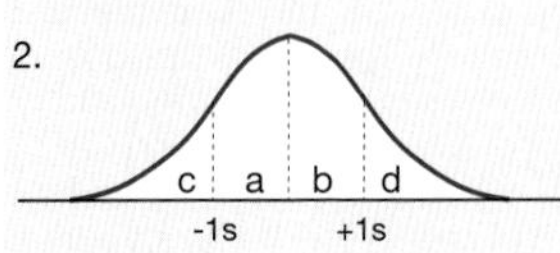

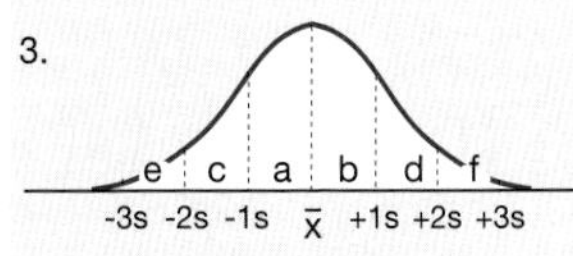

1 $$\bar{x} = \frac{\sum x}{n}$$ (Mean)

2 $$\bar{x} = \frac{\sum fx}{n}$$

3 $$\sigma^2 = \frac{\sum (x - \mu)^2}{N}$$ (Variance)

4 $$s^2 = \left(\frac{\sum (x - \bar{x})^2}{n}\right)\left(\frac{n}{n - 1}\right)$$

5 $$s^2 = \frac{\sum (x - \bar{x})^2}{(n - 1)}$$

6 $$s = \sqrt{\frac{\sum f (x - \bar{x})^2}{n - 1}}$$ (Standard deviation)

A. Normal distribution of a trait

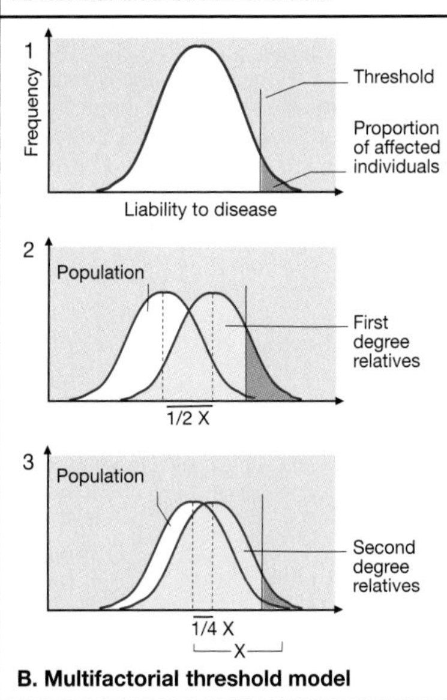

B. Multifactorial threshold model

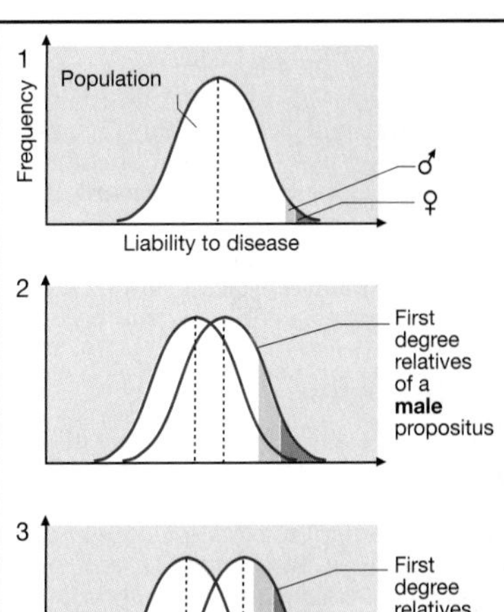

C. Different thresholds

Distribution of Genes in a Population

A population can be characterized by its distribution of genes, i.e., the frequency of alleles at different gene loci. The field of population genetics is concerned with the distribution patterns of alleles and the factors that alter or maintain their frequencies. A detrimental allele that causes disease in an individual might be expected to disappear sooner or later from a population. For the reasons given below, this is usually not the case. The frequency of an allele in a population is related to the distribution of genotypes among the offspring of different parental genotypes.

A. Expected frequency of genotypes in the children of parents with different genotypes

As shown on p. 132, a gene locus with two alleles (a dominantly inherited allele **A** and a recessively inherited allele **a**) can show six possible combinations (1–6). Each of these has an expected distribution of genotypes in their offspring according to the Mendelian laws. If one parent is homozygous and the other heterozygous, as in combinations 2 and 4, a 1:1 distribution of genotypes will result in the offspring (0.5 AA and 0.5 Aa). If both parents are heterozygous (combination 3), three genotypes will occur among their children in a ratio of 1:2:1 (0.25 AA, 0.5 Aa, and 0.25 aa). With combinations 1, 5, and 6 the respective offspring are of one genotype.

B. Allele frequency

Population genetics is concerned with the frequencies of different genotypes in a population. The frequency with which an allele is present at a given gene locus in a given population is called allele frequency (or also, gene frequency). The concept of allele frequency refers exclusively to the frequency of the allele in a population, and not directly to the frequency of the individual genotypes.

For a gene locus with two possible alleles **A** and **a**, the only possible genotypes are **AA** or **Aa** or **aa**. The frequency of the two alleles together (p the frequency of **A** and q the frequency of **a**) must be 100% (1.0). If two alleles **A** and **a** are equally frequent (each 0.5), they have the frequency of $p = 0.5$ for the allele **A** and $q = 0.5$ for the allele **a** (1); i.e., $p + q = 1$. The frequency distribution of the two alleles in a population follows a simple binomial relationship: $(p + q)^2 = 1$. Accordingly, the distribution of genotypes in the population corresponds to $p^2 + 2pq + q^2 = 1.0$. The expression p^2 corresponds to the frequency of the genotype **AA**; the expression 2pq corresponds to the frequency of the heterozygotes **Aa**; and q^2, the homozygotes **aa**.

When the frequency of an allele is known, the frequency of the genotype in the population can be determined. For instance if the frequency p of allele **A** is 0.6 (60%), then the frequency q of allele **a** is 0.4 (40%). Thus, the frequency of the genotype **AA** is 0.36; of **Aa**, 2 x 0.24 = 0.48; and of **aa**, 0.16 (2). And conversely, if genotype frequency has been observed, the allele frequency can be determined. If only the homozygotes **aa** are known (e.g., they can be identified due to an autosomal recessive inherited disease), then q^2 corresponds to the frequency of the disorder. From $p = 1 - q$ the frequency of heterozygotes (2pq) and of normal homozygotes (p^2) can also be determined.

	Genotype of parents	Genotype of offspring
1	AA and AA	1.0 AA
2	AA and Aa	0.50 AA 0.50 Aa
3	Aa and Aa	0.25 AA 0.50 Aa 0.25 aa
4	Aa and aa	0.50 Aa 0.50 aa
5	AA and aa	1.0 Aa
6	aa and aa	1.0 aa

A. Expected frequency of genotypes in children of parents with different genotypes

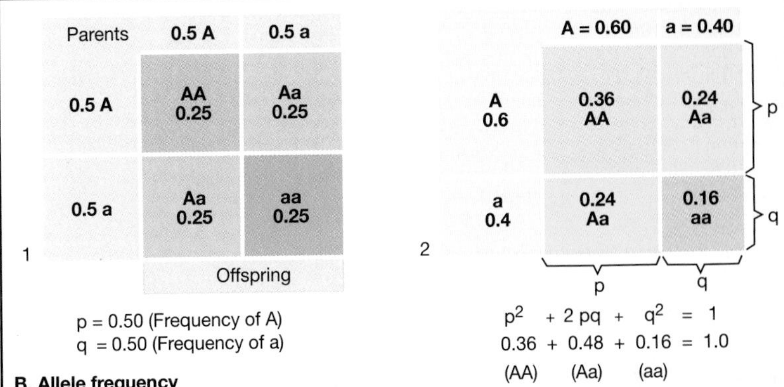

B. Allele frequency

Hardy–Weinberg Equilibrium

When an allele in the homozygous (or in some cases, heterozygous) state leads to a disorder that due to its severity and the time of onset prevents the individual from reproducing, it represents a selective disadvantage for that individual. Although it could be assumed that such a mutant allele would eventually disappear from a population, this is not the case. Rather, a frequency equilibrium develops according to principles and conditions elucidated in 1910 by the English mathematician Hardy and the German physician Weinberg (Hardy–Weinberg equilibrium principle).

A. Constant allele frequency (Hardy–Weinberg equilibrium)

An allele (here referred to as allele **a**) that leads to a severe disorder in the homozygous state is mainly present in a population in the undetectable heterozygous state. Only homozygotes (**aa**) come to attention due to illness. The heterozygotes (**Aa**) remain unrecognized and can not be distinguished by phenotype from the homozygotes **AA**. The frequency of affected individuals (homozygotes **aa**) is determined by the frequency of allele **a** (corresponding to q). The frequency of the three genotypes is determined by the binomial relationship $(p + q)^2 = 1$, where p represents the frequency of allele **A**, and q the frequency of allele **a** (see p. 148). The homozygous alleles (**aa**) eliminated in one generation by illness are replaced by new mutations. An equilibrium is reached between the elimination due to illness and the mutation frequency. A change in mutation frequency is accompanied by a change in the frequency of the corresponding allele. However, an equilibrium will again be reached in one generation.

B. Some factors influencing allele frequency

The Hardy–Weinberg equilibrium principle is valid only under certain conditions: Above all, it applies only when the population is sufficiently large and mating is random (panmixia). Selection for a particular allele in mating (e.g., in heterozygotes as opposed to homozygotes) would lead to an increase in the corresponding allele frequency. In such a population, a particular allele would become more frequent than would be predicted by the equilibrium. Thus, if preferential mating of certain genotypes occurs (assortative mating), a shift in the frequency of the corresponding allele will result. The Hardy–Weinberg equilibrium, therefore, is valid only with random mating.

A change in mutation frequency can change the frequency of an allele. However, after a single alteration of the mutation frequency, equilibrium will occur at a new level within one generation.

In a small population, the proportion of genotypes may be shifted due to casual fluctuations. The resulting frequencies of matings of different genotypes then lead to alteration in the allele frequency (genetic drift). An allele that is very rare or not present in a population can be introduced by migration and spread out as the population grows (founder effect).

Finally, a population may experience a drastic reduction in size, followed by an increase in the number of individuals. An allele that was previously rare in this population may by chance subsequently become relatively common as the population expands again. Such fluctuations have been observed in human populations.

(Photographic detail from "Coney Island, 1938, photograph by Weegee," hilma hilcox, Fotofolio).

A
A
a
a
Aa
Aa
New mutations
aa Homozygote
Elimination by disease

Alleles **a** are eliminated from the population owing to severe illness in homozygotes **aa,** but are replaced by new mutation. An equilibrium results.

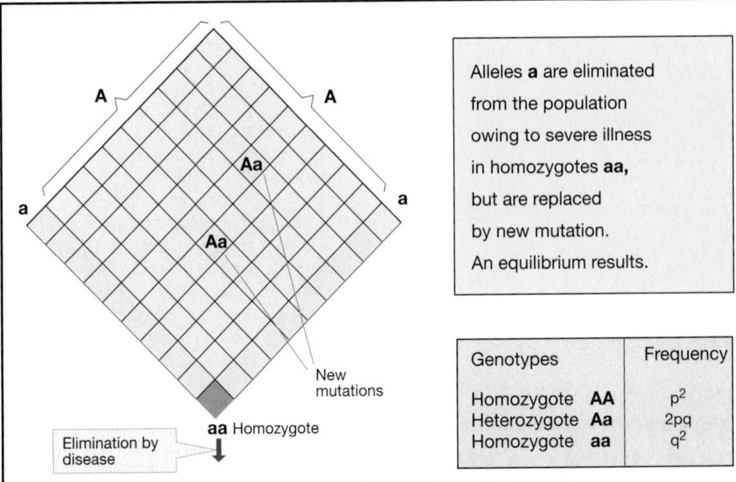

Genotypes		Frequency
Homozygote	**AA**	p^2
Heterozygote	**Aa**	2pq
Homozygote	**aa**	q^2

A. Constant allele frequency (Hardy-Weinberg equilibrium)

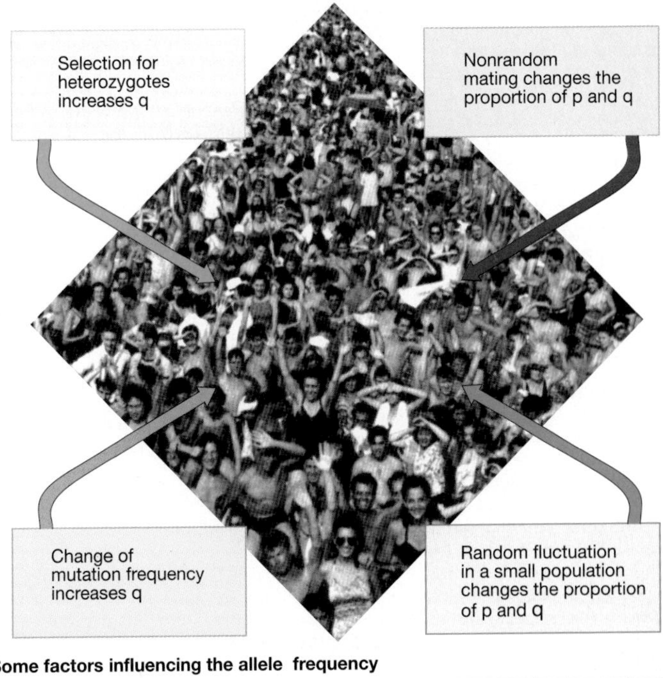

Selection for heterozygotes increases q
Nonrandom mating changes the proportion of p and q
Change of mutation frequency increases q
Random fluctuation in a small population changes the proportion of p and q

B. Some factors influencing the allele frequency

Consanguinity and Inbreeding

Parental consanguinity (being "of the same blood") means that the parents of an individual are genetically related, i.e., they have one or more ancestors in common. This increases the probability that the two related individuals may carry an identical mutation as a recessive allele inherited from a common ancestor. Mating between blood relatives is called inbreeding. The degree of inbreeding depends on the closeness of their relationship. Although inbreeding is not common in human populations, there are exceptions in certain regions and ethnic groups.

A. Simple types of consanguinity

The mating of very closely related individuals, i.e., brother/sister (1), half-siblings, father/daughter, mother/son, aunt/nephew (3), or an uncle/niece mating, etc., is referred to as incest. More frequent types of consanguineous unions in man are between first (2) or second cousins. The degree of relationship of two individuals can be expressed as the coefficient of relationship, **r**. It expresses the proportion of genes in the two persons that are identical by descent. Identity by descent means that identical alleles are copies of one particular allele inherited from an ancestor. The coefficient of relationship, **r**, is 1/2 for sibs, 1/8 for first cousins, and 1/4 for an uncle and niece (see also below). It can be determined from the number of generational steps connecting each common ancestor (here, A and B) to the consanguineous persons of interest (C and D in 1; G and H in 2; F and D in 3 in the corresponding path diagrams on the right). For each step, the probability of transmission of a given gene is 1/2 (50%). Individuals E and F in the path diagram of 2 and individual E in 3 indicate that homozygosity may also arise by transmission from a nonconsanguineous ancestor (homozygosity for independent alleles). The rarer a recessive disease, the greater the probability that the parents of an affected person are consanguineous.

B. Coefficient of relationship

The figure shows a four-generation pedigree connecting individuals C and D to their common ancestors, A and B. (It is assumed that A and B are not consanguineous themselves). The probability of transmission of a given allele is 1/2 for each generational step. Since the total number of steps connecting individuals A and C is 4, the probability of transmission of a given allele from A to C is $(1/2)^4$. The same applies to transmission from A to D. The same allele can only arrive at both individuals C and D if all steps (each with probability 1/2) have occurred uninterrupted; therefore, the probabilities of the steps are multiplied together. The total number of steps fom A to C and to D is 8; thus, the probability of transmission of a given allele from A to both C and D is $(1/2)^8$ or 1/256. Since individual B is also an ancestor of both C and D, the same applies to transmission from B to C and D. The coefficient of relationship, **r**, the sum of the probabilities of transmission from each common ancestor, is $(1/2)^8 + (1/2)^8 = 1/128$.

The coefficient of inbreeding (**F**) is the probability that two alleles at one locus in an offspring of related persons are identical by descent. In this diagram, **F** would apply to an offspring (not shown) of individuals C and D. **F** = 1/2 **r** of the parents, but it can also be determined from the number of direct steps from a common ancestor through both parents to their offspring (only one of the two parent-to-offspring steps being included), counted for each common ancestor.

By identifying regions in the genome that are homozygous in different patients in an inbred population, clues for the possible location of a gene can be obtained (homozygosity mapping).

Reference

Bittles AH, Neel JV. The costs of human inbreeding and their implications for variations at the DNA level. Nature Genet. 1994; 8: 117–121.

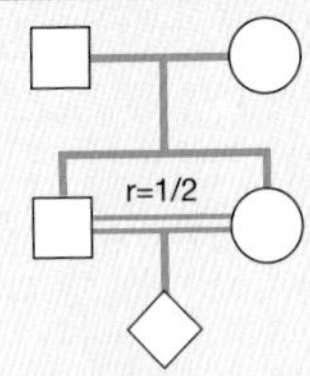

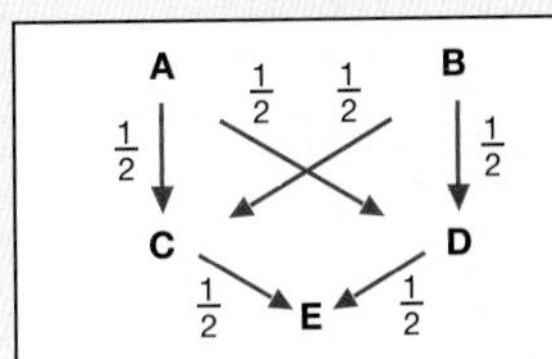

1. Brother/sister mating

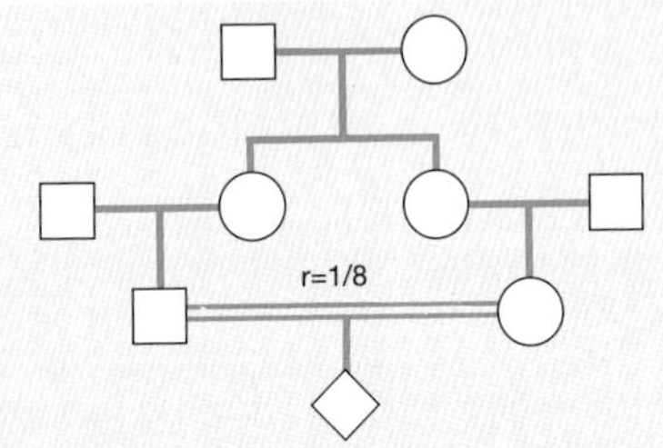

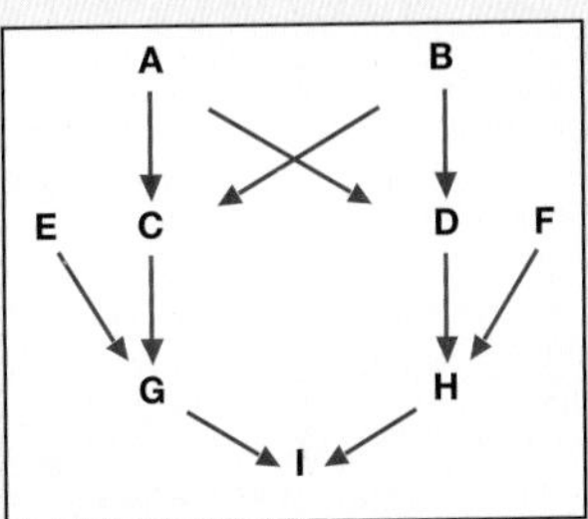

2. First cousins

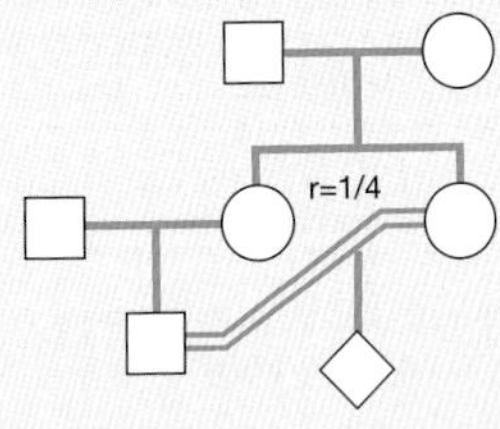

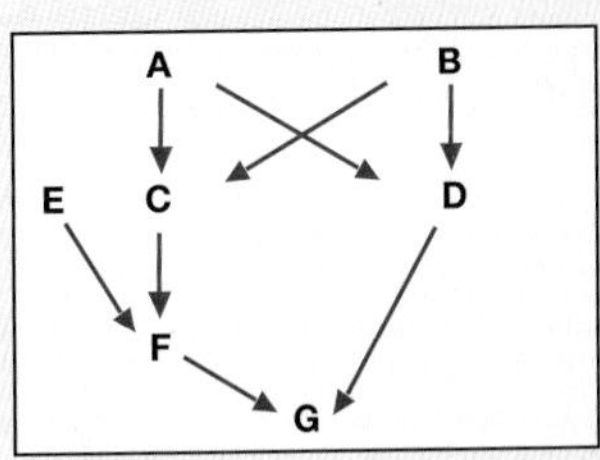

3. Nephew/aunt

A. Simple types of consanguinity

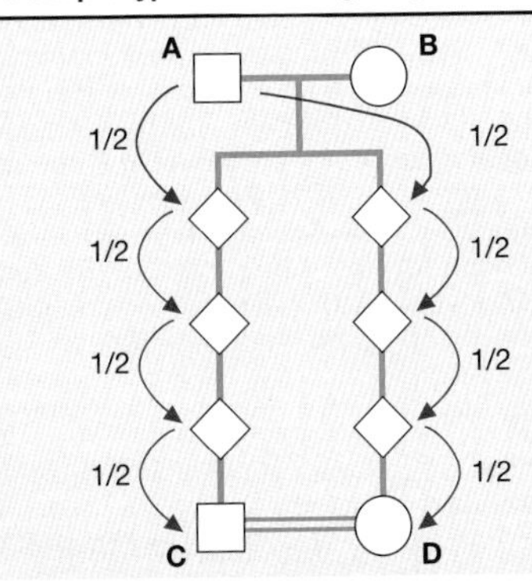

$$A \longrightarrow C = \left(\frac{1}{2}\right)^4$$

$$A \longrightarrow D = \left(\frac{1}{2}\right)^4$$

$$A \longrightarrow C \text{ and } D = \left(\frac{1}{2}\right)^8$$

$$B \longrightarrow C \text{ and } D = \left(\frac{1}{2}\right)^8$$

$$r = \left(\frac{1}{2}\right)^8 + \left(\frac{1}{2}\right)^8 = \frac{1}{128}$$

(Probability of descent from a common ancestor)

B. Coefficient of relationship (r)

Twins

Twin, triplet, or quadruple pregnancies regularly occur in many species of animals. In humans, about 1 in 80 pregnancies results in twins. They can arise from a single fertilized egg (monozygotic twins, MZ) or from two different eggs (dizygotic twins, DZ). Monozygotic twins are genetically identical.

Comparisons of monozygotic and dizygotic twins can yield information about the proportion of genetic influences on a complex trait. Comparisons of monozygotic twins who have been raised together with those who were raised apart are especially informative. They allow insight into the role of genetic factors as opposed to environmental influences. In a large study, the Minnesota Twin Study (Bouchard TJ, et al. Science. 1990; 250:223–228), monozygotic and dizygotic twins who had been separated since infancy were compared. In a series of complex tests, the conclusions about personality, temperament, nonprofessional interests, social adjustment, and other variables of personality demonstrated that monozygotic twins raised apart, compared with those raised together, were very similar in spite of their different environmental influences. About 70% of the variance of mental ability (different variants of intelligence quotient determined) was associated with genetic variation.

A. Types of twins

Monozygotic twins arise by separation at a very early stage of development. They always have a common placenta (monoplacental). They may lie in two amniotic cavities (diamniotic) or in a common amniotic cavity (monoamniotic). Dizygotic twins (DZ) always have their own amniotic cavity (diamniotic). They may each have their own placenta (diplacental), or they may share a common placenta.

B. Pathologic conditions in monozygotic twins

Incomplete separation (so-called Siamese twins) or connections between the blood circulations lead to pathologic conditions in monozygotic twins. A relatively frequent form of incomplete separation is thoracopagus (1), in which the twins are joined to various extents at the thoracic region. The blood circulation of twins can be connected via the common placenta (formation of a shunt) (2). This can lead to one twin receiving less blood or even bleeding to death. Especially severe malformations result from incompletely formed organs, e.g., absence of the heart in one twin (acardius) (3). If one of the twins dies very early in pregnancy, it may disappear by the time the other twin is born (vanishing twin).

C. Concordance of some traits in monozygotic (MZ) and dizygotic (DZ) twins

When twins show the same trait, they are said to be concordant; when they differ, discordant. Comparisons of the rate of concordance allow conclusions about the relative role of genetic factors in the etiology of complex traits such as some malformations, predisposition to diseases, height, blood pressure, etc.. However, definitive conclusions about genetic causes can rarely be reached from concordance rates alone.

D. Biochemical differences between monozygotic and dizygotic twins

Dizygotic and monozygotic twins differ not only with respect to certain disorders or physical or psychological traits, but also biochemically. Due to genetic differences, many chemical substances used in therapy are metabolized or excreted at different rates, owing to different activities of corresponding enzymes. An example is the rate of excretion of c-phenylbutazone, as studied by E. S. Vesell. This substance is excreted at the same rate in identical twins, whereas the rates of excretion differ between dizygotic twins or siblings.

Reference

Phelan, MC: Twins. In: Human Malformations and Related Anomalies. Vol. II. Stevenson RE, Hall JG, Goodman RM, eds. Oxford Monographs on Medical Genetics No. 37, Oxford Univ. Press 1993: 1047–1079.

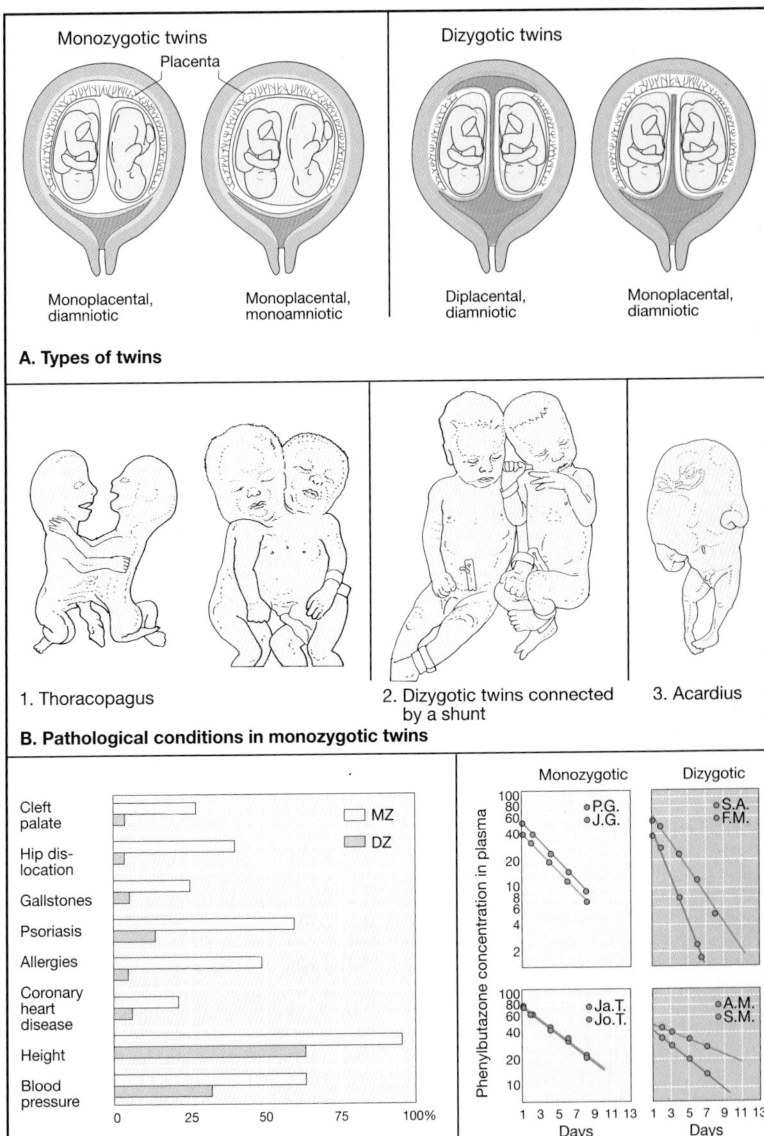

A. Types of twins

B. Pathological conditions in monozygotic twins

(after Connor & Ferguson-Smith, 1991)

C. Concordance of some traits in monozygotic (MZ) and dizygotic (DZ) twins

D. Excretion rate of a drug (data of E. S. Vesell)

Polymorphism

In nature, organisms of the same species usually differ in some aspects of their appearance. The differences are genetically determined and are referred to as polymorphisms. At many gene loci, two or more alleles may occur (genetic polymorphism). Genetic polymorphism is defined as the occurrence in a population of two or more genetically determined alternative phenotypes due to different alleles, whereby the least frequent allele cannot be maintained by repeated mutation alone. A gene locus is defined as polymorphic if the rare allele(s) has (have) a frequency of at least 0.01 (1%), and as a result, heterozygotes for such alleles occur with a frequency of at least 2%. Polymorphism can be observed at the level of the whole individual (phenotype), in variant forms of proteins and blood group substances (biochemical polymorphism), morphological features of chromosomes (chromosomal polymorphism), or at the level of DNA in differences of nucleotides (DNA polymorphism).

A. Polymorphism of the phenotype

An impressive example of phenotypic polymorphism is the color pattern on the wing sheaths of the Asian ladybug (*Harmonia axyridis*) (1). In the area of distribution, extending from Siberia to Japan, multiple variants can be distinguished (F. J. Ayala, Mechanisms of Evolution, Scientific American, September 1978). The different color combinations are due to different alleles of the same gene (1). In the California king snake (*Lampropeltis getulus californiae*) (2), color patterns differ to such an extent within the same species that they would seem to represent different species. Man (*Homo sapiens*) occurs with different forms of facial features, skin pigmentation, eye color, height, and other traits. The genes determining these traits are the result of adaption to different climatic and geographic conditions as man migrated out of Africa to the different continents. They only simulate extensive genetic differences between different ethnic populations; extensive genetic differences at the individual level actually are much greater than at the population level.

B. Polymorphism in relation to environmental conditions

A polymorphism is considered as neutral if the presence or absence of a certain allele does not confer any advantage or disadvantage. A polymorphism may represent an advantage for the population. A polymorphic population is likely to contain individuals who are better prepared for certain environmental conditions or changes thereof than if there were genetic uniformity. Most living organisms show geographical variation, i.e., there are differences in the frequencies of alleles between populations of the same species in different areas. The differences may be gradual and reflect adaption to environmental conditions. For example the average height of the yarrow plant on the slopes of the Sierra Nevada in California decreases as the altitude increases (1). By comparing the growth of plants whose seeds were obtained at different altitudes and sown in one garden, it can be shown that the average height of plants at the different altitudes is genetically determined (illustration after N. A. Campbell, Biology, 1990). The peppered moth (*Bison betularis*) is usually light-colored and hard to see on the normally light birch bark and is therefore protected (2). At the end of the last century, many darkly colored peppered moths were observed near Liverpool (England). This is an industrially caused genetic adaption phenomenon. On lichen-free darkly stained bark in the industrial area, the dark insects are hardly noticeable. Since the color-adapted type has better chances of survival, they are found in those areas in greater numbers than the nonadapted insects. (Origin of illustrations: F. J. Ayala [A1], J. H. Tashjian, San Diego Zoo [A2], N.A. Campbell, Biology, 1990 [B1], Laßwitz and K. Jäkel, Okapia, Frankfurt [B2]).

1. Asiatic beetle *Harmonia axyridis*

2. California king snake *Lampropeltis getulus californiae*

A. Polymorphism of the phenotype

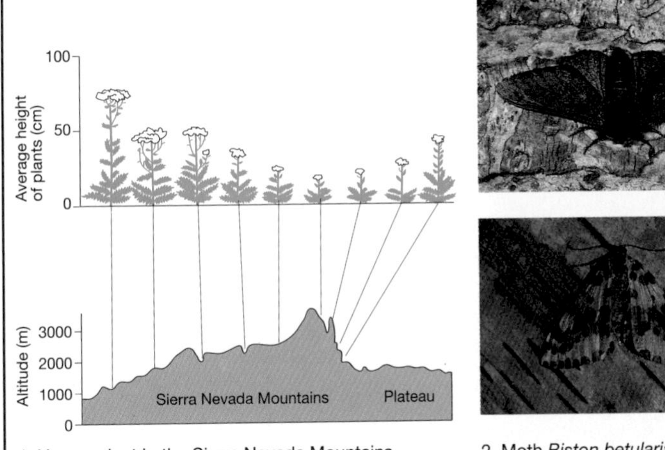

1. Yarrow plant in the Sierra Nevada Mountains

2. Moth *Biston betularia*, light and dark type

B. Polymorphism in relation to environmental conditions

Biochemical Polymorphism

Generally, polymorphism is not perceptible by phenotype. Rather it is detected by laboratory methods. Individual differences in the sequence of nucleotide bases of DNA can be determined. If a difference in the sequence leads to a change of the codon, a different amino acid will be incorporated at the corresponding site. This can be demonstrated by analyzing the gene product.

A. Recognition of polymorphism by means of gel electrophoresis

Polymorphism of a protein (gene product) can be demonstrated by gel electrophoresis when one variant form differs from the others by the presence of an amino acid with a different electrical charge. In this case, the allelic forms of the gene product can be distinguished due to their different speeds of migration in an electrical field (electrophoresis). When a gene product consists of two subunits (dimeric protein), heterozygous individuals will show three bands. A polymorphism that does not lead to a change in electrical charge cannot be identified in this manner.

B. Demonstration of polymorphism in gene products

Here, polymorphism is shown to be frequent in three typical gene products, the enzymes phosphoglucomutase, malate dehydrogenase, and acid phosphatase, in different species of *Drosophila*. Each of the diagram sections 1, 2, and 3 shows a starch gel electrophoresis of 12 fruit flies, each electrophoresis being specifically stained for the respective protein. With phosphoglucomutase (1) different migration speeds are observed in homozygous individuals as opposed to heterozygous (2, 4, 10). Malate dehydrogenase is a dimeric protein, so that heterozygous individuals (4, 5, 6, 8) show three bands. Acid phosphatase (3) shows a complex pattern because four alleles are involved. (Figure after Ayala and Kiger, 1984).

C. Frequency of polymorphism

In a study of average heterozygosity in *Drosophila willistoni* (1), 17.7% of 180 gene loci were found be heterozygous. The average heterozygosity is the proportion of heterozygous individuals in a population with reference to the number of analyzed loci. This is determined by adding the proportion of heterozygotes per total number of individuals per gene locus. The best insight into the considerable frequency of polymorphism is obtained from direct analysis of DNA. Hypervariable regions are found in many areas of the genome. Here, digestion with a restriction enzyme produces several different patterns due to individual differences in the size of the DNA fragments. In (2), the polymorphism at one locus near the H chain of the immunoglobulin J region in 16 individuals is shown (R. White et al., Cold Spring Harbor Symp. 1986; Quant Biol. 1986; 51: 29). Hypervariable regions lead to an individual pattern for each person (DNA fingerprint).

D. Genetic diversity and evolution

A population of genetically relatively homogeneous individuals is less likely to be able to adapt to changing environmental conditions than a genetically diverse species. Two populations of *Drosophila serrata* kept for 25 generations in separate closed bottles with limited availability of space and nourishment were compared. The population with the greater genetic diversity adjusted better to the environmental conditions (Figure after Ayala, 1980).

Reference

Beaudet AL, Scriver CR, Sly WS, Valle D. Genetics, biochemistry, and molecular basis of variant human phenotypes. In: Scriver CR, et al, eds. The Metabolic Bases of Inherited Disease. New York: McGraw-Hill; 1995: 53–118.

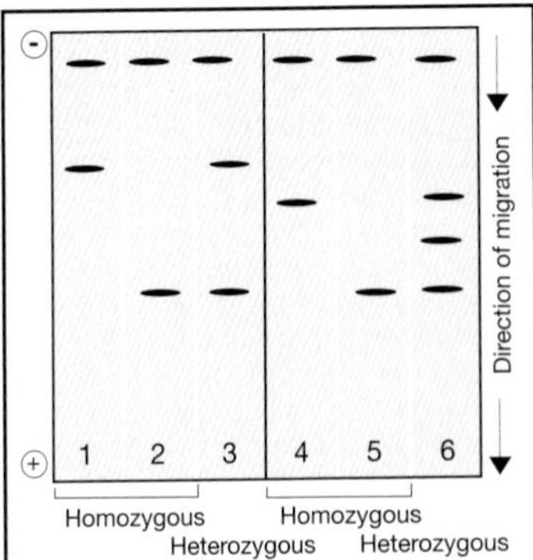

A. Recognition of polymorphism by gel electrophoresis

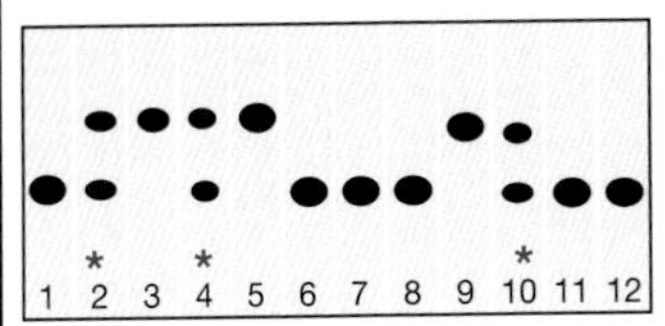

1. Phosphoglucomutase in *Drosophila pseudoobscura*

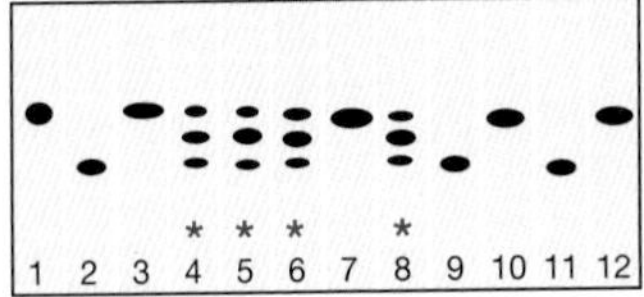

2. Malate dehydrogenase in *Drosophila equinoxialis*

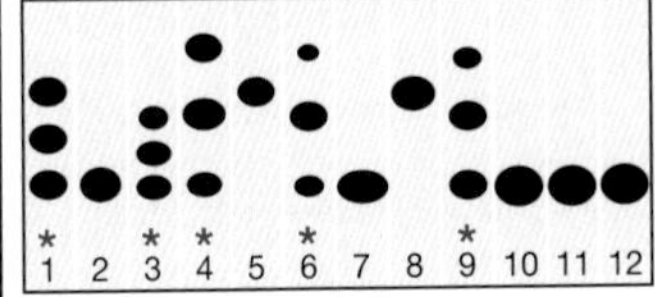

3. Acid phosphatase in *Drosophila equinoxialis*

(* = Heterozygous fruit flies)

B. Demonstration of polymorphism in gene products

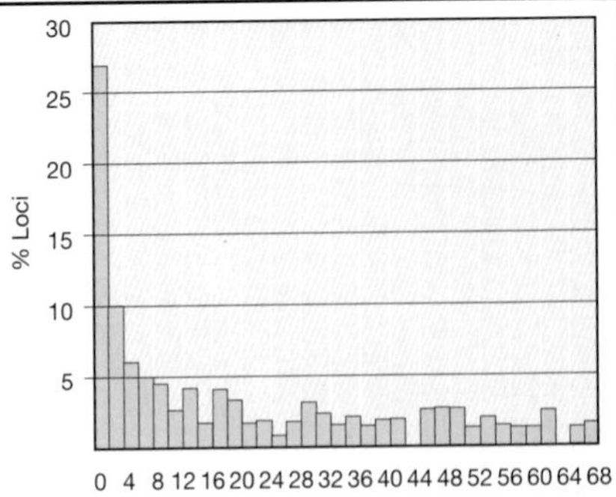

1. High-average heterozygosity (17.7 % of 180 loci)

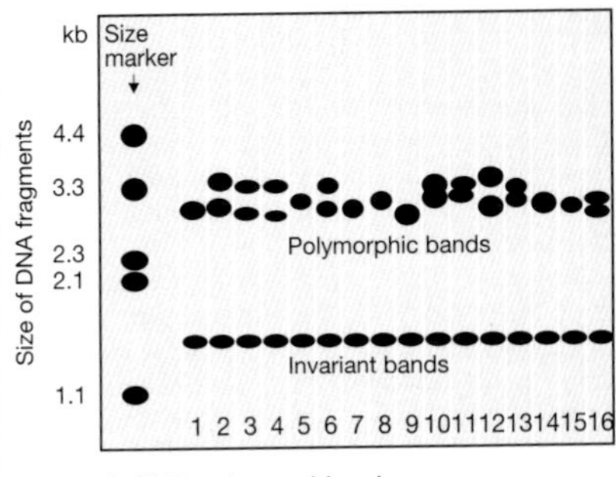

2. DNA polymorphism in man (16 individuals)

C. Frequency of polymorphism

Two populations of *Drosophila serrata*

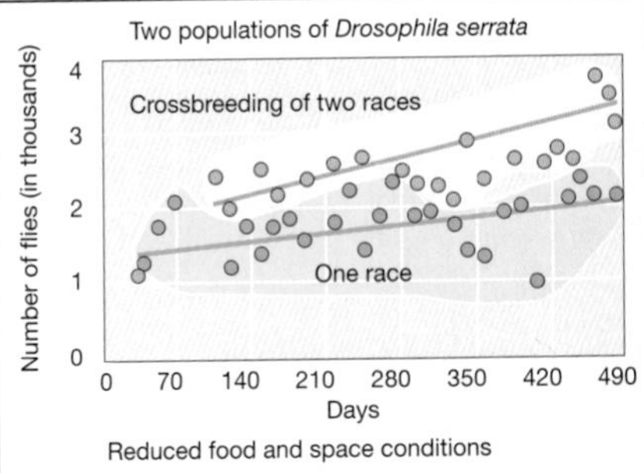

Reduced food and space conditions

D. Genetic diversity and evolution

Geographical Distribution of Genes

Within a species, alleles at a given gene locus may occur with different frequencies in different geographical regions. This can be due either to random differences in frequencies (genetic drift) or to selection. Selection means that one allele confers an advantage compared with another (selective advantage) with the result that the individual may have a slightly higher probability of survival in that particular environment.

A. Different frequencies of hereditary disorders, e.g., in Finland

Genetically determined diseases in humans (hereditary diseases) occur with different frequencies in different regions of the world. For example, the distribution of the grandparents of persons affected with three autosomal recessive inherited diseases in Finland shows an increased regional clustering because each mutation has arisen in a different region of Finland: (1) congenital flat cornea, in the western part of the country; (2) the Finnish type of congenital nephrosis (a severe renal disorder), in the southern part; and (3) diastrophic skeletal dysplasia, in other southern regions. There is no known basis for a selective advantage of these mutant alleles in the heterozygotes to explain their frequencies. The different distribution merely reflects the place and relative point in time of the mutation. The mutations must have arisen after Finland was settled, since they are unusually frequent there, but rare elsewhere. On the whole, diseases do not occur more frequently in the Finnish population than in other populations of the world. Similar examples are found in many other regions of the world and in other populations.

B. Distribution of malaria and frequency of heterozygotes due to selective advantage

The distribution of malaria in the Mediterranean region, West Africa, Sub-Saharan Africa, and East Africa including the Nile Valley and Asia (1) is similar to the frequency of heterozygotes for three forms of genetic disease: sickle cell anemia, thalassemia (see p. 278), and glucose-6-phosphate dehydrogenase deficiency (a red-cell disorder). Heterozygotes for these three genetically determined diseases have a selective advantage here. They are relatively protected because conditions in the blood of heterozygotes are less favorable for the malaria parasite than in the blood of normal homozygotes (see p. 283). Glucose-6-phosphate dehydrogenase deficiency (2) is an X-chromosomal disorder that leads to severe illness in affected males (hemizygotes). In the same way, homozygotes for sickle cell anemia (3) and different forms of a disorder of hemoglobin formation (thalassemia, 4) are very severely affected, whereas heterozygotes are not or are only mildly affected. This has led to a striking increase of heterozygotes for the mutant allele in areas infested with malaria. Protection for the heterozygotes occurs at the cost of affected homozygotes, but has a positive effect for the population as a whole, since the total frequency of affected is reduced. Nevertheless, the price of this advantage is considerable. More than 100 million people suffer from malaria in Africa and Asia. The sickle cell mutant has independently arisen at least four times in different regions and has become established there due to selective advantage. The situation is similar for thalassemia (see p. 284).

References

Cavalli-Sforza LL, Menozzi P, Piazza A. The History and Geography of Human Genes. Princeton: University Press; 1994.

Norio R, Nevanlinna HR, Perheentupa J. Hereditary diseases in Finland. Ann Clin Res. 1973; 5: 109–141.

Norio R. Diseases of Finland and Scandanavia. In: Rothschild HR, ed. Biocultural Aspects of Disease. New York: Academic Press; 1981: 359–415.

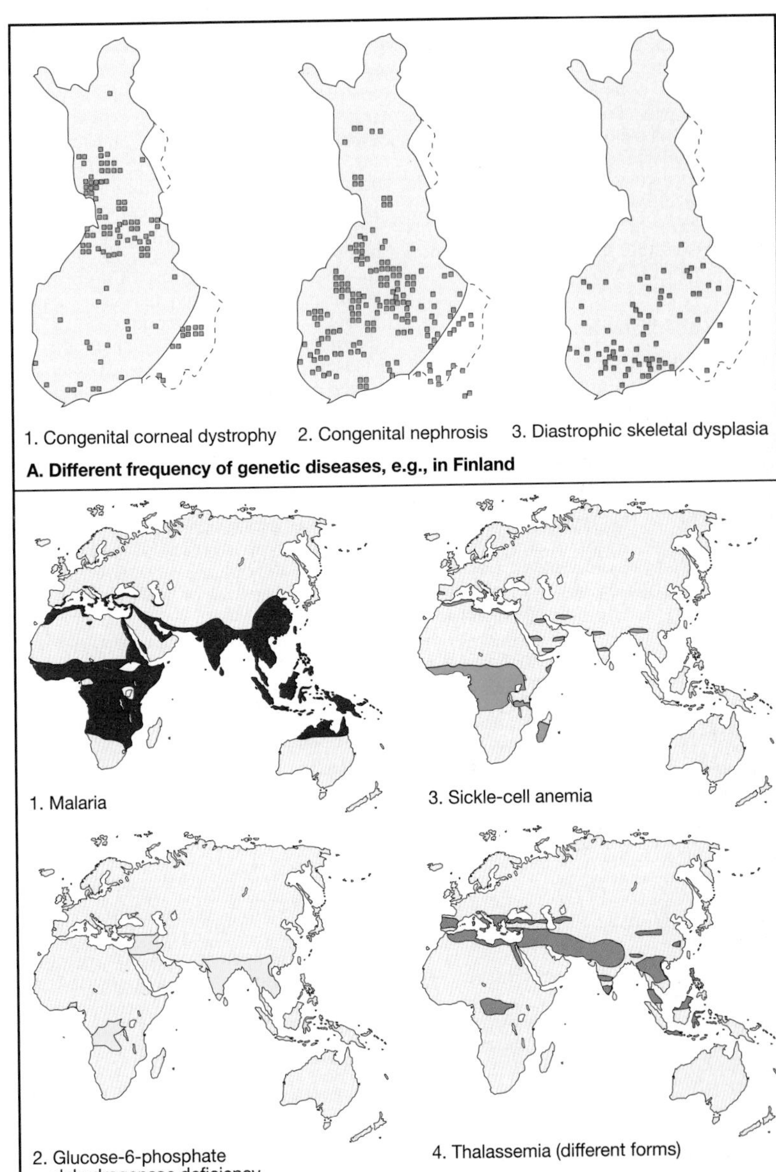

A. Different frequency of genetic diseases, e.g., in Finland

B. Distribution of malaria and frequency of heterozygotes by selective advantage

Nucleosomes

The total length of DNA in the haploid chromosome complement of mammals corresponds to about 1 meter. Since the total length of the human haploid chromosome complement in metaphase corresponds to about 115 μm, DNA in metaphase chromosomes must be shortened 10 000-fold by suitable packing. This occurs in discrete units, the nucleosomes, consisting of DNA and histones.

A. DNA and histones

Histone proteins (histones) are relatively simple proteins with a high proportion of positively charged amino acids (lysine and arginine), which enable them to bind firmly to the negatively charged DNA double helix. There are five types of histone molecules: H1, H2A, H2B, H3, and H4. Except for H1, they occur in exactly equal numbers. Histones are evolutionarily highly conserved. That is, they differ very little among the different organisms. There is an especially close spatial relationship between DNA and histones in the AT-rich regions of the minor grooves. Here, the DNA double helix is most easily bent.

B. Nucleosome

A nucleosome is made up of histones and two and a half turns of DNA double helix with an invariant number of the nucleotide pairs (146 base pairs). Two each of the histones H2A, H2B, H3, and H4 form a rounded octamer of about 6 x 11 nm in dimension. DNA has contact with the histones of a nucleosome in the order H2A, H2B, H4, H3, H3, H4, H2B, and H2A. The genes for histones are arranged one after the other in a tandemlike series and contain no introns. Their nucleotide base sequences are the most evolutionarily conserved of all genes.

C. Chromatin structure

DNA in the cell nucleus is tightly bound to proteins, with about equal amounts of histone and nonhistone proteins. The total complex of DNA and DNA-binding proteins in the cell nucleus of eukaryotic organisms is called chromatin. In electron micrographs, chromatin appears in formations with different densities. In some regions, it forms dense and compact structures, corresponding to a tightly folded arrangement of the nucleosomes. In other regions, the chromatin is somewhat spread out so that discrete individual structures can be recognized, corresponding to a partial folding of nucleosomes. Chromatin in an electron micrograph may also appear loosely wound and composed of individual units. In these segments, the nucleosomes are not folded on each other. Between nucleosomes the DNA is not bound to histones as in the partially folded arrangement. In the partially folded arrangement, DNA is bound to a fifth type of histone, histone H1. The dense segments of chromatin are called heterochromatin; the less dense, euchromatin. Genes in heterochromatin are generally not active. However, the folding of DNA in chromatin is not a means of transiently controlling the expression. As a rule the permanently inactivated genes are found here, in the completely folded chromatin segments.

D. Chromatin segments

The chromatin of the nucleus consists of segments of nucleosomes condensed to different degrees. DNA wound around a nucleosome is inactive and unreactive. In the DNA segments lying between series of nucleosomes, sequence-specific DNA-binding proteins are found. They regulate the expression (activity) of genes (see pp. 206,164). (Figures after B. Alberts et al., Molecular Biology of the Cell. 2nd ed., 1989).

References

Lewin B. Genes V. Oxford: University Press; 1994.

Watson JD, et al. Molecular Biology of the Gene. 3rd ed. 1987.

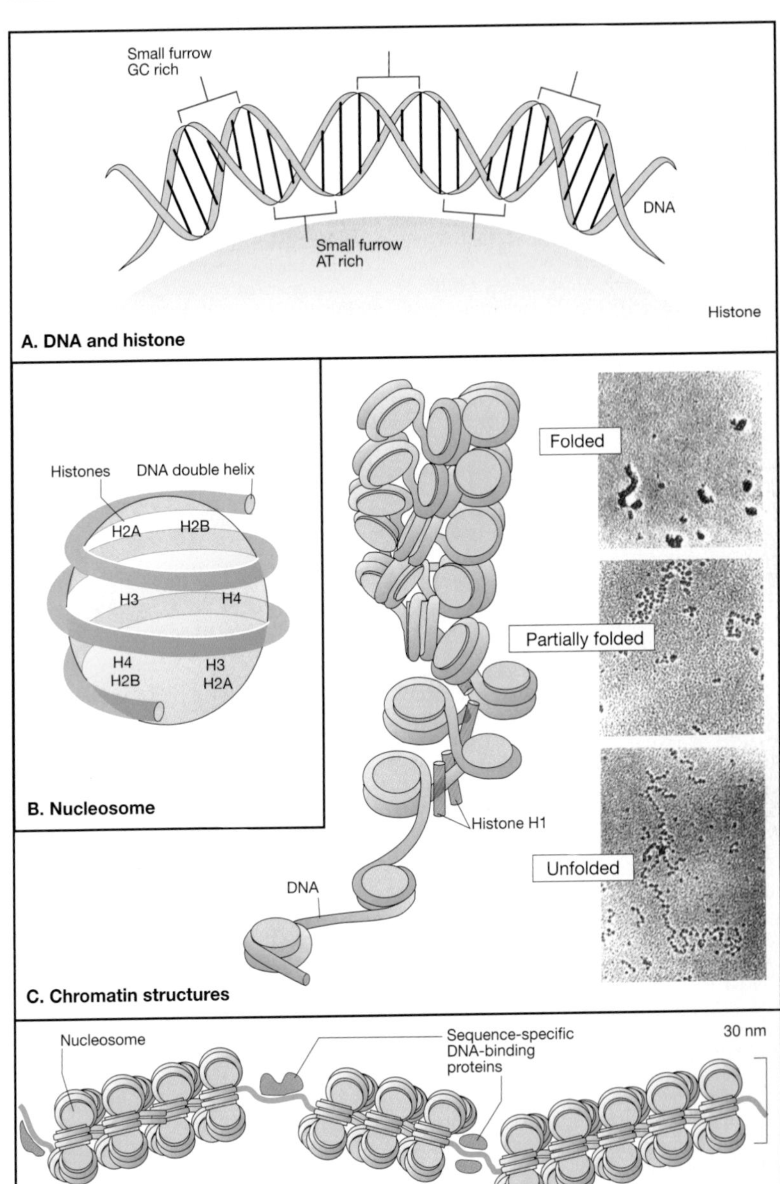

Small furrow
GC rich
DNA
Small furrow
AT rich
Histone
A. DNA and histone
Histones
DNA double helix
H2A
H2B
H3
H4
H4
H2B
H3
H2A
B. Nucleosome
Folded
Partially folded
Unfolded
Histone H1
DNA
C. Chromatin structures
Nucleosome
Sequence-specific
DNA-binding
proteins
30 nm
D. Chromatin segment

DNA in Chromosomes

In eukaryotic organisms the DNA is contained in chromosomes. Chromosomes consist of DNA and its closely associated proteins, which are histones and nonhistone proteins in about equal amounts. The chromatin of the cell nucleus is composed of chromosomes in interphase. Chromosomes are visible as individual structures only during mitosis. The haploid genome in humans contains 3 billion base pairs (3 x 10^9) in 22 autosomes and two sex chromosomes (X and Y).

A. DNA in metaphase chromosomes

A gene is a nucleotide sequence that serves as the functional unit for the formation of a complementary RNA molecule (transcription unit). A chromosome consists of a single, very long DNA molecule that contains a series of genes. It is folded and coiled in a complex manner. The spatial arrangement is secured by the nucleosomes.
In eukaryotic cells, the DNA is closely associated with histones to form characteristic structures, the nucleosomes. In addition, there are hundreds of sequence-specific DNA-binding proteins that recognize short DNA segments. Their binding is an important means of regulating gene activity (gene expression).
Chromosomal DNA is folded and packed in an efficient manner. Schematically, several levels of packing of DNA in a metaphase chromosome can be differentiated. The figure (above) shows a segment of DNA double helix from part of a chromatin segment. The chromatin segment consists of several nucleosomes connected by so-called linker stretches of DNA (see p. 162). The folding of a chromatin segment produces a chromatin fiber of 30 nm diameter, visible on electron microscopy. It consists of a series of tightly packed nucleosomes. These in turn form a part of a chromosome segment of about 300 nm diameter in the extended stage. A further packing level is represented by a thickened segment of a metaphase chromosome. This is called a condensed chromosomal segment. This segment corresponds to only a small part of one chromatid of a metaphase chromosome. As described on p. 170, each metaphase chromosome consists of two chromatids (sister chromatids), one of these chromatids being the result of DNA replication in the S phase. No transcription occurs in the chromatin of mitotic chromosomes. RNA synthesis is halted with the onset of chromosome condensation. (Figure after B. Alberts et al. Molecular Biology of the Cell. 2nd ed., 1989).

Reference

Tyler–Smith C, Willard HF. Mammalian chromosome structure. Current Opinion Genet Develop. 1993; 3: 390–397.

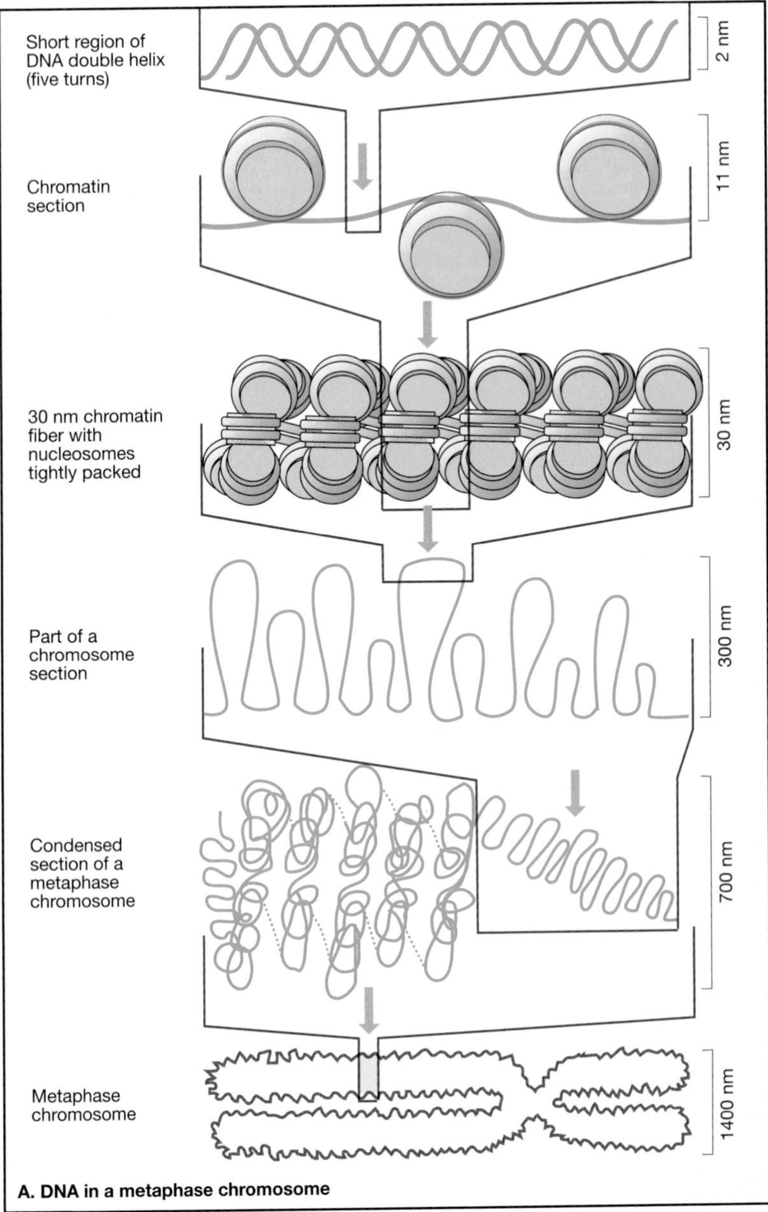

A. DNA in a metaphase chromosome

Polytene Chromosomes

In certain cells of insects, chromatin structures are visible under the light microscope as individual looplike domains. These cells and the chromosomes in them are greatly enlarged by prior DNA synthesis without cell division. That is, they contain about a thousand times more DNA than usual. The multiple copies of homologous chromosomes remain side by side in these cells and form a gigantic polytene chromosome. If the multiple copies of a polytene chromosome did not occur side by side, but rather as discrete chromosomes, the result would be referred to as polyploidy. Since there is a direct transition from polytene to polyploid chromosomes, the polytene chromosomes probably correspond in structure to normal chromosomes. Polytene chromosomes are readily visible under the light microscope because they are so large and lie precisely next to each other.

A. Polytene chromosomes in the salivary glands of Drosophila larvae

Under the light microscope, polytene chromosomes show readily recognized patterns of alternating dark and light bands. The light bands are called interbands. A polytene chromosome results from ten cycles of replication without division into daughter chromosomes. Thus, there are about 1024 (2^{10}) identical chromatid strands, which lie strictly side by side. About 85% of the DNA in polytene chromosomes is located in bands, and about 15% in interbands. The chromatin in the darkly stained bands is condensed to a much greater degree than the chromatin in the interbands. This can be attributed to an especially high degree of folding. Depending on the size, an individual band contains between 3000 and 300 000 nucleotide base pairs. Altogether, the Drosophila genome contains about 5000 bands and the same number of interbands. They have been numbered to produce a polytene chromosome map. A micrographic detail of a polytene chromosome from a drosophila salivary gland shows the characteristic band pattern. The dark bands are the result of chromatin condensation in the large interphase polytene chromosomes. These bands must be differentiated from those visible only in metaphase after special staining techniques in the banding patterns of the karyotypes of eukaryotic organisms (see p. 172). Intensely stained chromosomal segments correspond to a high degree of packing and are genetically inactive (heterochromatin); less tightly packed segments stain less distinctly and correspond to segments with genetic activity (euchromatin).

B. Functional stages in polytene chromosomes

Polytene chromosomes form structures that correlate with the functional state: During the larval development of drosophila, a series of expansions (puffs) appear in temporal stages in the polytene chromosomes. Chromosome puffs are decondensed, expanded segments that represent active chromosomal regions, i.e., regions that are being transcribed. The location and duration of the puffs reflect different stages of larval development (1). The incorporation of radioactively labeled RNA has been used to demonstrate that RNA synthesis, a sign of gene activity (transcription), occurs in these regions (2). (Figures after B. Alberts, Molecular Biology of the Cell. 2nd ed., 1989 and J. D. Watson et al., Molecular Biology of the Gene. 3rd ed., 1987).

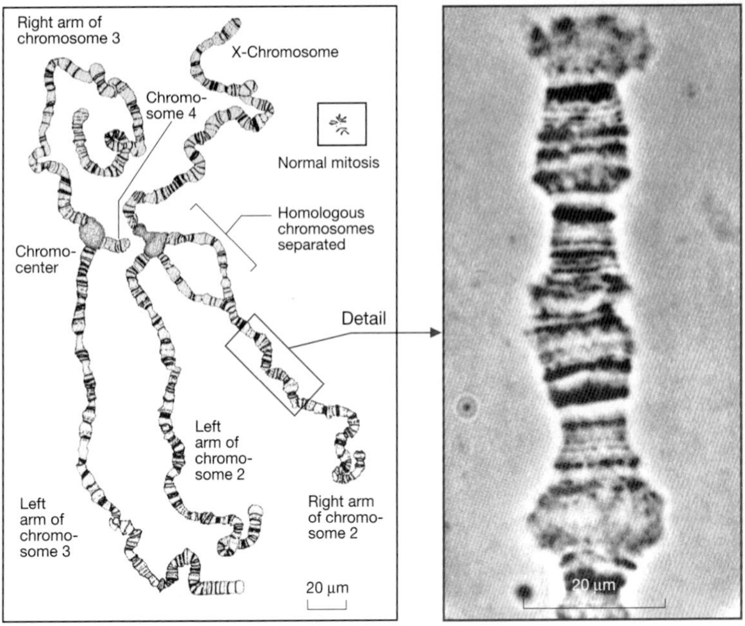

Right arm of chromosome 3
X-Chromosome
Chromo-some 4
Normal mitosis
Homologous chromosomes separated
Chromo-center
Detail
Left arm of chromo-some 2
Left arm of chromo-some 3
Right arm of chromo-some 2
20 µm
20 µm

A. Polytene chromosomes in salivary glands of Drosophila larvae

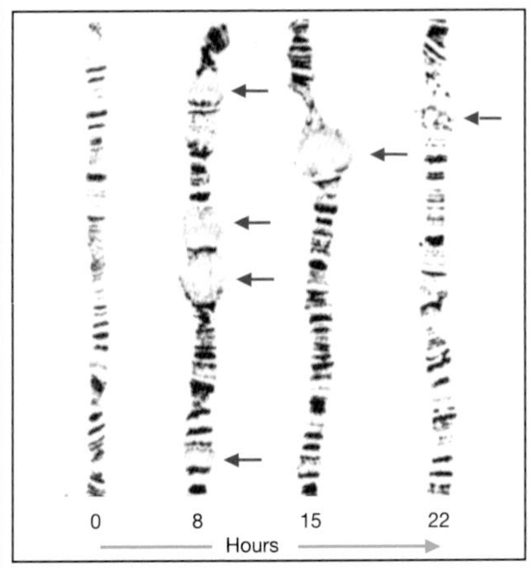

1. Formation of puffs (arrows)

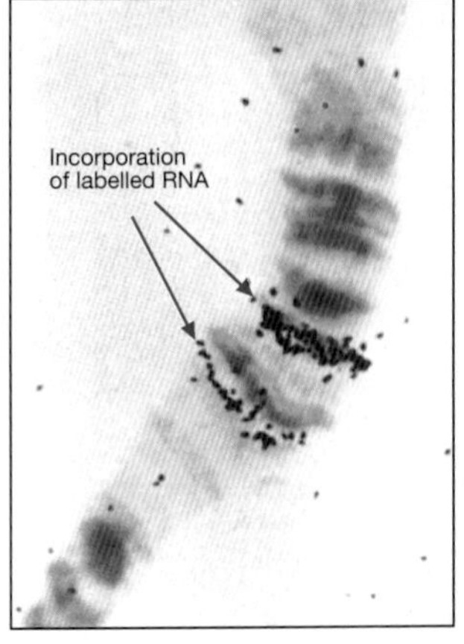

2. Evidence of gene activity

B. Functional stages in polytene chromosomes

DNA in Lampbrush Chromosomes

Usually, the chromatin in interphase nuclei is so densely packed that single chromatin threads are not directly visible. An exception is in the interphase nuclei of certain cells, especially immature amphibian oocytes with paired meiotic chromosomes. These chromosomes show RNA synthesis and form unusual chromatin loops. They are covered with newly transcribed RNA and are packed in dense RNA–protein complexes. These are visible under the light microscope as the so-called lampbrush chromosomes.

A. Chromosome structure in amphibian oocytes (lampbrush chromosomes)

The paired chromosomes of oocytes in meiosis consist of numerous chromatin loops arranged along an axis. Chiasma formation is visible at various locations.

Each segment of a lampbrush chromosome consists of a series of chromatin loops, originating from an axis. There, a condensed structure, the chromomere, is visible. Any given chromatin loop always contains the same DNA sequence and is formed in the same manner as the oocyte grows. Each loop corresponds to a particular unit of unfolded chromatin, which has unraveled and is transcriptionally active. Each cell contains four copies of a loop, two per chromosome (four-strand structure). Transcription occurs either along the whole loop or at parts of a loop.

The chromatin loops of a chromosome are paired, mirror-image structures. Each corresponds to the loop of a sister chromatid. The chromomere at the base of the loops consists of dense chromatin of the two sister chromatids, while the chromatin loops themselves are not condensed.

Lampbrush chromosomes in amphibian oocytes are unusually large compared with mitotic chromosomes (about 400 μm long, as opposed to at the most 10 μm) as shown in the micrograph (phase contrast exposure) (Joseph G. Gall, On the submicroscopic structure of chromosomes. Brookhaven Symp. Biol. 1956; 8: 17–32).

At the beginning of meiosis, when DNA replication is complete, the homologous pairs lie immediately next to each other and form characteristic structures composed of four chromatids.

Lampbrush chromosomes are distinguished by an especially high rate of RNA transcription. Most of the RNA transcripts are longer than in other chromosomes.

References

Alberts B, et al. Molecular Biology of the Cell. 2nd ed. 1989.

Callan HG. Lampbrush chromosomes. Proc Royal Soc Lond (Biol). 1982; 214: 417–448.

Traut W.: Chromosomen. Klassische und molekulare Cytogenetik. Heidelberg: Springer 1991.

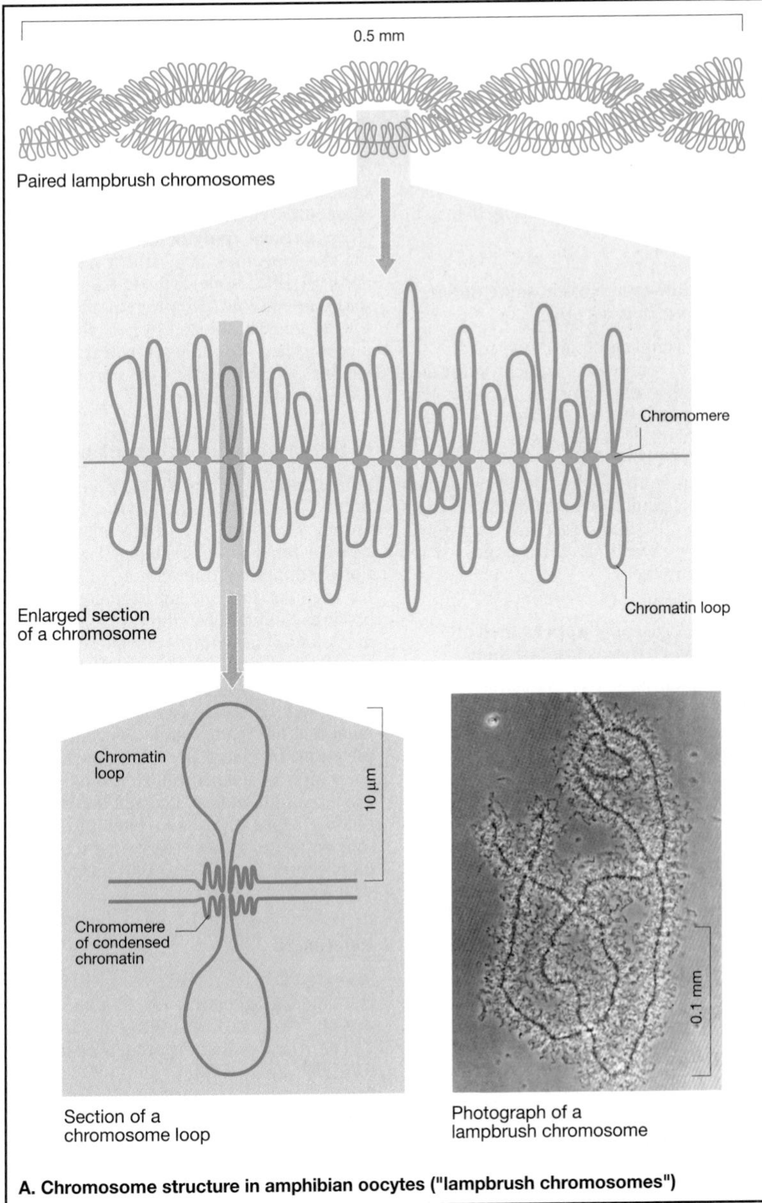

A. Chromosome structure in amphibian oocytes ("lampbrush chromosomes")

Metaphase Chromosomes

Chromosomes are visible as separate structures only during mitosis. In interphase, the chromosomes in chromatin cannot be individually differentiated. Electron microscopic studies have yielded some insight into chromosomal structure. In metaphase chromosomes, DNA is packed about 10 000 times more densely than in interphase.

A. A histone-free chromosome under the electron microscope

When certain proteins, especially histones, are removed from chromosomes, the chromosomal skeleton becomes visible under the electron microscope (photographs from J.R. Paulson and U.K. Laemmli. Cell. 1977; 12: 817–828) (1). Such a structure is surrounded by numerous darkly stained threads. A higher magnification (2) shows that this is a single continuous thread. It corresponds to the DNA double helix.

B. The microscopic appearance of metaphase chromosomes of man

With an approximately 1000-fold enlargement, the metaphase chromosomes of man and other vertebrates can be readily recognized under the light microscope as individual rodlike structures. A metaphase is shown here at about 2800-fold magnification. The chromosomes differ from each other in length, in the size and arrangement of their transverse light and dark bands (banding pattern), and in the point of attachment of the spindle (centromere), which is recognizable as a constriction. In prometaphase, the chromosomes are longer than in metaphase and show more bands. Thus, for certain purposes chromosomes are also studied in prometaphase.

C. Types of metaphase chromosomes

Depending on the location of its centromere (point of attachment of the spindle during mitosis), a chromosome can be distinguished as submetacentric, metacentric, acrocentric, or telocentric. The centromere divides a submetacentric chromosome into a short arm (p arm) and a long arm (q). In metacentric chromosomes, the short and long arms are about the same length. Acrocentric chromosomes show a dense appendage called a satellite (not to be confused with satellite DNA) at the end of the short arm. Satellite size differs for each acrocentric chromosome of an individual (chromosomal polymorphism). Telocentric chromosomes have neither a short arm nor a satellite. None of the human chromosomes are telocentric, whereas all chromosomes appear telocentric in the house mouse, *Mus musculus.* However, it is debatable whether telomeric chromosomes actually exist as defined.

D. Simple structural aberrations

A functionally relevant deviation from the normal structure is called a structural aberration. This is to be differentiated from chromosomal polymorphism. Loss (deletion) or doubling (duplication) of a particular segment may occur. A deletion may occur at the end of a chromosome (terminal deletion) or within a chromosomal segment (interstitial deletion). Prerequisite for a terminal deletion is one break; for an interstitial deletion, two breaks. A segment that has been doubled is called a duplication. In metaphase chromosomes, an aberration is seen in both chromatids because as a rule it has occurred before the S phase. Duplications and deletions represent opposite, and in some respects complementary, aberrations of chromosomal structure.

References

Rooney DE, Czepulkowski BH, eds. Human Cytogenetics. A Practical Approach. Vol. I, Constitutional Analysis. 2nd ed. The Practical Approach Series. Oxford: IRL Press; 1992.

Rooney DE, Czepulkowski BH, eds. Human Cytogenetics. A Practical Approach. Vol. II, Malignancy and Acquired Abnormalities. 2nd ed. The Practical Approach Series. Oxford: IRL Press; 1992.

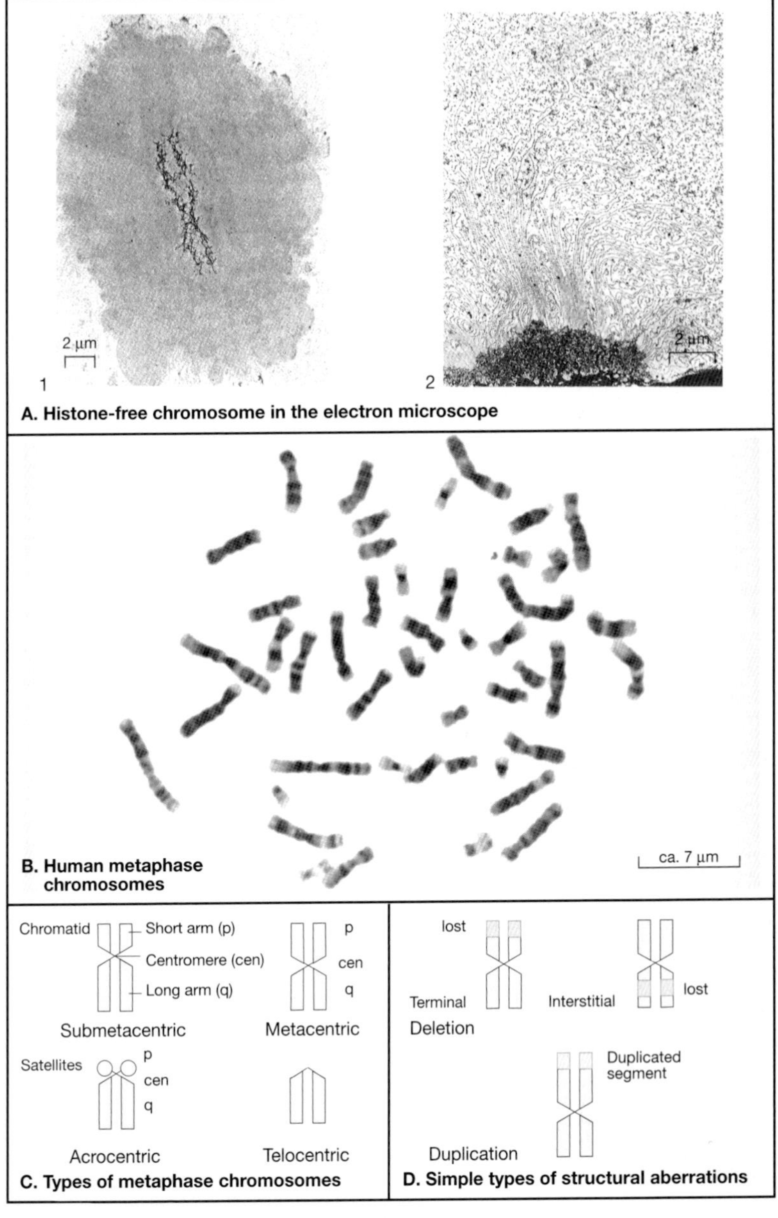

A. Histone-free chromosome in the electron microscope

B. Human metaphase chromosomes

C. Types of metaphase chromosomes

D. Simple types of structural aberrations

Karyotype

Karyotype refers to the arrangement of chromosomes in homologous pairs. They are arranged and numbered according to a convention. The karyotype is characteristic for each species. However, the term karyotype can also be applied to an individual or even to a single cell.

A. The karyotype of man

Man (*Homo sapiens*) has 22 pairs of chromosomes (autosomes) and in addition, either two X chromosomes, in females, or an X and a Y chromosome, in males (karyotype resp. 46,XX or 46,XY). In front of the comma, the karyotype formula gives the total number of chromosomes present, and after the comma, the composition of the sex chromosomes. The 22 pairs of autosomes in man are divided into seven groups (A–G) (see pp. 175–177).

B. Karyotype of the mouse *(Mus musculus)*

The standard karyotype of the mouse consists of 19 chromosome pairs in addition to the X and Y chromosomes. All chromosomes except the X and Y are telocentric and of similar size (1). However, they differ in their banding patterns, characteristic for each chromosome pair, and therefore are individually distinct. Certain strains of mice may show variants of the karyotype (2). These variants arise from fusion of certain of the chromosomes. In the example shown here, only chromosome pairs 1, 15, 19, and X correspond with those of the standard karyotype, while the others consist of fused chromosomes, e.g., chromosomes 4 and 2, chromosomes 8 and 3, etc. Structural rearrangements of the karyotype occurred with the separation of different species in evolution (figure from W. Traut: Chromosomen. Klassische und molekulare Cytogenetik. Heidelberg: Springer; 1991; photographs by H. Winking, Lübeck).

C. Flow cytometry karyotype in man

Because of their different lengths, metaphase chromosomes can also be presented in a flow-cytometry-based karyotype. With this method, individual chromosomes, stained and passed by a laser light source, give signals corresponding to their sizes. Although there are overlaps, e.g., between the similarly sized human chromosomes 9–12, or with chromosomes 1 and 2, a distribution pattern of light impulses based on individual chromosome sizes is obtained. The size of the X chromosome lies between those of chromosomes 8 and 7; the size of the Y chromosome as a rule corresponds to that of a chromosome 22, although Y chromosome sizes may differ considerably. Because of the technical expenditure required and the unsharp resolution, flow cytometry is not frequently used for practical diagnosis. However, it is of advantage in certain studies. (Figure from J.M. Connor & M.A. Ferguson-Smith: Essential Medical Genetics. 3rd ed. London: Blackwell Scientific Publications; 1991).

Reference

Buckle VJ, Kearney L. New methods in cytogenetics. Current Opin Genet Develop. 1994; 4: 374–382.

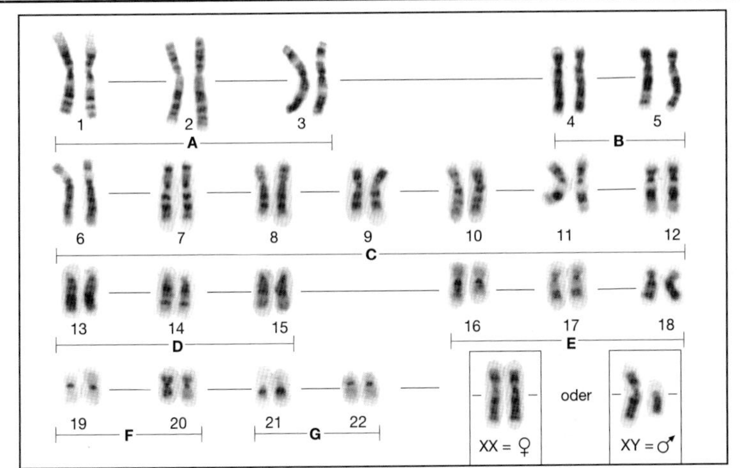

A. Karyotype of man

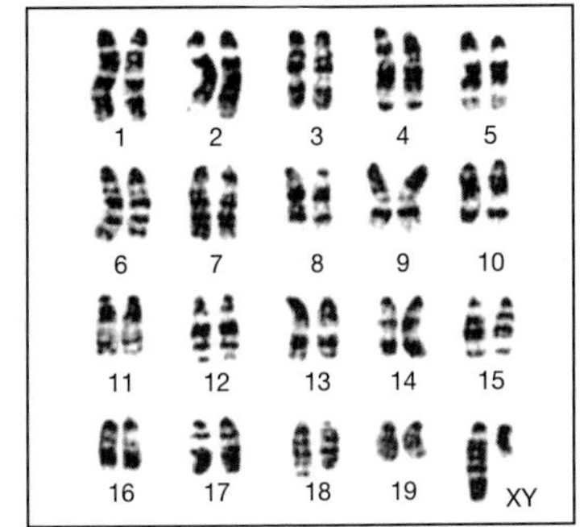

1. Standard

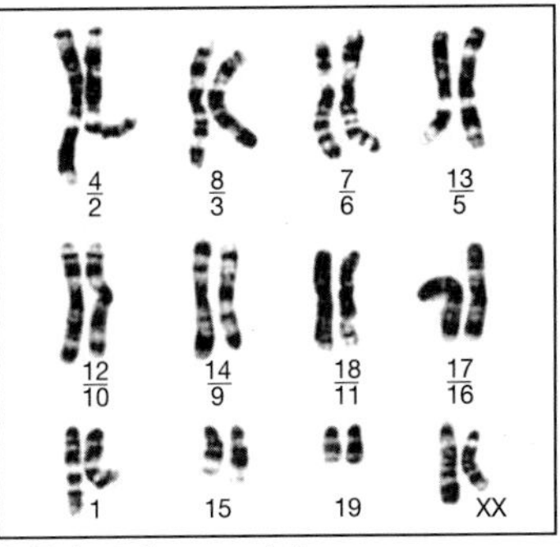

2. Variants in a population with fused chromosomes

B. Karyotype of the mouse *(Mus musculus)*

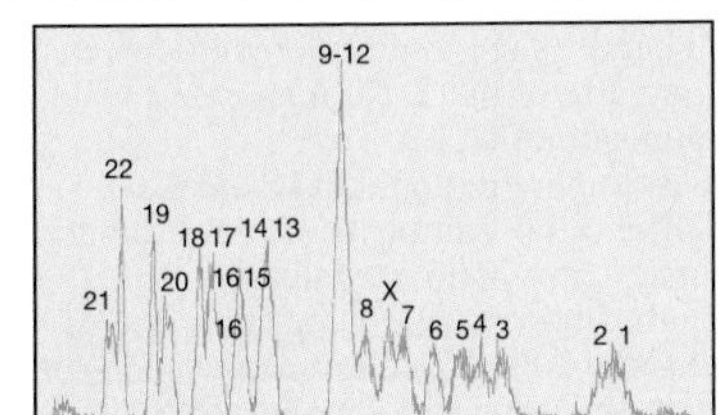

1. Normal female

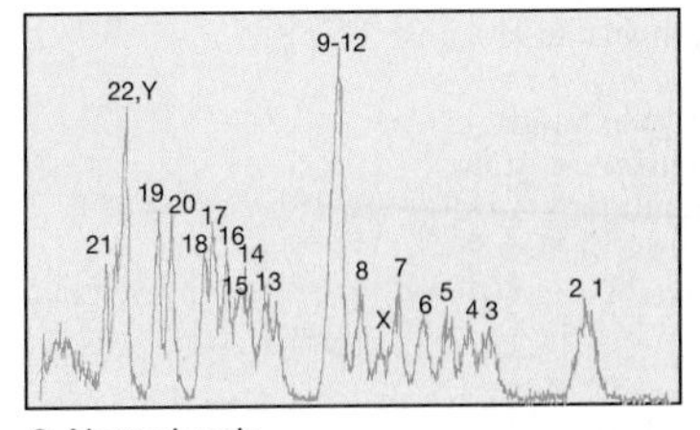

2. Normal male

C. Flow cytometry karyotype

G- and R-Band Patterns of the Metaphase Chromosomes of Man

With certain preparation and staining procedures, it is possible to distinguish light and dark bands in metaphase and prometaphase chromosomes under the light microscope. Each chromosome pair and each chromosome segment of sufficient length has a recognizable band pattern, specific for the respective chromosome. The basic types of bands are G bands (Giemsa stain induced) and R bands (reverse bands). A procedure used for fluorescence microscopy produces quinacrine-induced bands (Q bands). Their pattern corresponds to that of G bands. Pages 175 and 177 show photographs of human metaphase chromosomes with G bands (left) and R bands (right). Between them, the corresponding G- and R-band patterns are presented diagrammatically. Starting at the centromere and moving distally, each chromosome is divided into defined regions and bands. The dark bands on the photograph are black on the diagram; light bands are white. Only the G-band-system numbering is shown. About 550 individually distinguishable bands of the haploid chromosome complement can be recognized in the illustration (550-band stage).

A. The metaphase chromosomes of man (pairs 1–22, X and Y)

The 22 pairs of autosomes (1–22) are divided into seven groups. Chromosome pairs 1–3 form group A (chromosomes 1 and 3 are metacentric; chromosome 2 is submetacentric); chromosomes 4 and 5 form group B; 6–12 group C; etc. The acrocentric chromosomes in humans belong to groups D and G. The chromosomes of group F are metacentric. The possibility of identifying every chromosome region and many bands makes it possible to establish breakpoints and to state the localization of genes according to region and band number on each chromosome. If a band in a good preparation can be further differentiated, subdivisions can be defined by decimals. For instance, band 31 includes bands 31.1, 31.2, and 31.3. (Picture sources: Figures on p. 175 and 177 correspond to Figure 5 from ISCN 1985, An International System for Human Cytogenetic Nomenclature, 1985, DG Harnden & HP Klinger, eds.; published through Cytogenetics and Cell Genetics, Karger, Basel, 1985, and in Birth Defects: Original Article Series, vol. 21, no. 1, March of Dimes Birth Defects Foundation, New York, 1985. The G-band photographs are from U. Francke, Cytogenet. Cell Genet. 1981; 31: 24, and the R-band photographs from Camargo and Cervenka. Am J. Hum Genet. 1982; 34: 757.)

The main types of chromosome bands

Banding methods	*Type*	*Principal use*
Trypsin-induced Giemsa stain	G	Differentiates light and dark bands
AT-specific fluorochrome (quinacrine, Hoechst 33258)	Q	Light fluorescence in the region of dark G bands, some centromere regions, distal long arm of the Y chromosome
Reverse bands	R	Opposite of G
Centromere stain	C	Centromere region darkly stained
Bromodeoxyuridine (BrdU) for two cell cycles	SCE	Differential staining of sister chromatids (sister chromatid exchanges)
Distamycin A–DAPI	DA/DAPI	Light fluorescence in the short arm of chromosome 15, centromere regions of 1, 9, and 16; distal long arm of Y
Silver nitrate stain	NOR	Short arms of all acrocentric chromosomes
Giemsa 11	G11	Centromere of chromosome 11

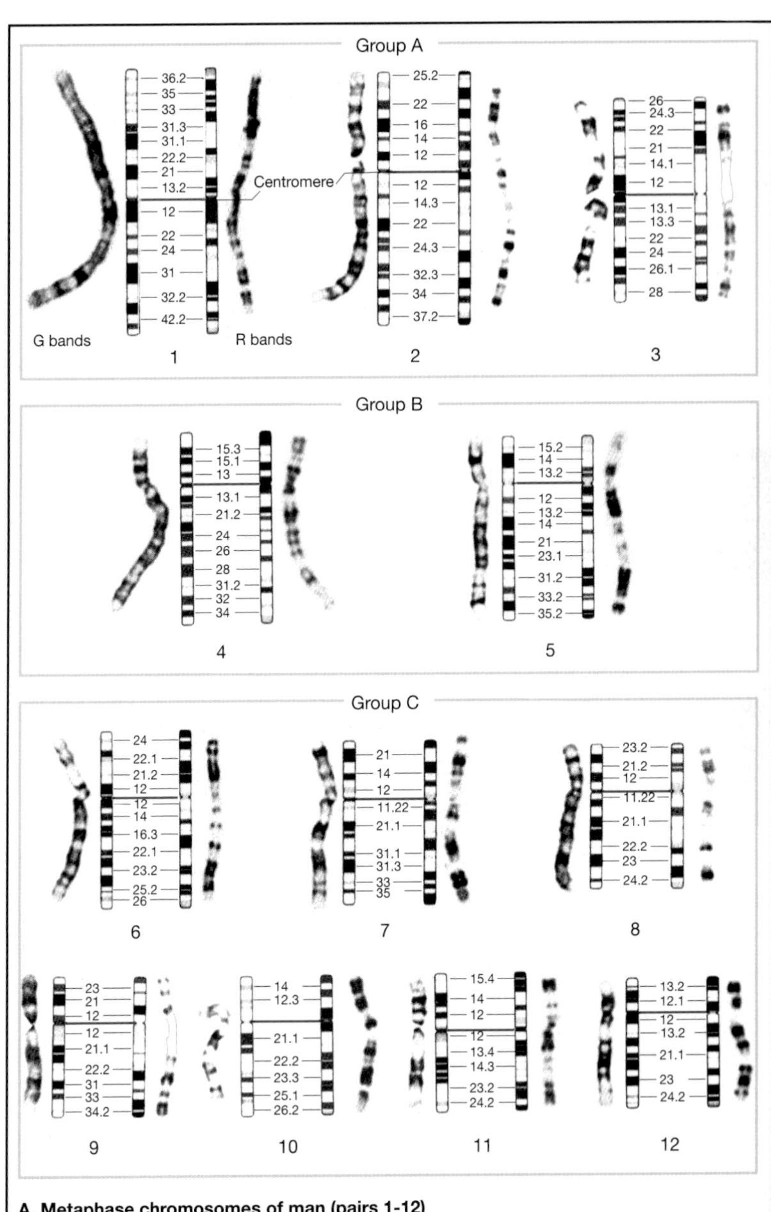

A. Metaphase chromosomes of man (pairs 1-12)

Designation of Chromosomal Aberrations

The following list of chromosome abbreviations includes the main types of chromosomal aberrations in man, abbreviated according to the International System for Human Cytogenetic Nomenclature (ISCN 1978).

46,XX	Normal female karyotype with 46 chromosomes (two X chromosomes)
46, XY	Normal male karyotype with 46 chromosomes (an X and a Y chromosome)
47,XXY	Karyotype with 47 chromosomes (two X chromosomes, a Y chromosome)
47,XY,+21	Karyotype with 47 chromosomes (X and Y chromosomes); an additional chromosome 21 (trisomy 21)
13p	Short arm of chromosome 13
13q	Long arm of chromosome 13
13q14	Region 1, band 4 of the long arm of chromosome 13
13q14.2	Subband 2 of 13q14
2q–	Shortening of the long arm of chromosome 2
del(2)	Deletion in chromosome 2
del(2)(q21–qter)	Deletion in chromosome 2 of region 2, band 1 (2q21) of the long arm to the end (telomere) of the long arm (qter)
inv(4)	Inversion in chromosome 4
inv(4)(p11q21)	Inversion in chromosome 4, 4p11 to 4q21 (pericentric inversion)
dup(1)	Duplication in chromosome 1
inv dup(1)	Inverted duplication in chromosome 1
inv dup(2)(p23–p24)	Inverted duplication of the bands p23 to p24 in chromosome 2
r(13)	Ring-shaped chromosome 13 (deletion implied)
i(Xq)	Isochromosome for the long arm of an X chromosome
dic(Y)	Dicentric Y chromosome
idic(X)	Isodicentric X
t(2;5)	Reciprocal translocation between a chromosome 2 and a chromosome 5
t(2q–;5q+)	Reciprocal translocation between a chromosome 2 and a chromosome 5; the long arm of 2 has been shortened and that of 5 has become longer
t(2;5)(q21;q31)	Reciprocal translocation with the breakpoints in q21 of chromosome 2 and q31 of chromosome 5
der(2)	A ("derivative") chromosome derived from chromosome 2
t(13q14q)	Translocation of the centric fusion type, of the long arms of a chromosome 13 and a chromosome 14. It results in a single chromosome
ins(5)	Insertion into chromosome 5
ins(5;2) (p14;q22q32)	The segment q22 to q32 of a chromosome 2 has been inserted into region p14 of a chromosome 5
fra(X)(q27.3)	X chromosome with a fragile site at position q27.3

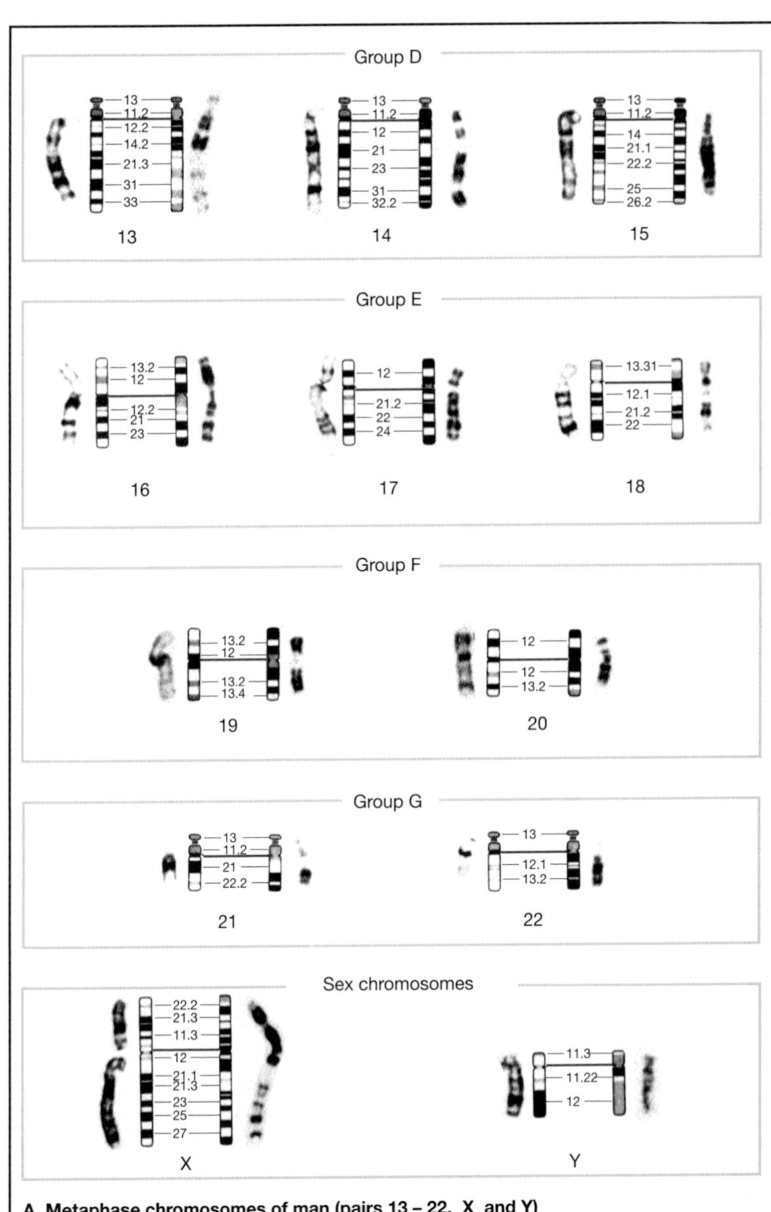

A. Metaphase chromosomes of man (pairs 13 – 22, X and Y)

Preparation of Metaphase Chromosomes for Analysis

Since chromosomes are visible as individual structures under the light microscope only during metaphase (or under special conditions and for certain purposes, during prometaphase), every chromosome analysis requires dividing cells. In vivo, only bone marrow cells contain a sufficient proportion of cells in mitosis. Thus, in vivo chromosome analysis of cells is limited to bone marrow. All other procedures for analyzing chromosomes in mitosis require culturing of suitable cells (cell culture). Most commonly, lymphocytes from blood are cultured for chromosome preparations. In addition, fibroblasts from a piece of skin can be propagated in cell culture for analysis (see p. 114). However, since this procedure is somewhat elaborate and time-consuming, it is used only for certain purposes.

A. Chromosome analysis from blood

For the cell culture, either peripheral blood is used directly or lymphocytes are isolated from peripheral blood (T lymphocytes). A sample of about 2 ml peripheral blood is needed. The blood is prevented from clotting by use of a heparinized syringe, since clumping of the blood cells precludes culturing (the proportion of heparin to blood is about 1:20). Peripheral blood or isolated lymphocytes are placed in a vessel with culture medium. Mitosis is generally stimulated with phytohemagglutinin, a protein from plants that unspecifically stimulates lymphocytes to divide. The culture requires about 72 hours at 37 °C for cells to divide. Lymphocyte cultures are suspension cultures; i.e., the cells divide in culture medium without attaching to the culture vessel. Cell division is arrested and the culture terminated by adding a suitable concentration of a colchicine derivative (colcemid) 2 hours prior to harvest. Colcemid interrupts mitosis during metaphase, so that a relative enrichment of cells in metaphase results. Cell preparation is carried out as follows: the culture solution is centrifuged; the cell sediment is placed in a hypo-osmolar KCl solution (0.075 molar), incubated for about 20 minutes, and centrifuged again. The resulting cell sediment is placed in fixative. The fixing solution is a mixture of methyl alcohol and glacial acetic acid in a ratio of 3:1. Usually the fixative is changed two to three times with subsequent centrifugation. After that, the fixed cells are taken up in a pipette and dropped onto a slide. The preparation is stained, and the slide covered with a cover glass. At this point the cells are ready for analysis: Suitable metaphases are located under the microscope with about 100 x magnification and are subsequently examined at about 1250 x magnification. During direct analysis with the microscope, the number of chromosomes and the presence or absence of all chromosomes and recognizable chromosome segments are noted. Since the preparation procedure itself may induce deviations from the normal chromosome number or structure in some cells, more than one cell must be analyzed. Depending on the purpose of the analysis, between 5 and 100 metaphases (usually 10–15) are examined. Some of the metaphases are photographed with the microscope and can subsequently be cut out of the photo (karyotyping). In this way a karyotype can be obtained from the photograph of a metaphase.

The time needed for a chromosome analysis varies depending on the problem, but is usually 3–4 hours. Analysis and karyotyping time can be shortened by computer procedures.

References

Therman E, Susman M. Human Chromosomes. Structure, Behavior, and Effects. 3rd ed., New York–Heidelberg: Springer; 1993.

Verma RS, Babu A. Human Chromosomes. Manual of basic techniques. New York: Pergamon Press; 1989.

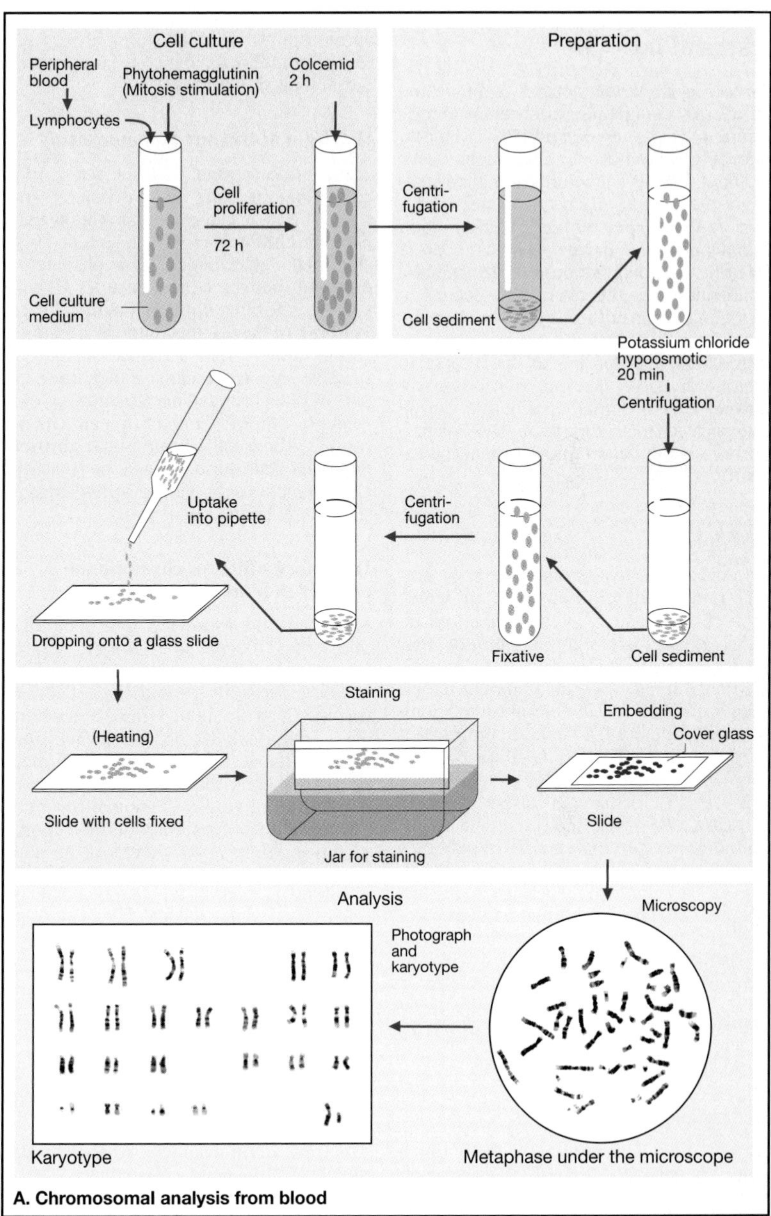

A. Chromosomal analysis from blood

Numerical Chromosome Aberrations

Deviation from the normal chromosome number in a single pair of chromosomes is referred to as aneuploidy. In humans, numerical chromosome aberrations occur in about 1 in 400 newborns. An abnormality of the number of chromosomes occurs as a result of their abnormal distribution (nondisjunction) during meiosis I or II (meiotic nondisjunction). With meiotic nondisjunction, the aberration occurs in all cells of a resulting organism. Nondisjunction in meiosis I and meiosis II can be differentiated according to the respective meiotic divisions (see pp. 108, 342). Abnormal chromosomal distribution during mitosis leads to an aberration in only a proportion of the cells (chromosomal mosaicism).

A. Triploidy

Triploidy refers to a deviation from the normal number of chromosomes in which each chromosome is present threefold instead of twofold. With tetraploidy, four copies of each chromosome are present. Triploidy arises when an abnormal oocyte with a double (46,XX) chromosome complement instead of a haploid complement (23,X) is formed. After fertilization by a normal spermatocyte, triploidy (69,XXX or 69, XXY) of maternal origin arises. In this case, two of the three complete sets of chromosomes are maternal. Triploidy may also arise during spermatogenesis, resulting in an abnormal spermatozoon that does not contain the normal haploid chromosome complement, but rather the diploid (46,XY). In this case, the triploidy (69,XXY) is of paternal origin (see p. 344). A further cause of triploidy is dispermy, or fertilization of a normal egg by two normal sperm.

B. Aneuploidy

In a trisomy (1), only one of the chromosomes is present threefold; all other chromosome pairs are normal. Rarely, two different trisomies occur, for two different chromosomes (double aneuploidy). If one chromosome of a pair is missing, it is referred to as monosomy (2).

C. Origin of trisomy and monosomy

The result of normal meiosis (see p. 108), consisting of two cell divisions (not shown here), is a normal distribution and a haploid chromosome complement. With abnormal distribution (nondisjunction either in meiosis I or in meiosis II), one gamete is formed with an additional chromosome, whereas the other is missing a chromosome. After fertilization, the respective zygote contains either three copies of one chromosome (trisomy) or only a single chromosome of a pair (monosomy). Abnormal chromosome distribution can occur during oogenesis (maternal nondisjunction) or during spermatogenesis (paternal nondisjunction).

D. Abnormalities of chromosome number in humans

In humans, the following autosomal trisomies may occur in liveborn infants: trisomy 13 with a frequency of 1 in 5000 newborns, trisomy 18 with 1 in 3000, and trisomy 21 with about 1 in 650 newborns (1). Additional X or Y chromosomes occur in about 1 in 800 newborns, much more frequently than the autosomal trisomies. But unlike the autosomal trisomies, they usually do not lead to defined clinical pictures. Triple X (47,XXX) or an additional Y chromosome (47, XYY) are generally not clinically apparent. On the other hand, monosomy X (3) leads to the clinical picture of Turner syndrome (see p. 344), and XXY causes Klinefelter syndrome.

References

Therman E, Susman M. Human Chromosomes: structure, behavior, and effects. 3rd ed. Heidelberg: Springer, 1993.

Traut W. Chromosomen. Klassische und molekulare Genetik. Heidelberg: Springer; 1991.

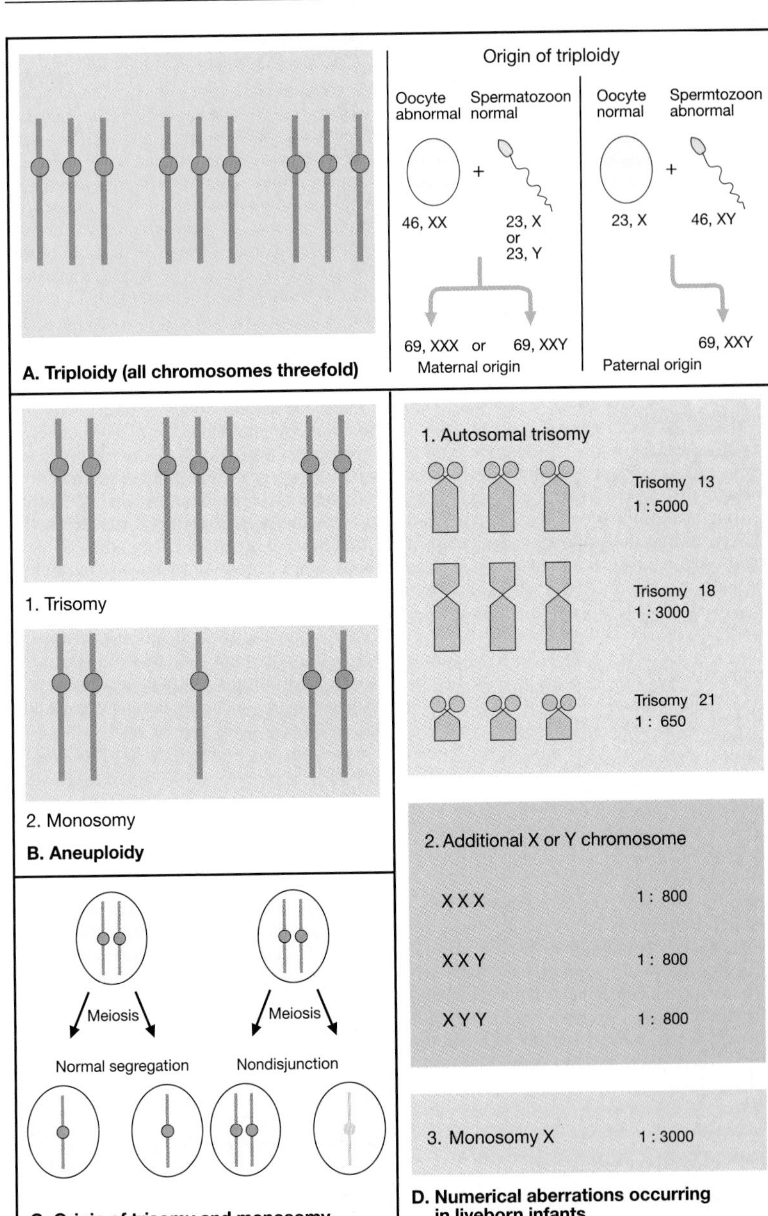

Origin of triploidy
Oocyte abnormal
Spermatozoon normal
Oocyte normal
Spermtozoon abnormal
46, XX
23, X or 23, Y
23, X
46, XY
69, XXX or 69, XXY
Maternal origin
69, XXY
Paternal origin
A. Triploidy (all chromosomes threefold)
1. Trisomy
2. Monosomy
B. Aneuploidy
Meiosis
Meiosis
Normal segregation
Nondisjunction
C. Origin of trisomy and monosomy
1. Autosomal trisomy
Trisomy 13
1 : 5000
Trisomy 18
1 : 3000
Trisomy 21
1 : 650
2. Additional X or Y chromosome
XXX 1 : 800
XXY 1 : 800
XYY 1 : 800
3. Monosomy X 1 : 3000
D. Numerical aberrations occurring in liveborn infants

Translocation

Translocation refers to an exchange of chromosome segments. A translocation can arise by centric fusion of two acrocentric chromosomes (Robertsonian translocation) or by exchange between two chromosomes (reciprocal translocation). With centric fusion, two complete chromosomes are involved; with reciprocal translocation, only a part of each of the two chromosomes is exchanged. In a translocation it is important to determine the breakpoints in each of the chromosomes involved.

A. Centric fusion of acrocentric chromosomes

Chromosome 14 and chromosome 21 (1) are the most frequently involved in fusions (about 1 in 1000 newborns). By fusion of the long arm of chromosome 21 (21q) and the long arm of chromosome 14 (14q), a chromosome t(14q21q) is formed (2). The satellite-carrying short arms of both chromosomes are lost, but this is insignificant. When germ cells (gametes) are formed, deviation from the normal chromosome number may result (3). Since chromosome 14 and chromosome 21 pair during meiosis, the following possible gametes may result: chromosome 14 alone (no chromosome 21), one chromosome 14 and one chromosome 21 (normal), the chromosome 14 fused to chromosome 21 (balanced), or the fused chromosome and one chromosome 21 (altogether two chromosomes 21).

After fertilization, the corresponding zygotes contain either only one chromosome 21 (unviable monosomy 21), a normal chromosome complement, a balanced chromosome complement with the fused chromosome, or a gamete with altogether three chromosomes 21 (trisomy 21). In the latter case, the clinical disorder Down syndrome (formerly called mongolism) results.

B. Reciprocal translocation

A reciprocal translocation is an exchange of chromosomal material between two chromosomes. Since usually no chromosomal material is lost or added with a reciprocal translocation, it does not cause clinical signs (i.e., it is balanced). However, carriers of a reciprocal translocation may form gametes with unbalanced chromosome complements. During meiosis, the chromosomes involved in the reciprocal translocation take part as usual in the homologous pairing of meiosis I. Each of the chromosomes not involved in the translocation pairs with its homologous partner that is involved in the translocation. This leads to the formation of a characteristic quadriradial configuration of the involved chromosomes. When these four chromosomes separate (segregation) during anaphase of meiosis (see p. 108), one of three possibilities may occur: With alternate segregation, one gamete receives the two normal chromosomes, and the other gamete the chromosomes involved in the translocation, i.e., it is balanced.

With nonalternate segregation (neighboring or adjacent chromosomes), the two chromosomes on the left go into one gamete and the two chromosomes on the right, into the other (adjacent–2). With the other possibility, the upper chromosomes go into one gamete, and the lower two into the other (adjacent–1). In each of the last two cases, an unbalanced distribution of the involved chromosome segments results. For example, after adjacent–2 segregation, gametes receive a partial duplication of the chromosome segment marked with red and a partial deficiency of the segment marked with blue (left pair of chromosomes) or a partial duplication of the blue segment and a partial deficiency of the red (duplication/deficiency). Different types of disorders result depending on the chromosome segments involved.

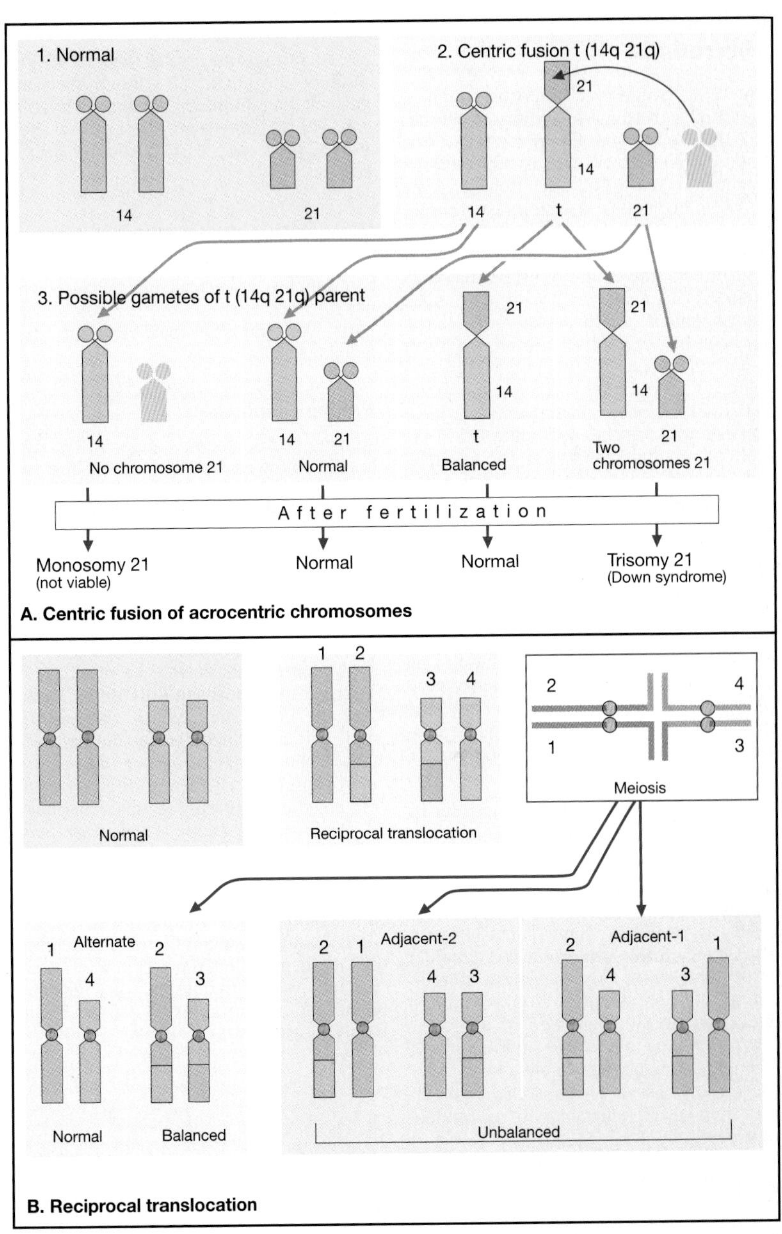

A. Centric fusion of acrocentric chromosomes

B. Reciprocal translocation

Structural Chromosomal Aberrations

Structural changes in chromosomes can be classified according to cytological types and their effect on the phenotype. The main cytological types are translocation (exchange) (see p. 182), deletion (loss, see p. 170), inversion, insertion, isochromosome, dicentric chromosome, and ring chromosome (see below). According to their effects, they can be differentiated as balanced or unbalanced. With a balanced rearrangement, no chromosomal material has been lost or gained. In this case, there is no effect on the phenotype. In unbalanced aberrations, chromosomal material either has been added (partial duplication) or lost (partial deficiency). Simple types of structural aberration, such as deletion and duplication are shown on p. 170.

A. Inversion

An inversion is a 180-degree change in direction of a chromosomal segment. Prerequisite for every inversion is a break at two different sites, followed by reunion of the inverted segment. Depending on whether the centromere is involved, a pericentric inversion (the centromere lies within the inverted segment) and a paracentric inversion can be differentiated.

B. The consequences of crossing-over in the inverted region

With homologous pairing during meiosis, an inversion loop is formed in the region of the inversion. When the inverted segment is relatively large, crossing-over may occur in this region. In the daughter cells, one chromosome may show a duplication (e.g., of segments A and B) and a deficiency (of segment F) (3), while the other chromosome shows deficiency of segments A and B and duplication of segment F (4). These chromosome segments are not balanced (aneusomy by recombination).

C. Isochromosome

An isochromosome arises when a normal chromosome (1) divides transversely instead of longitudinally, so that it is composed of two long arms (2) or of two short arms (3). In each case, the other arm is missing.

D. Dicentric chromosome

A dicentric chromosome contains two centromeres. It is unstable because it is usually torn apart during mitosis and its parts are divided between the two daughter cells.

E. Ring chromosome

A ring chromosome arises after two breaks followed by a joining of the two opposite ends. The distal segments are lost. Therefore, a ring chromosome is unbalanced.

F. Consequences of a ring chromosome

A ring chromosome is unstable because a break with reattachment ("crossing-over") of the chromatids during the prophase of mitosis usually leads to difficulties. In this case, a large ring-shaped chromosome with two centromeres arises during metaphase and telophase. Since the centromeres migrate in different directions during anaphase, the ring becomes disrupted. If this does not occur strictly symmetrically, two daughter cells will result with certain segments either missing (deficiency) or duplicated (duplication). In the example, one daughter cell with a deficiency of segment 4 and one daughter cell with a duplication of segment 4 are formed. Not infrequently, ring chromosomes are lost completely and a monosomy results.

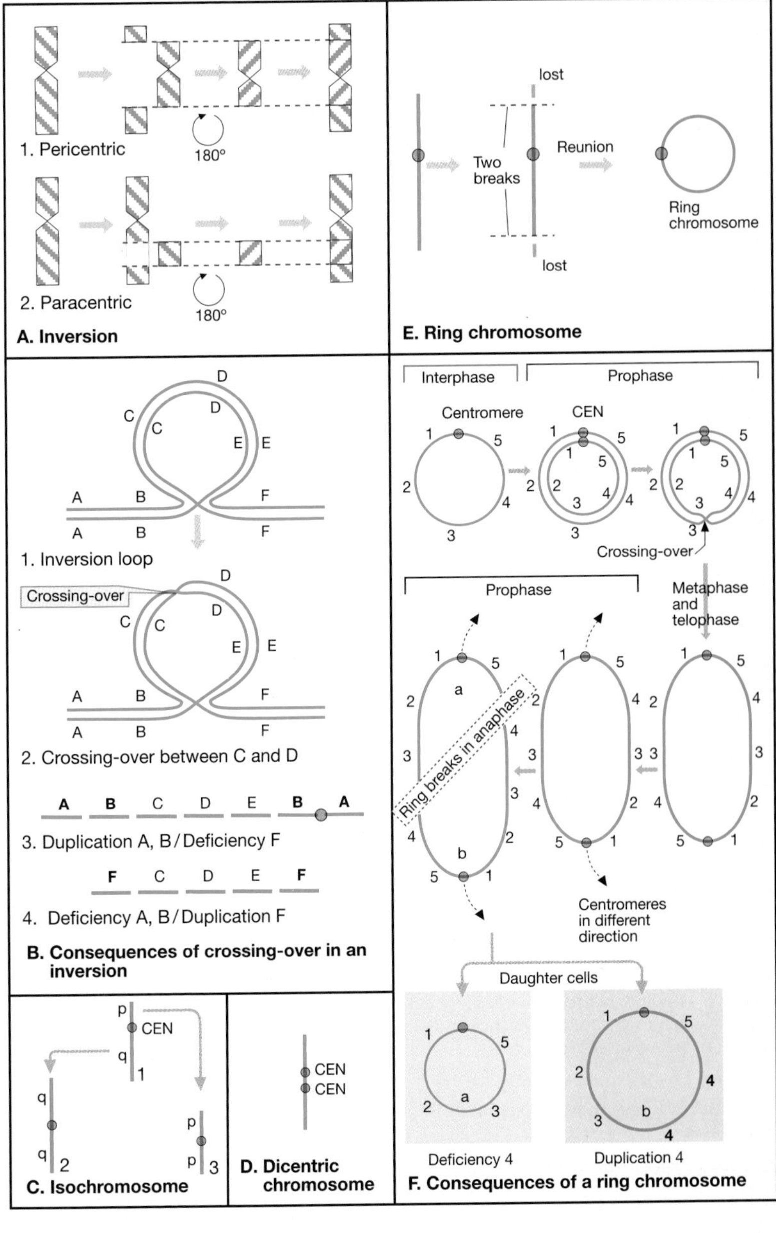
1. Pericentric
180°
2. Paracentric
180°
A. Inversion
lost
Two breaks
Reunion
Ring chromosome
lost
E. Ring chromosome
1. Inversion loop
Crossing-over
2. Crossing-over between C and D
A B C D E B A
3. Duplication A, B / Deficiency F
F C D E F
4. Deficiency A, B / Duplication F
B. Consequences of crossing-over in an inversion
Interphase
Prophase
Centromere
CEN
Crossing-over
Prophase
Metaphase and telophase
Ring breaks in anaphase
Centromeres in different direction
Daughter cells
Deficiency 4
Duplication 4
F. Consequences of a ring chromosome
CEN
C. Isochromosome
CEN
CEN
D. Dicentric chromosome

Correlation of Structure and Function in Chromosomes

In a chromosome, regions with special structures and functions can be differentiated. The centromere is the point of attachment of the microtubuli of the mitotic spindle (kinetochore). The telomeres at the ends contain no genes and have a special structure.

A. Heterochromatin and euchromatin

In 1928 Emil Heitz observed that certain parts of the chromosomes of a moss (*Pellia epiphylla*) remain thickened and deeply stained during interphase, as chromosomes otherwise do only during mitosis. He named these structures heterochromatin, as opposed to euchromatin, which becomes invisible during late telophase and subsequent interphase. Functionally, heterochromatin is defined as a region in which few or no active genes lie and in which repetitive DNA sequences occur. When active genes become located close to the heterochromatin, they usually become inactivated (position effect–variegation). (Figure from E. Heitz: Das Heterochromatin der Moose. I. Jahrb Wiss Bot. 1928; 69: 762–818).

B. Characteristic regions of a chromosome

The centromere and telomeres contain repetitive DNA sequences. They are evolutionarily conserved because they are important for chromosome stability. The segments located between the telomeres and centromere consist of trypsin-sensitive light and trypsin-resistant dark G bands. The light G-band areas of DNA form loops in which the protein-coding genes lie. The DNA loops are bound to a protein matrix at special attachment sites.

C. Model of a chromosome segment in interphase

A three-dimensional model of a chromosome segment shows that the constitutive heterochromatin (C band) in the centromere region is very tightly wound. In the light G bands, the euchromatin is relatively loosely, and in the dark G bands, somewhat more tightly packed. (With kind permission of the author L. Manuelidis: View of metaphase chromosomes, Science *250:* 1533–1540, 14th December 1990, Copyright 1990 by the AAAS).

D. Constitutive heterochromatin (C bands) in the centromeric region

The constitutive heterochromatin in the centromeric region can be specifically stained (C bands). The distal half of the long arm of the Y chromosome is also C-band positive. The centromeric heterochromatin in chromosomes 1, 9, and 16 and in the long arm of the Y chromosome in humans is polymorphic. The lengths of the heterochromatic segments in one or more of these regions may vary between different individuals. (From A.S. Verma and A. Babu: Human Chromosomes. New York: Pergamon Press; 1989).

E. Functional attributes of the euchromatin regions

The light and dark G bands differ in functional respects. An average G band contains around 1.5 megabases (Mb) DNA.

References

Bickmore WA, Sumner AT. Mammalian chromosome banding: an expression of genome organization. Trends Genet. 1989; 5: 144–148.

Passarge E. Emil Heitz and the concept of heterochromatin: Longitudinal chromosome differentiation was recognized fifty years ago. Am J Hum Genet. 1979; 31: 106–115.

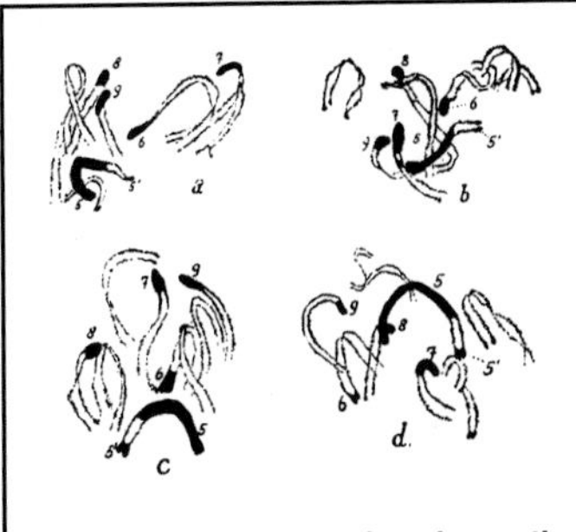

A. Heterochromatin and euchromatin

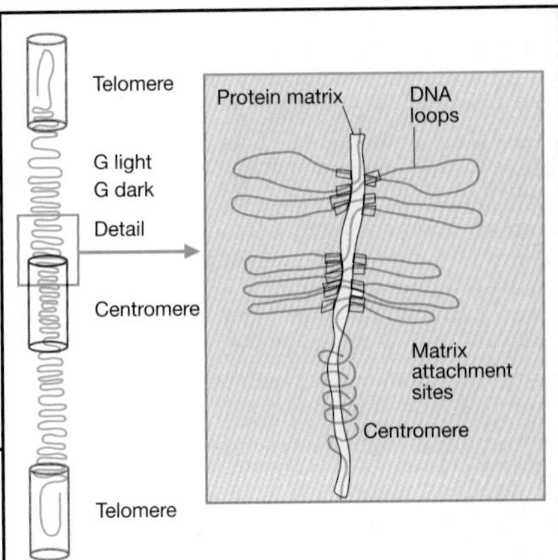

B. Special parts of a chromosome

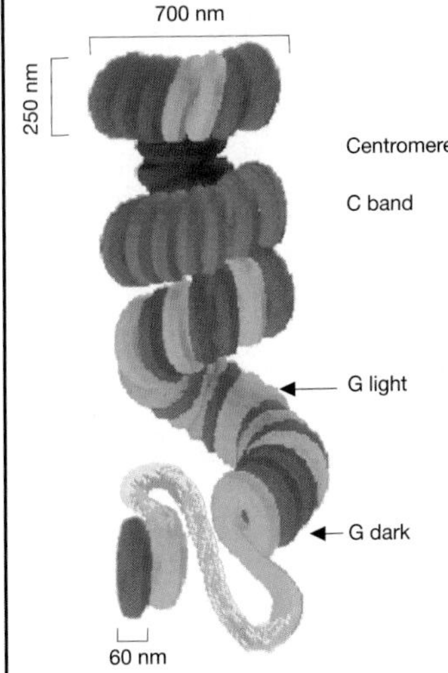

C. Model of a chromosomal segment in interphase

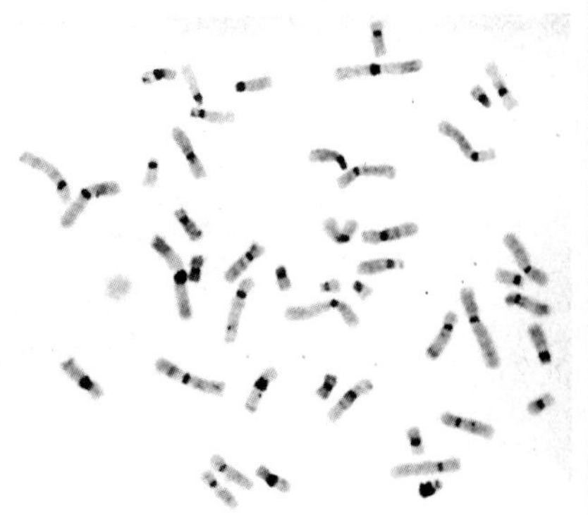

D. Constitutive heterochromatin (C bands) at the centromeres

Light G bands

GC-rich
Fluorescence with G-specific fluorochromes, e.g., olivomycin
Early replicating
Gene-rich
Alu repeats
SINE repeats (short interspersed repetitive DNA sequences)
Z-DNA conformation possible

Dark G bands

AT-rich
Fluorescence with AT-specific fluorochromes, e.g., quinacrine
Late replicating
Gene-poor
LINE repeats (long interspersed repetitive sequences)
HMG-1 nonhistone proteins bound to AT-rich areas
Minisatellites

E. Functional properties in euchromatin

Special Structure at the Ends of a Chromosome: the Telomere

The telomeres are located at both ends of eukaryotic chromosomes. The concept of telomere, introduced in 1938 by H. J. Muller, was initially applied to function. Without telomeres, chromosomes are unstable and can combine with other chromosomes to form dicentric or ring chromosomes. Recent findings demonstrate that telomeric DNA sequences differ in structure and function from other DNA sequences and that they are of fundamental biological significance.

A. Replication problem at the ends of linear DNA

Since new DNA can be formed only in the 3' direction (see p. 38), an unfilled gap should arise at each 5' end of a new double-stranded DNA molecule after the RNA primer is removed. DNA remains an unstable single strand in this region. In consecutive replication cycles (cell divisions), the new strand becomes shorter and shorter. This finally leads to the loss of functionally important sequences (genes).

B. G-rich repetitive sequences in the telomere region

A chromosome contains G-rich repetitive sequences in the telomere region. Hairpin-like folding and G-G pairing serve to stabilize the single-stranded DNA. These sequences are highly conserved in evolution, which underlines their functional importance.

C. Replication model of the telomere

Replication in the telomere region differs from that of remaining chromosomal DNA: Breaks in the C-rich sequences of the distal 3' to 5' strand (1) lead to the formation of an overhanging single strand (5' to 3')(2). Replication of the 5' to 3' strand does not include this 12 to 16 base-pair single strand (3). When replication is completed, a double strand of DNA is formed by folding back on the single strand, and the remaining gap is then closed by DNA synthesis (5). The 3' to 5' strand is replicated to the end in the regular manner (4). Then DNA replication characteristic for the telomere takes place (6). The 5' to 3' strand becomes elongated by means of an RNA-containing enzyme (RNA telomerase), which serves as template for synthesis of new DNA. Through this enzyme, the individual nucleotides of the telomere-specific repeats are added. After the 5' to 3' end has been extended, this single strand folds into a hairpinlike terminal loop (7). The result is two unshortened chromosome ends (5 and 7).

D. General structure of a telomere

In the terminal 6–10 kb of a chromosome, telomeric sequences and telomere-associated sequences can be differentiated (1). The telomere-associated sequences contain autonomously replicating sequences (ARS). The telomere sequences consist of about 250 to 1500 G-rich repeats (ca. 9 kb). They are highly conserved within a species (2) (see p. 192). The repeats differ among different classes of organisms, but the general telomere motif is similar. It consists of from one to four adenine/thymine pairs and from one to eight G/G pairs. Telomerase activity is essential for protozoans and yeast. In vertebrates it occurs mainly in germ cells, whereas somatic cells possess little or no telomerase activity. This explains the cell-division-dependent decrease of telomere length. This is viewed as being related to the aging process. Recent findings indicate that unlike normal cells, tumor cells have telomerase activity.

References

Blackburn EH. Structure and function of telomeres. Nature. 1991; 350: 569–573.

Haber DA. Telomeres, cancer, and immortality. N Engl J Med. 1995; 332: 955–956.

Zakian VA. Structure and function of telomeres. Ann Rev Genet. 1989; 23: 579–604.

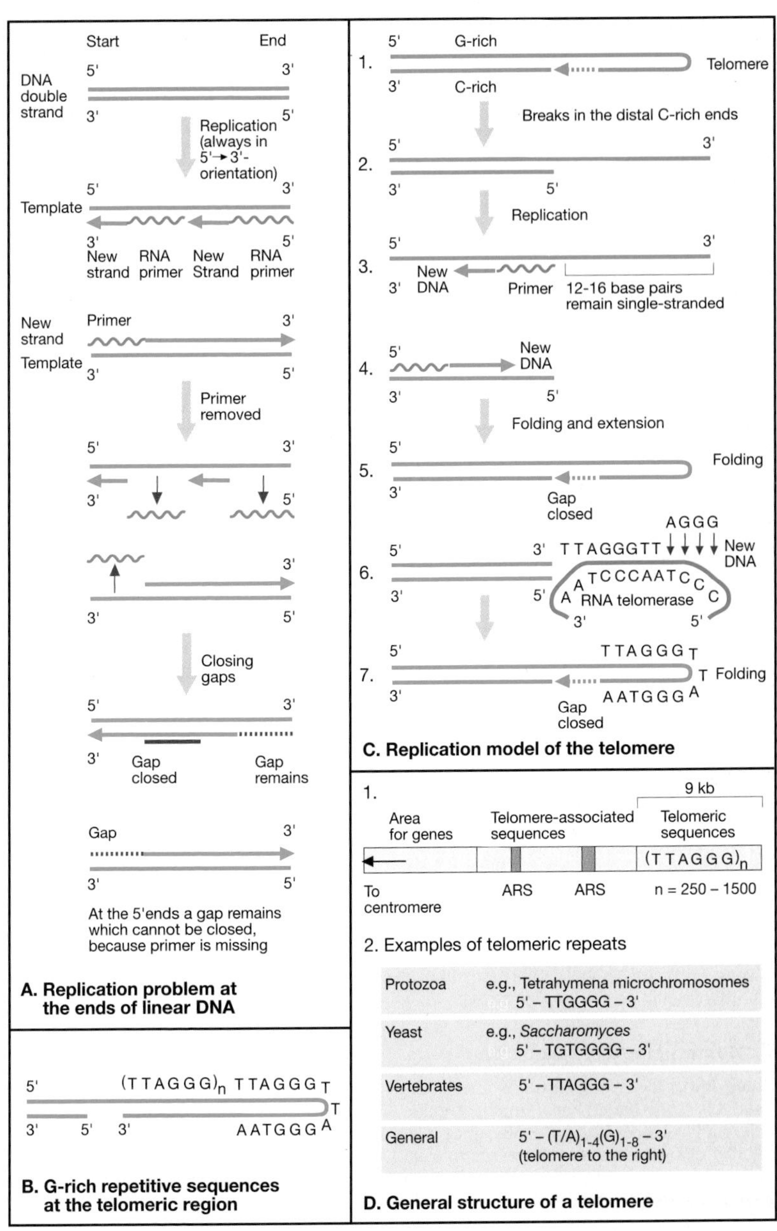

A. Replication problem at the ends of linear DNA

B. G-rich repetitive sequences at the telomeric region

C. Replication model of the telomere

D. General structure of a telomere

X-Chromosome Inactivation

Although somatic cells of female mammals contain two X chromosomes, the activities of enzymes coded for by the X chromosome are not higher in female cells than in male cells, with only one X chromosome. This is due to inactivation of most of the genes on one of the two X chromosomes. Thus, in any given cell in female mammals, genes of only one of the two X chromosomes are expressed (principle of X-chromosome inactivation, formerly the Lyon hypothesis, Mary F. Lyon, 1961).

A. X chromatin

In 1949, Barr and Bertram observed a stainable appendage in the nucleus of nerve cells of female (1 and 3) but not of male cats (2). The authors named this structure "sex chromatin." Similar structures were found in somatic cells of other female mammals; these included drumsticks in peripheral blood leukocytes (4) and small peripheral bodies in the nuclei of fibroblasts and of oral mucosa cells (5) in humans. Since each of these structures represents one of the two X chromosomes, they are now referred to as X chromatin. (Figures 1–3 from M.L.Barr & E.G. Bertram, Nature. 1949; 163: 676).

B. Dosage compensation

The activities of most of the enzymes coded for by genes on the X chromosome (A: glucose-6-phosphate dehydrogenase; B: phosphoglucomutase; C: hypoxanthine guanine-phosphoribosyl transferase, HGPRT) do not differ among individuals with different numbers of X chromosomes, since only one X chromosome is completely active.

C. Mosaic pattern of expression

Since one of the two X chromosomes is extensively inactivated in female somatic cells, only one of the two genes at a given X-chromosomal locus is expressed in each cell. As a result, heterozygotes show a mosaiclike distribution of cells expressing just one of the two alleles. In mice, X-chromosomal coat-color mutants show a mosaic of light- and dark-colored coat patches (1, after M. W. Thompson. Canad J Genet Cytol. 1965; 7: 202–213). In humans, a similar distribution of normal and absent sweat pores is seen in female heterozygotes for hypohidrotic ectodermal dysplasia. The sweat pores of hemizygotes are not developed or are extremely underdeveloped (hypohidrosis). Fingerprints of female heterozygotes show areas with normal sweat pores (black points) and areas with absent sweat pores (2a and 2b, from E. Passarge & E. Fries, Nature New Biology. 1973; 245: 58–59). In cell cultures from female heterozygotes for X-chromosomal HGPRT deficiency (HGPRT-, hypoxanthine-guanine-phosphoribosyl-transferase deficiency) the various colonies are either $HGPRT^-$ or $HGPRT^+$ (3) because each colony has been derived from a single cell (from B.R. Migeon, Am J Hum Genet. 1971; 23: 199–200, with kind permission of the author and publisher).

D. Pattern of X inactivation

One of the two X chromosomes becomes inactivated during early embryonic development. In each cell, either the maternal (X^M) or the paternal (X^P) X chromosome is inactivated. The inactivation is stable in all daughter cells and is not reversible. It results in a mosaic of paternally and maternally expressed X-chromosomal alleles, with an average 1:1 distribution. Deviation from a 1 : 1 distribution, with preferential inactivation of one particular X chromosome in more than 50% of cells, is possible.

E. Exceptions to X inactivation

Some genes on the inactivated X chromosome escape inactivation: genes on the extreme end of the short arm and at a few other loci on the short and long arms. X-chromosome inactivation is controlled by an X-inactivation center (XIC), which is expressed only on the inactive chromosome.

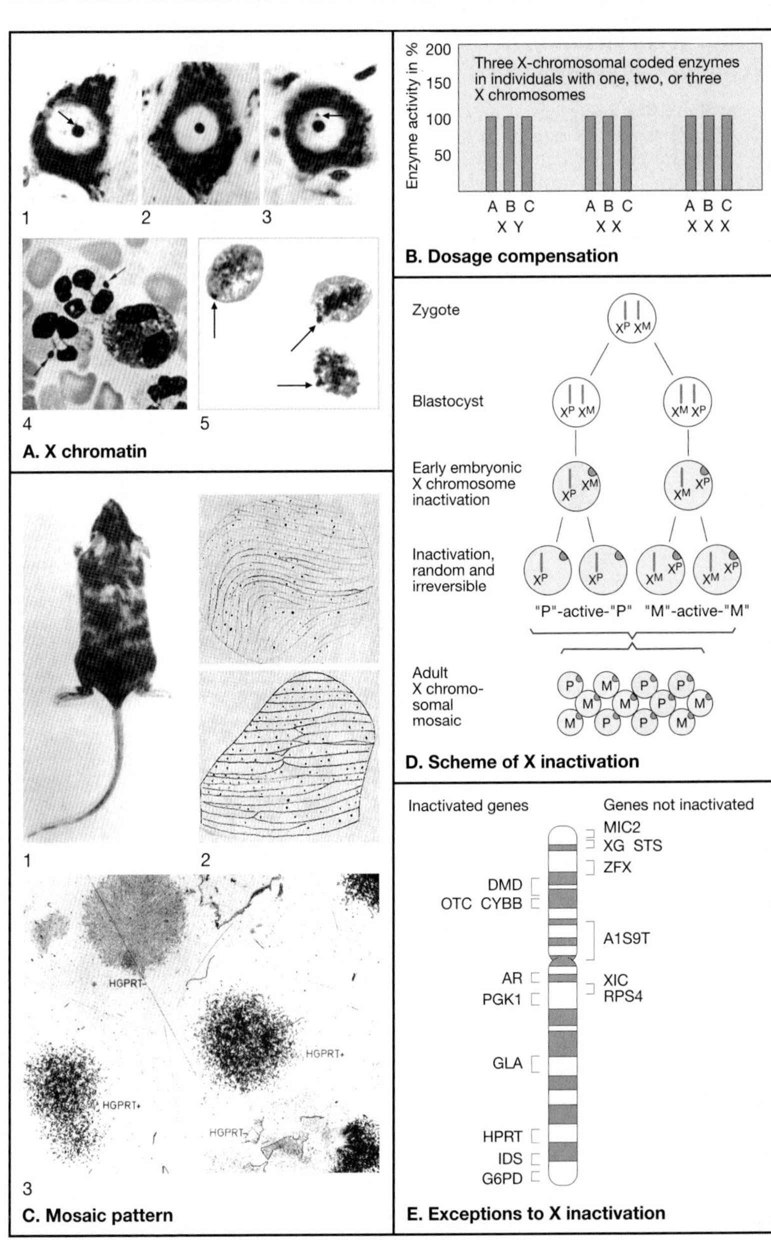

1
2
3
4
5
A. X chromatin
Enzyme activity in %
200
150
100
50
Three X-chromosomal coded enzymes in individuals with one, two, or three X chromosomes
A B C
X Y
A B C
X X
A B C
X X X
B. Dosage compensation
Zygote
Blastocyst
Early embryonic X chromosome inactivation
Inactivation, random and irreversible
"P"-active-"P"
"M"-active-"M"
Adult X chromosomal mosaic
D. Scheme of X inactivation
HGPRT-
HGPRT+
HGPRT+
HGPRT-
C. Mosaic pattern
Inactivated genes
Genes not inactivated
MIC2
XG STS
ZFX
DMD
OTC CYBB
A1S9T
AR
XIC
RPS4
PGK1
GLA
HPRT
IDS
G6PD
E. Exceptions to X inactivation

In Situ Hybridization in Metaphase and Interphase

In situ hybridization refers to procedures that demonstrate DNA sequences directly on chromosome preparations (in situ). Since resolution is relatively good (about 12 x 10^7 base pairs), the exact regional localization of a sequence on its corresponding chromosome can be determined.

A. Principle of in situ hybridization

Cells in metaphase or interphase are fixed on a slide and denatured to change the double-stranded DNA (1a) into single-stranded DNA (2). The metaphase or interphase preparation is then hybridized (3) with DNA sequences that are complementary to the region of interest and that have been labeled with biotin (1b). The hybridization site is made visible by means of a primary antibody against biotin; this antibody is bound to a fluorochrome (4), e.g., fluorescein isocyanate (FITC). Since the primary signal is quite weak, a secondary antibody (e.g., avidin) bound to biotin is attached (5). A further primary antibody can then be attached to the secondary antibody (6). This amplifies the signal, which can then be demonstrated by bright fluorescence under the light microscope.

B. Demonstration of the Philadelphia translocation in chronic myeloid leukemia (CML)

The Philadelphia translocation (1) in chronic myeloid leukemia (CML) (see p. 270) can be demonstrated in metaphase (2) and in interphase (3) by means of in situ hybridization. When a probe for the *BCR* gene is used in interphase, the normal signal consists of two fluorescing dots, one dot on each chromosome 22. (On good preparations of metaphase chromosomes, one dot is seen over each chromatid and appears as a double dot on a chromosome.) When the probe includes the breakpoint of the translocation, three signals are visible: the largest over the normal chromosome 22, a small one over the *BCR* sequences remaining in the distal long arm of a chromosome 22 (22q), and another small one over the sequences translocated to the distal long arm of chromosome 9 (photographs by kind permission of T. Cremer, Heidelberg; from: C. Lengauer et al.: Metaphase and interphase cytogenetics with Alu-PCR-amplified yeast artificial chromosome clones containing the BCR gene and the proto-oncogenes c-raf-1, c-fms, and c-erbB-2. Cancer Research. 1992; 2590–2596).

C. Translocation 4;8

This preparation shows the translocation of part of the long arm of a chromosome 8 to the short arm of a chromosome 4 in a patient with Langer–Giedion syndrome. The hybridization was done with a 170-kb YAC (yeast artificial chromosome) that spans the breakpoint of the translocation in the 8q24 region. Three fluorescent signals result: over the normal chromosome 8, over the part of 8q24 translocated to chromosome 4, and over the sequences remaining on chromosome 8. Chromosomes 4 and 8 were hybridized with alphoid probes, which are specific for the centromere region. (Preparation from H.J. Lüdecke, Essen).

D. Telomere sequences in metaphase chromosomes

This illustration shows part of a human metaphase chromosome after in situ hybridization with telomere sequences (see p. 188). Each chromosome shows four signals, one over each end (telomere) of each chromatid because the telomeric sequences are the same for all chromosomes. (Photographs from Robert K. Moyzes, Los Almos National Laboratory, with kind permission of the author. From: Scientific American, August 1991; 34–41).

Reference

Tkachuk DC, et al. Detection of *bcr-abl* fusion in chronic myelogenous leukemia by in situ hybridization. Science. 1990; 250: 559–562.

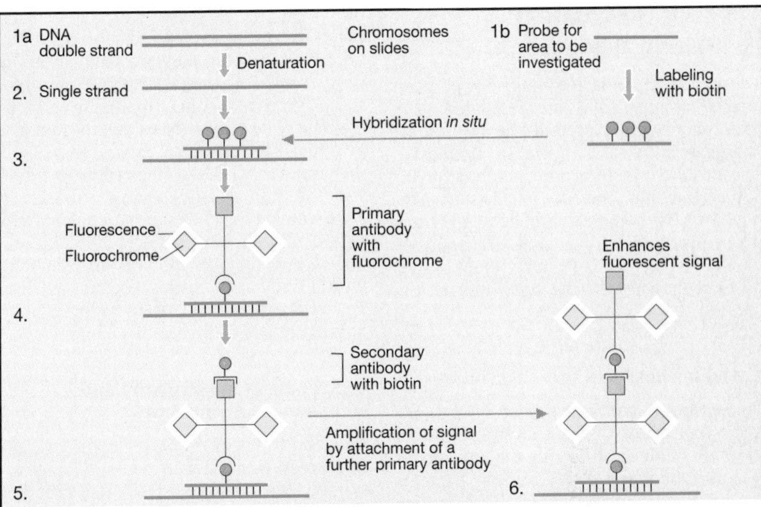

A. Principle of *in situ* hybridisation

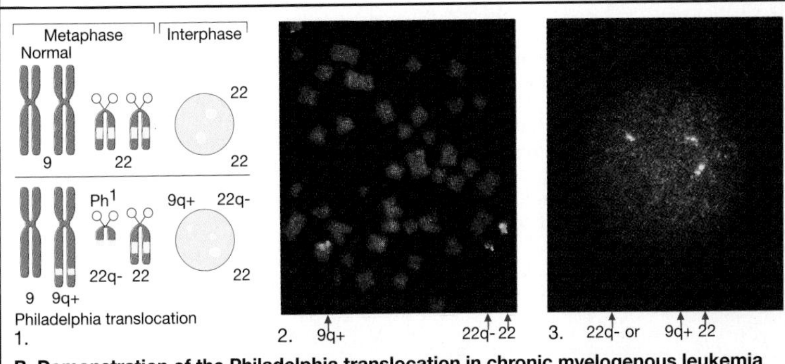

B. Demonstration of the Philadelphia translocation in chronic myelogenous leukemia

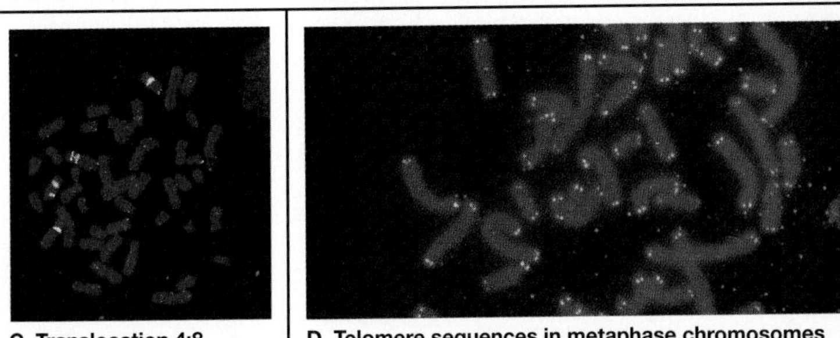

C. Translocation 4;8

D. Telomere sequences in metaphase chromosomes

The Cell Nucleus and Ribosomal RNA

The cell nucleus is the main center from which cell functions are regulated. The principal function of cells is to produce endogenous proteins (protein synthesis). The proteins, in turn, are required for innumerable vital processes, such as the catalyzation of complex biochemical reactions, production of energy, transport of molecules, etc. Cells from different tissues differ with respect to the genes that are expressed.

A. Cell nucleus and protein synthesis

Transcription and processing of the primary transcripts (RNA splicing) occur in the cell nucleus. RNA in the nucleus is bound to nuclear RNA-binding proteins for stabilization. The mature RNA is then released from the nucleus into the cytoplasm, where it serves as a template for forming polypeptides during translation. The whole process, from transcription to the completion of translation, is referred to as protein synthesis. For translation to occur, the mRNA must be bound to ribosomes. Ribosomes are complex protein structures made up of numerous subunits, which in turn are the products of individual genes (ribosomal genes).

B. Nucleolus and the synthesis of ribosomes

Ribosomes are synthesized in the nucleolus. The nucleolus is a morphologically and functionally specific region in the cell nucleus. The numerous subunits of a ribosome are formed by the transcription of a 45S rRNA precursor molecule. The term 45S refers to the size of the molecule, expressed as the sedimentation rate during ultracentrifugation (large S values indicate a large, small S values a small molecule). After the 45S rRNA precursor has been transcribed in the nucleolus, a large ribonucleoprotein particle is formed. Ribosomal proteins synthesized in the cytoplasm and nuclear RNA proteins are part of this complex. The rRNA genes lie next to each other in groups on certain chromosomal segments; in man, they are located on the short arms of the acrocentric chromosomes, the nucleolus-organizing regions (NORs). The only rRNA gene located in a different chromosomal region in man is the gene for 5S RNA. Each ribosome consists of a small and a large subunit. In eukaryotes, the small subunit contains 18S rRNA, and the large subunit 5.8S, 5S, and 28S rRNA; the 40S and 60S subunits bind with a number of further proteins to form an 80S ribosome (the sedimentation values are not additive).

C. Schema of the structure and components of ribosomes

The ribosomes of prokaryotic and eukaryotic organisms differ in size and composition of their ribosomal RNA and proteins, but are quite comparable in their basic structure. The size of a completed ribosome in prokaryotes is 70S, in eukaryotes 80S. Ribosomes consist of a large subunit (50S in prokaryotes, 60S in eukaryotes) and a small subunit (30S in prokaryotes, 40S in eukaryotes). The large subunit of prokaryote ribosomes contains 5S and 23S rRNA; that of eukaryotes contains 5S, 28S, and 5.8S rRNA. The small subunits each contain only one small rRNA. The numerous proteins of the large subunits give the ribosomes a characteristic composition and three-dimensional structure.

Reference

Alberts B, et al. Molecular Biology of the Cell. 3rd ed. 1995.

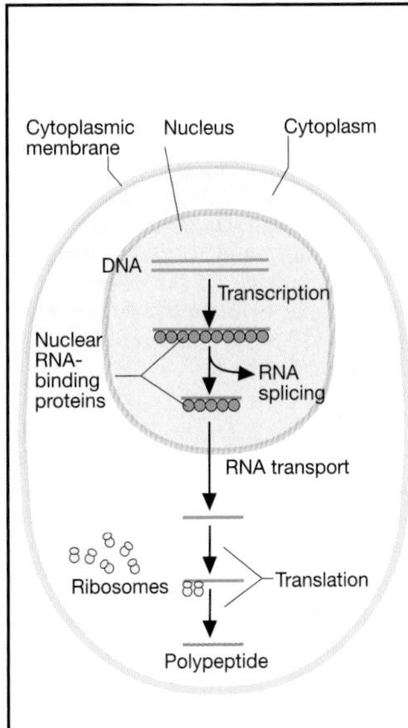

A. Nucleus and protein synthesis

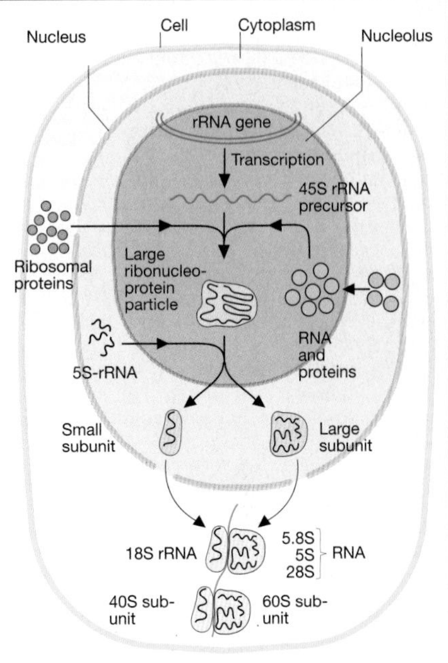

B. Nucleolus and synthesis of ribosomes

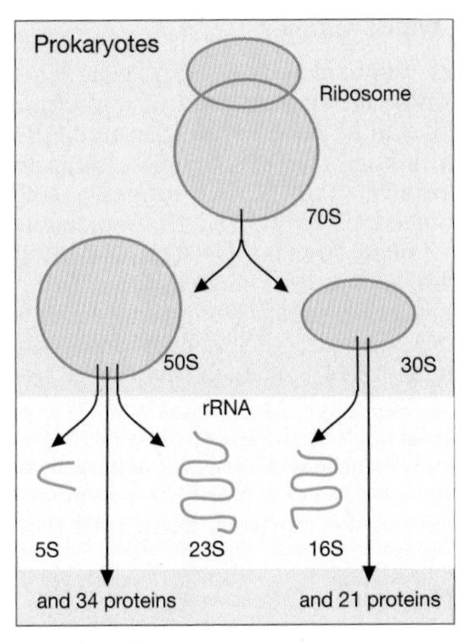

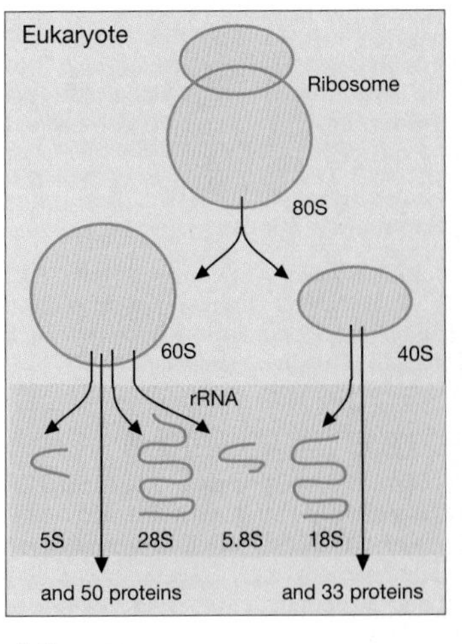

C. Overview of the structure and components of ribosomes

Transcription

The transcription of one strand of DNA into a complementary RNA molecule is the first step in gene expression. Multiple proteins (called transcription factors) form a transcription complex, which binds to DNA. Although there are differences in transcription in prokaryotes and eukaryotes, much of the basic process is the same. Transcription is catalyzed by RNA polymerase. RNA polymerase in *E. coli* has five subunits (two α, two β, and one sigma), each coded for by its own genes. RNA polymerases in eukaryotes are complex (see p. 202). Eukaryotic RNA polymerase consists of three different enzymes, which transcribe different types of genes.

A. Transcription by means of RNA polymerase

At initiation (1), RNA polymerase binds to the DNA double helix. At this point the DNA starts to unwind and RNA synthesis begins with elongation (2). As the polymerase moves along the DNA, mRNA is formed. The DNA rewinds behind the polymerase. At termination (3), the RNA polymerase is removed from the DNA. At this point, the formation of the unstable primary transcript is completed. Since it is unstable, it is immediately translated in prokaryotes and modified (processed) in eukaryotes (see p. 50).

B. Polymerase binding site

Bacterial RNA polymerase binds to a specific region of about 60 base pairs of the DNA. Several active centers can be identified (not shown here).

C. Promoter of transcription

Transcription must begin at a specific position of DNA, just upstream (at the 5' end) of a gene. This transcription initiation site is called a promoter. A promoter is a short nucleotide sequence of DNA that regulates the onset of transcription by binding to RNA polymerase. Two distinct promoter regions can be recognized above the transcription starting point. These sequences are evolutionarily highly conserved (consensus sequences). In prokaryotes, a promoter with a consensus sequence consisting of six base pairs, TATAAT (also called a Pribnow box after its discoverer) is located 10 base pairs above the starting point; another region of conserved sequences, TTGACA, is located 35 base pairs above the gene (at the 5' end). These sequences are referred to as the –10 box and the –35 box respectively (the term "box" is derived from the sequence identity or similarity in all genes). In eukaryotes, the location and the sequences of the promoters differ slightly from those of the prokaryotes (see p. 202).

D. A transcription unit

A transcription unit is all of the DNA sequences in a given segment that are used in transcription. It begins at the promoter and ends at the terminator. The region around the promoter at the 5' end is designated proximal; around the terminator at the 3' end, distal.

E. Determination of the starting point of transcription

One way to identify an active gene is to determine the starting point of transcription. This can be done by comparing the RNA formed and the DNA template. After transcription, the RNA formed (single-stranded) is hybridized to a complementary single strand of DNA (RNA/DNA hybridization). An endonuclease (S1 nuclease) that cleaves only single-stranded DNA degrades the nonhybridized single strand of DNA, while the hybridized strand is protected (RNA protection assay). Subsequently, the RNA can be removed and the transcribed DNA segment analyzed (e.g., size determined or sequenced). (Figures after Singer & Berg, 1991).

Reference

Rosenthal N. Regulation of gene expression. N Eng J Med. 1994; 331: 931–933.

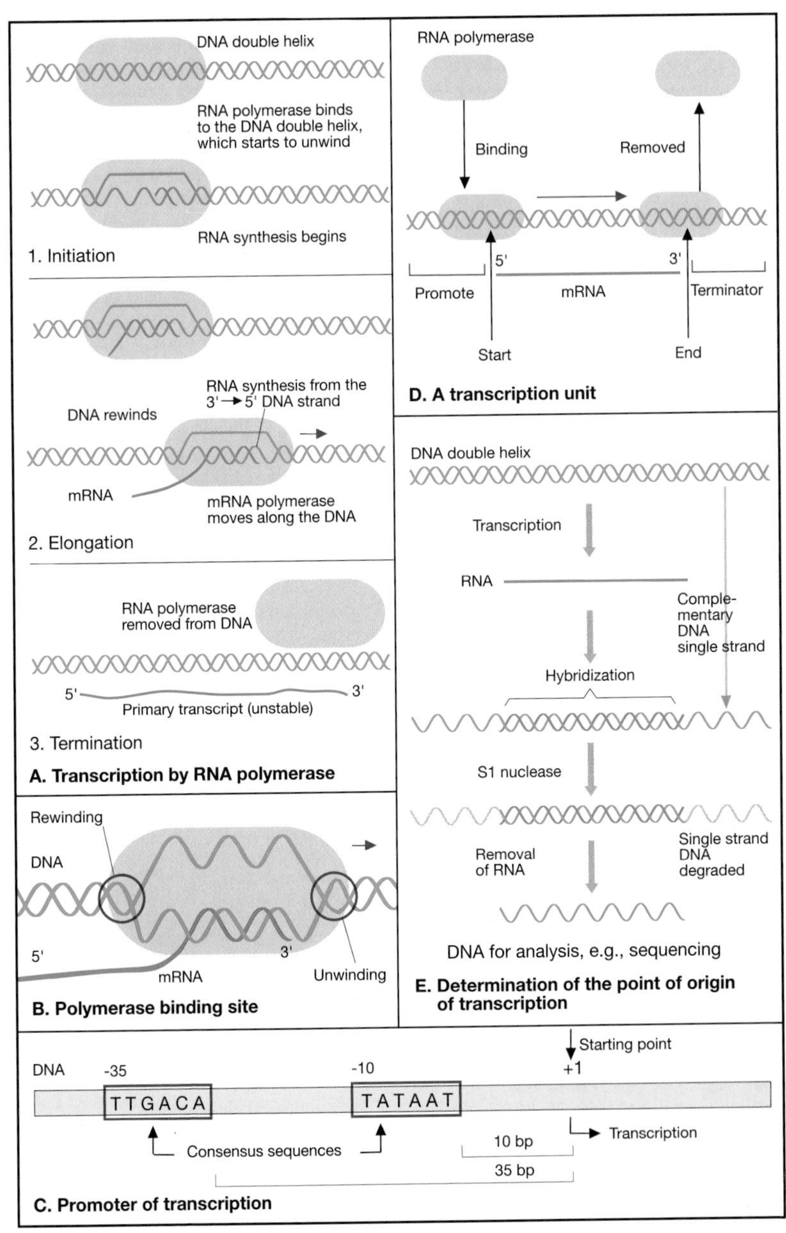

A. Transcription by RNA polymerase

B. Polymerase binding site

D. A transcription unit

E. Determination of the point of origin of transcription

C. Promoter of transcription

Control of Gene Expression in Bacteria by Induction

The regulation of gene expression is a basic function of prokaryotic and eukaryotic organisms. Prokaryotic organisms rely entirely on their ability to adapt rapidly to changes in external conditions. When available, substances usually not present in the nutrient medium must be degraded and no longer need to be synthesized by the organism itself. On the other hand, substances not present must be synthesized by the cell. The control of gene expression occurs at different levels. Regulator proteins may act as repressors (suppression of RNA polymerase activity) or as activators (inducing RNA polymerase activity). Control of prokaryotic genes is often facilitated in that functionally related genes usually lie together and therefore can be regulated together (operon) (see p. 350).

A. Induction of enzymes in bacteria

The presence of certain substances in the nutrient medium induces the synthesis of enzymes for their utilization in bacteria. An example in *Escherichia coli (E. coli)* is the activation of three enzymes for lactose catabolism by lactose. Within 10 minutes after lactose is added to the nutrient medium, the enzymes β-galactosidase, β-galactoside permease, and β-galactoside transacetylase increase manifold. Beta-galactosidase is the enzyme that splits lactose into galactose and glucose (2).

B. The lactose operon in *E. coli*

A series of genes whose regulation is coordinated is called an operon. Three structural genes that code for the synthesis of lactose-degrading enzymes (genes lac Z, lac Y, and lac A) form the lactose operon (lac operon). These three genes are regulated by a promoter at the 5' end and are transcribed by a common mRNA (polycistronic transcript). Normally they show little activity because a lac repressor inhibits lac mRNA synthesis. The former is the gene product of the lac I regulator gene. *E. coli* can use lactose as its sole source of carbon and energy because large amounts of β-galactosidase can be synthesized within a short time.

C. Control of the lac operon

The three structural genes of the lac operon, lac Z, lac Y, and lac A, are controlled by means of a repressor protein that binds to the promoter/operator region (P–O). When the repressor is bound to the P–O region, RNA polymerase cannot bind to the promoter region. Transcription is blocked (1), and the three gene products are not formed. The lac operon is activated when a β-galactoside binds to one of the subunits of the repressor (2) and inactivates it. RNA polymerase can then bind to the promoter region and transcription can begin.

D. Gene-regulating nucleotide sequences

Promoter and operator regions partially overlap. When there is a positive signal to express the lac operon, a catabolite activator protein (CAP) binds to cyclic adenosine monophosphate (cAMP); this complex binds to a specific nucleotide sequence above (in the 5' direction) the lac promoter. The RNA-polymerase binding site contains specific nucleotide sequences (promoter elements), one lying 35 nucleotide base pairs (bp) and another 10 bp 5' (upstream) from the start codon for translation of the lac Z gene. The gene-regulating effect of these nucleotide sequences was determined by observing that mutations in this region led to inactivation of the gene.

(Figures redrawn from Stryer L, Biochemistry. 3rd ed. 1989).

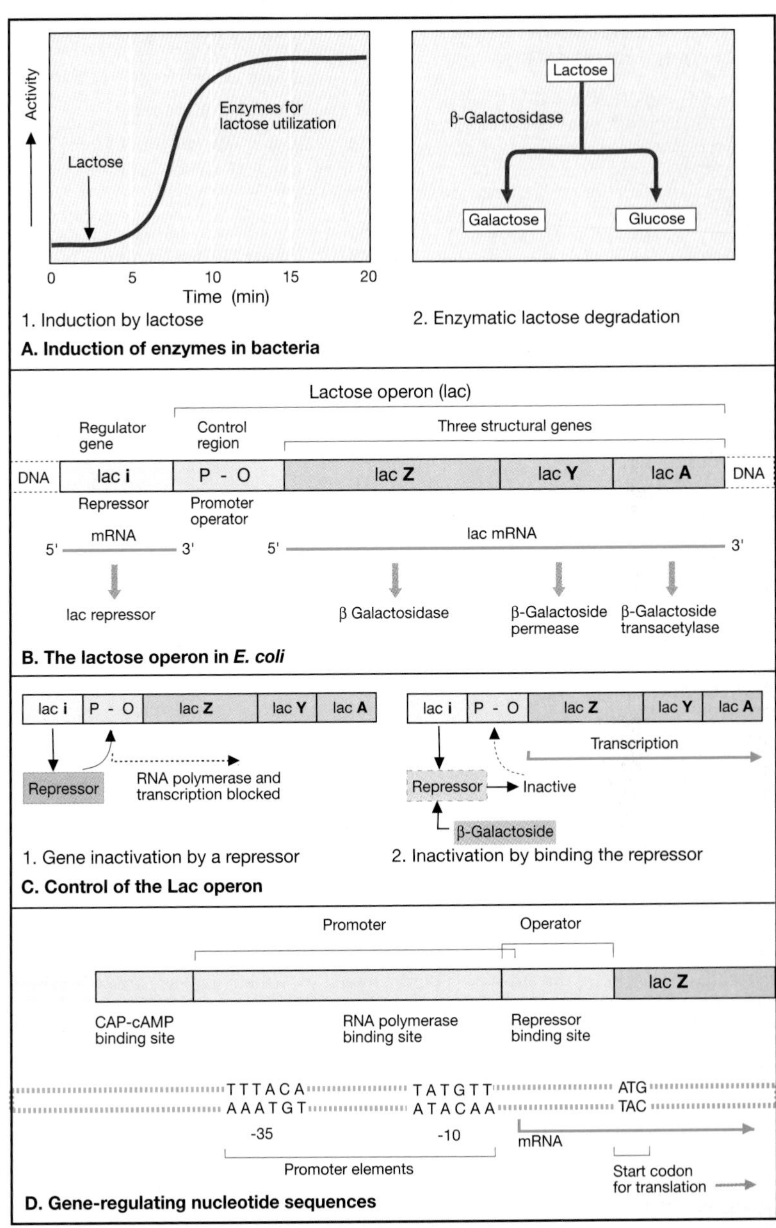
Activity
Lactose
Enzymes for lactose utilization
0 5 10 15 20
Time (min)
1. Induction by lactose
Lactose
β-Galactosidase
Galactose
Glucose
2. Enzymatic lactose degradation
A. Induction of enzymes in bacteria
Lactose operon (lac)
Regulator gene
Control region
Three structural genes
DNA
lac i
P - O
lac Z
lac Y
lac A
DNA
Repressor
Promoter operator
mRNA
5' 3'
lac mRNA
5' 3'
lac repressor
β Galactosidase
β-Galactoside permease
β-Galactoside transacetylase
B. The lactose operon in E. coli
lac i
P - O
lac Z
lac Y
lac A
Repressor
RNA polymerase and transcription blocked
1. Gene inactivation by a repressor
lac i
P - O
lac Z
lac Y
lac A
Transcription
Repressor
Inactive
β-Galactoside
2. Inactivation by binding the repressor
C. Control of the Lac operon
Promoter
Operator
lac Z
CAP-cAMP binding site
RNA polymerase binding site
Repressor binding site
TTTACA
AAATGT
TATGTT
ATACAA
ATG
TAC
-35
-10
mRNA
Promoter elements
Start codon for translation
D. Gene-regulating nucleotide sequences

Control of Gene Expression in Bacteria by Repression

If a gene is usually expressed (is active), it is said to be constitutive. Gene expression in bacteria can vary considerably depending on the presence or absence of certain substances in the nutrient medium. An important mechanism to control transcription is a signal (termination signal) that can terminate transcription or translation. It lies between the promoter and the beginning of the first structural gene and is called the attenuator (attenuation of translation).

A. Regulation of synthesis of the amino acid tryptophan in *E. coli*

Tryptophan is an essential amino acid in eukaryotic organisms. Bacteria can synthesize tryptophan, but will do so only when it is not present in the nutrient medium (1). If tryptophan is added to the medium, enzyme activity for tryptophan biosynthesis decreases within about 10 minutes (2).

B. Biosynthesis of tryptophan by means of five enzymes and five genes

Tryptophan is synthesized from chorismate via four intermediates; this occurs in five steps, regulated by five enzymes. The enzymes are coded for by five genes (tryp A–E) (CdRP = carboxyphenyl-amino-deoxyribulose-phosphate).

C. Tryptophan operon in *E. coli*

The tryptophan operon in *E. coli* consists of these five genes and their regulating sequences. The latter include promoter and operator, a leader sequence, and attenuator sequences. Translation of the five structural genes results from a continuous trp operon mRNA. In this, leader sequences coded for by L-sequence genes are connected in series. The attenuator sequences are part of the L-sequences. When tryptophan is present in the medium, translation of trp leader RNA is discontinued in the region of an attenuator sequence, before reaching the first structural gene.

D. The role of the attenuator

The weakening (attenuation) of the expression of the tryptophan operator is controlled by a sequence of about 100–140 base pairs after (in the 3' direction) the starting point of transcription (tryptophan mRNA leader). In the presence of tryptophan, the trp mRNA leader is interrupted in the region of an attenuator sequence (1), and translation does not take place. In the absence of tryptophan, translation is continued. The trp leader peptide contains two tryptophan residues (2). When tryptophan is deficient, translation is delayed and a stop signal weakened.

E. Stop signal by means of mRNA conformation

The two tandem tryptophan codons (UGG–UGG) contained in the trp operon transcript are crucial for the expression of the trp operon. If tryptophan is present, transcription progresses rapidly. In two areas, intramolecular base pairing occurs due to inverted repeats: one shortly before position 90 and the other between positions 114 and 130. The result is a hairpinlike loop with a stop codon UAA at position 141 (terminator hairpin). Thus, translation is stopped here. With tryoptophan deficiency, translation is delayed in the region of the two codons for tryptophan (UGG), and intramolecular base-pairing occurs only at one point. The terminator loop is not formed and the stop codon remains single stranded and is weakened (attenuated). Therefore translation is continued over the complete, about 7000-nucleotide-long trp mRNA. Attenuation is a frequent mechanism for regulating gene expression in bacteria.

(Figures redrawn from Stryer L, Biochemistry. 3rd ed. 1989).

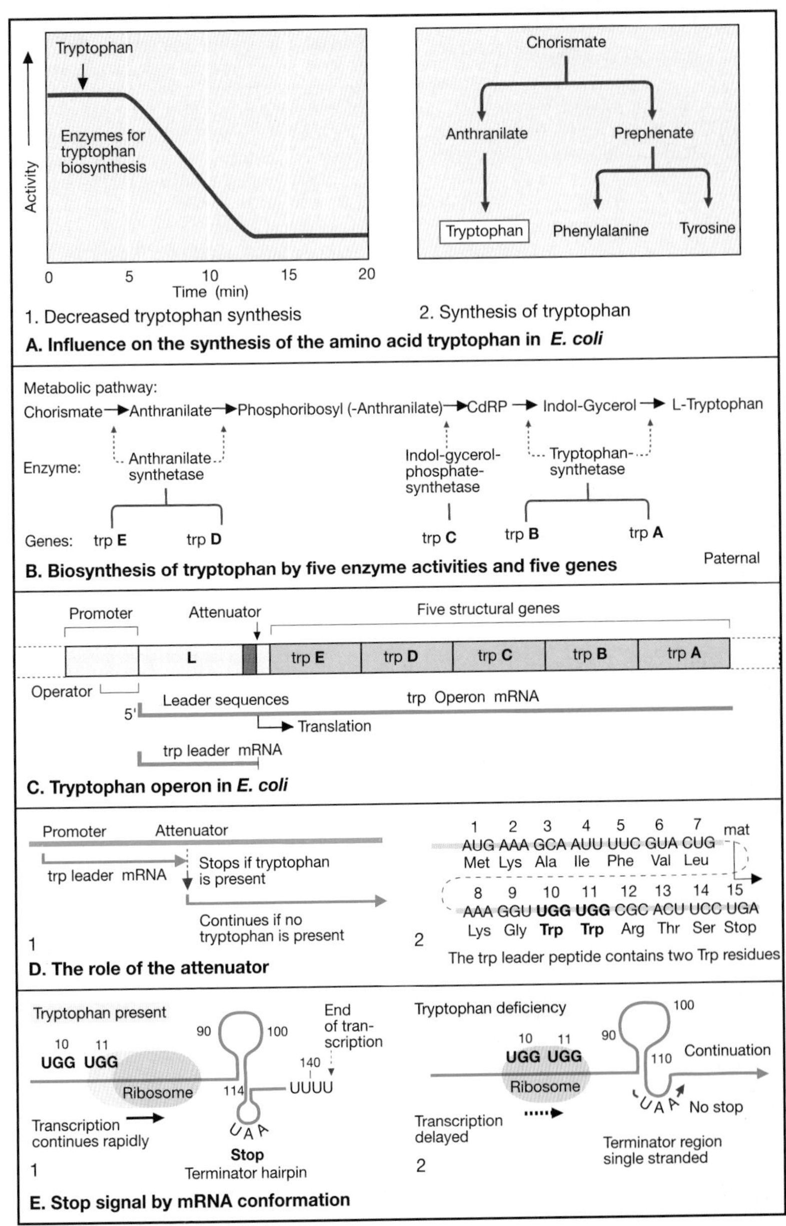

Tryptophan
Activity
Enzymes for tryptophan biosynthesis
0 5 10 15 20
Time (min)
1. Decreased tryptophan synthesis
Chorismate
Anthranilate
Prephenate
Tryptophan
Phenylalanine
Tyrosine
2. Synthesis of tryptophan
A. Influence on the synthesis of the amino acid tryptophan in E. coli
Metabolic pathway:
Chorismate → Anthranilate → Phosphoribosyl (-Anthranilate) → CdRP → Indol-Gycerol → L-Tryptophan
Enzyme:
Anthranilate synthetase
Indol-gycerol-phosphate-synthetase
Tryptophan-synthetase
Genes: trp E trp D trp C trp B trp A
B. Biosynthesis of tryptophan by five enzyme activities and five genes
Paternal
Promoter
Attenuator
Five structural genes
L
trp E
trp D
trp C
trp B
trp A
Operator
5'
Leader sequences
trp Operon mRNA
Translation
trp leader mRNA
C. Tryptophan operon in E. coli
Promoter
Attenuator
trp leader mRNA
Stops if tryptophan is present
Continues if no tryptophan is present
1
D. The role of the attenuator
1 2 3 4 5 6 7
mat
AUG AAA GCA AUU UUC GUA CUG
Met Lys Ala Ile Phe Val Leu
8 9 10 11 12 13 14 15
AAA GGU UGG UGG CGC ACU UCC UGA
Lys Gly Trp Trp Arg Thr Ser Stop
2
The trp leader peptide contains two Trp residues
Tryptophan present
10 11
UGG UGG
Ribosome
90
100
End of transcription
140
114
UUUU
UAA
Stop
Terminator hairpin
Transcription continues rapidly
1
Tryptophan deficiency
10 11
UGG UGG
Ribosome
90
100
110
Continuation
UAA
No stop
Transcription delayed
Terminator region single stranded
2
E. Stop signal by mRNA conformation

Control of Transcription

Control of the transcription of a gene is prerequisite for proper gene function. Transcription is controlled at promoters and other DNA sequences (enhancers) outside of the actual coding region. Transcription control in prokaryotes and eukaryotes differs in some respects and corresponds in others. Characteristic for the control of expression of all genes is a region lying in the 5' direction above the coding sequence, the so-called promoter region (see pp. 196, 198).

A. Promoter region

In prokaryotes, two important areas in the promoter region are 35 and 10 nucleotide base pairs above (in the 5' direction) the starting point of transcription. These sequences, the –35 box and the –10 box (Pribnow box), are called promoters. Similar sequences occur in all living organisms (see p. 196); they are of fundamental importance for gene function (2). Analysis of mutations in the region of the regulating sequences (promoter region) has shown that certain regions are extremely sensitive to base substitution (mutation), an indication of their importance. Although some mutations have only mild effects on transcription control, most have marked effects (2).

B. Transcription units

In prokaryotes, genes may occur singly, and be individually controlled (1), or several genes that code for functionally related proteins may be controlled together in an operon (2). Ribosomal genes have their own transcriptional units (3). They contain information for the synthesis of 16S ribosomal RNA (rRNA), 23S rRNA, 5S rRNA, and tRNA along with their corresponding regulatory regions. Most prokaryote transcriptional units are for polycistronic mRNA. From the latter, several corresponding functional proteins are translated together. The primary transcripts are neither modified nor processed, but instead are directly translated into several polypeptides. In contrast, the individual eukaryotic transcripts that code for multiple RNAs are specifically spliced after transcription before mature, stable RNA products are formed. The regions that code for the various RNAs contain interspersed sequences (spacer sequences) that are eliminated during the processing of the rRNAs and tRNAs.

C. RNA polymerases

The initiation of transcription by RNA polymerase is similar in eukaryotes and prokaryotes. In *E. coli*, two control regions 10 and 35 base pairs upstream (5' direction) of the transcription starting point can be distinguished (1). The 3' end of a gene contains sequences that regulate the termination of transcription (terminator). Eukaryotes have three RNA polymerases: pol I, pol II, and pol III (2). They transcribe three different classes of genes, which are provided with different specific transcription control signals. These regulatory sequences consist of relatively short DNA sequences. Each motif is a binding site for a specific protein, a transcription factor. The distance of a controlling region from the starting point for transcription varies. Polymerase I is responsible for ribosomal genes; polymerase II, for polypeptides and some RNAs; and polmerase III, for tRNAs, 5S RNAs, and some smaller RNAs. RNA polymerase II is extremely sensitive to α-amanitin, the poison of the mushroom *Amanita phalloides*.

(Figure after Singer & Berg, 1991).

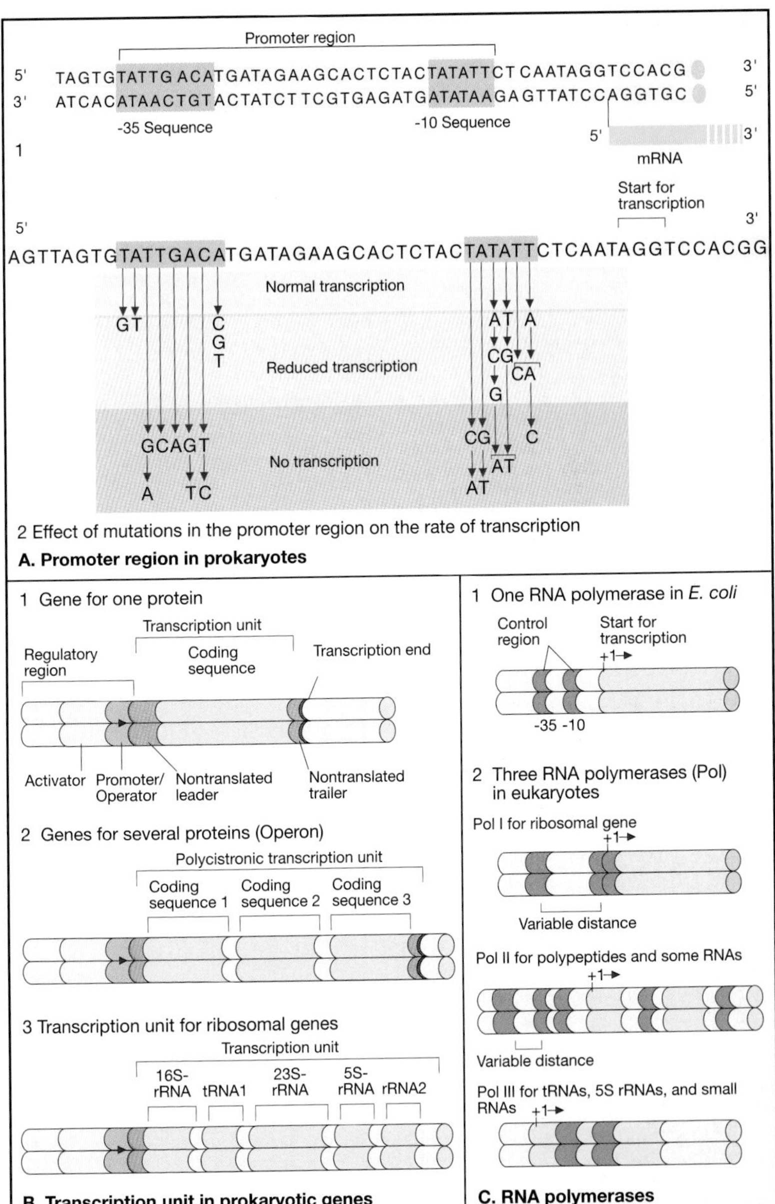
Promoter region
5' TAGTGTATTG ACATGATAGAAGCACTCTACTATATTCT CAATAGGTCCACG 3'
3' ATCACATAACTGTACTATCT TCGTGAGATGATATAAGA GTTATCCAGGTGC 5'
-35 Sequence
-10 Sequence
5' mRNA 3'
1
Start for transcription
5' AGTTAGTGTATTGACATGATAGAAGCACTCTACTATATTCTCAATAGGTCCACGG 3'
Normal transcription
Reduced transcription
No transcription
2 Effect of mutations in the promoter region on the rate of transcription
A. Promoter region in prokaryotes
1 Gene for one protein
Transcription unit
Regulatory region
Coding sequence
Transcription end
Activator
Promoter/ Operator
Nontranslated leader
Nontranslated trailer
2 Genes for several proteins (Operon)
Polycistronic transcription unit
Coding sequence 1
Coding sequence 2
Coding sequence 3
3 Transcription unit for ribosomal genes
Transcription unit
16S-rRNA
tRNA1
23S-rRNA
5S-rRNA
rRNA2
B. Transcription unit in prokaryotic genes
1 One RNA polymerase in E. coli
Control region
Start for transcription
+1
-35 -10
2 Three RNA polymerases (Pol) in eukaryotes
Pol I for ribosomal gene
+1
Variable distance
Pol II for polypeptides and some RNAs
+1
Variable distance
Pol III for tRNAs, 5S rRNAs, and small RNAs
+1
C. RNA polymerases

Transcription Control in Eukaryotes

Transcription in eukaryotes and prokaryotes differs in two main respects. In eukaryotes, one gene codes for a single polypeptide (monocistronic transcription unit), and the initial transcript is processed into mature messenger mRNA. This involves intron splicing (see p. 50) and substantial modification of the ends of the primary transcript.

A. Prototype of a eukaryotic structural gene

A structural gene is a gene that codes for a polypeptide gene product. It can be divided into sections involved in transcription (transcription unit) and regulatory sequences. Regulatory sequences are located both upstream (the 5' direction) and downstream (the 3' direction) of the gene. In addition, internal regulatory sequences may occur in introns. Some regulatory sequences are located far from the gene. Together with the promoter (see p. 196), they are required to regulate transcription.

B. Prototype of mature eukaryotic mRNA

Mature eukaryotic mRNA is produced from its precursor RNA by the removal of introns, addition of a 5' cap at the 5' end, and addition of numerous adenine nucleotides at the 3' end (polyadenylation). A noncoding sequence (5' leader) is located in front of the translation start signal (AUG), and a trailer sequence, at the 3' end in back of the translation stop signal (UAA). Both addition of the 5' cap and polyadenylation involve enzymatic reactions.

C. 7-Methyl-guanosine cap

The translation of eukaryotic mRNA is similar to that of prokaryotic mRNA, with two distinct differences: (1) transcription and translation occur at different locations in the eukaryotic cell: transcription occurs in the cell nucleus, and translation in the cytoplasm; (2) the 5'- and 3'-ends of eukaryotic mRNA have special structures. The structure at the 5' end is called a cap. Through the action of guanosine-7-methyl transferase, guanosine is bound by a triphosphate bridge to the first and second ribose groups of the precursor mRNA chain. The guanosine is methylated in position 7, as are the two initial ribose residues at the beginning of the RNA chain. Except for the mRNAs transcribed by DNA viruses, eukaryotic mRNA usually contains a single protein-coding sequence (monocistronic messenger).

D. Polyadenylation at the 3' end

Eukaryotic termination signals have been less well recognized than the regulators of gene activity at the 5' end. Eukaryotic primary transcripts are split by a specific endonuclease shortly after the sequence AAAUAA. Subsequently, about 100–250 adenine nucleotides are attached to the 3' end of the transcript by means of a poly-A-polymerase (polyadenylation). The poly-A end binds to a protein. All mRNAs, except those that code for histone proteins, possess a poly-A terminus.

(Figure after Singer & Berg, 1991).

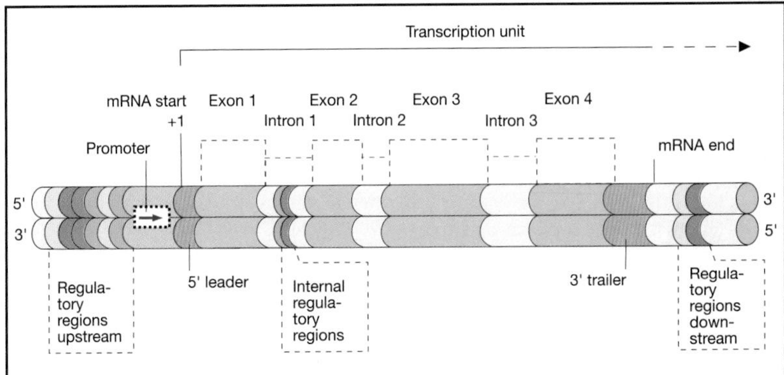

A. Prototype of a eukaryotic structural gene

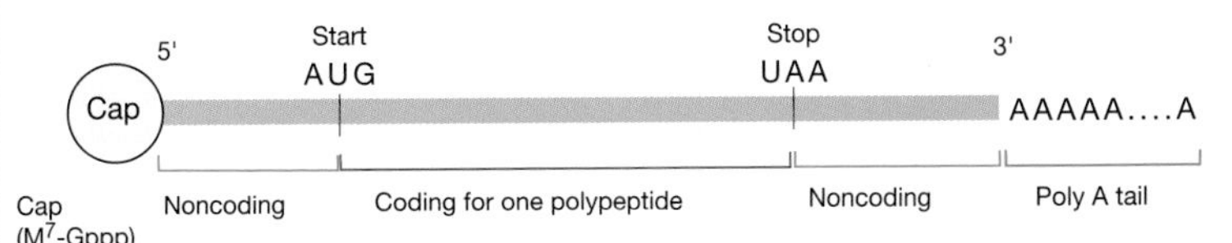

B. Prototype of a mature eukaryotic mRNA

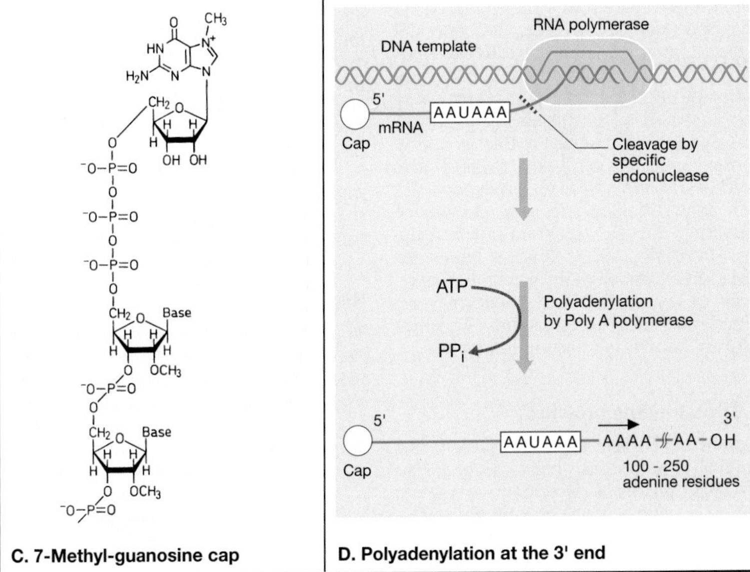

C. 7-Methyl-guanosine cap

D. Polyadenylation at the 3' end

Transcription Factors: DNA-Binding Proteins

The control of gene expression is of immediate importance for cell function and survival. In prokaryotes, it is required for functional adjustment to the given environmental conditions. In eukaryotes, the regulation of transcription is required for the differentiated development and function of different tissues. Control of gene expression is accomplished by transcription factors, proteins that interact with DNA. There are three basic groups of regulatory DNA motifs, short DNA sequences that regulate transcription: (1) sequences that establish the exact beginning of translation, (2) DNA segments that regulate the end, or termination, and (3) DNA sequences near the promoter that have specific effects on gene activity (repressors, activators, enhancers, and others).

A. General effect of transcription factors

The RNA-polymerase II activity needed to transcribe eukaryotic genes for polypeptides and some RNAs (see p. 202) does not by itself produce adequate mRNA. RNA polymerase alone leads to disorganized initiation of mRNA synthesis at several sites, and the rate of synthesis is poor (1). However, if cell nucleus extract is added, optimal mRNA synthesis results, with exact starts and sufficient production (2). Cell nuclei contain adequate amounts of activating factors (diagram after Knippers, 1990, Fig. 12.14). Since at least some of these factors do not lie in neighboring regions of a given DNA segment, they are called trans-activating factors. Together, they are known as transcription factors.

B. DNA-binding proteins

The best characterized motif for DNA binding is helix–turn–helix. In the DNA-binding domains of the protein, two neighboring α helices of the protein are stereochemically arranged so that one of the helices can form a tight bond with six bases of a major groove of DNA (recognition site). The other helix forms the stabilizing region (diagram after Singer & Berg, 1991). Here, the geometrical configuration, not the amino acid sequence, is critical.There are many transcription factors that can be differentiated according to structure and DNA-binding site. One class of transcription factors comprises proteins that bind to promoter elements of the CCAAT box ("CAT box") type. These include CAAT-binding transcription factors (CTF), CCAAT-binding proteins (CBP), enhancer and CBP-binding proteins (E/CBP), nuclear factor I (NFI), and others.

C. Transcription factors with zinc-finger motif

Some DNA-binding proteins contain characteristic structures called zinc fingers because of their zinc content and fingerlike form. A zinc atom is bound to a pair of cysteines (Cys) and a pair of histidines (His) in such a way that a fingerlike loop of about ten amino acids is formed. Their side chains can react with DNA. A number of transcription factors, such as the mammalian Sp1 factor, steroid receptor, genes for LDL-lipoprotein acceptor, estrogen receptor, and others contain zinc fingers. There are at least three types of zinc fingers; they differ in the number and position of zinc-bound cysteine and histidine residues.

(Figure A adapted from Knippers 1990; Figures B and C, from Singer and Berg 1991).

References

Tjian R, Maniatis T. Transcriptional activation: A complex puzzle with few easy pieces. Cell. 1994; 77: 5–8.

Tjian R. Molecular machines that control genes. Scientific Amer. February 1995; (Vol. 272): 38–45.

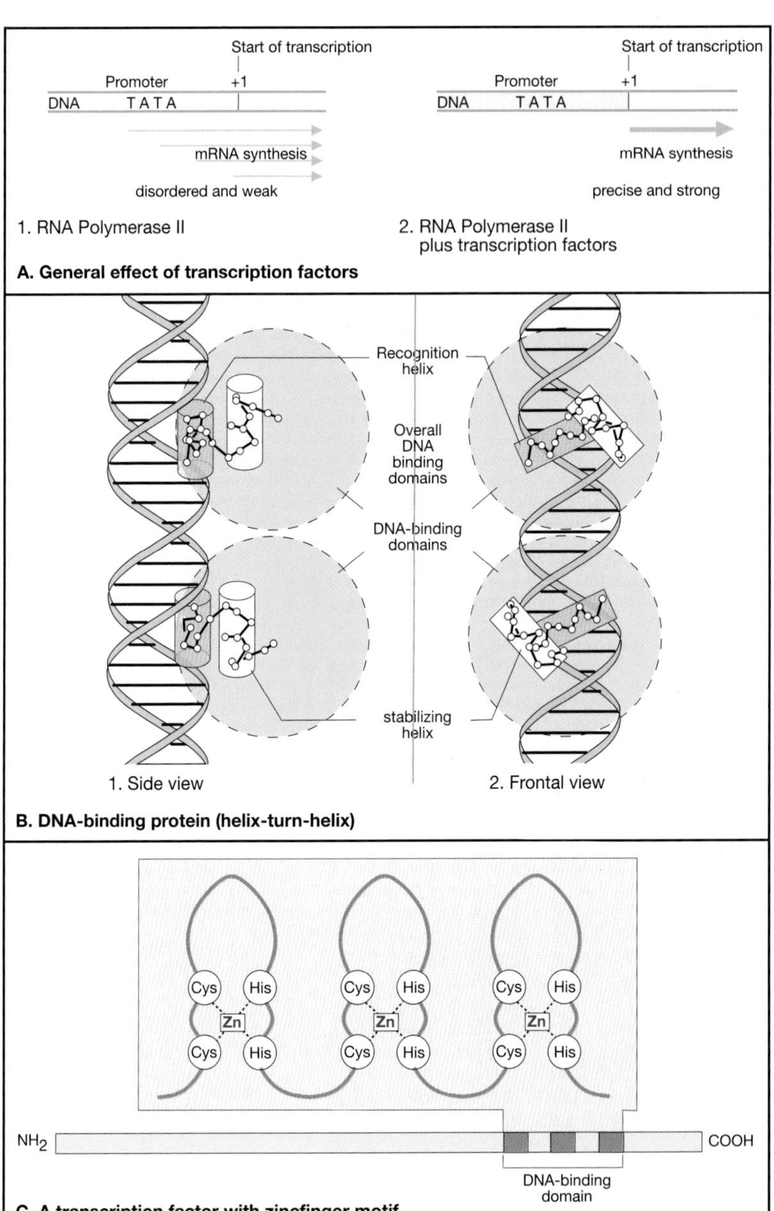

A. General effect of transcription factors

B. DNA-binding protein (helix-turn-helix)

C. A transcription factor with zincfinger motif

Transcription Factors II

A number of transcription factors, such as C/EBP (CAT/enhancer-binding protein), and *c-myc,* and *c-fos* (see p. 262) contain an unusual structural motif, the so-called leucine zipper. As the name suggests, its structure resembles that of a zipper. Some genes are controlled by specific steroid hormones.

A. Transcription factors with leucine zipper

In a leucine-zipper protein, a DNA-binding region and the leucine zipper can be distinguished (1). The essential element of a leucine zipper is the periodic repetition of four leucine residues on a relatively long α-helical segment of the protein (2). The α-helical sections are structured so that each leucine is separated from the next by seven other amino acids (amino acids 2–7). In this way, the four leucines on each side of the helix can form a type of zipper (3). The leucine-zipper domains in mammalian and yeast transcription factors are different for each gene family.

B. Activation by steroid hormone binding

Transcriptional enhancers are regulatory regions of DNA that increase the rate of transcription. Their spacing and orientation vary relative to the starting point of transcription. The enhancer is activated by binding to a hormone–receptor complex. This activates the promoter, and transcription begins (active gene). Numerous important genes in mammalian development are regulated by steroids (steroid-responsive transcription). The latter include: glucocorticoids and mineralocorticoids, the steroids of glycogen and mineral metabolism; sex hormones, which function in embryonic sex differentiation and control of reproduction; and others. Normal bone development and function are under the control of steroidlike vitamin D. Another steroidlike hormone is retinoic acid, an important regulator of differentiation during embryogenesis (morphogen). These hormones initiate their physiological effects by associating with corresponding steroid-specific transcellular receptors (hormone–receptor complex).

C. Evidence of a protein-binding region in DNA

Protein-binding regions in DNA represent regulatory areas; thus, their analysis can yield some insights into gene regulation. Protein-binding DNA regions can be demonstrated in several ways. With band-shift analysis (1), protein-bound and non-protein-bound DNA fragments are differentiated using gel electrophoresis. In an electrical field where speed of migration is according to size, a DNA fragment that is part of a DNA–protein complex migrates more slowly than a free DNA fragment of the same size. The DNA–protein complex is found at a different position ("band shift"). DNA footprinting (2) is another procedure for identifying protein-binding sites on DNA. The principle of DNA footprinting is that a protein-bound DNA region, e.g., the polymerase–promoter complex, is protected from the effects of a DNA-cleaving enzyme (DNAse I). Previously isolated DNA is cut up into different fragments by DNAse, and the fragments are sorted according to size by gel electrophoresis. Since the DNA protein-binding region is protected from DNAse cuts (DNAse I protection experiment), DNA bands from the binding region are missing ("footprint").

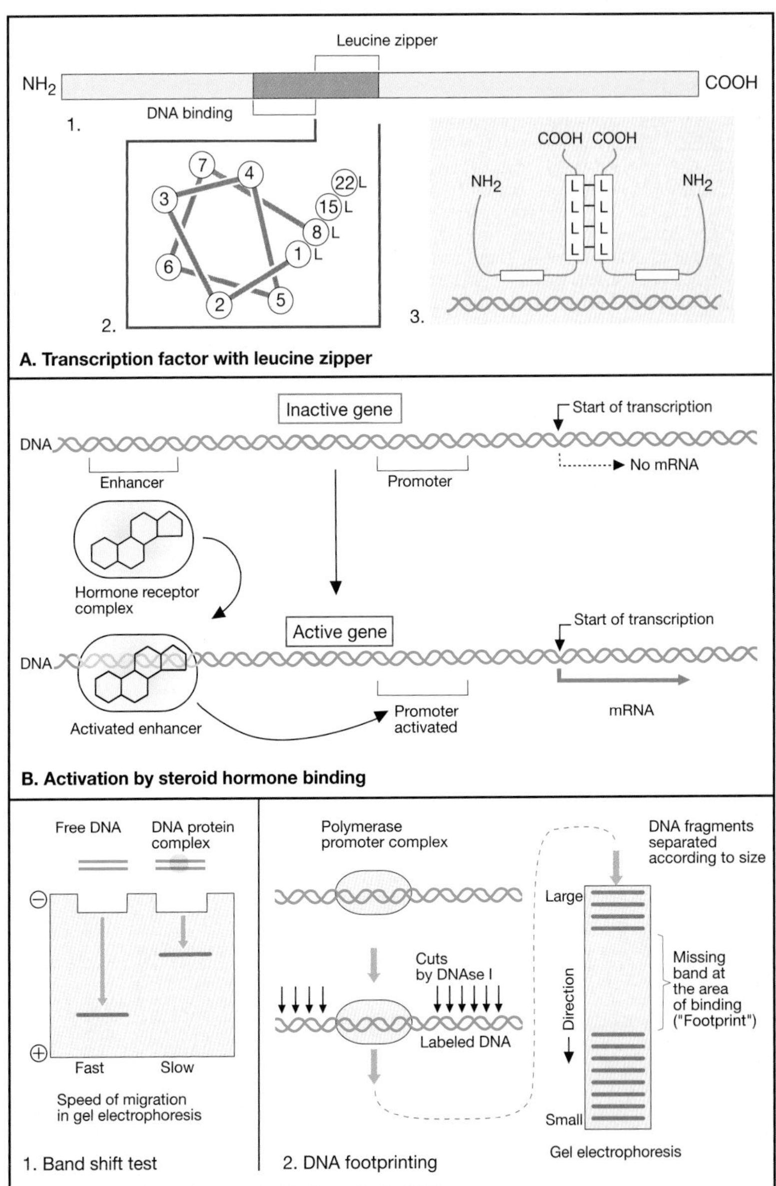

A. Transcription factor with leucine zipper

B. Activation by steroid hormone binding

C. Demonstration of a protein-binding site in DNA

Inhibitors of Transcription and Translation

A number of natural and artificial substances are able to inhibit transcription or translation. They can be used to treat cancer or as antibiotics to treat infections. Although most substances are unspecific and not suitable for therapeutic purposes, some are very specific and therefore important for the understanding of transcription and translation or for therapy. Basically, one can distinguish whether an agent interferes with transcription or with translation.

A. Insertion of actinomycin D between a GC base pair

Actinomycin D is a complex polypeptide produced by a species of streptomyces bacteria. It consists of a phenoxazone ring with two symmetrical side chains (1). It acts by becoming intercalated between two neighboring GC base pairs in double-stranded DNA. Viewed from the side (2), the inserted actinomycin D molecule is seen very distinctly within the DNA double helix. In the view from above (3) the actinomycin D molecule forms a narrow layer within the DNA double helix, bound by the two neighboring GC base pairs. The degree of inhibition by actinomycin D varies greatly. High concentrations of actinomycin D block replication, whereas low concentrations suffice to inhibit transcription. (Figure from Singer & Berg, 1991)

B. Puromycin imitates an aminoacyl tRNA

Puromycin, a polypeptide from *Streptomyces alboniger*, blocks polypeptide synthesis in the ribosomes of pro- and eukaryotes. Its action is based on the structural similarity with an aminoacyl tRNA. An aminoacyl tRNA is a tRNA molecule with an amino acid attached to its 3' end. Normally a peptide bond is formed by peptidyl transferase between the amino group of the incoming aminoacyl tRNA at the A (aminoacyl) position and the carboxyl group of the peptidyl tRNA at the P (peptidyl) position. The structure of puromycin resembles that of aminoacyl tRNA, but since the amino acid is bound via an N atom, puromycin cannot be attached to the A position in the ribosome. The resulting polypeptidyl–puromycin adduct is removed from the ribosome, and the protein synthesis ends prematurely.

C. Inhibitors of protein synthesis

Numerous naturally occurring and artificially produced substances inhibit protein synthesis by inhibiting transcription or certain phases of translation. Some have clinical significance as antibiotics; others are toxicologically significant. An example for the specificity of some inhibitors is α amanitin, a dicyclic octapeptide of the fungus (mushroom) *Amanita phalloides*. In very low concentrations, it binds to RNA polymerase II and thereby blocks the formation of precursor mRNA in eukaryotes. In contrast, RNA polymerase I is insensitive to this toxin, and polymerase III binds to it only in higher concentrations.

(Data after Singer & Berg, 1991).

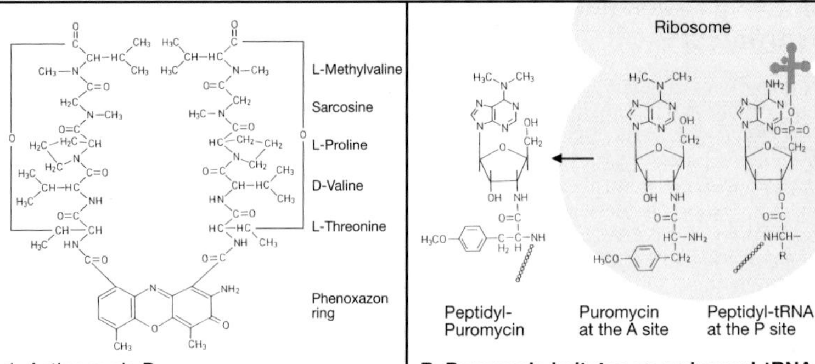

1. Actinomycin D

B. Puromycin imitates an aminoacyl-tRNA

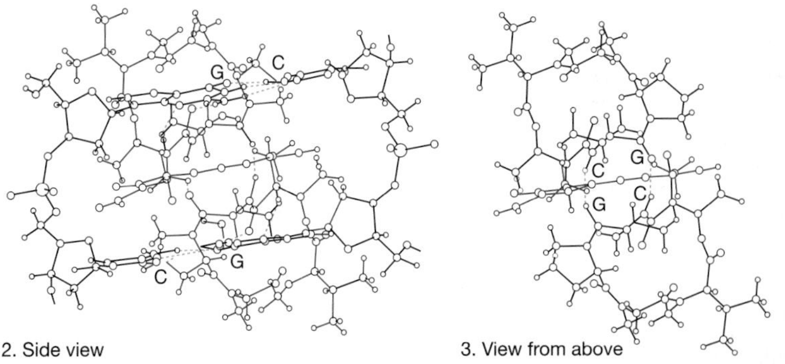

2. Side view

3. View from above

A. Intercalation of actinomycin D between a GC base pair

In Prokaryotes (Examples) :	
Actinomycin	Binding between adjacent G - C bases in DNA (Intercalation)
Chloramphenicol	inhibits peptidyltransferase of the 70S ribosome
Erythromycin	binds to the 50S particle and arrests synthesis of the 70S ribosome
Neomycin	binds to the 30S ribosomal subunit and inhibits binding of a tRNA
Puromycin	premature chain termination
Rifamycin	inhibits RNA synthesis by binding to the β subunit of the RNA polymerase holoenzyme
Streptomycin	as erythromycin
Tetracycline	inhibits binding of tRNA to the 30S ribosomal subunit

In Eukaryotes (Examples) :	
α-Amanitin	inhibits polymerase II
Chloramphenicol	inhibits peptidyltransferase of mitochondrial ribosomes
Cycloheximide	inhibits peptidyltransferase
Diphteria toxin	inhibits initiation factor 2 and translocation

C. Inhibitors of protein synthesis

Gene Expression and DNA Methylation

The expression of genes in vertebrates is often correlated with the degree of methylation of cytosine within and near the gene-regulating regions. As a rule, an active gene is associated with decreased or absent methylation in these regions (hypomethylation). Unexpressed gene regions contain relatively many methyl groups (hypermethylation). DNA methylation may be stable for many generations of cell division.

A. Inactive DNA is methylated in cytosine-rich sequences

Methyl groups in eukaryotic DNA easily attach to cytosine residues at position 5, giving rise to 5-methylcytosine (5-meC). Therefore, methylation is found predominantly in CG-rich sequences. 5-Azacytosine, a pyrimidine base analog of 5-azacytidine, cannot be methylated at position 5 because this position contains a nitrogen atom instead of a carbon atom. Thus, methylation can be reversed by incorporating 5-azacytosine. However, gene expression is not generally regulated by methylation. This follows from the absence of methylated DNA in *Drosophila*.

B. Methylation-sensitive restriction enzyme

The methylation pattern of DNA can be determined by using a methylation-sensitive restriction enzyme. Such an enzyme can cleave DNA only when the recognition sequence is not methylated. When it is methylated, the recognition site is abolished and no cut results.

C. GC-rich segments in DNA

About 25–75% of DNA contains C-rich segments. Here, methylation is especially frequent.

D. Recognition of a methylated DNA segment

A methylated DNA segment can be recognized in a Southern blot. The restriction endonuclease *Hpa*II cleaves its recognition sequence CCGG only if it is not methylated. *Msp*I cuts the same sequence both when it is methylated and when it is unmethylated. If a given DNA segment contains two CCGG recognition sequences, one of which is methylated, three fragments are formed after the action of *Msp*I, but only two after the action of *Hpa*II. The methylated recognition site cannot be cleaved by *Hpa*II; therefore, only one fragment results in this area.

E. Nonmethylated GC-rich segment at the 5' end of an expressed gene

Most of the CG dinucleotides in vertebrate DNA are highly methylated. About 1% of vertebrate DNA yields multiple small fragments of, on average, 120-base-pair CpG islands (cytosine–phosphate–guanine, CpG) after digestion with the restriction endonuclease *Hpa*II. Since *Hpa*II is a methylation-sensitive enzyme, these must be nonmethylated DNA fragments (HTF, *H*paII *t*iny *f*ragments). These nonmethylated islands are located at the 5' ends of expressed genes. A CpG island is a segment of CG dinucleotides about 1–2 kb long. Mammalian genomes contain about 30 000 such CpG islands, separated from each other by an average of about 100 kb. Their presence is generally assumed to be an indication of the beginning of a structural gene. However, some genes do not contain CpG islands, such as those for the β-globin family, growth hormone, insulin, α interferon, and others.

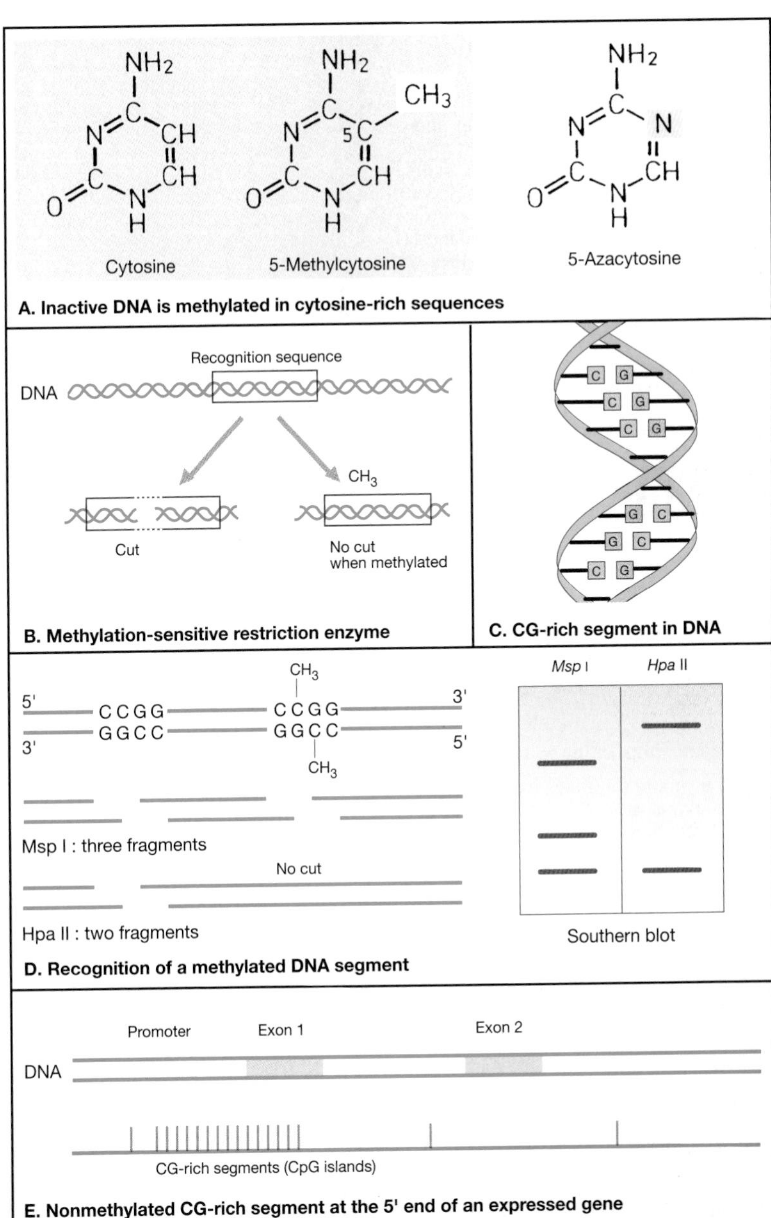
NH2
CH3
N
CH
C
O
H
5
Cytosine
5-Methylcytosine
5-Azacytosine
A. Inactive DNA is methylated in cytosine-rich sequences
Recognition sequence
DNA
CH3
Cut
No cut
when methylated
B. Methylation-sensitive restriction enzyme
C G
G C
C. CG-rich segment in DNA
5'
3'
C C G G
G G C C
Msp I : three fragments
No cut
Hpa II : two fragments
Msp I
Hpa II
Southern blot
D. Recognition of a methylated DNA segment
Promoter
Exon 1
Exon 2
DNA
CG-rich segments (CpG islands)
E. Nonmethylated CG-rich segment at the 5' end of an expressed gene

Genomic Imprinting

Certain regions of the maternal and paternal genomes are not functionally equivalent. Gene loci in such regions may show differences in activity with respect to their parental origin. Thus, for certain genes, the paternal and maternal genomes contribute to the developing embryo in different ways. This parent-of-origin specific difference in regulation of a gene locus is referred to as genomic imprinting.

A. Uniparental disomy

Uniparental disomy (UPD) is the presence of two homologous chromosomes (or chromosome segments) from only one parent and the absence of the corresponding chromosome from the other parent. For example, a mouse that received two chromosomes 11 from its mother will be smaller than normal. If it received two chromosomes 11 from its father, it will be larger than normal. The term isodisomy refers to two copies of the same parental chromosome; heterodisomy means that one each of two homologous chromosomes comes from one parent. Uniparental disomy has been shown to cause abnormalities in man (see p. 340).

B. Genomic imprinting

Genomic imprinting may involve a gene locus of either maternal or paternal origin. In this example, imprinting of a locus on one of the maternal chromosomes is shown. In somatic cells of both females and males, the paternal locus on homologous chromosomes is active, whereas the maternal locus is inactive. In germ cells, the state of imprinting is reset. Thus, the zygote will have the normal pattern in imprinting. Genomic imprinting and uniparental disomy have been demonstrated for several diseases in humans, such as Prader–Willi syndrome and Angelman syndrome on chromosome 15q (see p. 350).

C. Parent-of-origin differences in methylation pattern

One explanation for genomic imprinting is that it is a result of differences in methylation. If a gene of maternal origin is methylated (inactive), but that of the paternal is not, only the paternal gene will be expressed. The imprinting is reset in the germ line. In this case, both homologous genes become methylated in the female germ line, and both become demethylated in the male germ line. The result is an allelic difference in methylation pattern of an imprinted gene in the zygote.

D. Imprinting of distal chromosome 7 in mouse embryos

Genomic imprinting has been especially well demonstrated for the distal third of mouse chromosome 7. If the distal thirds of both chromosomes 7 in mouse embryos come from the father, the mice develop larger than normal controls. On the other hand, if the distal thirds of both chromosomes 7 come from the mother, the embryos are too small and die on the 16th day of development. This is due to differences in imprinting of the gene for insulin-like growth factor 2 (*Igf2*), which lies in this region and is important in embryonal development. The distal segment of mouse chromosome 7 is homologous with the distal third of the short arm of chromosome 11 in humans (11p15). In humans, uniparental disomy in this area predisposes to a syndrome characterized by overgrowth and an increased risk to develop a particular type of cancer of the kidney (Wilms tumor) (Beckwith–Wiedemann syndrome).

(Figure after Ferguson-Smith et al. 1991).

References

Ferguson-Smith AC, et al. Embryological and molecular investigations of parental imprinting on mouse chromosome 7. Nature. 1991; 351: 667.

Nicholls RD. New insights reveal complex mechanisms involving genomic imprinting. Am J Hum Genet. 1994; 54: 733–740.

Wilson GN, Hall JG, Cruz de la F. Genomic imprinting: Summary of an NICHD conference. Am J Med Genet. 1993; 46: 675–680.

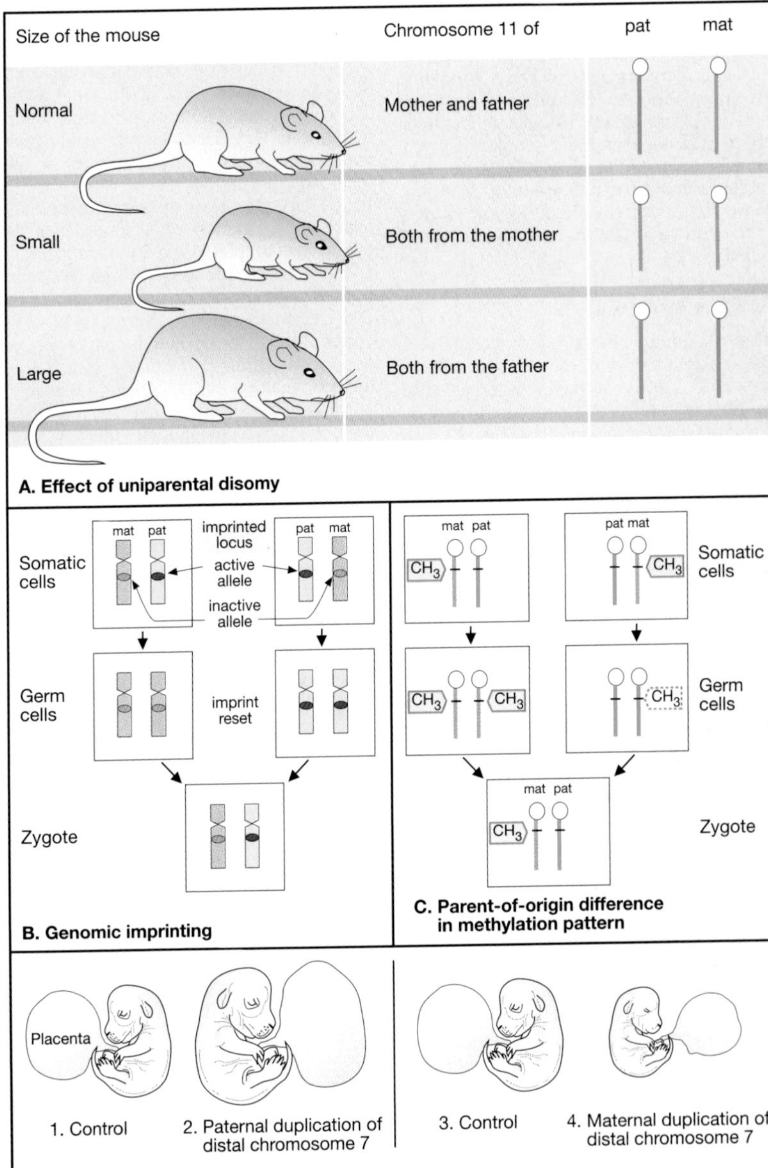

A. Effect of uniparental disomy

B. Genomic imprinting

C. Parent-of-origin difference in methylation pattern

1. Control
2. Paternal duplication of distal chromosome 7
3. Control
4. Maternal duplication of distal chromosome 7

D. Imprinting of the distal part of chromosome 7 in mouse embryos

Alternative DNA Structures

Gene expression and transcription can be influenced by changes of DNA topology. However, this type of control of gene expression is relatively universal and nonspecific. Thus, it is more suitable for permanent suppression of transcription, e.g., in genes that are expressed only in certain tissues or are active only during the embryonic period and later become permanently inactive.

A. Three forms of DNA

The DNA double helix does not occur as a single structure, but rather represents a structural family of different types. The original classic form, determined by Watson and Crick in 1953, is B-DNA. The essential structural characteristic of B-DNA is the formation of two grooves, one large (major groove) and one small (minor groove). There are at least two further, alternative forms of the DNA double helix, Z-DNA and the rare form A-DNA. While B-DNA forms a right-handed helix, Z-DNA shows a left-handed conformation. This leads to a greater distance (0.77 nm) between the base pairs than in B-DNA and a zigzag form (thus the designation Z-DNA). A-DNA is rare. It exists only in the dehydrated state and differs from the B form by a 20-degree rotation of the perpendicular axis of the helix. A-DNA has a deep major groove and a flat minor groove.

B. Major and minor grooves in B-DNA

The base pairing in DNA (adenine–thymine and guanine–cytosine) leads to the formation of a large and a small groove because the glycosidic bonds to deoxyribose (dRib) are not diametrically opposed. In B-DNA, the purine and pyrimidine rings lie 0.34 nm apart. DNA has ten base pairs per turn of the double helix. The distance from one complete turn to the next is 3.4 nm. In this way, localized curves arise in the double helix. The result is a somewhat larger and a somewhat smaller groove.

C. Transition from B-DNA to Z-DNA

B-DNA is a perfect regular double helix except that the base pairs opposite each other do not lie exactly at the same level. They are twisted in a propellerlike manner. In this way, DNA can be easily bent without causing essential changes in the local structures.

In Z-DNA the sugar–phosphate skeleton has a zigzag pattern; the single Z-DNA groove has a greater density of negatively charged molecules. Z-DNA may occur in limited segments in vivo. A segment of B-DNA consisting of GC pairs can be converted into Z-DNA when the bases are rotated 180 degrees. Normally, Z-DNA is thermodynamically relatively unstable. However, transition to Z-DNA is facilitated when cytosine is methylated in position 5 (C5). The modification of DNA by methylation of cytosine is frequent in certain regions of DNA of eukaryotes. There are specific proteins that bind to Z-DNA, but their significance for the regulation of transcription is not clear.

References:

Stryer L. Biochemistry. 3rd ed. 1988.

Watson JD, et al. Molecular Biology of the Gene. 3rd ed. 1987.

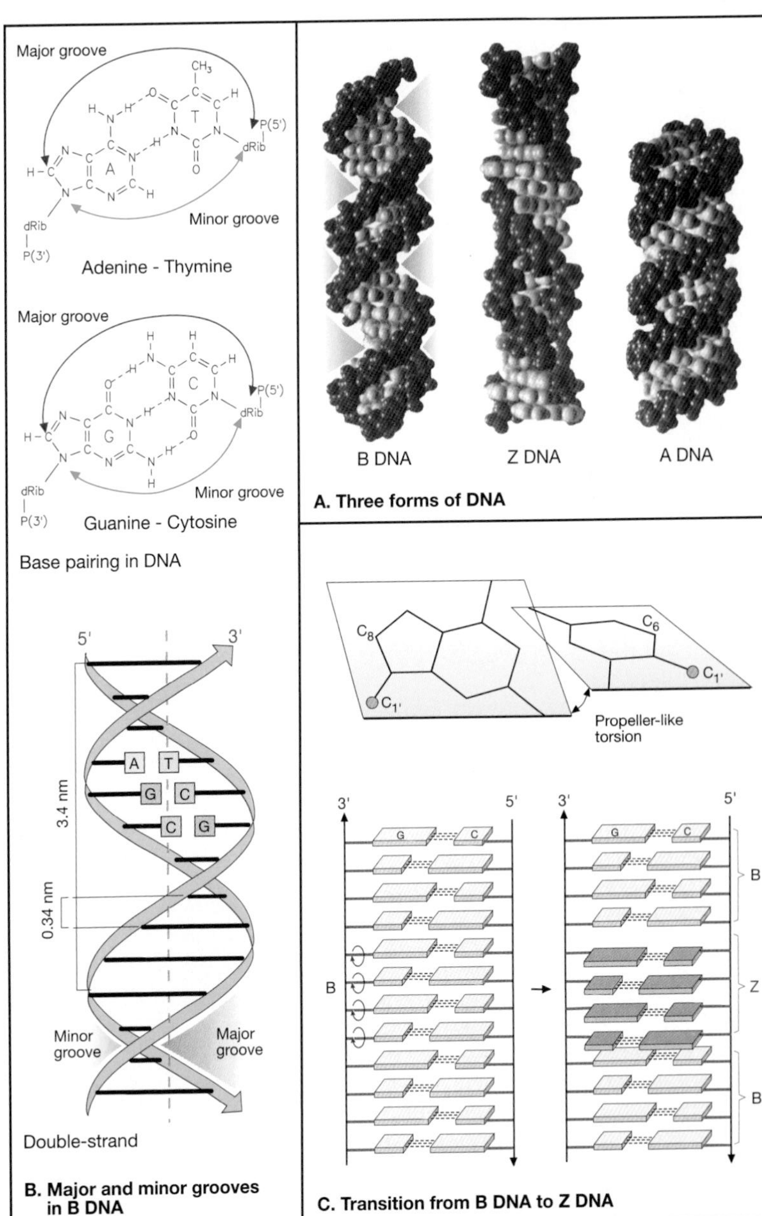
Major groove
Minor groove
dRib
P(5')
P(3')
Adenine - Thymine
Major groove
Minor groove
Guanine - Cytosine
Base pairing in DNA
5'
3'
3.4 nm
0.34 nm
A T
G C
C G
Minor groove
Major groove
Double-strand
B. Major and minor grooves in B DNA
B DNA
Z DNA
A DNA
A. Three forms of DNA
C8
C6
C1'
Propeller-like torsion
3'
5'
G C
B
Z
C. Transition from B DNA to Z DNA

Genetics and Medicine

Types of Cell Surface Receptors

Specific receptors on cell surfaces (and in the nucleus or cytosol) convey cell-to-cell signals into the cells and the functional answers. The basic structures of their genes are similar because they have been derived from a relatively small group of ancestral genes. How they bind to the ligand (the signal-releasing molecule) and the functional answer of the cell are specific. When a ligand binds to a receptor, a series of reactions is initiated that alters the function of the cell. Receptors with direct and indirect ligand effects can be distinguished.

A. Cell surface receptors with direct ligand effect

Many hormones cannot pass through the plasma membrane; instead, they interact with cell surface receptors. Their effects are direct and very rapid. With ligand-activated (or ligand-gated) ion channels (1), binding of the ligand to the receptor changes the conformation of the receptor protein. This causes an ion-specific channel in the receptor protein to open. The resulting flow of ions changes the electric charge of the cell membrane. Receptors with ligand-activated protein kinase (2) further activate a substrate protein. Most protein kinases phosphorylate tyrosine (tyrosine kinase), serine, or threonine by transferring a phosphate residue from adenosine triphosphate (ATP), which is then converted to adenosine diphosphate (ADP). Other receptors mediate the removal of phosphate from a phosphorylated tyrosine side chain by means of their phosphatase activity (3). With one important type of receptor, ligand binding activates guanylate cyclase (4), which catalyzes the formation of cyclic guanosine monophosphate (cGMP) from guanosine triphosphate (GTP). The cGMP functions as a second messenger and brings about a rapid change of activity of enzymes or nonenzymatic proteins. Removal or degradation of the ligand reduces the concentration of the second messenger and ends the reaction. (Diagrams after J. Darnell, H. Lodish, D. Baltimore, Molecular Cell Biology. 2nd ed. 1990).

B. Hormones with immediately effective cell reactions

Important examples of hormones that function as ligands are amino acid derivatives, arachidonic acid derivatives, and many peptide hormones. Epinephrine, norepinephrine, and histamine act directly and very rapidly. Peptide hormones such as insulin or adrenocorticotropic hormone (ACTH) initially occur as precursor polypeptides, which are split by specific proteases to form active molecules. Some peptide hormones are coded for by a common gene; differential RNA splicing of the transcript of this gene produces different precursors for translation. (Abbreviations used: ACTH = adrenocorticotropic hormone, FSH = follicle-stimulating hormone, LH = leutinizing hormone, TSH = thyroid-stimulating hormone) (Figure data after Darnell J, Lodisch H, Baltimore D. Molecular Cell Biology. 2nd ed. New York: Scientific American Books; 1990).

C. Cell surface receptors with indirect ligand effect

Many cell surface receptors act indirectly. When they bind to a ligand they induce a series of intracellular activation steps. This reaction system consists of a receptor protein, a protein (G protein) bound to a guanosine residue, and an enzyme to be activated. Ligand binding alters the receptor protein and activates the G protein (2). This moves to the effector, e.g., an enzyme complex (3), and activates it (4). In this way, a second messenger is formed that triggers further reactions in the cell, e.g., cyclic adenosine monophosphate (cAMP) by means of the enzyme adenylate cyclase (see p. 226).

References

Darnell T, Lodish H, Baltimore D. Molecular cell biology. 2nd ed. New York: Scientific American Books; 1990.

Watson JD, Gilman M, Witkowski J, Zoller M. Recombinant DNA. 2nd ed. New York: Scientific American Books; 1992.

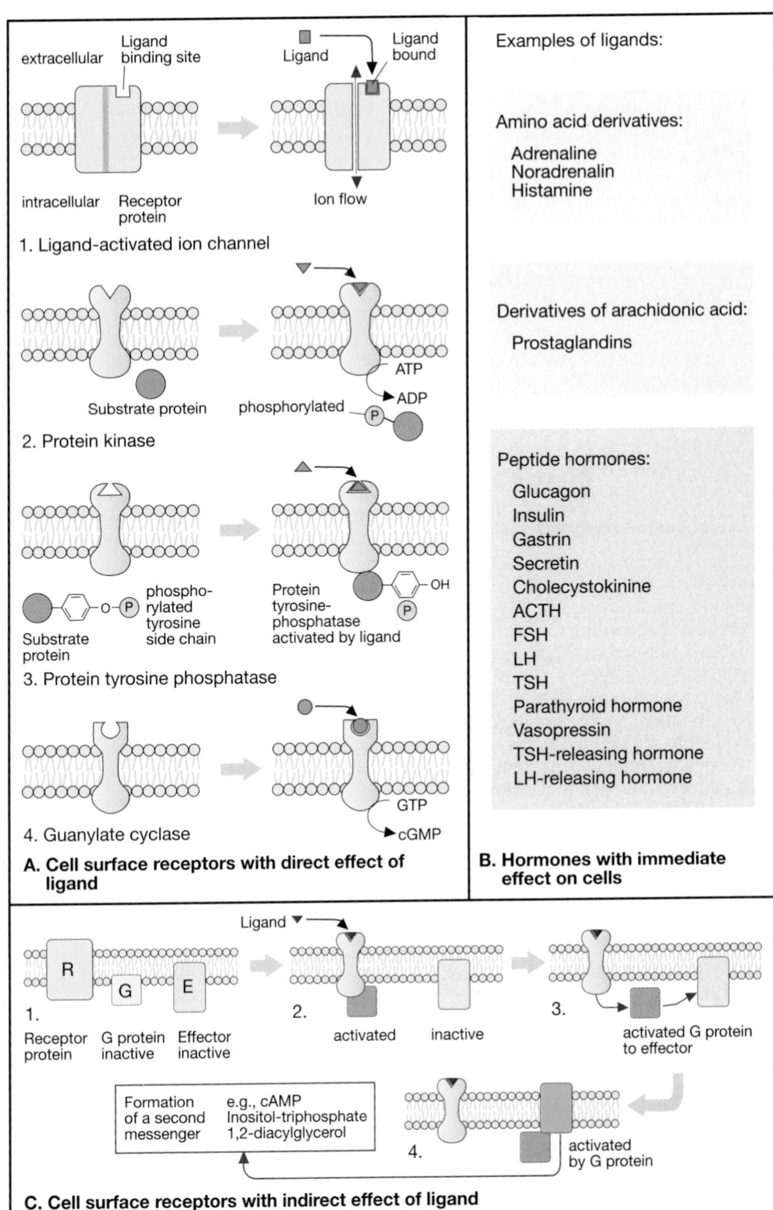
extracellular
Ligand binding site
Ligand
Ligand bound
intracellular
Receptor protein
Ion flow
1. Ligand-activated ion channel
ATP
ADP
Substrate protein
phosphorylated
P
2. Protein kinase
phospho-rylated tyrosine side chain
Substrate protein
Protein tyrosine-phosphatase activated by ligand
OH
3. Protein tyrosine phosphatase
GTP
cGMP
4. Guanylate cyclase
A. Cell surface receptors with direct effect of ligand
Examples of ligands:
Amino acid derivatives:
Adrenaline
Noradrenalin
Histamine
Derivatives of arachidonic acid:
Prostaglandins
Peptide hormones:
Glucagon
Insulin
Gastrin
Secretin
Cholecystokinine
ACTH
FSH
LH
TSH
Parathyroid hormone
Vasopressin
TSH-releasing hormone
LH-releasing hormone
B. Hormones with immediate effect on cells
Ligand
R
G
E
1.
Receptor protein
G protein inactive
Effector inactive
2.
activated
inactive
3.
activated G protein to effector
Formation of a second messenger
e.g., cAMP
Inositol-triphosphate
1,2-diacylglycerol
4.
activated by G protein
C. Cell surface receptors with indirect effect of ligand

Receptors of Neurotransmitters

Impulses are relayed between nerve cells or between nerve and muscle cells by various transmitter molecules (neurotransmitters). Their effects are further relayed by receptors in the cell membrane. Receptors can be differentiated according to their structure, which in turn determines their specificity.

A. Acetylcholine as a neurotransmitter

Cholinergic synapses convey the nerve impulse from one nerve cell to another or from a nerve cell to a muscle cell (motor endplate). Acetylcholine leads to postsynaptic depolarization through the release of potassium ions (K^+) and the uptake of sodium ions (Na^+). This process is regulated by an acetylcholine receptor.

B. Acetylcholine receptors

The acetylcholine receptors are of two genetically and functionally different types. Pharmacologically they can be differentiated according to the effects of nicotine and muscarine. The nicotine-sensitive acetylcholine receptor is an ion channel for potassium and sodium. It consists of five subunits: two α, one β, one γ, and one δ (1). Acetylcholine binds as a ligand to the two α subunits. Each subunit consists of four transmembrane domains (2). Each subunit is coded for by a gene (3). These genes have similar structures and nucleotide base sequences. The ligand-gated ion canal is an example of direct transport without an intermediate carrier. A mutation in the second transmembrane region has been shown to change the ion selectivity from cations to anions (G.-L. Galzi. Nature. 1992; 359: 500).
The muscarine-sensitive type of acetylcholine receptor is a protein that contains seven transmembrane subunits (4). Since each exists in the form of an α helix, it is referred to as a seven-helix transmembrane protein. The amino end (NH_2) lies extracellularly; the carboxy end (COOH), intracellularly. The transmembrane domains are connected by intracellular and extracellular polypeptide loops (4). Different domains of the whole protein are distinguished (5) according to location and the relative proportion of hydrophilic and hydrophobic amino acids. The amino end and the carboxy end each form a domain just like the intracellular (a–c) and extracellular portions (d–f). The transmembrane domains located within the plasma membrane (1–7) consist for the most part of hydrophobic amino acids. The structure of the gene product corresponds to the general structure of the gene (6). The different domains are coded for by individual exons. The DNA nucleotide sequences within functionally similar domains are similar.
The seven-helix transmembrane motif occurs in many receptors. The general structures of the genes and of the gene products are very similar, but they differ in their specificity of binding to other functionally relevant molecules (G proteins). They play a role not only as neurotransmitters but also in the transmission of light, smell, and taste.

(Figures based on JD Watson et al., 1992).

Reference

Watson JD, Gilman M, Witkowski J, Zoller M. Recombinant DNA. 2nd ed. New York: Scientific American Books; W H Freeman; 1992.

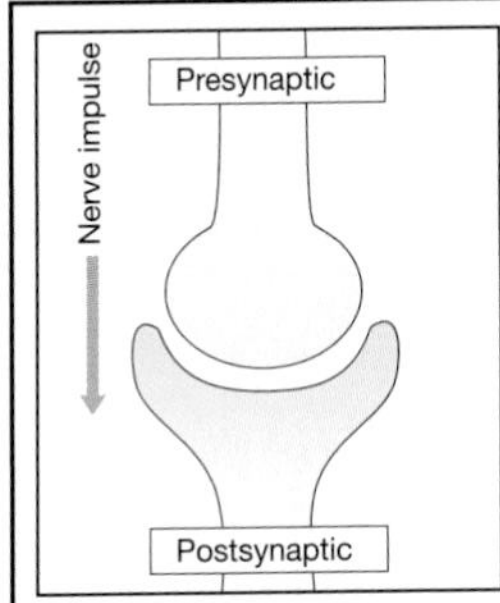

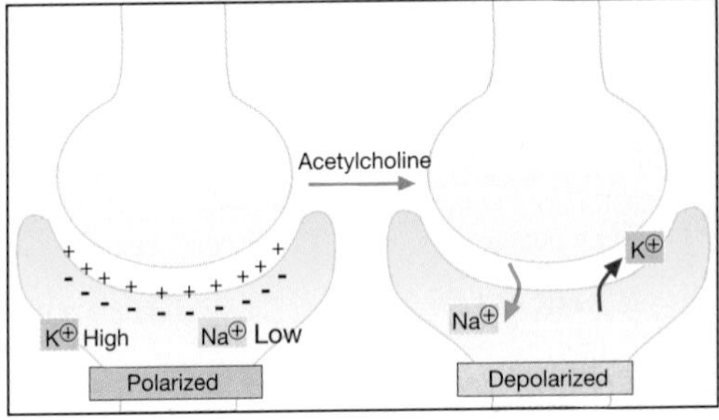

Cholinergic synapsis (nerve/nerve or nerve/muscle)

A. Acetylcholine as neurotransmitter

Two types of acetylcholine receptors

Cation-specific channel in muscle of vertebrates (nicotine-sensitive)

Seven-helix transmembrane protein bound to G-proteins (muscarine-sensitive)

1. Acetylcholine binds to α-subunits (ligand binding)

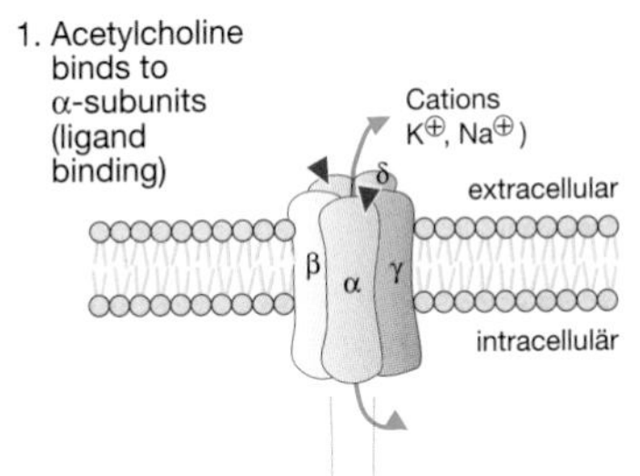

2. Each subunit has four transmembrane domains

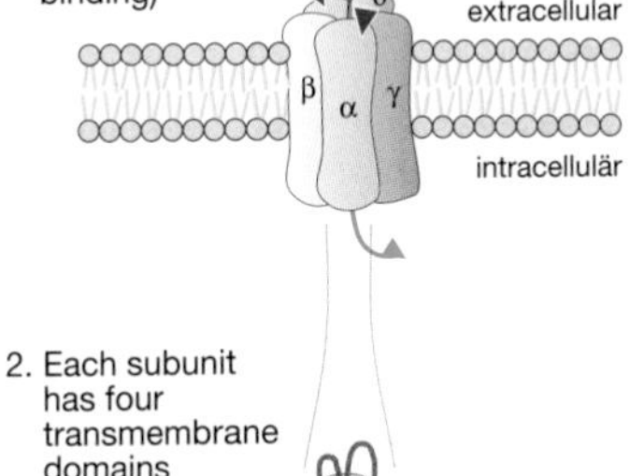

4.

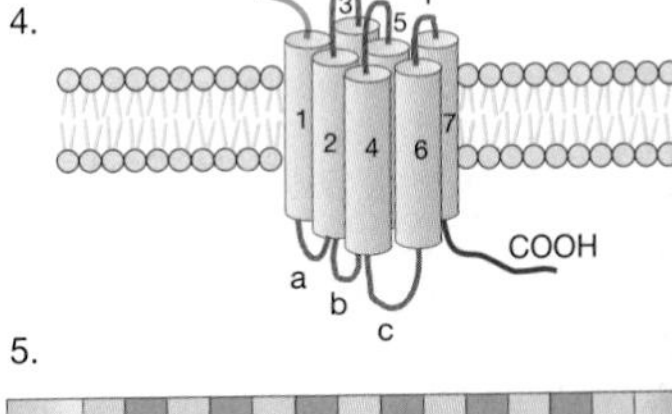

5.

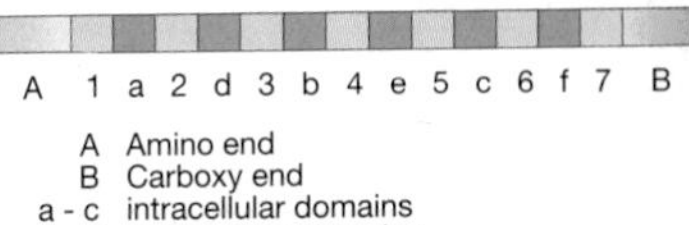

A Amino end
B Carboxy end
a - c intracellular domains
d - f extracellular domains
1 - 7 transmembrane domains (hydrophobic)

3. One gene for each subunit:

2 for α-subunits
1 for β-subunit
1 for γ-subunit
1 for δ-subunit

6. Gene structure (diagram)

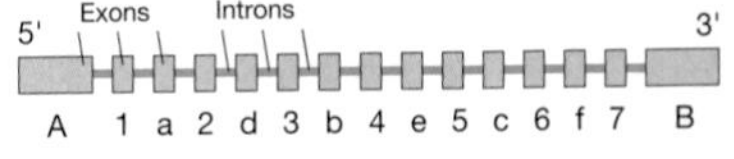

The different domains are encoded by individual exons

B. Acetylcholine receptor

Structural Motif of Transmembrane Signal Transmitters

Functional signals between cells are received by transmembrane proteins as signal transmitters. During evolution, relatively simple precursor genes for such proteins gave rise to multiple structurally and functionally related genes. Their corresponding proteins serve to transmit ions (sodium, potassium, calcium, chloride, and others), as neurotransmitters, and for perception of light and odors, etc. Cloning of these genes has yielded insight into the variety of functions of transmembrane signal transmitters. Their general structure can be traced back to an evolutionarily conserved ancestral molecule.

A. Transmembrane structure of volt-gated ion channels

The direct flow of ions across the cell membrane is regulated by ion channels. The transmembrane proteins, composed of several domains, are arranged so that they form pores that can be opened and closed. The simplest model is the potassium channel (1). This membrane-bound polypeptide contains six transmembrane domains. The amino and the carboxy ends of the protein lie within the cell. Changes in cell membrane potential or voltage cause the channel to open (or close) in order to initiate (or terminate) a brief flow of ions. Domain four, which is composed of polar amino acids, is crucial for the flow of ions. Sodium and calcium ion channels consist of four subunits (2) of similar structure, each resembling a potassium channel. The similarity is due to the common evolutionary origin of their genes. The four subunits of the sodium channel (3) are positioned to form a very narrow porelike passage, much narrower than a potassium channel, through the plasma membrane. Ion transport is brought about by membrane depolarization (3 and 4). (Figure after J.D. Watson et al., 1992).

B. Seven-helix structure of transmembrane signal transmitters

Indirect transmission of signals is more frequent than the direct transport of ions or ligand-gated impulse transmission. Here, the transmembrane protein is involved only in the first step of signal transmission. Further steps follow. An especially common structural motif is a transmembrane protein containing seven units in the form of α helices within the plasma membrane. The amino end is extracellular; the carboxy end, intracellular. Different oligosaccharide side chains are usually bound to the extracellular domains. The intracellular domains have binding sites for other molecules involved in signal transmission. The seven-helix motif is the characteristic structure of G-protein-binding receptors (p. 226). As the G proteins themselves, these receptors and their genes form a large family with a long evolutionary history. In yeast, they serve to discern the pheromones of the mating types; in higher organisms they are the basis for transmitting signals of vision, smell, and taste. (Figure redrawn from L. Stryer, Biochemistry, 1995).

C. A receptor with two transmembrane protein chains, α and β

The receptor for γ-aminobutyric acid (GABA) utilizes two transmembrane protein subunits, α and β. Both the amino and the carboxy ends are extracellular. The two chains are coded for by different genes. Several oligosaccharide side chains are present on the extracellular side. The β chain contains a phosphorylation site for cAMP-dependent protein kinase.

Reference

Sabatini DD, Adesnik MB. The biogenesis of membranes and organelles. In: Scriver CR, Beaudet AL, Sly WS, Valle D, eds. The Metabolic and Molecular Bases of Inherited Disease. 7th ed. New York: McGraw-Hill; 1995: 459–553.

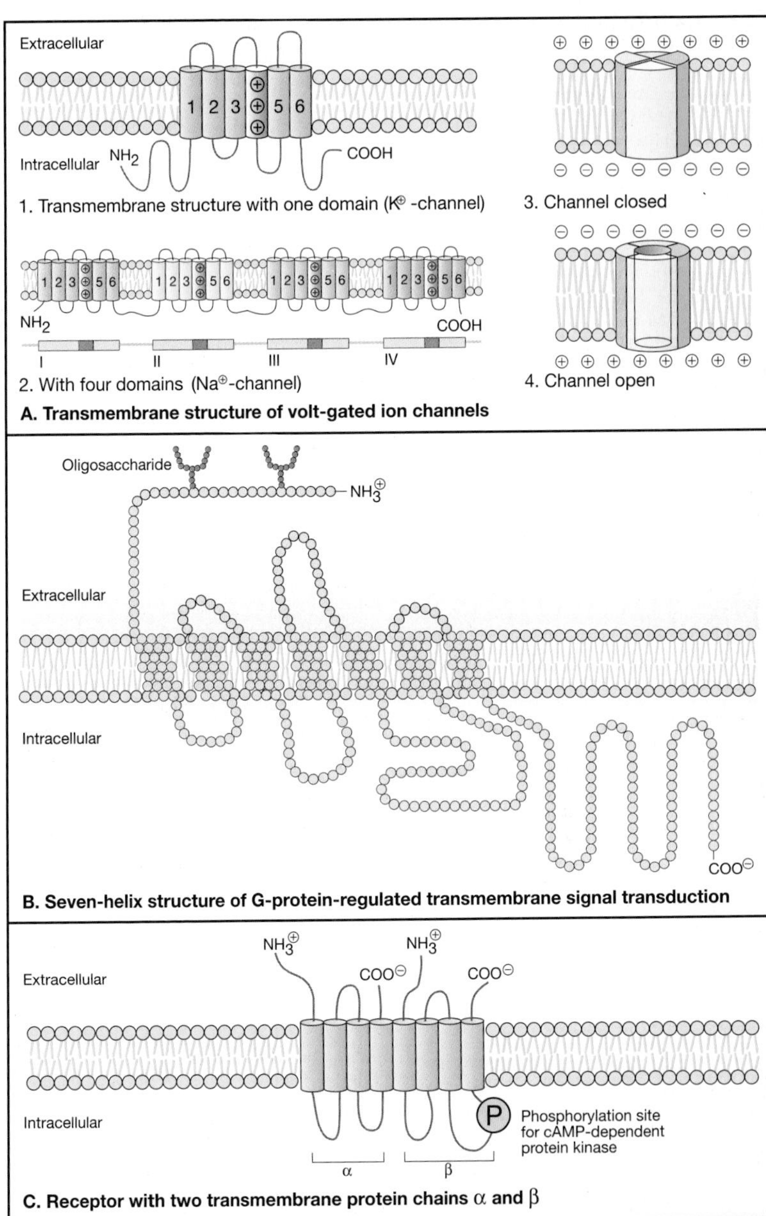

A. Transmembrane structure of volt-gated ion channels

B. Seven-helix structure of G-protein-regulated transmembrane signal transduction

C. Receptor with two transmembrane protein chains α and β

Transmembrane Transmission of Signals by G Proteins

The indirect transmission of signals into the cell is mediated by transmembrane proteins, which traverse the cell membrane. A first messenger, e.g., a hormone like adrenaline, triggers an intracellular reaction by binding to a specific receptor. This leads to activation of a second messenger, which in turn initiates a series of reactions that result in a change of cell function. Many of the genes for the different proteins involved in the indirect transmission of signals are known.

A. Stimulatory G protein (G_s) and hormone–receptor complex

There are many endogenous messengers (hormones) with their own specific receptors. First the hormone binds to the receptor (formation of a hormone–receptor complex). The intracellular transmission of signals is mainly carried out by special guanine-nucleotide-binding proteins, or G proteins. By binding to guanosine triphosphate (GTP, a nucleotide composed of guanine, a sugar, and three phosphate groups), the G protein becomes activated and initiates further reactions. G proteins consist of three subunits: α, β, and γ. The α subunit (stimulatory G protein, G_s) binds to the effector protein. Immediately thereafter, G_α is inactivated (GTPase) by hydrolysis of GTP to GDP (guanosine diphosphate). This transforms the G protein back into an inactive form (G_i).
Several human diseases due to defective G protein or a defective G-protein receptor are known (Clapham DE. Mutations in G protein-linked receptors: novel insights on disease. Cell. 1993; 75: 1237–1239).

B. Four hormone classes

Four principal classes of hormones can be differentiated: (1) amino acid derivatives such as epinephrine and epinephrine derivatives, (2) polypeptides such as glucagon, (3) steroids such as cortisol and its derivatives, and (4) fatty acid derivatives such as the prostaglandins.

C. Formation and hydrolysis of cAMP

The key reaction is the formation of cyclic adenosine monophosphate (cAMP) from adenosine triphosphate (ATP) by means of adenylate cyclase. Intracellular cyclic AMP transmits the activation initiated by the hormone–receptor complex without a molecule having passed through the plasma membrane. cAMP is responsible for many physiological reactions. It becomes inactivated when converted into adenosine monophosphate (AMP) by phosphodiesterase. cGMP (cyclic guanosyl monophosphate) functions in the same manner as cAMP to initiate an intracellular reaction.

D. G-protein cycle to activate adenylate cyclase

When a hormone binds to its specific receptor, a structural change occurs (1). This activates the α subunit of the G protein, which separates from the β and γ subunits (2). The stimulatory G protein (G_s-α) binds to the effector protein, usually adenylate cyclase, and activates it (3). cAMP is then formed from ATP, while GTP is hydrolyzed to GDP at the G-α subunit. This inactivates the effector protein and the formation of cAMP is terminated. Thus, the signal is of very short duration, and the initial conditions are rapidly restored. Several toxins exert their activity by interrupting this cycle. For example, cholera toxin inhibits inactivation of the G_s-α protein so that adenylate cyclase remains activated and large amounts of sodium and water are lost through the intestinal mucous membranes.

References

Bourne HR, Sanders DA, McCormick F. The GTPase superfamily: conserved structure and molecular mechanism. Nature. 1991; 349: 117–127.

Linder ME, Gilman AG. G-Proteins. Scient Amer. July, 1992: 36–43.

Watson JD, Gilman M, Witkowski J, Zoller M. Recombinant DNA. 2nd ed. Scientific American Books, New York: W H Freeman; 1992.

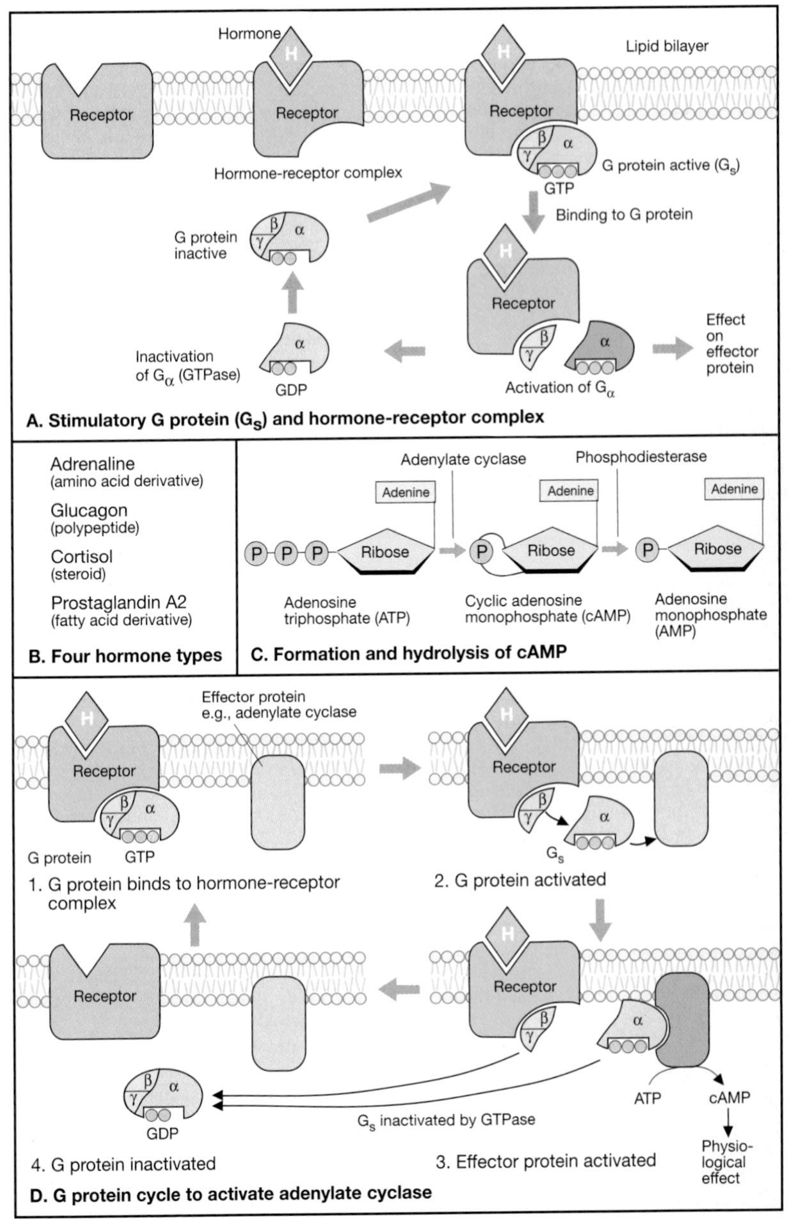

A. Stimulatory G protein (G_S) and hormone-receptor complex

B. Four hormone types

C. Formation and hydrolysis of cAMP

D. G protein cycle to activate adenylate cyclase

Rhodopsin as a Photoreceptor

The human retina contains about 110 million rod cells for vision in the dark and 6 million cone cells for color vision in the light. They contain photoreceptors that convert light into a nerve impulse. Rhodopsin is the photoreceptor for weak light. The light-transmitting system consists of numerous components coded for by genes that are similar in structure and function to genes for other transmembrane signal-transmitting molecules.

A. Rod cells

A rod cell consists of an outer segment with a photoreceptor region and an inner segment comprising cell nucleus and cytoplasm with endoplasmic reticulum, Golgi apparatus, and mitochondria. The outer segment contains about 1000 discs with rhodopsin molecules in the membrane. In the periphery, the approximately 16-nm-thick discs are folded by the protein peripherin. (Diagram after L. Stryer, 1995).

B. Photoactivation

In 1958, George Wald and co-workers discovered that light isomerizes 11-*cis*-retinal (1) very rapidly into all-*trans*-retinal (2), a form that practically does not exist in the dark (approx. 1 molecule /1000 years). The light-induced structural change is so great that the resulting atomic motion can trigger a reliable and reproducible nerve impulse. The absorption spectrum of rhodopsin (3) corresponds to the spectrum of sunlight, with an optimum at a wave length of 500 nm. Although vertebrates, arthropods, and mollusks have anatomically quite different types of eyes, all three phyla use 11-*cis*-retinal for photoactivation.

C. Light cascade

Photoactivated rhodopsin triggers a series of enzymatic steps (light cascade). First, a signal-transmitting protein of visualization, transducin, is activated by photoactivated rhodopsin. Transducin belongs to the G-protein family, i.e., it can assume an inactive GDP and an active GTP form. GTP activates phosphodiesterase. This very rapidly hydrolyzes cGMP and lowers the cGMP concentration in cytosol, which leads to closure of the sodium ion channels. Immediately thereafter, phosphodiesterase is inactivated by means of a G-protein cycle.

D. Rhodopsin

Rhodopsin is a seven-helix transmembrane protein with binding sites for functionally important molecules such as transducin, rhodopsin kinase, and arrestin on the cytosol side. The binding site of the light-sensitive molecule (chromophore) is lysine in position 296 of the seventh transmembrane domain. The light-absorbing group consists of 11-*cis*-retinal. The amino end of rhodopsin is located in the disc interspaces, and the carboxy end on the cytosol side. About half of the molecule is contained in the seven transmembrane hydrophobic domains; one fourth, in the disc interspaces; and one fourth, on the cytosol side.

E. cGMP as transmitter in the vizualization process

The light cascade ends with rapid hydrolysis of cGMP, the internal transmitter in visualization. This leads to rapid closure of the sodium ion channels and hyperpolarization of the membrane to initiate a nerve impulse, which is transmitted as a signal to the brain.

References

Stryer L. Molecular basis of visual excitation. Cold Spring Harbor Symp Quant Biol. 1988; 53: 283–294.

Stryer L. Biochemistry. 4th ed. New York: WH Freeman; 1995.

Schoenlein RW, et al. The first step in vision: Femtosecond isomerization of rhodopsin. Science. 1991; 254: 412–415.

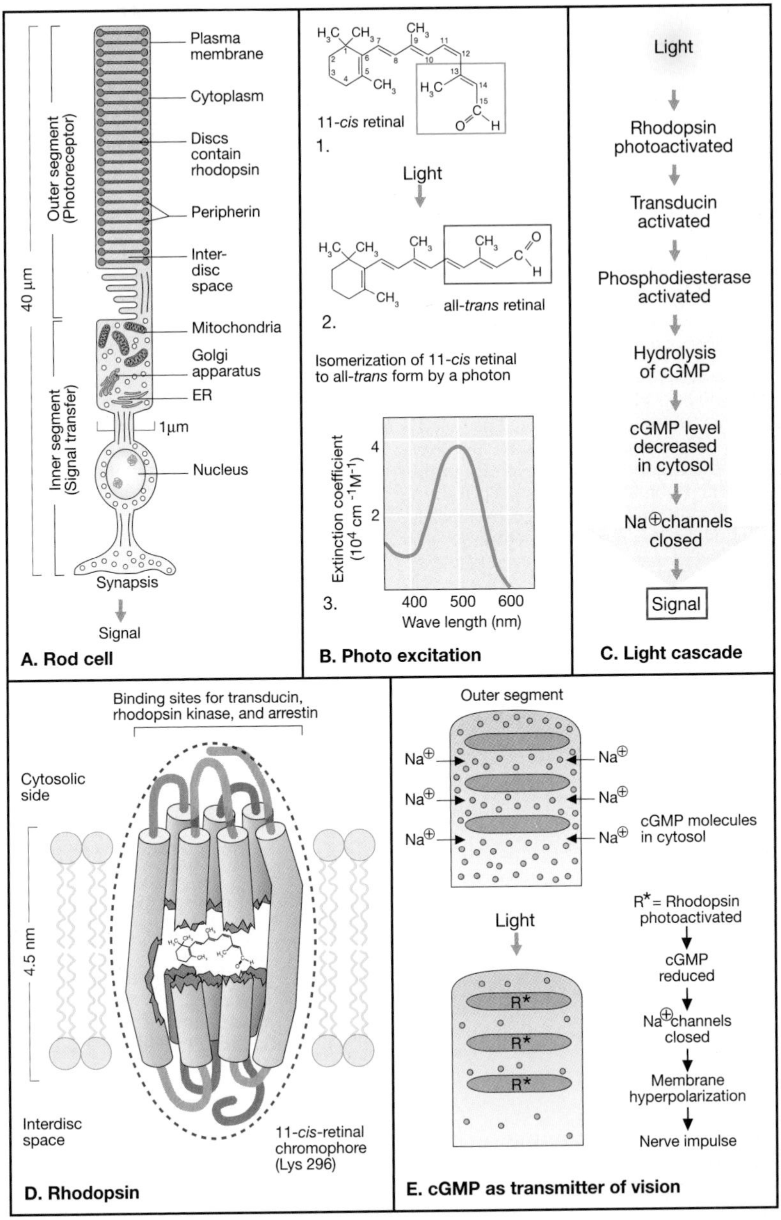

Plasma membrane
Cytoplasm
Discs contain rhodopsin
Peripherin
Inter-disc space
Mitochondria
Golgi apparatus
ER
1µm
Nucleus
Outer segment (Photoreceptor)
Inner segment (Signal transfer)
40 µm
Synapsis
Signal
A. Rod cell
11-cis retinal
1.
Light
all-trans retinal
2.
Isomerization of 11-cis retinal to all-trans form by a photon
Extinction coefficient (10⁴ cm ⁻¹M⁻¹)
4
2
400
500
600
Wave length (nm)
3.
B. Photo excitation
Light
Rhodopsin photoactivated
Transducin activated
Phosphodiesterase activated
Hydrolysis of cGMP
cGMP level decreased in cytosol
Na⊕channels closed
Signal
C. Light cascade
Binding sites for transducin, rhodopsin kinase, and arrestin
Cytosolic side
4.5 nm
Interdisc space
11-cis-retinal chromophore (Lys 296)
D. Rhodopsin
Outer segment
Na⊕
cGMP molecules in cytosol
Light
R*
R* = Rhodopsin photoactivated
cGMP reduced
Na⊕channels closed
Membrane hyperpolarization
Nerve impulse
E. cGMP as transmitter of vision

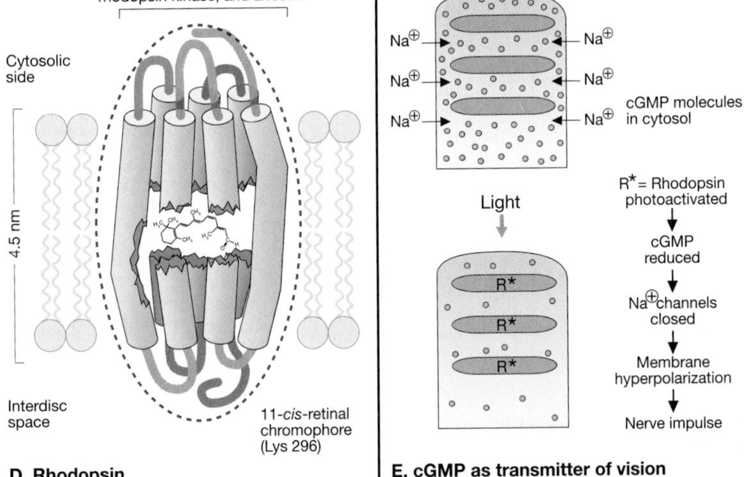

Mutations in the Rhodopsin Gene Causing Retinitis Pigmentosa

Retinitis pigmentosa (RP) is a genetically heterogeneous group of diseases that lead to pigmental degeneration of the retina and progressive blindness. Numerous mutations in the rhodopsin gene have been shown to be the cause of different forms of RP. Mutations in other genes coding for proteins of the light cascade may also cause retinitis pigmentosa.

A. Retinitis pigmentosa

The fundus of the eye shows distinct displacement of pigmentation, with irregular hyper- and depigmentation. The papilla (optic disk) shows waxy yellow discoloration. The loss of vision, especially in dim light (night blindness), proceeds from the periphery to the center at different rates depending on the form of the disease until only a very narrow central visual field remains (photograph from E. Zrenner, Tübingen).

B. Point mutation in codon 23

The first point mutation demonstrated in the rhodopsin gene (T. P. Dryja et al., Nature. 1990; 343: 364) was a transversion from cytosine to adenine in codon 23. This changed the codon CCC, for proline (Pro), into CAC, for histidine (His). Since the proline in position 23 occurs in more than ten related G-protein receptors, it must be very important for normal function.

C. Mutations in rhodopsin

The gene locus for rhodopsin (RHO) in man lies on the long arm of chromosome 3 in region 2, band 1.4 (3q21.4). Dominant and autosomal recessive inherited mutations have been demonstrated in humans. Most mutations lead to the exchange of an amino acid, although deletions may also occur. Of the 348 amino acids of rhodopsin, 38 are identical (invariant) at various positions in vertebrates. More than 80 different mutations are known for autosomal dominant inherited RP. An increasing number of mutations are recognized to cause autosomal recessive RP. In addition, mutations of at least two other gene loci have been recognized to lead to retinitis pigmentosa: mutations in the gene for peripherin on the short arm of chromosome 6 in humans (6p) and a locus in the centromeric region of chromosome 8. Other photoreceptor genes that are candidates as disease loci are the α and β subunits of phosphodiesterase (PDE).

D. Demonstration of a mutation in codon 23 by means of oligonucleotides after PCR

This pedigree with autosomal dominant inherited retinitis pigmentosa due to mutation in codon 23 (1) includes 13 affected individuals in three generations (affected females, black circles; affected males, black squares). Using polymerase chain reaction (PCR) (see p. 72), Dryja et al. (1990) demonstrated the mutation in amplified fragments of exon 1 (2). The normal oligonucleotide corresponds to the normal sequence between codons 26 and 20. The mutant sequence of the oligomere "RP" contains the mutant sequence CAC. All affected individuals gave a hybridization signal with the RP oligomer (2) (II-2 and III-4 were not examined), whereas unaffected individuals did not (see p. 70 for demonstration of a point mutation with oligonucleotides).

References

Barkur SS. Retinitis pigmentosa and related disorders. Am J Med Genet. 1994; 52: 467–474.

Dryja TP. Retinitis pigmentosa. In: Scriver CR et al., eds. The Metabolic and Molecular Bases of Inherited Disease. 7th ed. New York: McGraw-Hill; 1995: 4297–4309.

McInnes RR, Bascom RA. Retinal genetics: a nullifying effect for rhodopsin. Nature Genetics. 1992; 1: 155–157.

Wright AF. New insights into genetic eye disease. Trends Genet. 1992; 8: 85–91.

A group of hereditary diseases with degeneration of the retinal pigment

Night blindness

Progressive loss of vision

Frequency about1:3500

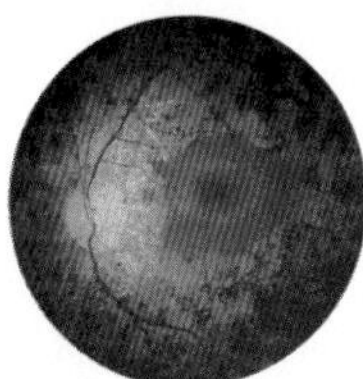

Typical fundus with pigment changes, narrow vessels, and pale, waxy optical nerve

Frequency of the different genetic forms

25% autosomal dominant
20% autosomal recessive
8% X-chromosomal
47% Mode of inheritance uncertain in an individual patient

Important diagnostic signs

Fundus:
narrow vessels
pale optic nerve
macular changes
widened light reflex
pigment epithelium changes
electroretinogram silent

Secondary changes in the anterior chamber:
vitreous body changes
cataract
myopia

A. Retinitis pigmentosa

normal — C T A G

mutant — C T A G

C A T — Tyr 20
G A G — Glu 21
C T T — Phe 22
C C C — Pro 23 → mutant: C A C His
C G A — Ser 24
C G C — Arg 25
A

B. Point mutations in codon 23

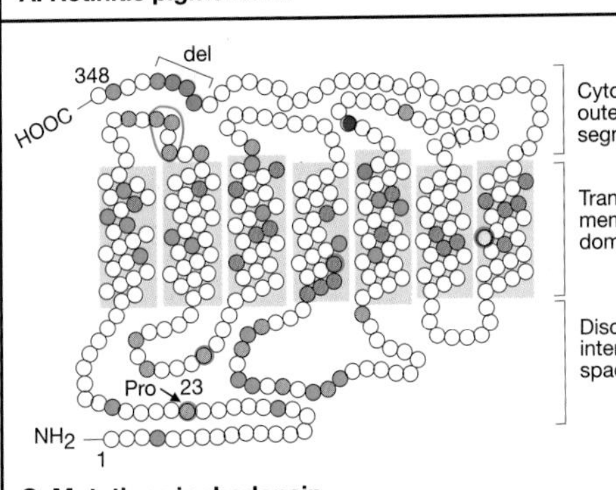

= Invariant amino acid in different vertebrates
= Retinal binding site
= Point mutation*
= Deletions*
= Point mutation in autosomal recessive RP
— = gt → tt Intron 4-donor splice site mutation

* in autosomal dominant retinitis pigmentosa

C. Mutations in rhodopsin

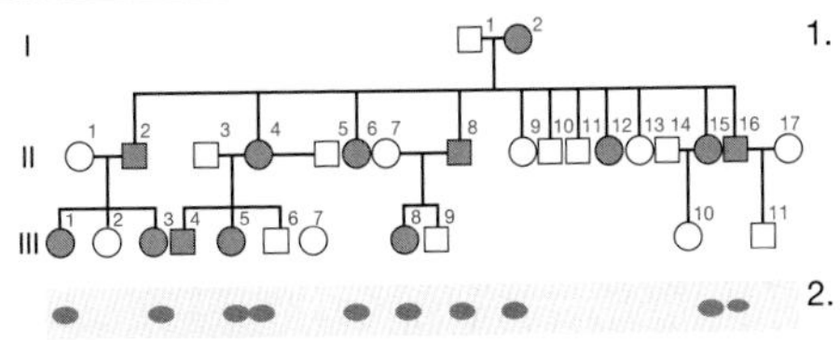

1. Pedigree with autosomal dominant retinitis pigmentosa due to mutation in codon 23

2. Autoradiogram of hybridization of amplified DNA fragments in codon 23 with oligomer 3'-CATGAGCTTCACCGACGCA-5' for the mutant sequence

D. Demonstration of a mutation in codon 23 by oligonucleotides after PCR

Color Vision

As suggested by Thomas Young in 1902, color vision in humans is mediated by three receptor types in the cone cells of the retina, one each for blue, green, and red.

A. Genes for photoreceptor proteins in cones

The gene for the blue receptor is autosomal; the genes for the red and green receptors are X chromosomal. The absorption spectra of the three receptors show maxima of 426 nm for blue, about 530 for green, and about 550 for red. The red receptor was recently discovered to be polymorphic, with two somewhat different absorption maxima at 552 and 557 nm.

B. Evolution of the genes for visual pigment photoreceptors

The photoreceptor genes arose from a single ancestral gene (protogene). The rhodopsin–transducin pair is found in invertebrates and is at least 700 million years old. The blue receptor is almost as old as rhodopsin, about 500 million years. The separation into a receptor for green and one for red must have occurred only about 30 million years ago, after the Old World and New World apes separated, since man and the Old World apes have three cone pigments whereas New World apes have two.

C. Structural similarity of the visual pigments

In 1986, J. Nathans and co-workers sequenced the genes for color photoreceptors and observed marked structural similarities, especially of the green and red receptor genes. Here the gene products (the receptors) are shown and their similarities compared. The dark dots indicate variant amino acids; the light dots are identical amino acids, given in percentages.

D. Polymorphism in the photoreceptor for red

A.G. Motulsky and co-workers (J. Winderickx et al., 1992) demonstrated variant codons in three regions of the red receptor gene (1). Serine was found at position 180 in 60% of the investigated males; alanine in 40%. Position 230 showed polymorphism of isoleucine (Ile) and threonine (Thr); position 233, alanine (Ala) and serine (Ser) (2). Differences in red color perception could be demonstrated by the color-mixing test procedure of Raleigh (3).

E. Normal and defective red–green vision

One gene for red and one to three genes for green lie close together on the long arm of the X chromosome in humans (1). Since the sequences of these genes are very similar, unequal crossing-over is not infrequent (2). Intergenic crossing-over leads to loss (green blindness) or duplication; intragenic crossing-over, to a hybrid gene (red blindness). Green blindness results from loss of a gene for the green receptor; red blindness, from a defective or absent red receptor. With red–green blindness, neither a normal red nor a normal green receptor is present. About 1% of all men are red–green blind and about 2% green blind. About 8% show weakness in differentiating red from green.

References

Motulsky AG, Deeb SS. Color vision and its genetic defects. In: CR Scriver et al., eds. The Metabolic and Molecular Bases of Inherited Disease. 7th ed. New York: McGraw-Hill; 1995: 4275–4295.

Nathans, J, Thomas D, Hogness DS. Molecular genetics of human color vision: the genes encoding blue, green, and red pigments. Science. 1986; 232: 193–202.

Neitz M, Neitz J. Numbers and ratios of visual pigment genes for normal red–green color vision. Science. 1995; 267: 1013–1016.

Winderickx J, et al. Polymorphism in red photopigment underlies variation in colour matching. Nature. 1992; 356: 431–433.

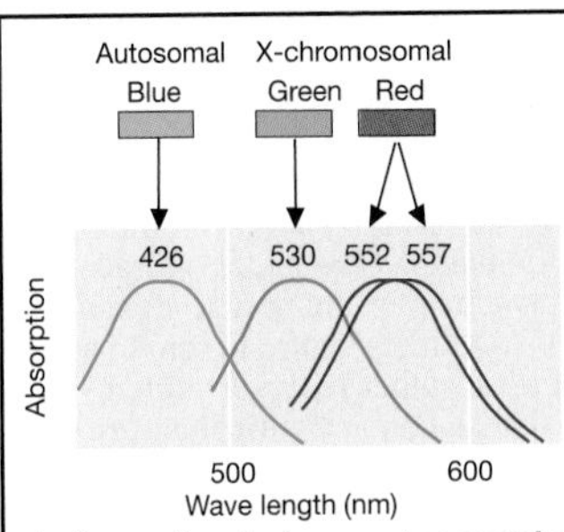

A. Genes for photoreceptor proteins in rods

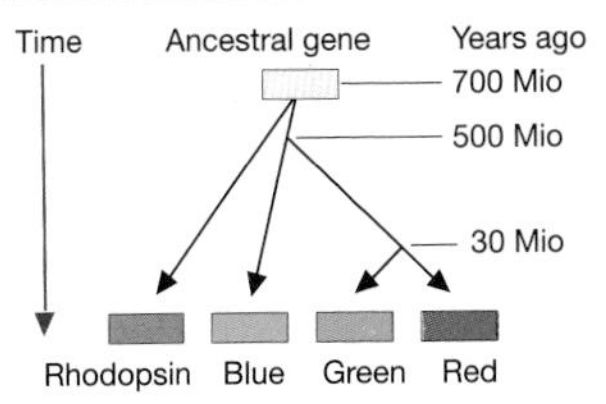

B. Evolution of genes for visual pigment photoreceptors

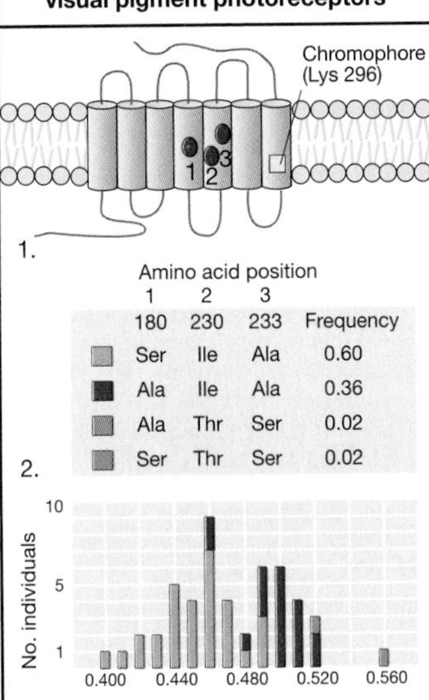

	1	2	3	
	180	230	233	Frequency
	Ser	Ile	Ala	0.60
	Ala	Ile	Ala	0.36
	Ala	Thr	Ser	0.02
	Ser	Thr	Ser	0.02

D. Polymorphism in the photoreceptor red

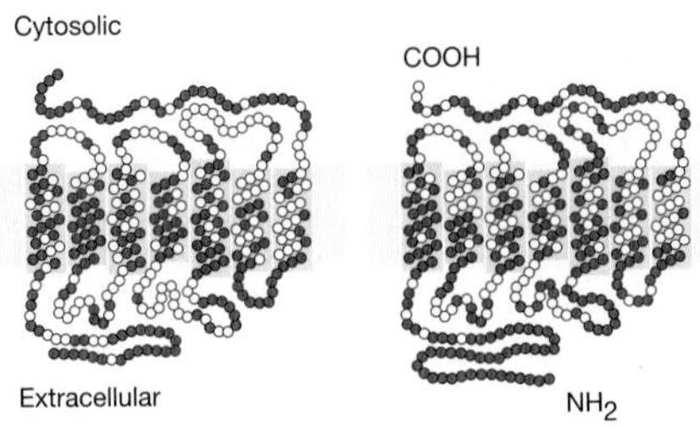

1. Blue/Rhodopsin 75% 2. Green/Rhodopsin 41%

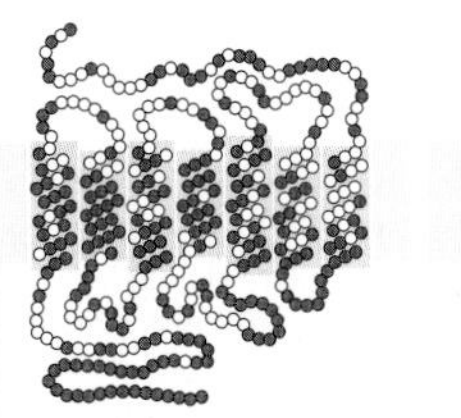

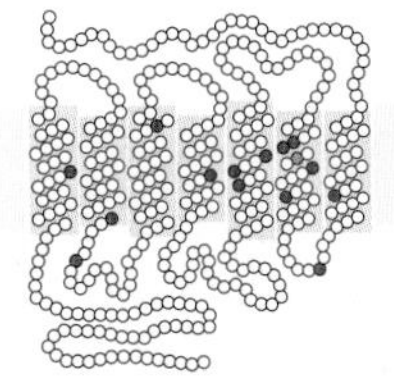

3. Green/Blue 44% 4. Green/Red 96%

C. Similar structure of visual pigments

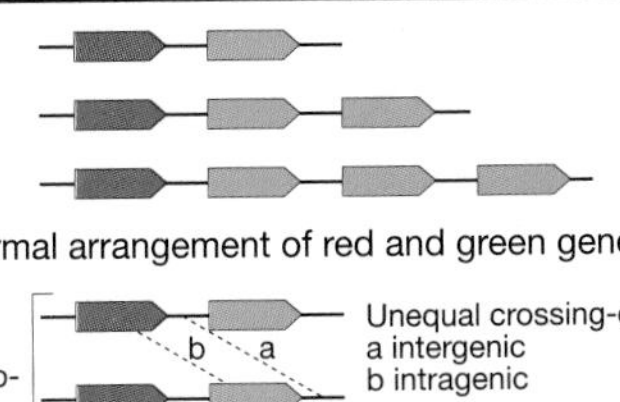

1. Normal arrangement of red and green genes

Maternal chromosomes

Unequal crossing-over
a intergenic
b intragenic

Green blindness
Normal
Red blindness
Red-green blindness
Red-green blindness
Red blindness

Deletion

2. Examples of different consequences of unequal crossing-over

E. Normal and abnormal red/green vision

Odorant Receptor Gene Family

Vertebrates can differentiate thousands of individual smells. Although their ability to distinguish differences in color is based on only three classes of photoreceptors, their sense of smell is regulated by a large multigene family of receptors that are highly specific for individual odorants. In fish, about 100 and in mammals, about 1000 genes code for specific olfactory receptors. These genes are expressed exclusively in the olfactory epithelium of the nasal mucous membrane.

A. Olfactory nerve cells in the nasal mucous membrane

The peripheral olfactory neuroepithelium of the nasal mucous membrane consists of three cell types: olfactory sensory neurons, whose axons lead to the olfactory bulb (1), supporting cells (2), and basal cells (3), which serve as stem cells for the formation of olfactory neurons during the individual's entire lifespan. Each olfactory neuron is bipolar, with olfactory cilia in the lumen of the nasal mucous membrane and a projection to the olfactory bulb, the first relay station of the olfactory system on the way to the brain.

B. Odor-specific transmembrane receptors and GTP-binding protein (G_s [olf])

Each receptor in the cilia of the olfactory neurons binds specifically to one odorant. Binding of the receptor activates adenylate cyclase via a specific GTP-binding protein (stimulatory G protein of the olfactory system, G_s [olf]). This opens a sodium ion channel and initiates a cascade of intracellular signals that result in a nerve signal, which is conveyed to the brain.

C. Olfactory receptor protein

The cloning of a large gene family from the olfactory epithelium of the rat (L. Buck & R. Axel, Cell. 1991 *65*: 175–187), demonstrated that a receptor protein contains seven transmembrane regions and shows marked structural homology with rhodopsin and β-adrenergic receptors. Unlike rhodopsin, the olfactory receptor proteins contain variable amino acids, especially in the fourth and fifth transmembrane domains. The third intracellular loop between transmembrane domains V and VI is relatively short (17 amino acids), in contrast to other receptor proteins. It is assumed that contact with the various G proteins takes place here (Buck and Axel, 1991).

D. Assignment of olfactory receptor RNA to neurons

A gene for the receptor of a given odorant is expressed in an individual in only a few neurons. J. Ngai et al. (Cell *72*: 657–666 and 667–680, 1993) classified individual olfactory neurons in the olfactory epithelium of the catfish *(Ictalurus punctatus)*. Only 0.5–2% of all olfactory neurons recognize a given receptor probe such as probe 202 (1) or 32 (2). Odors are distinguished in the brain according to which neurons are stimulated. The topographical position of each neuron is specific for each odorant.

E. Subfamilies within the multigene family

Amino acid sequences derived from partial nucleotide sequences of cDNA clones (F2–F24) (1) investigated by Buck and Axel (1991) were very variable, especially in transmembrane domains III and IV. Within subfamilies, there was homology due to conserved sequences (2). For example, F12 and F13 differ in only 4 of 44 positions (91% identical).

References

see p. 373.

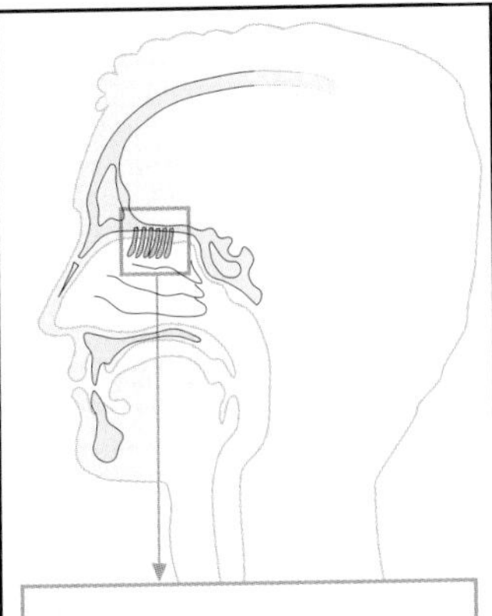

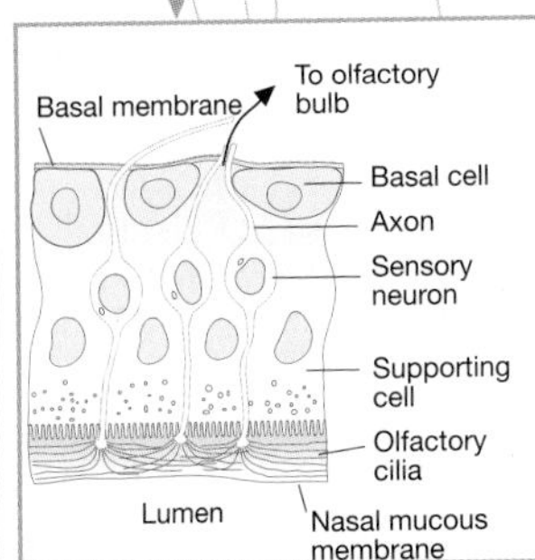

A. Olfactory nerve cells in the nasal mucous membrane

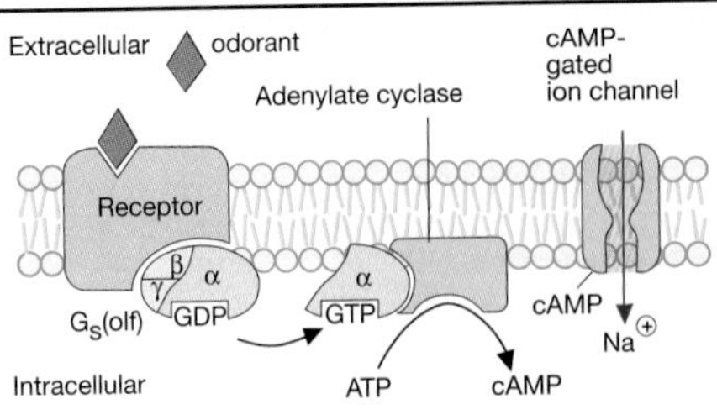

B. Odor-specific transmembrane receptor and GTP-binding protein [G_s(olf)]

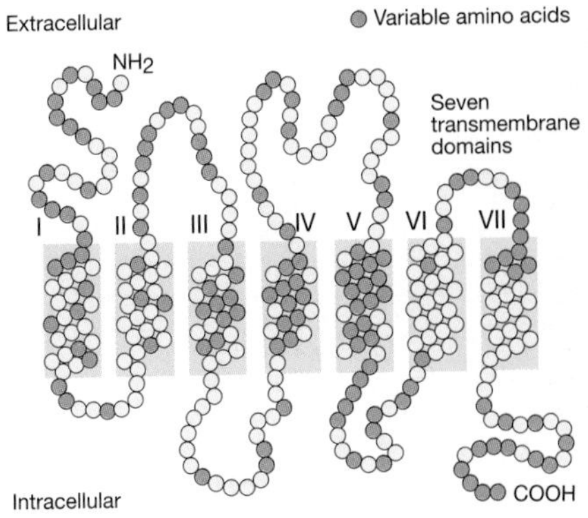

C. Olfactory receptor protein

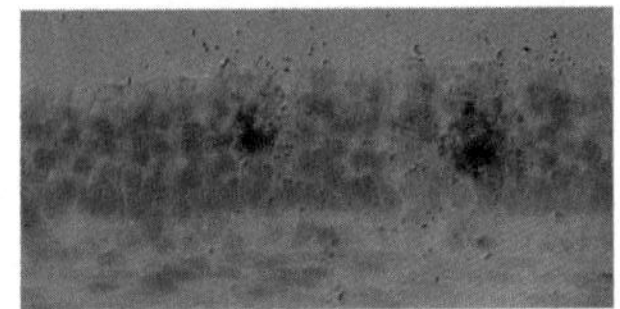

1. Receptor probe 202

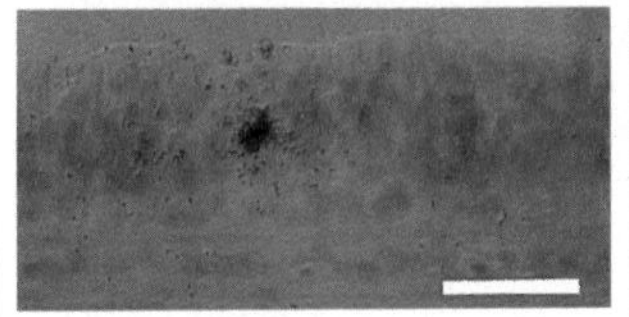

2. Receptor probe 32 25 µm

D. Correspondence of olfactory receptor RNA to neurons

cDNA clones		Transmembrane domain	
F2	RVNE	VVIFIVVSLFLVLPFALIIMSYV	RIVSSILKVPSSQGIYK
F3	FLND	LVIYFTLVLLATVPLAGIFYSYF	KIVSSICAISSVHGKYK
F5	HLNE	LMILTEGAVVMVTPFVCILISYI	HITCAVLRVSSPRGGWK
F6	QVVE	LVSFGIAFCVILGSCGITLVSYA	YIITTIIKIPSARGRHR
F7	HVNE	LVIFVMGGIILVIPFVLIIVSYV	RIVSSILKVPSARGIRK
F8	FPSH	LTMHLVPVILAAISLSGILYSYF	KIVSSIRSMSSVQGKYK
F12	FPSH	LIMNLVPVMLAAISFSGILYSYF	KIVSSIHSISTVQGKYK
F13	FPSH	LIMNLVPVMLAAISFSGILYSYF	KIVSSIRSVSSVKGKYK
F23	FLND	VIMYFALVLLAVVPLLGILYSYS	KIVSSIRAISTVQGKYK
F24	HEIE	MIILVLAAFNLISSLLVVLVSYL	FILIAILRMNSAEGRRK

1. Variable amino acid sequences

F12	FPSH	LIMNLVPVMLAAIISFSGILYSYF	KIVSSIHSISTVQGKYK
F13	FPSH	LIMNLVPVMLAAIISFSGILYSYF	KIVSSIRSVSSVKGKYK
F8	FPSH	LTMHLVPVILAAIISLSGILYSYF	KIVSSIRSMSSVQGKYK
I12	FPSH	LIMNLVPVMLGAIISLSGILYSYF	KIVSSVRSISSVQGKHK
F23	FLND	VIMYFALVLLAVVVPLLGILYSYS	KIVSSIRAISTVQGKYK
F3	FLND	LVIYFTLVLLATVVPLAGIFYSYF	KIVSSICAISSVHGKYK

2. Homology within subfamilies

E. Subfamilies within the multigene family

Developmental Mutants in *Drosophila melanogaster*

The embryonic development of an organism is determined by genes that are active only during specific phases. Analysis of developmental mutants of embryos of the fruit fly *Drosophila melanogaster* has provided insight into the genetic regulation of developmental processes. Early developmental phases in embryos of very different organisms are regulated by similar genes.

A. The segmental organization of the fruit fly *(Drosophila melanogaster)*

The development of a fruit fly from the fertilized egg cell to the segmented body of the adult organism takes 9 days. The larvae hatch after 1 day and pass through defined stages of embryonic development. The embryo forms a cocoon at 5 days and, after metamorphosis, emerges as a 2-mm-long adult fly. The head of the adult has three segments (C1–3), the thorax three (T1–3), and the abdomen eight segments (A1–8). A fruit fly has altogether 14 parasegments (P1–14), each corresponding to the last half of one and the first half of the next segment. The segmental organization is discernible in the larva.

B. Embryonic lethal mutations

Embryonic lethal mutations can be identified by appropriate crosses. One fourth of the progeny of heterozygous flies (A/a) for an embryonic mutant gene (a) are homozygotes (aa) for the mutant allele (1). If a mutation involves maternal genes only (maternal effect), progeny of female homozygotes (bb) are lethally affected (2). Maternal effect genes code for early gene products that determine the polarity of the embryo; see below (C).

C. Examples of developmental mutants

Many developmental mutations are known in *Drosophila melanogaster*. They can be classified into different hierarchical gene classes. The normal larva (wild type) consists of three head segments, three thorax segments, eight abdominal segments, and the tail end (1). A mutation for anterior maternal effect, *bicoid*, leads to the development of a larva without head or thorax (2). A mutation called *nanos* affects the posterior end of the early larva. *Gap* genes establish the basic pattern of segmental organization. Mutations of the *gap* genes lead to omissions (gaps) in the segmental construction of the larva. In the *Krüppel* mutant (3), all thoracic and the abdominal segements 1–5 are missing; in the *Knirps* mutant (4), abdominal segments 1–6 are absent. The genes for *pair-rule* determine the orientation and developmental fate of the 14 parasegments. Some mutations affect every second segment. With *even-skipped* (5), all even-numbered parasegments are missing. Mutation of the gene *fushi tarazu* leads to fewer than normal segments being formed (*fushi tarazu* is Japanese for too few segments). *Segment polarity* genes determine the polarity of each segment (7). There are more than ten *segment polarity* genes. *Homeotic selector* genes (8) determine the ultimate fate of each segment. With the mutant *antennapedia* (Ant), the antenna normally attached immediately under the eye is replaced by a leg (homeotic leg).

References

Gehring WJ, et al. Homeodomain-DNA recognition. Cell. 1994; 78: 211–223.

Kenyon C. If birds can fly, why can't we? Homeotic genes and evolution. Cell. 1994; 78: 175–180.

Lawrence PA. The Making of a Fly. The genetics of animal design. Oxford: Blackwell Scientific; 1992.

Nüsslein-Volhard C, Frohnhöfer HG, Lehmann R. Determination of anterior–posterior polarity in Drosophila. Science. 1987; 238: 1675–1681.

Watson JD, Gilman M, Witkowski J, Zoller M. Recombinant DNA. 2nd ed. New York: Scientific American Books, WH Freeman; 1992.

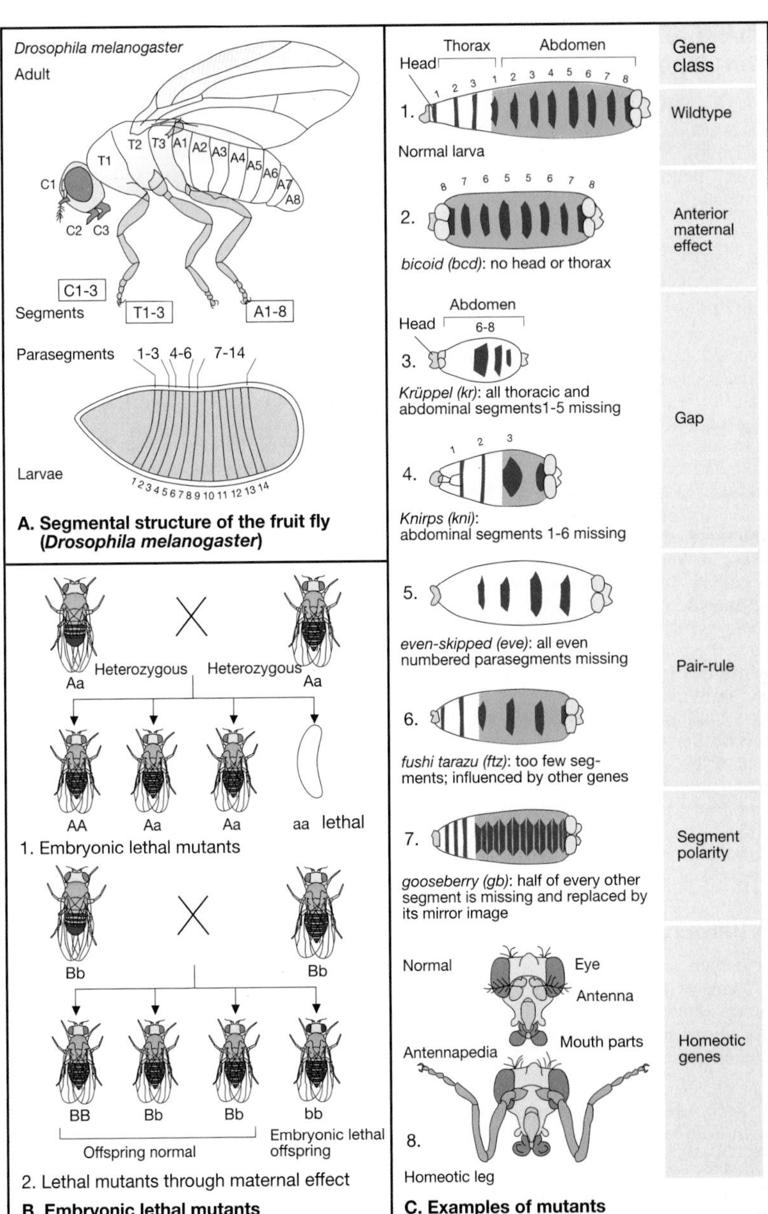

Drosophila melanogaster
Adult
T1
T2
T3
A1
A2
A3
A4
A5
A6
A7
A8
C1
C2 C3
C1-3
Segments
T1-3
A1-8
Parasegments 1-3 4-6 7-14
Larvae
1 2 3 4 5 6 7 8 9 10 11 12 13 14
A. Segmental structure of the fruit fly (Drosophila melanogaster)
Heterozygous
Heterozygous
Aa
Aa
AA
Aa
Aa
aa lethal
1. Embryonic lethal mutants
Bb
Bb
BB
Bb
Bb
bb
Offspring normal
Embryonic lethal offspring
2. Lethal mutants through maternal effect
B. Embryonic lethal mutants
Thorax
Abdomen
Head
Gene class
1.
Normal larva
Wildtype
2.
bicoid (bcd): no head or thorax
Anterior maternal effect
Abdomen
Head
6-8
3.
Krüppel (kr): all thoracic and abdominal segments1-5 missing
Gap
4.
Knirps (kni): abdominal segments 1-6 missing
5.
even-skipped (eve): all even numbered parasegments missing
Pair-rule
6.
fushi tarazu (ftz): too few segments; influenced by other genes
7.
gooseberry (gb): half of every other segment is missing and replaced by its mirror image
Segment polarity
Normal
Eye
Antenna
Mouth parts
Antennapedia
Homeotic genes
8.
Homeotic leg
C. Examples of mutants

Genes for Embryonic Development

The genes for embryonic development in *Drosophila* are organized in a functional hierarchy. Similar genes (in some regions even identical) occur in mice and in humans.

A. Hierarchy of developmental genes

Four independent systems control the embryonic development of *Drosophila melanogaster*. A gradient of maternally derived protein coded for by the *bicoid* gene (bcd) determines development in the anterior region (head); *nanos*, in the posterior region. They activate the *gap* genes. The three most important *gap* genes, *hunchback, Knirps,* and *Krüppel*, code for transcription factors of the zinc finger type (p. 206). *Hunchback* is expressed from the anterior part to about the middle of the embryo; *Krüppel*, in the region of thoracic segments 4–6 and the first six abdominal segments; *Knirps*, further posteriorly (parasegments 10–14). The *gap* genes induce the *pair rule* genes. *Segment polarity* genes determine the correct orientation of the individual segments. After segmentation is complete, *selector* genes determine the further development and the ultimate fate of the segments. They consist of three large gene complexes, the *antennapedia* complex (*ANT*-C), the *bithorax* complex (*BX*-C), and the *ultrabithorax* complex (*UBX*). Mutations of these gene complexes lead to unusual structures.

B. Bithorax mutation

A mutant for the *bithorax* complex (*BX*–C) causes the development of an additional thoracic segment with completely developed wings. (Photograph from Lawrence, after E. B. Lewis).

C. Structure of the *antennapedia* gene with homeobox

Antennapedia is a gene of the *antennapedia* complex *(ANT*-C), which is expressed in parasegments 5 and 6. It contains a segment of highly conserved DNA sequences in exon 8, the homeobox. Its sequence is identical in a wide variety of organisms, from drosophila to mammals. It codes for about 60 amino acids (homeodomain). The homeodomain contains four domains of a helical protein with DNA-binding properties and functions as a transcription factor (W. J. Gehring et al., 1990).

D. Homeotic genes in drosophila and Hox genes in mouse

Homeobox genes also occur in vertebrates. A series of genes with the same anterior to posterior orientation and corresponding to the ANT and BX complexes (*ANT*-C and *BX*-C) are found in the embryonic brain of the mouse (and in man). The temporal expression corresponds to this orientation.

E. Homeobox genes (HOX genes)

In humans and in mice there are four groups (clusters) of genes that correspond to the homeo genes of drosophila (HOX 1–4 in humans, Hox 1–4 in the mouse). Since mutations of these genes in mice lead to characteristic disorders, it is assumed that they are of clinical significance in humans also. The posterior end corresponds to the 5' end and the anterior end to the 3' end of the coding DNA.

References

Gehring WJ, et al. The structure of the homeodomain and its functional implications. Trends Genet. 1990; 6: 323–329.

Krumlauf R. Hox genes in vertebrate development. Cell. 1994; 78: 191–201.

Marx J. Homeobox genes go evolutionary. Science. 1992; 255: 399–401.

Reddihough G. Homing in on the homeobox. Nature. 1992; 357: 643–644.

Scott MP. A rational nomenclature for vertebrate homeobox (HOX) genes. Nucleic Acids Res. 1993; 21: 1687–1688.

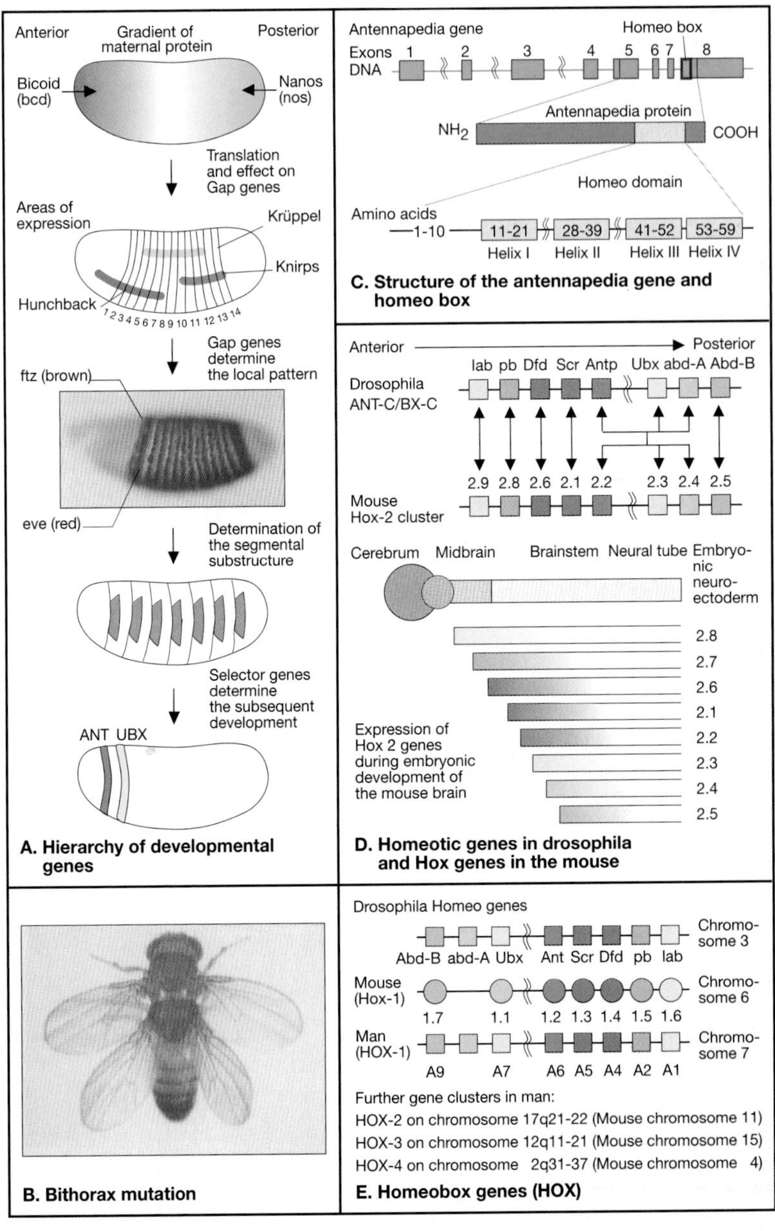

A. Hierarchy of developmental genes

B. Bithorax mutation

C. Structure of the antennapedia gene and homeo box

D. Homeotic genes in drosophila and Hox genes in the mouse

E. Homeobox genes (HOX)

Developmental Genes in a Plant Embryo

In plants, as in animals, the basic structural plan is genetically determined. In the plant *Arabidopsis thaliana* (common wall cress), nine genes with numerous alleles determine the organization of the plant embryo along an apical–basal longitudinal axis. They determine a radial pattern and the form (U. Mayer et al., 1991). Genetic regulation of the embryonic development of a plant can be inferred from the observed mutations. This example from plant genetics is included here because *Arabidopsis* is an important object of genetic studies.

A. Normal development and structure

Although the adult forms of flowering plants differ greatly, these plants undergo similar stages during their embryonic phases. *Arabidopsis* is especially well suited for genetic analysis because it is small and has a short generation time and a diploid genome. The basic structural plan can be understood as an axial and a radial pattern superimposed on each other. An octant stage, a globular stage, and a so-called heart stage can be differentiated before the seedling is formed. The regions A, C, and B of the octant stage correspond to the regions A, C, and B of the heart stage. Region A forms the cotyledon and the meristem; C, the hypocotyl region; and B, the root. The seedling consists of a set of identifiable structures including vessels (v), external epidermis (e), short meristem (s), cotyledons (c), and hypocotyl (h). In the heart stage, the essential organization of the plant is predetermined.

B. Deletions in the apical–basal pattern

The mutations can be induced by 0.3% ethyl methanesulfonate. Using complementation analysis, Mayer et al. (1991) determined mutations in three areas of the plant. These mutations affect the apical–basal pattern, the radial pattern, and the shape. Apical–basal delections involve one of several genes, each leading to a characteristic phenotype: apical deletion (*Gurke*), central deletion (*Fackel*), basal deletion (*monopteros*), and terminal deletion (*Gnom*).

C. Wild type

The normal structure of *Arabidopsis* results from two basic processes: formation of patterns (apical–basal and radial orientation) and morphogenesis through different cell forms and regional differences in cell division.

D. Phenotypes of embryonic mutants

The four mutant phenotypes in the apical–basal pattern are Gurke (9 alleles), Fackel (5 alleles), monopteros (11 alleles), and Gnom (15 alleles) (see figure section B). Deletions in the radial pattern lead to phenotypes Keule (9 alleles) and Knolle (2 alleles). Mutants of shape are Fass (12 alleles), Knopf (6 alleles), and Mickey (8 alleles). (Photos from U. Mayer et al., 1991). The *monopteros* gene (*mp*) is apparently very important for apical–basal development. However, it also has an indirect effect on the spatial arrangement of the apical structures. It is not necessary for root development (Berleth & Jürgens, 1993).

References

Berleth T, Jürgens G. The role of the monopteros gene in organising the basal body region of the Arabidopsis embryo. Development. 1993; 118: 575–587.

Mayer U, Torres Ruiz RA, Berleth T, Miséra S, Jürgens G. Mutations affecting body organization in the *Arabidopsis* embryo. Nature. 1991; 353: 402–407.

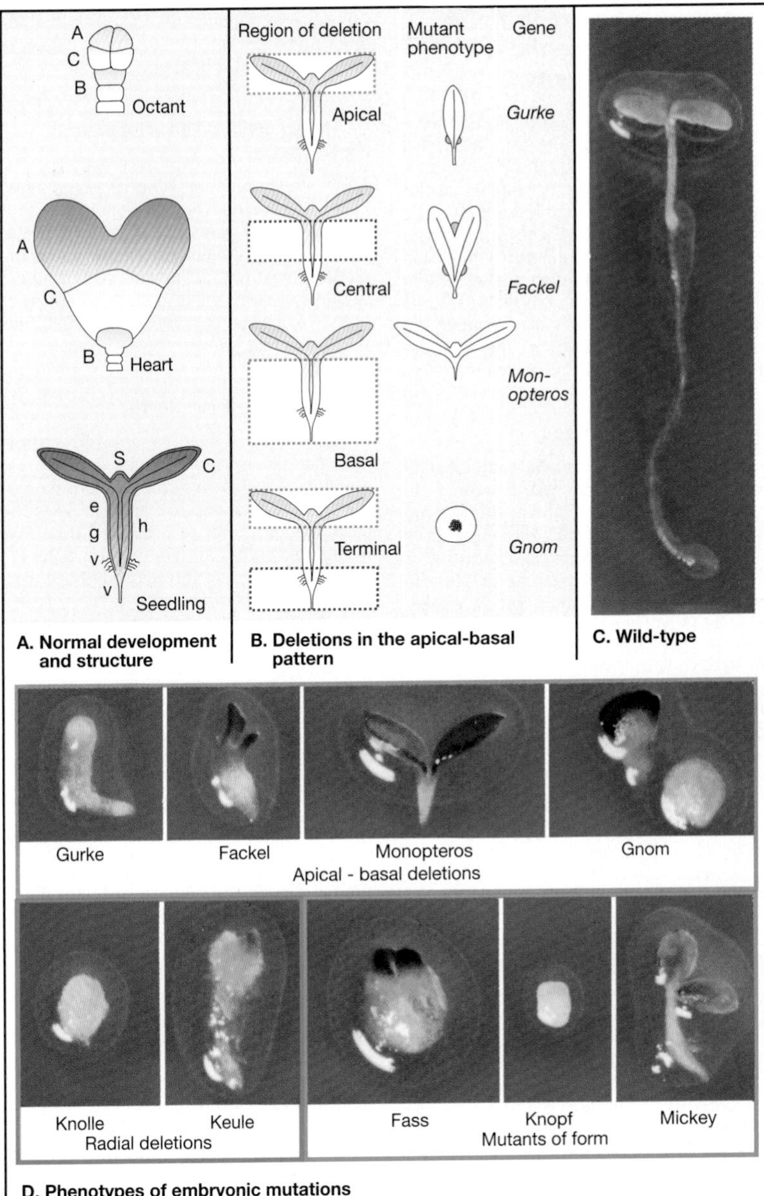

A. Normal development and structure

B. Deletions in the apical-basal pattern

C. Wild-type

D. Phenotypes of embryonic mutations

Developmental Program for Individual Cells in the Nematode *C. elegans*

Development from a zygote into an adult organism requires that the time and place of cell divisions be organized. Whereas cell origin in complex organisms cannot be determined, it is possible to analyze the origin of each individual cell in the nematode *Caenorhabditis elegans*. In 1965, Sydney Brenner introduced the genetic analysis of *C. elegans* as a model to investigate the interaction of genetic, anatomic, and physiological traits in the development of a relatively simple nervous system.

A. *Caenorhabditis elegans*

Caenorhabditis elegans is a small (about 1 mm long), transparent worm with a life cycle of about 3 days. The basic structure is a bilaterally symmetric elongated body of nerves, muscles, skin, and intestine. It exists as one of two sexes: hermaphrodite or male. Hermaphrodites produce eggs and sperm and can reproduce by self-fertilization. The adult hermaphrodite worm has 959 somatic cell nuclei; the adult male worm, 1031. In addition, there are 1000 to 2000 germ cells. The haploid genome consists of 8 x 10^7 nucleotide base pairs (about eight times more than that of the yeast *Saccacharomyces* and about half as many as that of *Drosophila*). Six homologous chromosome pairs contain around 3000 genes, many of which have been identified and sequenced. (Diagram from W.B. Wood 1988, after Sulston and Horvitz 1977).

B. Embryonic origin of the individual cells

The developmental pathway of each individual cell can be traced. The various tissues arise from six founder cells. Of the 959 adult cells, 302 are nerve cells. Except for cells of the intestine and germ line, differentiated tissues stem from different founder cells. Cells with similar function are not necessarily related, while cells with different functions may be of the same origin. Genetically established rules determine the fate of the two daughter cells at each cell division.

C. Mutations in the developmental control genes

Many developmental control genes have been identified from analysis of ethyl methanesulfonate-induced point mutations. Some mutants determine an incorrect cell type (e.g., Z instead of B); others divide too early or too late (division mutants) (resulting, e.g., in two times C, instead of B and C).

D. Programmed cell death

Programmed cell death (apoptosis) is a normal part of vertebrate and invertebrate development. During embryonic development of *C. elegans*, one in eight somatic cells regularly dies at a defined time and branching point (1). The photographs (2) show the death of a cell (cell p11.aap) over a time span of about 40 minutes (from Wood 1988, after Sulston and Horvitz 1977). Mutations that interfere with programmed cell death can lead to severe disturbances. Vaux et al. (Science. 1992; 258: 1955–1957) have shown that the human *BCL–2* gene corresponds to a gene in *C. elegans*. The human *BCL–2* gene codes for an inner mitochondrial membrane protein that acts by inhibiting cell loss by apoptosis in pro-B lymphocytes. Disruption of this gene on human chromosome 18 causes follicular lymphoma, a B-cell tumor. The *BCL–2* gene is a homolog of the *C. elegans ced–9* gene, a gene that prevents apoptosis.

References

Hengartner MO, Horvitz HR. *C. elegans* gene *ced–9* encodes a functional homolog of the mammalian proto-oncogene *bcl–2*. Cell. 1994; 76: 665–676.

W.B. Wood and the Community of *C. elegans* Researchers. The Nematode *Caenorhabditis elegans*. New York: Cold Spring Harbor; 1988: Monograph 17.

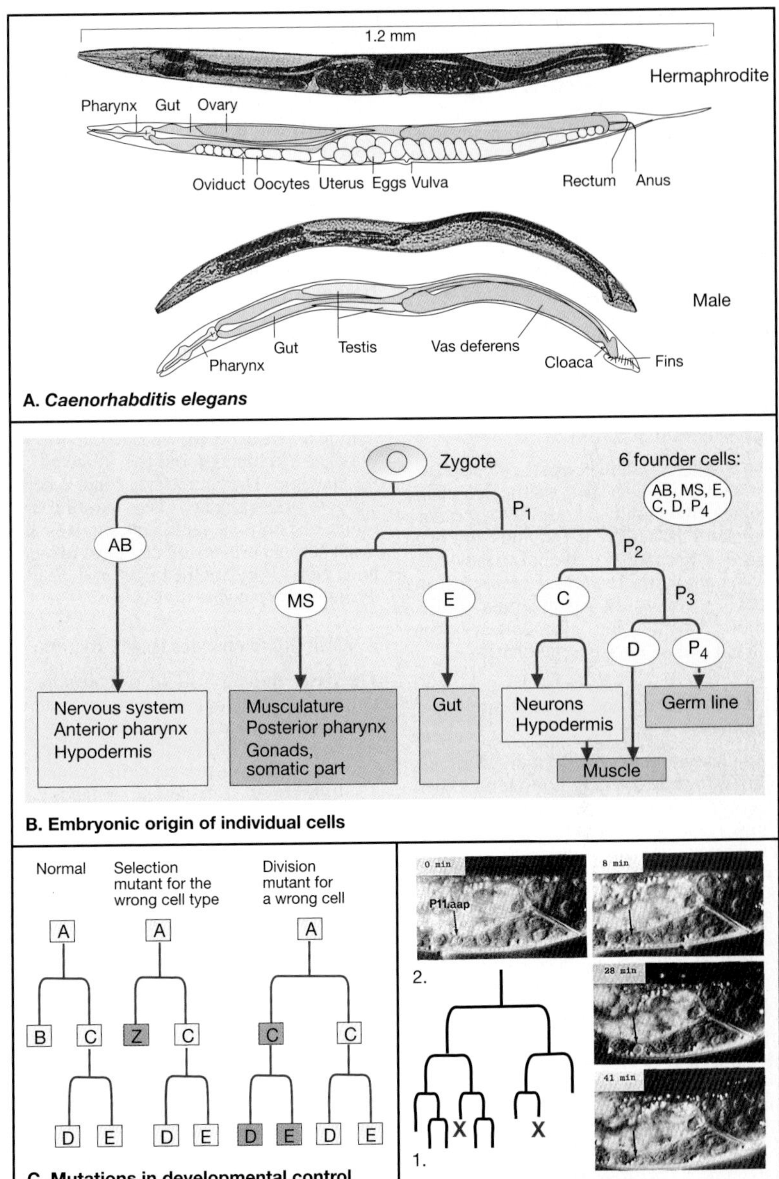

A. *Caenorhabditis elegans*

B. Embryonic origin of individual cells

C. Mutations in developmental control genes

D. Programmed cell death (X)

Components of the Immune System

The function of the immune system is to recognize and to eliminate invading foreign organisms. The immune response must be fast, specific, and aimed exclusively against foreign molecules or cells harboring an infective agent, but never against its own normal cells. The effector molecules must be prepared for contact with a variety of organisms and molecular structures. With these requirements, an extremely efficient, genetically complex immune system has evolved. Its development and function are regulated by genes of extraordinary diversity.

A. Lymphatic organs

The immune system consists of peripheral blood lymphocytes and of the lymphatic organs (lymphoid tissue). The primary lymphoid tissues are the thymus and bone marrow. Secondary lymphoid tissues are the lymph nodes in various regions of the body, especially the nasopharynx, axillae, groins, and intestines. The spleen is considered a secondary lymphatic organ.

B. Lymphocytes and the immune response

Lymphocytes carry the immune response. This was shown by the following experiment: The normal immune response of a mouse to an administered antigen (foreign organism, foreign molecular structure) was destroyed by a high dose of X irradiation, to which the immune system is especially sensitive. The immune response was then restored by lymphocytes of a genetically identical (from an inbred strain) unradiated mouse, but not by any other cells. Other cells are ineffective since only lymphocytes produce an immune response.

C. T cells and B cells

Lymphocytes exist as one of two functionally and morphologically different types, T lymphocytes and B lymphocytes. T lymphocytes undergo differentiation in the thymus during embryonic and fetal development, thus the designation T lymphocyte or T cell. B lymphocytes differentiate in various stages in the the bone marrow in mammals and in the bursa of Fabricius in birds (thus the designation B cells). A series of further differentiating steps take place in the lymph nodes (T cells) and in the spleen (B cells).

D. Cellular and humoral immune response

During the first phase of the immune response provoked by an antigen (e.g., a bacterium, virus, fungus, or foreign protein) there is rapid proliferation of B cells (humoral immune response). The B cells mature to plasma cells, which form free antibodies (immunoglobulins) directed at the antigen. The antibodies bind specifically to the antigens. The humoral immune response is rapid, but it does not reach foreign organisms that have invaded body cells. They are the target of the cellular immune response.

E. Antibody molecules (basic structure)

The basic structural motif of an antibody molecule (immunoglobulin) is a Y-formed protein composed of two heavy chains (H chains) and two light chains (L chains). They are held together at defined sites by disulfide bonds. L chains and H chains contain regions with variable and constant sequences of amino acids.

F. Antigen–antibody binding

A foreign molecule, the antigen, is recognized and firmly bound to a specific region of the antibody molecule, the antigen-binding site. Here, the amino acid sequence of the antibody molecule differs from one molecule to the next in three hypervariable regions. The result is the ability to bind a wide spectrum of different antigenic molecules. The three-dimensional structure of this region is known precisely, and important details of the binding process are understood (Diagram from B. Alberts et al., Molecular Biology of the Cell).

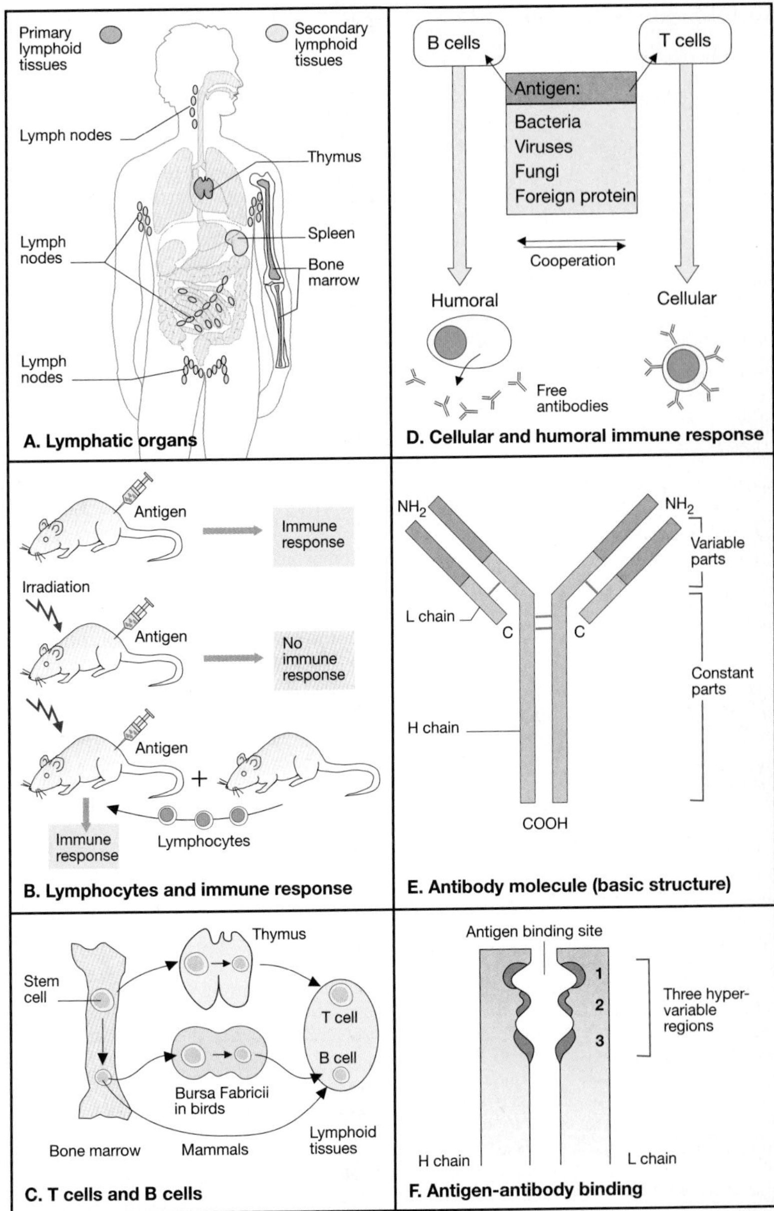

A. Lymphatic organs

D. Cellular and humoral immune response

B. Lymphocytes and immune response

E. Antibody molecule (basic structure)

C. T cells and B cells

F. Antigen-antibody binding

Immunoglobulin Molecules

Immunglobulins are the effector molecules of the immune system. They occur either as membrane-bound cell surface receptors or as free antibodies. The enormous diversity of their variable regions enables them to bind many very different antigens. Although they differ in details of their function, they share a relatively simple basic pattern, which is derived from a common ancestral molecule (p. 256).

A. Immunoglobulin G (IgG)

Immunoglobulin G is the prototype of antibody molecules produced by derivatives of B lymphocytes, the plasma cells. The molecule has two H chains and two L chains, held together by disulfide bonds. Each H chain has three constant regions (CH1, CH2, and CH3) and one variable region (V_H). Each H chain has a total of 446 amino acids, of which the first 109 belong to the variable region at the N-terminal end. Each L chain has one variable (V_L) and one constant (C_L) domain and consists of 214 amino acids. In the L chains also, the first 109 amino acids form the variable region. The variable domains of the H chains and the L chains form the antigen-binding sites. The three hypervariable regions within the V region of each chain are also called complementary determining regions (CDR) because the actual physical contact of molecules based on their complementary structure occurs in these regions. Each domain consists of about 110 amino acid residues. A jointlike area (hinge) between constant region 1 (CH1) and constant region 2 (CH2) of the heavy chain allows some flexibility of the molecule. The H chains are bound to each other and the H to the L chains by disulfide bridges (–S–S–). Furthermore, there are disulfide bridges within the constant and variable domains. The L chains are of one of two types, κ or λ. In addition to immunoglobulin G, there are other types of immunoglobulins, which differ from each other in the constant part of the H chain: IgA (C_α), IgD (C_δ), and IgE (C_ε). A very large immunoglobulin, IgM, is made up of five IgG subunits. The different types of H chains are referred to as isotypes.

B. T-cell receptor

Antigen receptors on the surfaces of T cells are heterodimers of covalently bound polypeptide chains, one α and one β chain. The basic structure of a cell-surface antigen receptor is similar to that of the secreted immunoglobulins except that the cell-surface receptors contain just one constant region. The β chain is the slightly larger chain. The constant regions of the α and β chains (C) each consist of 140–180 amino acids. The variable regions (V) consist of 102–109 amino acids and contain three hypervariable regions, like the immunoglobulin molecules. In addition to the α and β, genes for a T-cell-receptor γ and δ chain exist. The loci for the TCRα, γ, and δ chains in man lie together on chromosome 14; the locus for TCRβ lies on chromosome 13.

C. The different domains of an immunoglobulin molecule are coded for by different genes

Each immunoglobulin and receptor molecule is coded for by different DNA sequences, which belong to a large series of genes for the L chains (types κ and λ) and the H chain. Genes for the H chain are located on chromosome 14q32 in humans and chromosome 14 in mice. The genes for the κ light chain are located on the short arm of chromosome 2 (2p12) in humans and chromosome 6 in mice, and for the λ light chain on the long arm of chromosome 22 (22q11) in humans and on chromosome 16 in mice. The genes are rearranged in the developing B and T cells in a pattern that is different and specific in each cell, as seen in the next plate.

Reference

Strominger JL. Developmental biology of T cell receptors. Science. 1989; 244: 943–950.

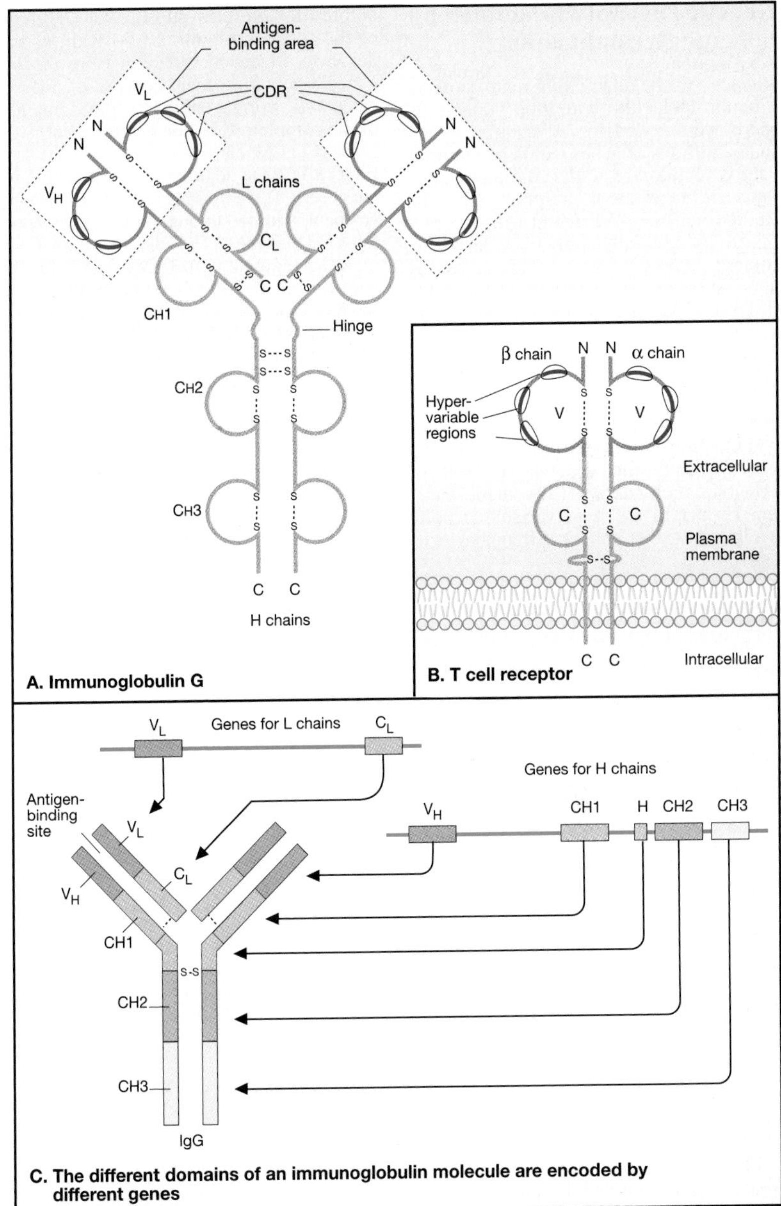

A. Immunoglobulin G

B. T cell receptor

C. The different domains of an immunoglobulin molecule are encoded by different genes

Genetic Diversity Generated by Somatic Recombination

If each of the many different immunoglobulin molecules with their variable regions were coded for by separate genes, many millions of genes would be required. This is not the case. Rather, during lymphocyte differentiation, a practically unlimited number of different cells are produced by recombination of a large but limited number of genes. This occurs by somatic recombination of genes during the differentiation of B cells and T cells. Antibody diversity arises by the following genetic mechanisms: (1) multiple DNA sequences of the germ-line genes for H and L chains can code for Ig molecules with different specificities; (2) somatic recombination of the various DNA segments greatly increases the number of possible combinations; (3) somatic mutations occur in the hypervariable regions, leading to further genetic differences.

A Somatic recombination during the formation of lymphocytes

Somatic recombination occurs within the genes for the L and the H chains during the maturation of B cells and within the genes for the four T-cell receptor chains in the maturing T cells. By means of this process, different coding DNA sequences (exons) of the various domains of the particular Ig molecule are in unique combinations in each cell. This provides each molecule with an antigen-binding specificity that differs from that of all other cells. Here, an example of the genetic processes during formation of an immunoglobulin H chain is shown. The exons (V1 to Vn) of the variable region (V) lie at the 5' end of the IgH locus. They are separated from each other by different lengths of noncoding DNA. A small exon (60–90 bp) that codes for a signal to initiate translation (leader or signal peptide L) lies more than 90 base pairs (bp) in the 5' direction of the V-region exons. Signal peptides guide the growing polypeptide into the lumen of the endoplasmic reticulum before they are cleaved off. The D genes of the constant region (C) lie at different distances in the 3' direction from the V genes. Each C segment consists of different exons, corresponding to the domains of the complete C region and different isotypes (C_μ, C_δ), etc.

First, a D gene segment and a J segment are joined (D–J joining). Next, this D–J segment is joined to one of the V segments (V–D–J joining). The combination of V, D, and J segments is different in each cell, with at least 25 000 different possible combinations due to the number of different segments present (100–125 V segments, 12 D segments, and 4 J segments). The joined VDJ segments form the primary RNA transcript. At this stage, noncoding segments (introns) are still present. In the example shown, a D2 is joined to a J1, but not with J2–4. J segments not directly connected to a D segment (here, J2–4) are removed. After the RNA is processed by splicing, the mature messenger RNA (mRNA) is formed as template for the translation of an H chain polypeptide. By further processing, such as removal of the leader segment (L) and glycosylation of the protein at certain sites, the definitive H chain is finally produced (of type **μ** in the example shown). Unlike the H chains, the L chains have no diversity (D) genes, so that a J and a V gene are directly joined by somatic recombination during DNA rearrangement in the lymphocytes. Thus, rearrangement of the genes for the H chain and two types of L chains in the lymphocyte DNA leads to a new combination of genes in each cell.

Reference

Abbas AK, Lichtman AH, Pober JS. Cellular and Molecular Immunology. 2nd ed. Philadelphia: WB Saunders; 1994.

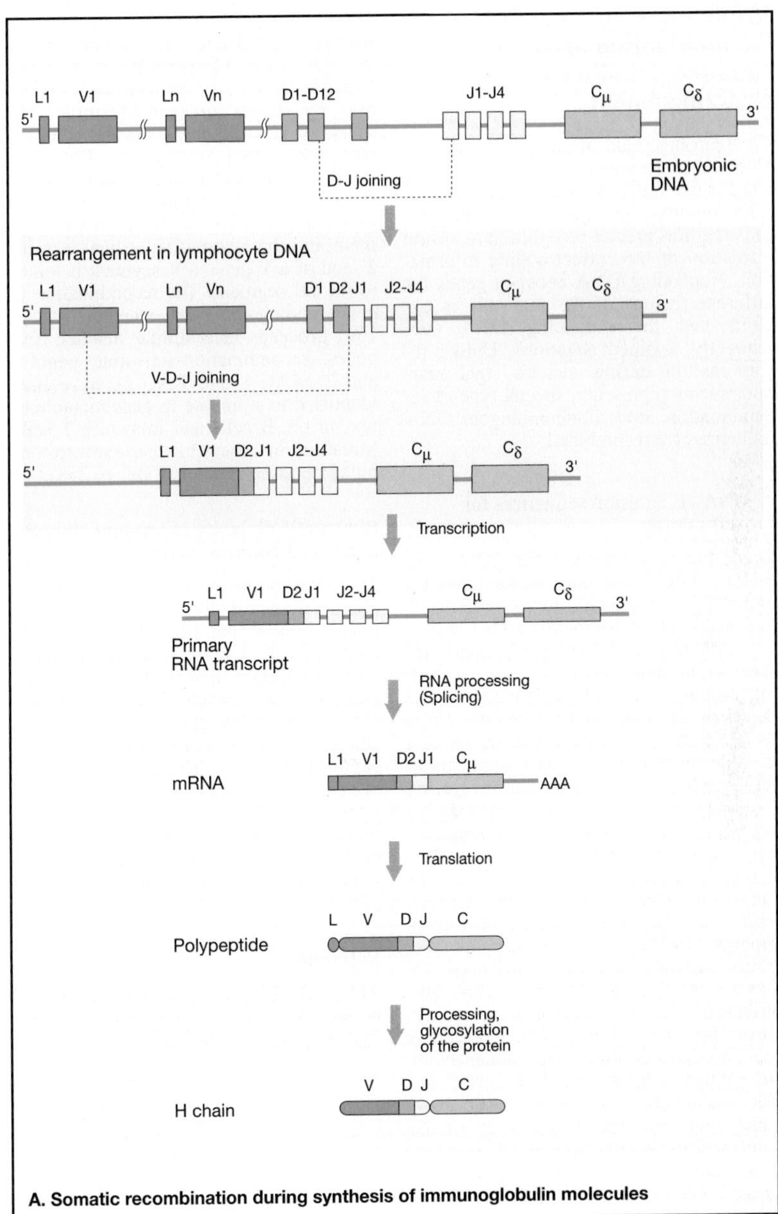

A. Somatic recombination during synthesis of immunoglobulin molecules

Mechanisms of Somatic Recombination in Immunoglobulin Gene Rearrangement

The rearrangement of genes for immunoglobulin molecules in immature B cells and for the T cell receptor in immature T cells involves an excision mechanism. This requires precise recognition to ensure formation of the correct coding information. Noncoding DNA between genes for different regions of the molecule is excised and the remaining DNA subsequently rejoined (ligation). Unlike recombination during meiosis, IgG rearrangements represent a special type of recombination, since nonhomologous DNA sequences are recombined.

A. DNA recognition sequences for recombination

Recombination between the DNA segments coding for an immunoglobulin molecule is mediated by a system of enzymes called recombinases. Their activities are varied and comprise lymphocyte-specific and more general activities including exonuclease and ligase functions. The enzymes are controlled by specific DNA recognition sequences. These are located in adjacent noncoding DNA segments at the 3' end of each V exon (exons for the variable region) and at the 5' end of each J segment. The D segments are flanked on both sides by recognition sequences. Recognition sequences are noncoding, but highly conserved DNA segments of seven base pairs (heptamer) or nine base pairs (nonamer). They are separated by precisely defined intervals, produced by spacers of 23 or 12 base pairs (bp). Upstream (5' direction) and downstream (3' direction) from a D segment, the spacers are 12 base pairs long. The sequences of the nucleotide base pairs of the spacers are not conserved. The characteristic rearrangement between neighboring signal sequences for Ig molecules and TCR receptors requires spacers of different lengths, i.e., 12 and 23 base pairs (so-called 12/23 rule). When an H chain is formed, nonhomologous pairing of the heptamer of a D segment and of a J segment occurs. These D and J segments are then joined (D–J joining) by means of recombination: The spacer of 12 or 23 base pairs and all of the intervening DNA forms a loop. This is excised, and the D and J segments are joined. By pairing and recombination of the recognition sequences at the 5' end of a DJ segment and the recognition sequence at the 3' end of a V gene, a V segment is joined to the DJ segment. The recombination of genes coding for T-cell receptors (see p. 254) proceeds in a similar manner. Two genes, recombination-activating genes 1 and 2 (RAG-1 and RAG-2), have been identified to stimulate Ig gene recombination in pre-B cells and immature T cells. Mutations in these genes cause severe combined immune deficiency (see p. 258).

B. Genetic diversity in immunoglobulin and T-cell-receptor genes

The total diversity, about 10^{18} possible combinations for all types of genes for immunoglobulins and T-cell receptors, is the result of different mechanisms. To begin with, different numbers of variable DNA segments are available for different chains (250–1000 for the H chain, 250 for the L chains, 75 for the α chain of the T-cell receptor TCR_{α}, etc.). The different D and J segments also multiply the number of possible combinations. Finally, DNA sequence changes (somatic mutations) occur regularly in the hypervariable regions, further increasing the total number of possible combinations.

Reference

Abbas AK, Lichtman AH, Pober JS. Cellular and Molecular Immunology. 2nd ed. Philadelphia: WB Saunders; 1994: 80.

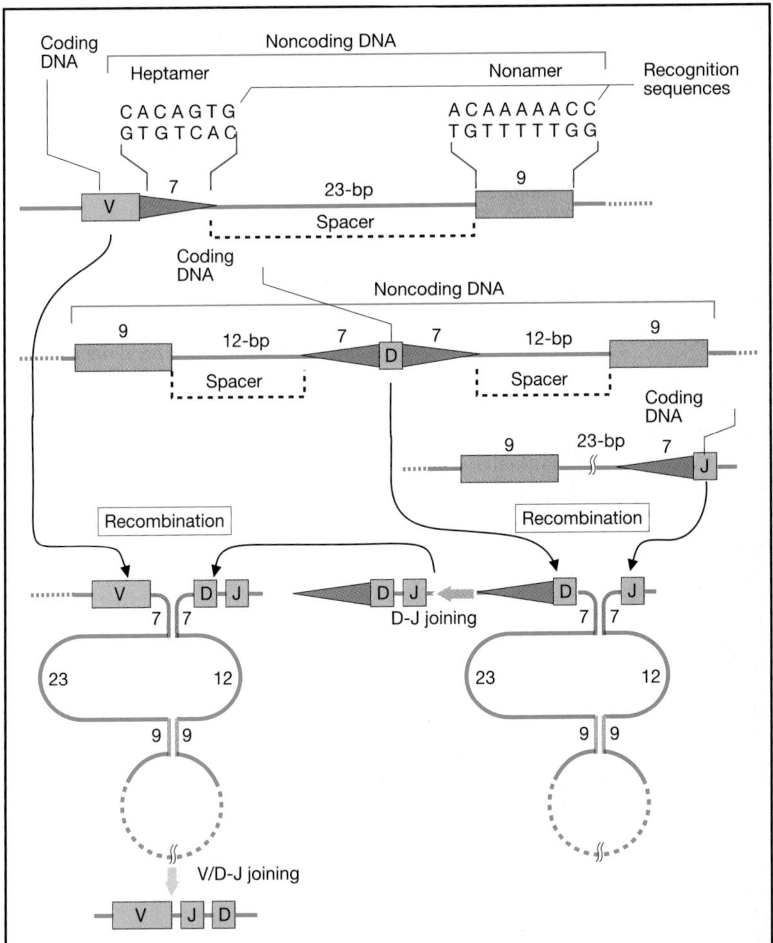

A. DNA recognition sequences for recombination

Mechanism	Immunoglobulin H chain	Immunoglobulin L chains	TCRαβ α	TCRαβ β	TCRγδ γ	TCRγδ δ
Variable domain	250 – 1000	250	75	25	7	10
Number of D segments	12	0	0	2	0	2
Number of J segments	4	4	50	12	2	2
Variable segment combinations	65 000 – 250 000		1825		70	
Total diversity	10^{11}		10^{16}		10^{18}	

B. Genetic diversity in immunoglobulin and T-cell receptor genes

Genes of the MHC Region

The MHC region (*m*ajor *h*istocompatability *c*omplex, MHC) is a region of highly polymorphic genes (about 10–50 alleles per locus). It spans about 3500 kb on the short arm of chromosome 6 in humans and on chromosome 17 in mice. Collectively, these genes are called immune response (Ir) genes. They are expressed on the surface of various cells. Their products can be demonstrated serologically or by cellular reactions in a mixed lymphocyte test. MHC genes control the immune response to antigen proteins by specific binding to T cells.

A. Basic structure of the MHC gene complex in humans and in mice

The gene loci of the MHC region are grouped into three classes (I–III). Class I in humans includes HLA-A, HLA-B, and HLA-C; in mice, D, L, and K of the H2 system. Many other loci also belong to this class, such as HLA-E to -J. Class II includes HLA-DP, -DQ, and -DR in humans and I-A and I-E (the letter I, not the Roman numeral) in mice. Strictly speaking, the class III genes do not belong to the MHC loci. They contain genes for different complement proteins and a few other genes. Here, alleles are designated according to a numerical system, e.g., HLA-A2, -B5, DR4, etc. The gene products of the alleles of the HLA system (*h*uman *l*eukocyte *an*tigens; also said by some to refer to *L*os *An*geles, where some of the first basic discoveries were made) can be demonstrated by the toxicity of serum of defined specificities to other leukocytes (serological cytotoxicity). Cytolysis occurs unless the specificities of the serum and of the cells being tested are identical. The gene products of the alleles of the HLA-D system are distinguished by the mixed lymphocyte test, based on lymphocyte proliferation as a reaction to foreign T cells.

B. Genomic organization of the MHC loci

The class II loci are located closest to the centromere. Each consists of a series of genomic subunits. The HLA-D region is about 900 kb long and contains a far greater number of loci than shown here. In addition, some have been renamed. For example HLA-DP has two α and two β genes, now called A1, A2 and B1, B2, respectively. The class III loci contain genes for complement factors C2 and C4, steroid 21-hydroxylase, and cytokines (tumor necrosis factor, TNFA; and lymphotoxin, LT, new designation TNFB). Several other class III genes are located here.

C. Class I and class II MHC molecules

Corresponding to the general organization of the individual class I and class II genes, the class I and class II molecules are basically similar. Both consist of two different polypeptide chains. In class I molecules, an MHC-coded α chain is associated with a non-MHC-coded β chain (β_2-microglobulin). The extracellular portion of the α chain consists of three domains, α3, α2, and α1, each with about 90 amino acids. Alpha1 and α2 form the highly polymorphic peptide-binding region; α3 and β_2-microglobulin structurally correspond to an immunoglobulinlike region. Elucidation of the crystalline structure of the class I MHC molecules showed that α1 and α2 interact to form a type of platform of eight-stranded β-folded proteins. The cleft formed between α1 and α2 (25 Å x10 Å x 11 Å) can bind a protein fragment consisting of 10–20 amino acids. Class II MHC molecules consist of two polypeptide chains, α and β, each with two domains, i.e., α1, α2 and β1, β2, each with about 90 amino acids and a transmembrane region of about 25 amino acids. As with the class I molecules, the peptide-binding regions (α1 and β1) are highly polymorphic. Unlike the β1 domain, α1 does not contain a disulfide bridge.

Reference

Trowsdale J. Genomic structure and function in the MHC. Trend Genet. 1993; 9: 117–122.

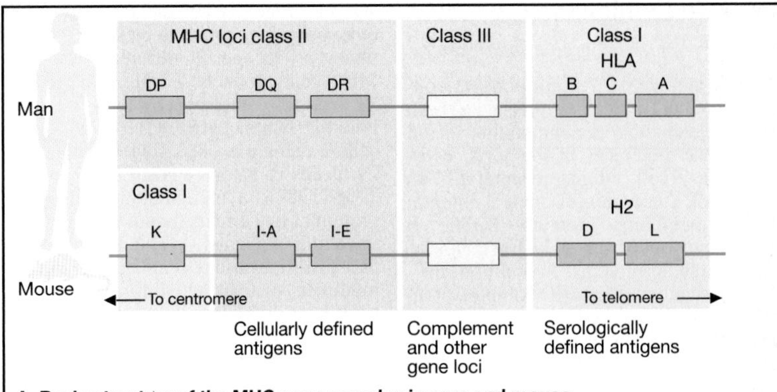

A. Basic structure of the MHC gene complex in man and mouse

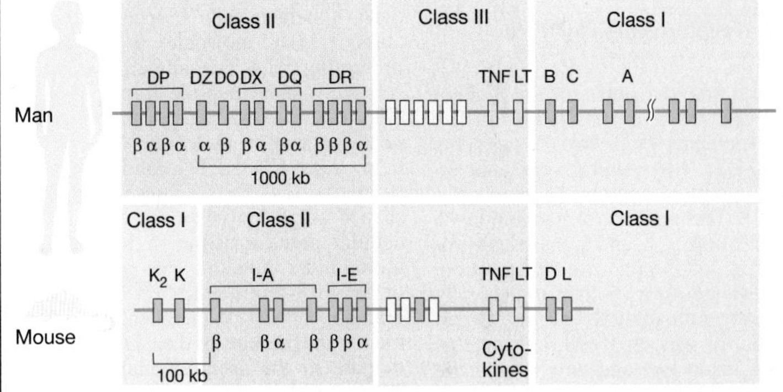

B. Genomic organization of the MHC loci

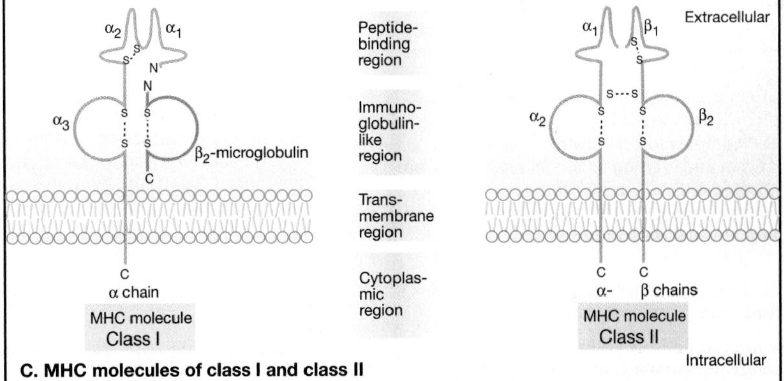

C. MHC molecules of class I and class II

T-Cell Receptors and the MHC System

The cell surface of a T lymphocyte contains receptor molecules that with high specifity recognize foreign antigens and cell-surface molecules of the MHC complex. The T-cell antigen receptor (TCR) consists of a complex of several integral plasma membrane proteins. Unlike B cells, T cells recognize only fragments of foreign antigen proteins. In addition, they bind physically to the MHC complex of an antigen-presenting cell. During maturation of the T cells in the thymus, T-cell gene segments are rearranged in a defined order by somatic recombination, similar to the formation of immunoglobulins.

A. T-cell receptor genes (TCR) in humans

In the germ line, the genes for the β chain of the T-cell receptor consist of 75–100 variable segments (V_β), two D segments (D-β1, D-β2), two joining segments (J-β1, J-β2), and two constant segments (C-β1, C-β2). The genomic organization of the genes for the α, δ, and γ chains is similar. During T-lymphocyte maturation, different segments are joined together by somatic recombination as during B-cell maturation. In a given T cell, only one of the two α-chain loci and only one of the two β-chain loci become functionally rearranged and expressed (allelic exclusion). As in the rearrangement of immunoglobulin genes, different mechanisms help produce diversity of the T-cell receptor genes. The genomic organization in humans and mice is very similar. The β-chain genes are located on chromosome 7 in humans and on chromosome 6 in mice. The α- and δ-chain gene loci are located on chromosome 14 in both humans and mice. The γ-chain genes lie on chromosome 7 in humans and on chromosome 13 in mice.

B. T-cell receptor binding to antigens and MHC surface proteins

Unlike B cells, T cells recognize and react to foreign protein antigens only when the antigens are attached to the surface of other cells. Two different classes of T lymphocytes recognize different types of MHC gene products. T cells with the ability to destroy other cells by cytolysis (cytolytic T lymphocytes, CTLs, or "killer cells") recognize class I MHC molecules by means of a coreceptor, CD8 (formerly T8). CD8 is a membrane-bound glycoprotein of two dimers (α and β), whose extracellular domains are held together by two disulfide bridges (either α-chain homodimers or α–β heterodimers). The extracellular aminoterminal domains are immunoglobulinlike. In general, the structure of CD8 resembles a box with rounded corners (Leahy et al., Cell. 1992; 68: 1145–1162; Praham P. Nature. 1992; 357: 538–539). The other class of T lymphocytes ("helper cells") specifically bind class II MHC molecules of the antigen-presenting cells. The antigen consists of a relatively small peptide fragment of foreign protein. With antigen presentation, a series of surface molecules lend physical help, e.g., CD3, CD4, and CD8. CD4 (formerly T4) is a coreceptor, a rod-shaped single polypeptide with an extracellular part consisting of four immunoglobulinlike domains. The gp120 protein of the HIV virus reacts with the second domain of CD4. A number of accessory membrane proteins of the CD3 system participate in the specific binding between TCR and MHC (TCR–CD3 complex).
(Drawings adapted from Abbas et al., 1994).

References

Amadou C, et al. Localization of new genes and markers to the distal part of the human major histocompatibility complex (MHC) region and comparison with the mouse: new insights into the evolution of mammalian genomes. Genomics. 1995; 26: 9–20.

Fugger L, et al. The role of human major histocompatibility complex (HLA) genes in disease. In: Scriver CR, et al., eds. The Metabolic and Molecular Bases of Inherited Disease. 7th ed. New York: McGraw-Hill; 1995: 555–585.

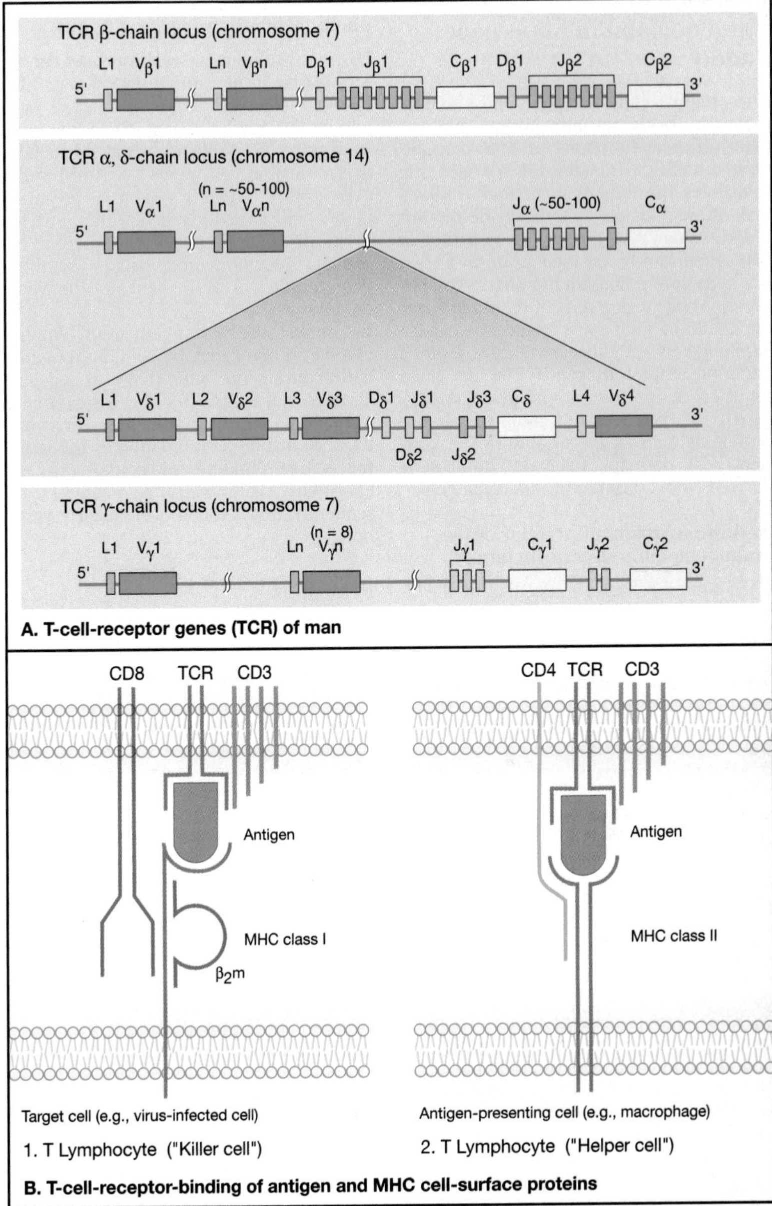

A. T-cell-receptor genes (TCR) of man

B. T-cell-receptor-binding of antigen and MHC cell-surface proteins

Evolution of the Immunoglobulin Supergene Family

The many cell surface and soluble molecules of the immune system that mediate different functions such as recognition, binding, or adhesion of specific molecules show many structural similarities. Some parts are found outside the immune system. As a group, they constitute a gene superfamily, derived from an ancestral gene common to all members. The homologies of the domains of their gene products and of their gene sequences can be explained by evolutionary origin from a common ancestral gene. The Ig gene family code for immunoglobulin domains, usually of about 70–110 amino acids homologous with an Ig variable (V) or constant (C) domain. Each Ig domain is derived from conserved DNA sequences.

A. Basic structure of proteins of the immunoglobulin supergene family

The immunoglobulin molecules of the T-cell receptors (TCR) and the class I and class II MHC molecules (1) are basically similar. They consist of variable Ig-like domains (V), constant Ig-like domains (C), or primordial Ig-like domains (H).
Although their genes are located on different chromosomes, the gene products form functional complexes with each other. Others, such as the V, D, and J gene segments of all antigen receptors and their genes for the C domain lie close together in gene clusters. In addition, genes for the MHC loci and for the two CD8 chains lie together. The basic structures of accessory molecules (2) such as CD2, CD3, CD4, CD8, and thymosine 1 (Thy 1) are relatively simple. Other members of the Ig superfamily (3) are the Fc receptor II (FcRII); polyimmunoglobulin receptor (pIgR), which transports antibodies through the membranes of epithelial cells; NCAM (neural cell adhesion molecules); and PDGFR (platelet-derived growth factor receptor) (3) (after T. Hunkapiller & L. Hood, Advances in Immunology. 1989; 44: 1–63).

B. Evolution of genes of the immunoglobulin supergene family

Distinct evolutionary relationships can be recognized in the homology of genes for Ig-like molecules. A precursor gene for a variant (V) and a constant (C) region must have arisen by duplication and subsequent diversification of a gene for a primordial cell surface receptor. Such a primitive gene could have looked like the gene for thymosine (Thy-1) or poly-Ig receptor. No somatic recombination occurs in these gene families or in the genes of the MHC complex.
In contrast, the rearrangement of lymphocyte germ line genes by somatic recombination during the maturation of B cells and T cells is the basis for the formation of immunoglobulins, T-cell receptors, and CD8. Somatic recombination of the genes for antigen-binding molecules was an enormous evolutionary advantage. Consequently, this is found even in early vertebrates.

References

Abbas AK, Lichtman AH, Pober JS. Cellular and Molecular Immunology, 2nd ed. Philadelphia: WB Saunders; 1994.

Hood L, Kronenberg M, Hunkapillar, T. T-cell antigen receptors and the immunoglobulin supergene family. Cell. 1985; 40: 225.

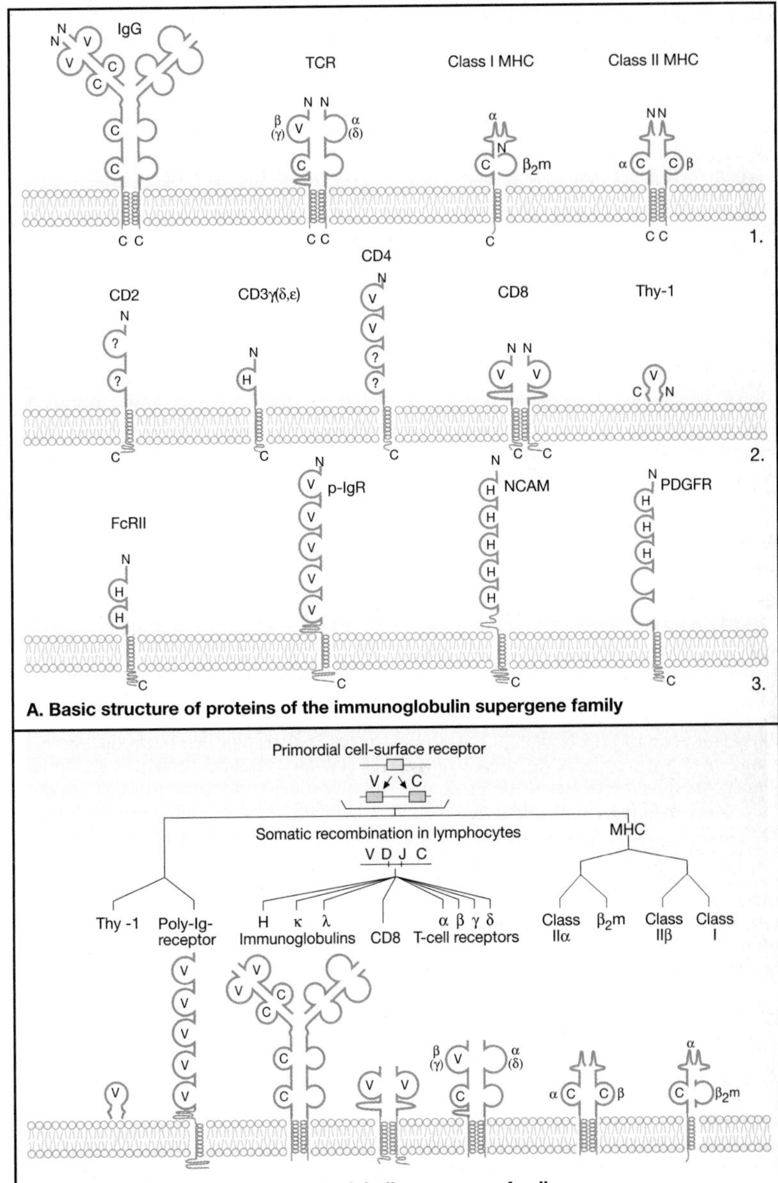

A. Basic structure of proteins of the immunoglobulin supergene family

B. Evolution of genes of the immunoglobulin supergene family

Hereditary and Acquired Immune Deficiency

Immune system protection against infectious organisms and their effects after successful invasion is a basic requirement for survival. Functional impairment of the immune system leads to severe illness, collectively called the immune deficiency diseases. The primary cause may be either congenital, due to a genetic defect (hereditary or primary immune deficiency), or acquired (secondary immune deficiency).

A. Examples of hereditary immune deficiency diseases

Many genetically determined defects of the immune system are known. A defect in the common precursor cell of B and T cells leads to a group of diseases called severe combined immune deficiency (SCID). In this group of diseases, both B cells and T cells are affected. This has correspondingly severe clinical consequences. At least one X-chromosomal and one autosomal recessive form are known. A disorder of B-cell maturation leads to the absence of free immunoglobulins (X-chromosomal agammaglobulinemia type Bruton). Other disorders concern B-cell differentiation (variable immune deficiency) or alteration of an isotype (isolated Ig isotype deficiency, such as IgA deficiency). An embryonic developmental disorder of the third and fourth branchial arches results in the absence or hypoplasia of the thymus and parathyroids, and in some cases of the aortic arch (DiGeorge syndrome). T lymphocytes cannot mature because the thymus is hypoplastic. Absence or hypoplasia of the parathyroids leads to severe dysfunction in calcium metabolism.

B. Example of acquired immune deficiency: HIV-1 infection

Acquired immune deficiency is a frequent accompanying sign of diseases and immunosuppressive medications and the sequel to a harmful dose of whole-body irradiation. In the last 10 years, the disease caused by HIV-1 and HIV-2 virus infection (AIDS, *a*cquired *i*mmune *d*eficiency *s*yndrome) has become very significant. It affects CD4-expressing T cells (T helper cells and macrophages). The HIV virus belongs to the lentivirus family. It is a retrovirus with an RNA genome of two identical RNA strands of about 9.2–9.7 kb located within the viral capsid protein (see p. 94). This is surrounded by a phospholipid membrane envelope derived from the host cell. Two virus-coded membrane glycoproteins, a transmembrane glycoprotein weighing 41 kD (gp41) and a glycoprotein weighing 120 kD (gp120) are bound to the envelope. The proteins gp120 and gp41 play a decisive role in HIV infection. Gp120 in primates (man and the anthropoid apes) has a high binding affinity for the CD4 receptors of T cells and mononuclear phagocytes. The first step in HIV infection is the high-affinity binding of the viral gp 120 to the CD4 receptor of the T helper cells. When gp120 binds to a CD4 receptor, the viral genome is taken into the cell. The viral RNA is transcribed into provirus DNA by reverse transcriptase. Virus production begins after the provirus DNA is integrated into the DNA of the T cell (infected cell). The new infectious virions are released from the cell by exocytosis (see p. 89). If the provirus DNA is not integrated into the host cell genome, a latently infected cell results. Virus production may be activated at a later time by, e.g., stimulation of the T-cell receptor or by cytokines. The infection leads to loss of $CD4^+$ T cells and to immune deficiency. In addition, the binding of $CD4^+$ T cells to MHC class II molecules of antigen-presenting cells is blocked.

References

Barrett DJ, et al. Antibody deficiency diseases. In: Scriver CR, et al., eds. The metabolic and molecular bases of inherited disease. 7th ed. New York: McGraw-Hill; 1995: 3879–3894.

Pantaleo GC, et al. The immunopathogenesis of human immunodeficiency virus infection. N Eng J Med. 1993; 328: 327–335.

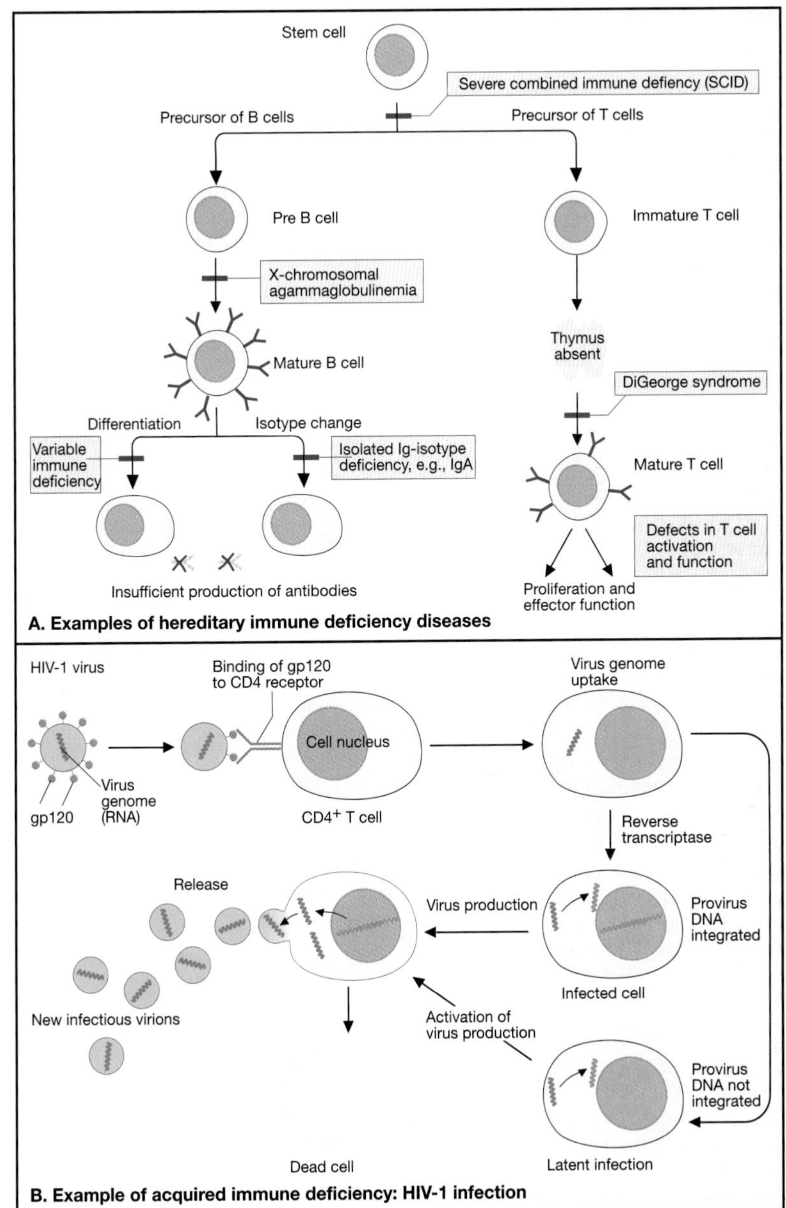

A. Examples of hereditary immune deficiency diseases

B. Example of acquired immune deficiency: HIV-1 infection

The Influence of Growth Factors on Cell Division

Development, differentiation, and the maintenance of vital functions require exact regulation of the time and location of cell divisions. Rapidly multiplying cells in embryonic tissues must be controlled, just like those in stationary phases in adult tissues. Rapid response to injury or to foreign antigens requires controlled cell division. Multicellular organisms have an extensive repertoire of genetically regulated mechanisms at their disposal for controlling cell division and tissue proliferation. As a group, they are referred to as growth factors. Every growth factor has a specific cell-surface receptor. Binding to the receptor initiates (or in some cases blocks) cell division. Most growth factors regulate only certain types of cells and tissues.

A. Control of cell division by growth factors

Basically, cell division (mitosis) can be controlled by stimulation or inhibition. In the absence of stimulation or with active inhibition, no mitosis occurs. Growth factors have an effect not only on specific types of cells, but also on defined phases of the cell cycle. The most frequently controlled phase of the cell cycle is the transition from G_0 to G_1. The growth factor group includes growth factors for epidermal cells (EGF), for nerve cells (NGF), for connective tissue or mesenchymal cells (fibroblasts, FGF), and for thrombus-forming cells in the inner lining (endothelium) of blood vessels (PDGF). Their stimulating effect may be opposed by an antagonistic effect (e.g., TGF, transforming growth factor or TNF, tumor necrosis factor). The function of each growth factor is mediated by a specific receptor.

B. Activation of a growth factor receptor

A growth factor receptor becomes activated by specific extracellular binding to the growth factor. The activated receptor in turn activates a substrate protein.

C. PDGF-receptor kinases have an effect on numerous substrates

A receptor such as the PDGF (*p*latelet-*d*erived *g*rowth *f*actor) receptor can have an effect on numerous substrates. Substrates of the PDGF receptors include the Ras protein (see **D**), the Src protein (the name is derived from the tumor, a sarcoma, in which it was first found), phospholipase C (a signal transmitter), and others.

D. Ras proteins as signal transmitters

The Ras proteins play a central role as signal transmitters. They belong to the group of G proteins (guanosine-residue-binding proteins with signal-transmitting functions, see p. 226). The binding of growth factor, e.g., PDGF, activates the Ras protein by stimulating the exchange of associated GDP (guanosine diphosphate) for GTP (guanosine triphosphate) and triggering a short time-limited signal that initiates cell division. The signal is terminated by inactivation of Ras by a GTPase-activating protein (GAP), which converts GTP into GDP. Mutation of the Ras protein or of GAP can remove the time limit of the cell-stimulating signals and result in an active condition with uncontrolled cell division. This can lead to a tumor with uncontrolled growth (malignancy). Several mutations have been defined in the pertinent genes.

(Diagrams adapted from JD Watson et al., Recombinant DNA, 1992).

Reference

Park M. Oncogenes: Genetic abnormalities of cell growth. In: Scriver CR, et al., eds. The Metabolic and Molecular Bases of Inherited Disease. 7th ed. New York: McGraw-Hill; 1995: 589–611.

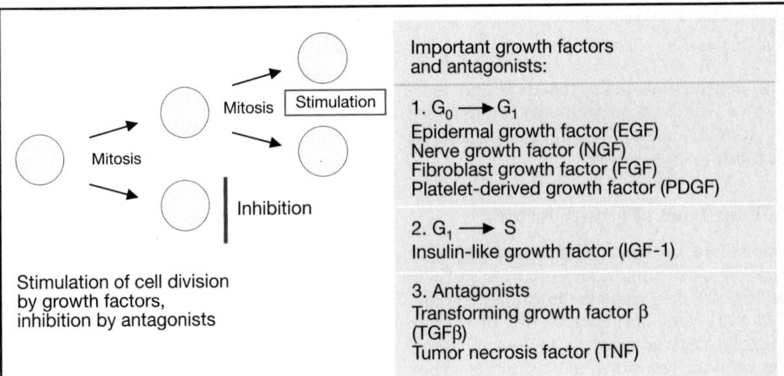

A. Control of cell division by growth factors

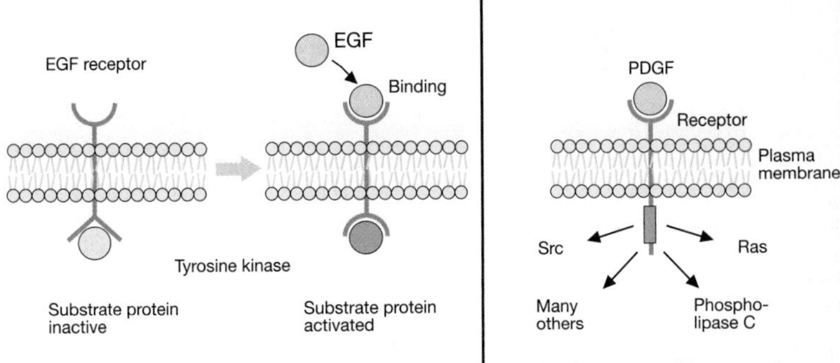

B. Activation of a growth factor receptor

C. PDGF-receptor kinases act via several substrates

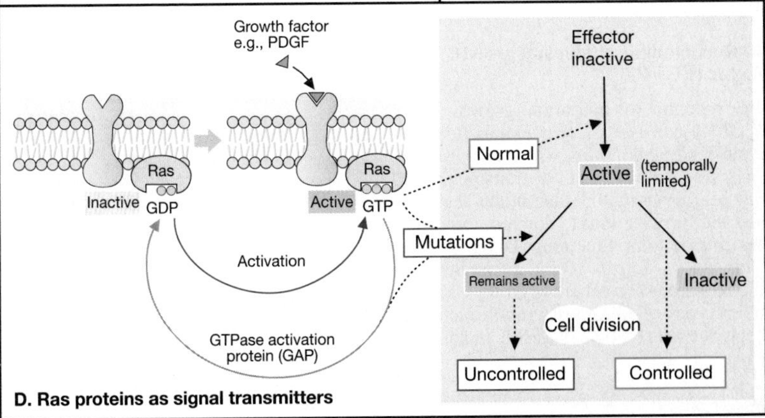

D. Ras proteins as signal transmitters

Growth Factors and Their Receptors

For mammalian cells, the time and location of cell division (growth) is controlled by growth factors, which either stimulate or inhibit cell division.

A. Four types of growth factors

Growth factors can be grouped into four classes: class I includes epidermal growth factor (EGF), growth hormone (somatotropin), and erythropoietin or PDGF (platelet-derived growth factor). Class II are cell-surface hormone receptors. They are often protein tyrosine kinases in the cytoplasmic domains of the receptor (see p. 226). Class III forms the large group of intracellular signal transmitters (transducers). They belong to different families, such as protein kinases, e.g., src or the ras proteins, including H-ras virus (Harvey murine sarcoma). Class IV are nuclear transcription factors. *Jun* codes for the protein AP1, which binds to DNA sequences in the promotors and enhancers of many genes. Further nuclear transcription factors are fos, myc (from myelocytomatosis in chickens), myb (from myeloblastosis in chickens), or N-myc. Others, such as protein p53 and retinoblastoma gene product (RB), are transcription factors that suppress cell proliferation (see pp. 268, 274).

B. Receptor for epidermal growth factor (EGFR)

The receptor for epidermal growth factor (EGFR) consists of two extracellular ligand-binding domains, which are cysteine-rich for stabilization; a transmembrane domain; a short juxtamembrane segment; and the tyrosine kinase domain, important for intracellular function. (Diagram after Branden & Tooze, 1991). In addition to the receptor shown here (FGFR1), several other types of fibroblast growth factor receptors exist (FGFR2, FGFR3, and others).

C. Domain structure of four growth factor receptors

The evolutionary relationship of certain growth factor receptors and their genes is evident in the similarity of their structures and nucleotide base sequences. The EGF receptor consists of 1186 amino acids. The different domains are coded for by matching exons in the gene. The receptor for insulinlike growth factor (IGF-1) shows structural similarities to the EGF receptor, although the former consists of two α and two β chains bound by disulfide (–S–S–) bridges. The thrombocyte-derived receptor for growth factor (PDGF, platelet-derived growth factor) shows similarity only in the protein-tyrosinase region. The extracellular domains 1–5, consisting of about 500 amino acids, structurally resemble immunoglobulin even though only 20% of their sequences are homologous. For stability, all ligand-binding extracellular domains except four contain the disulfide bridges characteristic of immunoglobulins. The product of the viral oncogene v-*erbB* resembles the EGF receptor, but the gene product lacks the largest of the extracellular domains of the EGF receptor and 34 amino acids of the C-terminal end. Because of this, the tyrosine-kinase region is altered so that uncontrolled cell division is induced after infection with the virus.

Genetic defects in different types of fibroblast growth factors (FGF3 and FGF2) have recently been found to be the cause of important genetic diseases such as achondroplasia, Apert syndrome, Crouzon syndrome, Pfeiffer syndrome, and others (R Shiang et al. 1994; F Rousseau et al. 1994; W Reardon et al. Nature Genet. 1994; 8: 98–103. M Muenke et al. Nature Genet. 1994; 8: 269–274).

References

Shiang R, et al. Mutations in the transmembrane domain of FGFR3 cause the most common form of dwarfism, achondroplasia. Cell. 1994; 78: 335–342.

Rousseau F, et al. Mutations in the gene encoding fibroblast growth factor receptor 3 in achondroplasia. Nature. 1994; 371: 252–254.

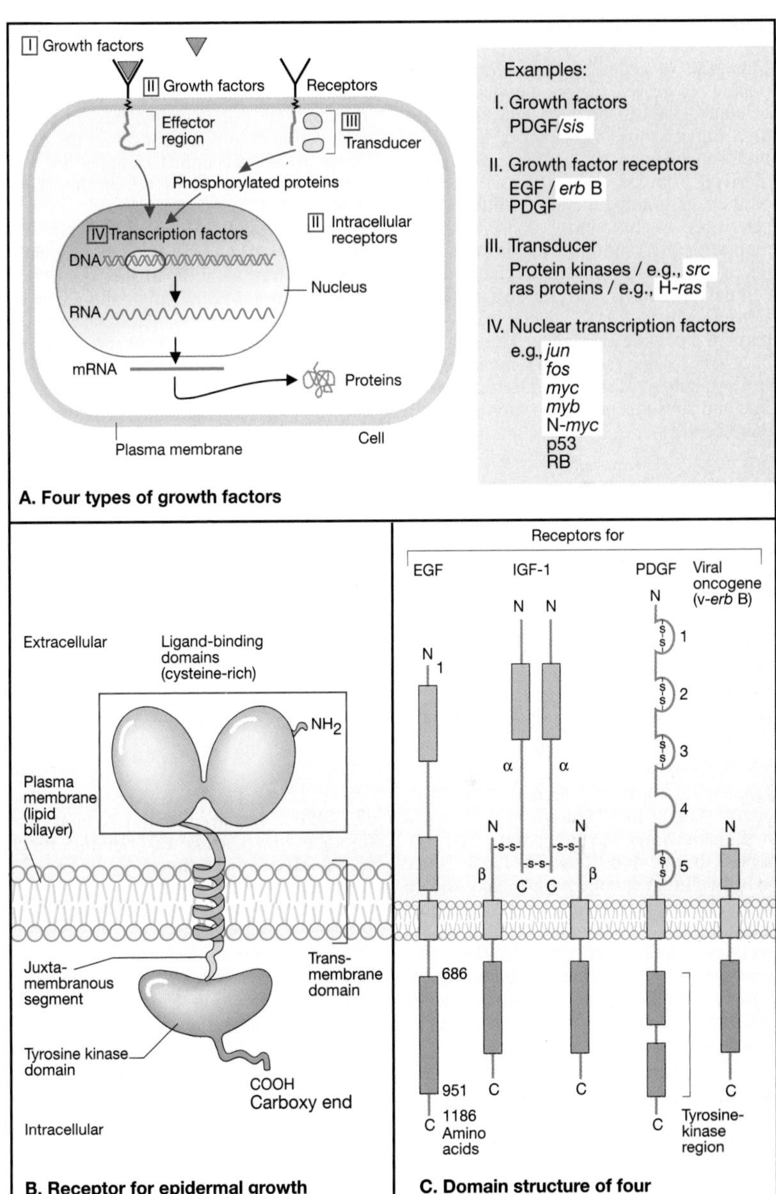

A. Four types of growth factors

B. Receptor for epidermal growth factor (EGF)

C. Domain structure of four growth receptors

Cellular Oncogenes

Mutations of genes for the different components of growth factors may lead to loss of control of cell division and eventually to a tumor; thus, these genes are called proto-oncogenes or cellular oncogenes (derived from Gk *oncos*, tumor). Because of their essential function, cellular oncogenes are evolutionarily conserved, i.e., their structure (and function) is very similar or identical in distantly related multicellular organisms. Certain viruses (retroviruses with an RNA genome) contain parts of a cellular oncogene. This is a viral oncogene, v-*onc* (see p. 94). Invasion of a cell by such a virus can lead to transformation and subsequent cell division of the affected cell.

A. Cellular and viral oncogenes

A typical retrovirus contains an RNA genome that codes for three genes or groups of genes: *gag* (group-specific antigen), *pol* (polymerase), and *env* (coat protein, *env*elope). As with all genes of higher organisms, a cellular oncogene (c-*onc*) consists of exons and introns with defined structure and sequence, as in, e.g., the gene *src* (src is derived from sarcoma, a tumor that is induced by a change in this gene). The virus may contain parts of the cellular oncogene (c-*scr*). This is designated viral oncogene (v-*src*) (Rous sarcoma virus). In chickens, it induces a malignant tumor (a sarcoma), first observed by Peyton Rous in 1911. Since many cellular oncogenes are also known in an altered, viral form, it is assumed that the viruses have integrated parts of the respective cellular oncogenes into their own genomes. Virus-induced tumors are known especially in chickens, rodents, and cats. In man, they do not play a special role in the induction of tumors.

B. Mechanisms of oncogene activation

A cellular oncogene controls cell division. It controls the time and location of the orderly proliferation of cells and tissues (normal growth). Genetic changes can lead to disorders of the regulation of cell division, increased proliferation of cells, and formation of a tumor. This can be traced back to relatively few mechanisms. A point mutation in a critical region of the gene can lead to disturbances in the regulation of cell division. Examples are mutations in codon 12 or 63 of the H-*ras* gene.

An inactive cellular oncogene may become activated when it is moved by chromosomal translocation to the vicinity of an active gene. In Burkitt lymphoma, an inactive gene is moved to the region of an active gene for the H or L chain of an immunoglobulin. In other cases, the breakpoint of the chromosome translocation may lie within a cellular oncogene and thereby affect its expression. An example is the Philadelphia translocation (see p. 270). Multiplication (amplification) of a gene is a further mechanism that can lead to altered (usually increased) expression.

C. Examples of cellular oncogenes and their proteins

The table shows examples of the about 60 known cellular oncogenes, their basic functions, a few tumors induced in man by mutation of the cellular oncogene (c-*onc*), and tumors induced in vertebrates by the homologous viral oncogene (v-*onc*). (Data from Darnell J, Lodisch H, Baltimore D. Molecular Cell Biology. 2nd ed., 1990. Cannon-Albright LA, et al.: Assignment of a locus for familial melanoma, MLM, to chromosome 9p13–p22. Science. 1992; 258:1148).

References

Levine AJ, Broach JR, eds. Oncogenes and cell proliferation. Current Opin Genet and Development. Feb. 1995; 5: 1–150.

Park M. Oncogenes: Genetic abnormalities of all growth. In: Scriver CR et al., eds. The Metabolic and Molecular Bases of Inherited Disease. 7th ed. New York: McGraw-Hill; 1995: 589–611.

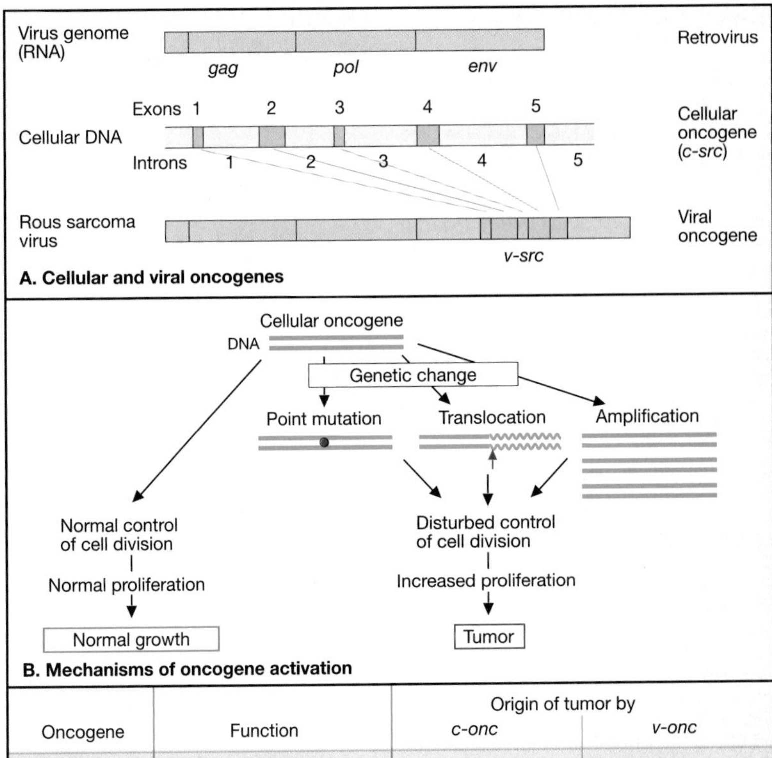

A. Cellular and viral oncogenes

B. Mechanisms of oncogene activation

Oncogene	Function	Origin of tumor by *c-onc*	Origin of tumor by *v-onc*
sis	One form of PDGF		Simian sarcoma
abl	Tyrosine protein kinase	Chronic myeloigenous leukemia	Abelson mouse leukemia
fes	Tyrosine protein kinase		Cat sarcoma
erbB	Epidermal growth factor	Mammary carcinoma Ovarian carcinoma	Chicken erythro-blastosis
src	Membrane-bound protein kinase		Rous chicken sarcoma
Ha-*ras* Ki-*ras* N-*ras*	Membrane-bound G proteins with GTPase activity	Different carcinomas, neuroblastoma	Harvey mouse sarcoma Kirsten mouse sarcoma
fos	Transcription factor AP1		Osteosarcoma in chicken
myb	Nuclear protein	Leukemias	Myeloblastosis in chicken
myc	Nuclear protein	Leukemias	Myelocytosis in chicken
N-*myc*	Nuclear protein	Neuroblastoma	

C. Examples of cellular oncogenes and their proteins

Tumor Suppressor Genes

Tumor suppressor genes are a group of genes that normally suppress cell division. Loss of normal function in both alleles leads to uncontrolled cell division and tumor growth. The presence of a normal allele suppresses the development of a tumor (tumor suppression). The loss of function must affect both alleles before an effect occurs (i.e., mutations of tumor suppressor genes are recessive on a cellular level). This requires two mutational events, under some circumstances more. Tumor suppressor genes correspond to the brakes of a car; cellular oncogenes, to the accelerator.

A. Tumor suppressor gene

In contrast to the cellular oncogenes, for which a change in one allele will alter normal function, both alleles of a tumor suppressor gene must lose their function before a tumor develops. The first event is usually a mutation by base exchange or deletion. The second event, affecting the other allele (allele 2), may also be a mutation, but the loss of function more often appears to be from loss of the chromosome after a faulty cell division (mitotic nondisjunction) or other mechanisms (e.g., mitotic recombination with gene conversion).

B. Loss of heterozygosity in tumor cells

Usually, in about half of the individuals who are heterozygous for DNA markers at the tumor suppressor gene locus of interest, the loss of one allele (event 2) can be demonstrated by Southern blot analysis. In contrast to normal somatic cells (blood), tumor cells contain only one allele (loss of heterozygosity, LOH). The remaining allele was most likely altered by mutation. By demonstration of LOH, the mutant allele can be identified. LOH is useful in diagnosis as an indication of the existence of a tumor suppressor gene.

C. Somatic and germinal mutation

The first mutation in a suppressor gene can either be present in the zygote (germinal mutation, i.e., germ cell mutation due to transmission from an affected parent or due to new mutation) or occur in a single cell of the corresponding tissue (somatic mutation). Loss of function of one allele (corresponding to event 1 in A) predisposes the cell to tumor development.

With a germinal mutation, all cells are predisposed. The tumor arises after loss of function of the second allele. When somatic mutation occurs in a single cell, loss of function of the second allele rarely affects the same cell. But with a germ cell mutation, loss of function of the second allele is frequent, since all cells carry the first mutation, i.e., are predisposed. With somatic mutation, the tumor occurs sporadically (is not hereditary) and arises unifocally from a single cell. In the hereditary form resulting from a germ cell mutation, several tumors may arise from different cells (multifocal tumor). The predisposition for the tumor in the hereditary form shows autosomal dominant inheritance.

D. Examples of tumor suppressor genes

Numerous types of tumors arise due to loss of function in both alleles of a tumor suppressor gene. Loss of heterozygosity (LOH) can be demonstrated in the tumor cells in about half of these patients.

References

Cavence W, White RL. The genetic basis of cancer. Scient Amer. March 1995; 50–57.

Skuse GR, Ludlow JW. Tumour suppressor genes in disease and therapy. Lancet. 1995; 345: 902–906.

Stanbridge EJ. Human tumor suppressor genes. Ann Rev Genet. 1990; 24: 615–657.

Weinberg RA. Tumor suppressor genes. Science. 1991; 254: 1138–1146.

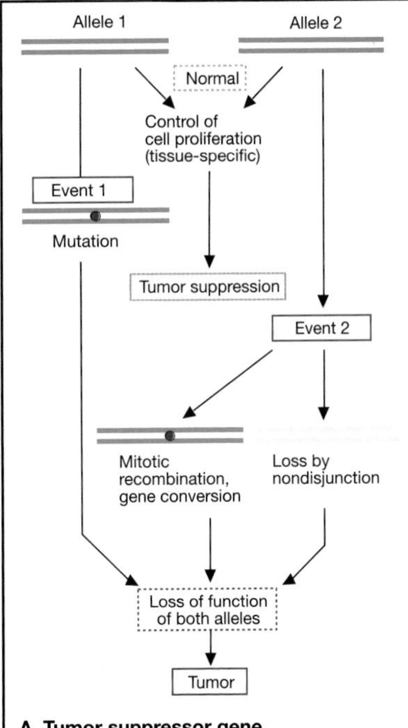

A. Tumor suppressor gene

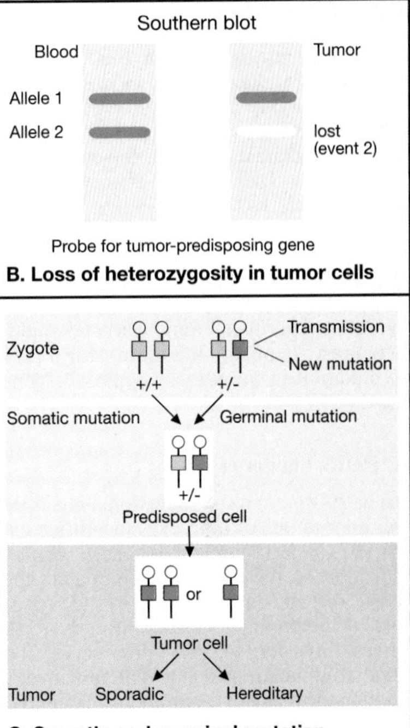

B. Loss of heterozygosity in tumor cells

C. Somatic and germinal mutation

Chromosomal localization	Type of tumor
1p	Melanoma; multiple endocrine neoplasia type 2; neuroblastoma, others
1q	Breast cancer (one form)
3p	Kidney cell carcinoma; von Hippel-Lindau disease
5q	Familial polyposis coli, colon carcinoma
9p	Familial melanoma
9q	Bladder carcinoma
10q	Astrocytoma, MEN type 2
11p	Wilms tumor and others
13q	Retinoblastoma; osteosarcoma
17p	Breast cancer, colon carcinoma, and others
17q	Neurofibromatosis Type 1
18q	Colon carcinoma
22q	Neurofibromatosis type 2; meningioma

D. Examples of tumor-suppressor genes (loss of heterozygosity in tumor cells)

Retinoblastoma

Retinoblastoma is a tumor that arises in the undifferentiated retina in one or both eyes during infancy or early childhood. With a frequency of about 1 in 12 000, it is the most frequent tumor of the eye in this age group. The normal retinoblastoma gene product has DNA-binding function and controls the cell cycle at the transition from the G1 to the S phase. The loss of function of both alleles of the gene results in a tumor. Retinoblastoma occurs in a hereditary (germ cell mutation) and a non-hereditary (somatic mutation) form due to mutation in the tumor suppressor gene *RB-1*.

A. Retinoblastoma

An early sign of a retinoblastoma is a light shimmer in one or both eyes (so-called cat eye) (1). The tumor appears as a whitish elevation on the retina (2). Untreated, the tumor spreads out within the eye (3) (photos: W. Höpping, Essen). In advanced stages, invasion of the optic nerve and other complications occur. The tumor may be limited to one eye or may occur in both eyes (unilateral or bilateral retinoblastoma). The tumor may arise at a single site, in somatic mutation (unifocal), or at several sites, in the hereditary form (multifocal). However, in clinical practice, the two forms often cannot be differentiated.

B. The retinoblastoma locus on chromosome 13

In humans, the locus for retinoblastoma (*RB-1*) is in region 1, band 4.1 to 4.2 (13q14.1–q14.2), on the proximal half of the long arm of chromosome 13.

C. The retinoblastoma gene *(RB-1)*

The retinoblastoma gene is a large gene of about 180 kb. It consists of 27 exons, which code for a small 4.7-kb transcript. Several genomic probes and cDNA probes are used for diagnosis. Restriction sites for *Hin*dIII form the most important map. Today, usually microsatellites are used for indirect DNA diagnosis of familial retinoblastoma.

D. Demonstration of a deletion

The patient (red square in the pedigree) has an interstitial deletion of one chromosome 13 with loss of the regions 13q14.1 and 13q14.2. A Southern blot analysis (digestion with *Hin*dIII and use of genomic probe H3-8) demonstrates the deletion at a molecular level. The affected boy inherited a 2.1-kb DNA fragment (allele 1) from his mother, but not the 1.5-kb fragment (allele 2) from his father. His sister has inherited allele 1 from their mother and allele 2 from their father. Thus, she is not at risk for retinoblastoma (findings of B. Horsthemke, Essen).

E. Demonstration of a point mutation

In the family shown (1), the mother (I-2) and a daughter (II-2) are affected. The mutation causes a C to T transition in the codon CAA (glutamine) at position 575. In this way, the stop codon TAA was formed (2) (findings of D. Lohmann, Essen). It results in a gene product that is about 300 amino acids too short and nonfunctional. In the daughter II-1, the tumor risk can be ruled out with certainty, since she does not have the mutation. She will not have to be subjected to frequent examination under general anesthesia for many years, which would otherwise be necessary with an a priori 50% risk.

References

Newsham IF, et al. Retinoblastoma. In: Scriver CR, et al., eds. The metabolic and molecular bases of inherited disease. 7th ed. New York: McGraw-Hill; 1995: 613–642.

Lohmann D, et al. Spectrum of small length germline mutations in the RB1 gene. Hum. Mol. Genet. 1994; 12: 2187–2193.

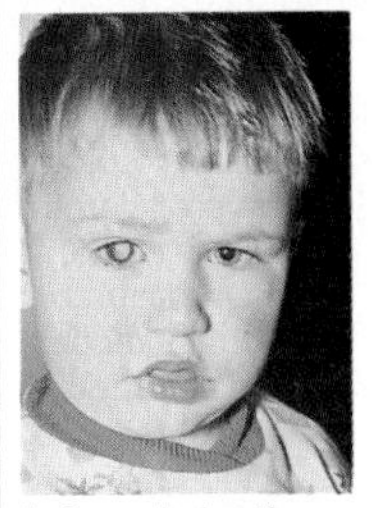
1. So-called cat's eye

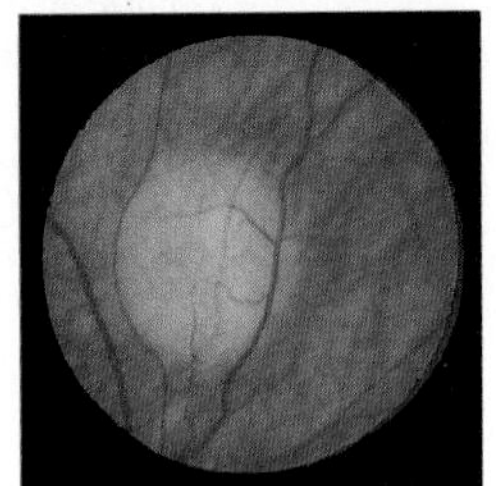
2. Tumor in the retina

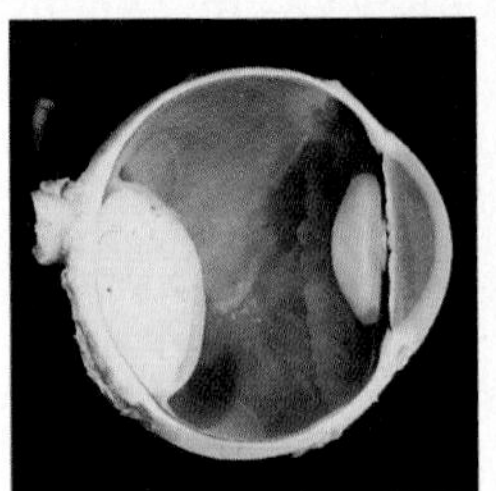
3. Large tumor in the eye

A. Retinoblastoma

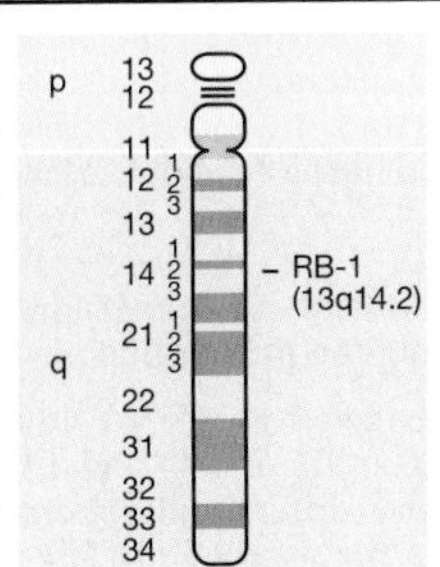

B. Retinoblastoma locus on chromosome 13

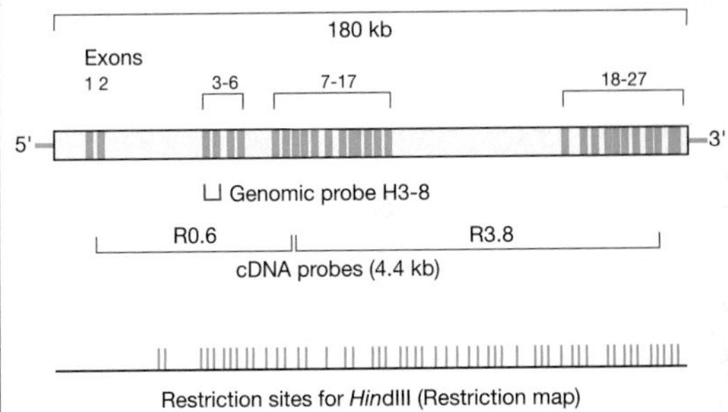

C. Retinoblastoma gene (RB-1)

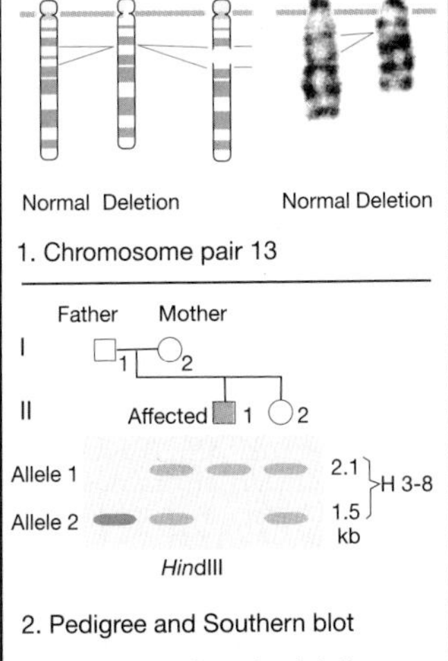

D. Demonstration of a deletion

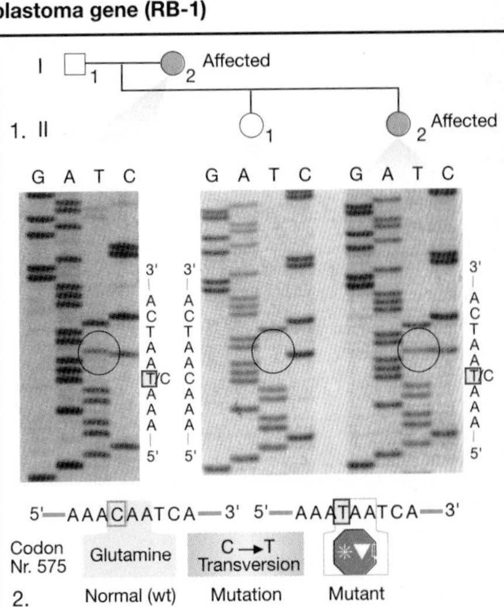

E. Demonstration of a point mutation

Tumor Origin Due to Chromosomal Translocation

Chronic myeloid leukemia (CML) is a malignant tumor that originates from a single cell of the bone marrow in adulthood. The number of myelocytes (white blood cells from the bone marrow) are greatly increased. The disease follows a chronic course. Acute crises develop intermittently and terminally. In about 90% of the patients, affected bone marrow cells contain a chromosome 22 with a shortened long arm (22q–, Philadelphia chromosome).

A. The Philadelphia chromosome (Ph^1) in different forms of leukemia

A Philadelphia chromosome is present in the bone marrow cells of most patients with the chronic form of the disease (CML). If it is not present, the illness progresses more rapidly than usual and has a poorer prognosis. In addition, the Philadelphia chromosome may be found in some acute leukemias (acute lymphocytic leukemia, ALL; acute myelocytic leukemia, AML) in adults and in children. Here, Ph^1 indicates a poor prognosis, whereas its absence is favorable.

B. The Ph^1 translocation [t(9;22)(q34;q11)]

The Philadelphia chromosome arises by reciprocal translocation between a chromosome 22 and a chromosome 9. The breakpoints are in 9q34 and 22q11. A good half of the long arm of a chromosome 22 is translocated to the long arm of a chromosome 9. A very small segment of the distal long arm of a chromosome 9 (9q34), not visible under the light microscope, is translocated to a chromosome 22. The Philadelphia chromosome (22q–) consists of the short arm and the proximal one-third of the long arm of a chromosome 22 and the small distal segment from the long arm of a chromosome 9. For demonstration of the Philadelphia translocation by in situ hybridization, see p. 192.

C. The Ph^1 translocation leads to the fusion of two genes

The breakpoints of the Ph^1 translocation are located in the *BCR* gene of chromosome 22 and in the *ABL* gene of chromosome 9. The translocation leads to the fusion of these genes. The exact locations of the breakpoints differ from patient to patient, but in the *BCR* gene they are limited to a small region of just 6 kb (thus the designation *BCR*, or *b*reakpoint *c*luster *r*egion). In CML, the breakpoints lie in exons 10–12 of the *BCR* gene; in acute Ph^1-positive leukemias (e.g., ALL), further in the 5' direction in exon 1 or 2. The breakpoint region in the *ABL* gene extends over 180 kb between exons 1a and 1b, which are separated by an intron of about 150 kb.

D. The gene fusion leads to changes in transcription and gene production

The *ABL* gene codes for mRNA transcripts of 7 kb (exon 1b, 2–11) and 6 kb (exon 1a, 2–11) by differential splicing; these in turn code for a protein of about 145 000 molecular weight ($p145^{abl}$). From the fusion of the two genes in CML, an 8.5-kb mRNA transcript results, which codes for a fusion protein of 210 000 mol. wt. ($p210^{bcr/abl}$). In the acute form of leukemia (ALL), a transcript results that codes for a fusion protein of 190 000 mol. wt. ($p190^{bcr/abl}$). In contrast to the normal protein, it has high tyrosinase activity. This results in uncontrolled cell division in the affected cells and tumor growth.

References

Cline MJ. The molecular basis of leukemia. N Engl J Med. 1994; 330: 328–336.

Hentze BM, Kulozik AE, Bartram CR. Einfuhrung in die medizinische Molekularbiologie. Grundlagen, Klinik, Perspektiven. Berlin: Springer; 1990.

Kurzrok R, Gutterman JU, Talpaz M. The molecular genetics of Philadelphia-positive leukemias. N Engl J Med. 1988; 319: 990.

Chronic myelogenous leukemia (CML)

Myelocytes in peripheral blood increased
Early chronic course
Adults affected
Origin from one myeloid cell
In ca 90% a Philadelphia chromosome (Ph^1)
(Translocation 22q to 9q)
Poor prognosis when Ph^1-negative

Other Ph^1-positive acute leukemias

Lymphocytes or myelocytes increased
Acute course
20% adults with ALL Ph^1-positive
2% adults with AML Ph^1-positive
5% children with ALL Ph^1-positive
Philadelphia translocation as in CML
Poor prognosis when Ph^1-positive

A. Philadelphia translocation (Ph^1) in different forms of leukemia

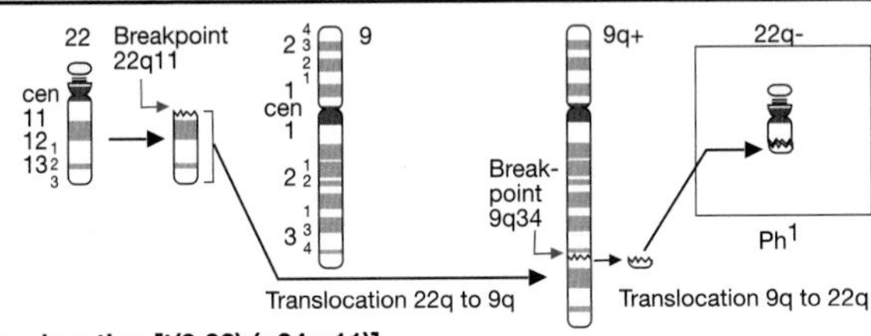

B. Ph^1 translocation [t(9;22) (q34;q11)]

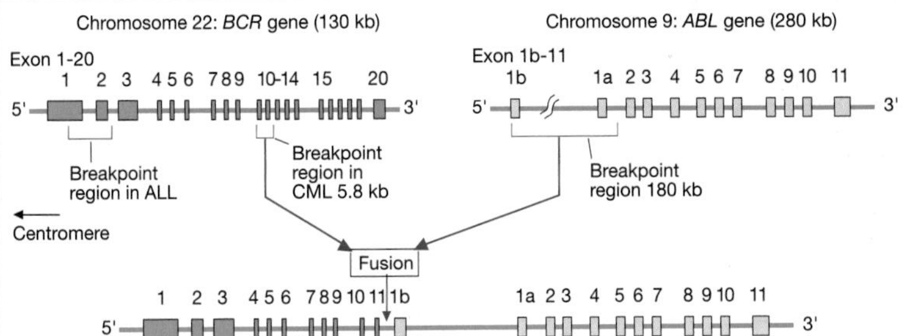

C. Ph^1 translocation causes fusion of two genes

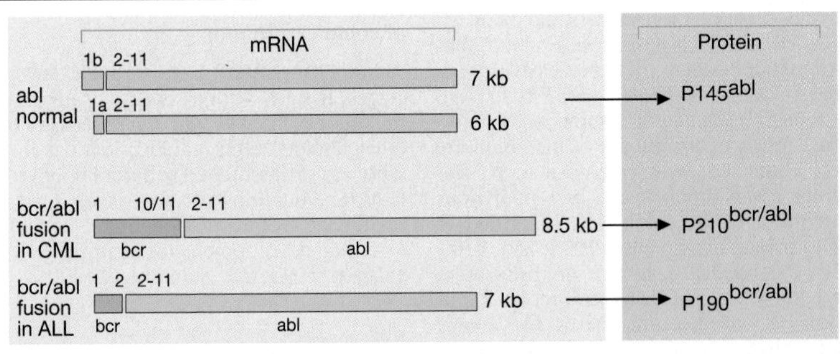

D. Gene fusion causes altered transcripts and gene products

Mutations in Intestinal Tumors (Colon Carcinoma)

Carcinoma of the colon (intestinal cancer) is one of the most frequent forms of malignancy in adults. The involvement of genetic factors in the etiology is implicated by a tendency to increased incidence in some families and by several forms of autosomal dominant hereditary polyposis coli (polyp formation in the large intestine) with a high risk for carcinoma of the colon.

A. Polyposis coli and colon carcinoma

Familial polyposis (FAP) is an autosomal dominant hereditary disease. In late childhood and early adulthood, up to 1000 and more polyps develop in the mucous membrane of the large intestine (colon) (1). Each polyp can develop into a carcinoma (2). Early recognition of this risk is important. In about 85% of affected persons, small hypertrophic areas that do not affect vision are present in the retina (3). (Photos 1 and 2 from U. Pfeifer, Institut für Pathologie der Universität Bonn; photo 3 from W. Friedl et al., Dtsch Ärztebl. 1991; 88: B–851, with kind permission of the authors).

B. Mutations at different gene loci in polyposis coli and carcinoma of the colon

At least six gene loci are involved in the development of carcinoma of the colon associated with polyposis coli. Somatic mutations may occur in two recessive oncogenes (Ras genes *KRAS1* and *KRAS2*) and in four dominant tumor-suppressor genes. Most forms of carcinoma of the colon are not associated with polyposis coli. Recently, genes predisposing to nonpolypous carcinoma of the colon were identified. P. Peltomäki et al. (Science. 1993; 260; 810–812) localized a gene that predisposes to hereditary nonpolypous colorectal carcinoma to chromosome 2p15–16. Tumor cells showed instability of dinucleotide repeats (CA) (L. A. Aaltonen et al., Science *260*: 812–816, 1993). Since no loss of heterozygosity (LOH) was found in tumor cells, this is not a tumor-suppressor gene.

C. The *APC* gene and distribution of mutations

The coding sequences of the *APC* gene are contained in 15 exons. The protein consists of 2844 mainly hydrophilic amino acids in a sequence without homology to other known genes (Groden et al., 1991; Joslyn et al., 1991). The loss of function of both of the alleles of the *APC* gene is thought to initiate development of a carcinoma of the colon (Ichii et al., 1992). Powell et al. (1992) found somatic mutations exclusively within the first 2000 amino acids, mainly within exon 15. A deletion at one particular site (codon 1309) predisposes to a tumor occurring at a young age (about 30 years).

D. Indirect DNA diagnosis in FAP

Linked DNA markers (RFLPs) near the APC locus (1) can be used for indirect DNA diagnosis. The alleles of three flanking marker pairs (K,k and E,e on the centromere side and A,a on the distal side) form the haplotypes, e.g., e–K–a and e–k–a in individual I-1 in the pedigree (2). The mutation-carrying haplotype must be e–K–a. Since individual III-2 has inherited this haplotype, he is at risk for the disease, whereas individual III-1 is not.

E. Several mutations in the production of colon carcinoma

Tumor formation goes through several stages. It starts with a somatic or germinal mutation in the APC gene. After loss of the other allele (LOH), an adenoma develops with less differentiated cells and polyp formation. Mutations in other genes lead to malignant transformation and eventually to tumor development (diagram after Vogelstein, 1990; E.J. Stanbridge, 1990).

References

see p. 373.

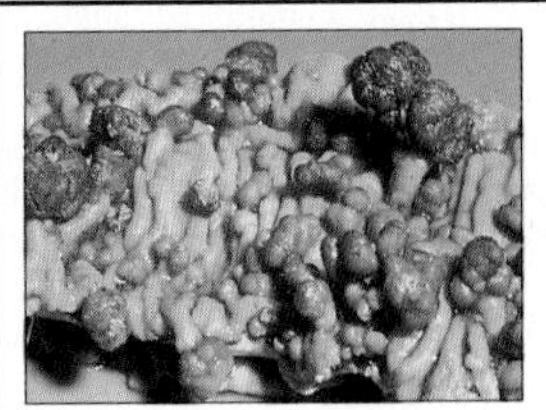
1.

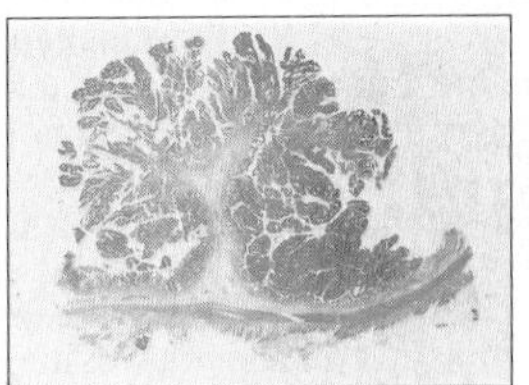
2.

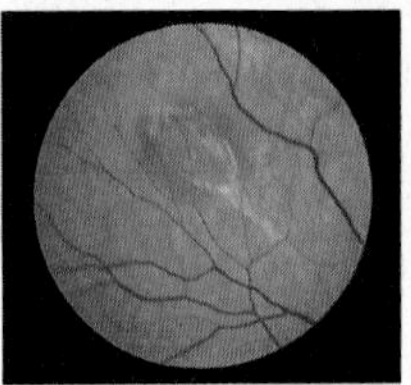
3.

A. Polyposis coli and colon carcinoma

McKusick no.	Disease/gene involved	Gene locus
190110	Colorectal adenoma/carcinoma (KRAS1)	6p12-11
190070	Colorectal adenoma/carcinoma (KRAS2)	12p12
114500	Isolated colon carcinoma (somatic mutation in the APC gene)	5q21
175100	Familial adenomatous polyposis coli (germinal mutation in the APC gene)	5q21
175100	APC with other tumors: Gardner syndrome (allelic)	5q21
159350	DNA sequences mutated in colon carcinoma (MCC)	5q21-22
191170	p53 gene (identical with McKusick *120460)	17p12-13
120470	Deletion in colorectal carcinoma (DCC)	18q21.3

B. Mutations at different gene loci in polyposis coli and colon carcinoma

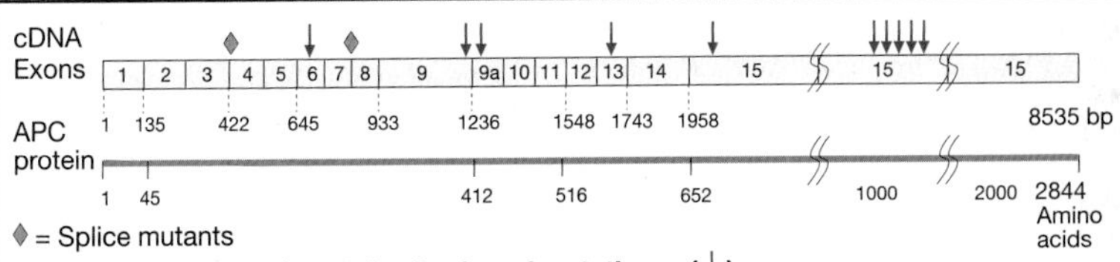

C. APC gene (scheme) and distribution of mutations (↓)

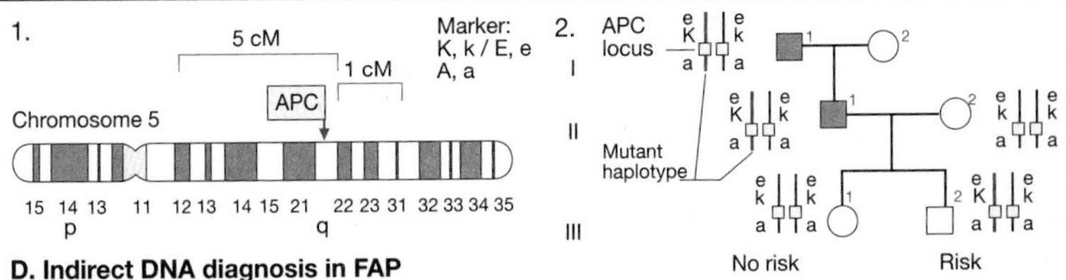

D. Indirect DNA diagnosis in FAP

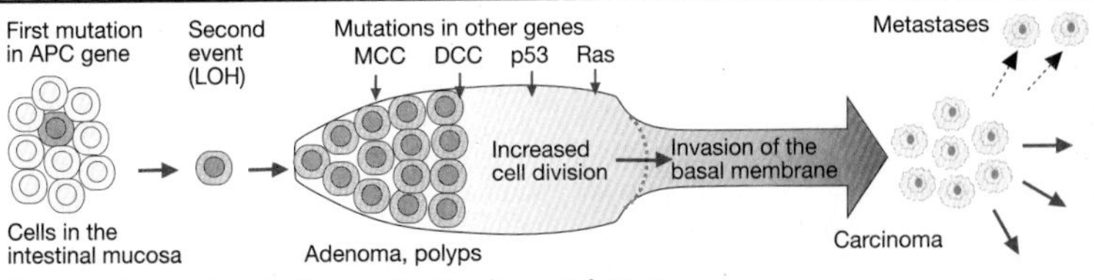

E. Several mutations in the origin of colon carcinoma

The *p53* Gene and Tumors

Mutations in the *p53* gene are among the most frequent genetic changes in cells of different malignant tumors in man. The *p53* gene is located on the short arm of chromosome 17 and codes for a nuclear protein of 53 000 daltons (53 kD, thus the designation p53). The p53 protein is a transcription factor and an important regulator of the cell cycle. When p53 is inactivated by mutation or by oncogene proteins from a DNA tumor virus, regulation of the cell cycle is disrupted and a tumor can develop.

A. The *p53* gene and gene product

The *p53* gene contains 393 codons. A region with transcription-activating (ACT) properties is located at the N-terminal end of its gene product. The HSP domain (coded for by codons 13–29) of mutant p53 can bind to heat-shock proteins. A serine residue at position 315 can become phosphorylated by the product of the *CDC2* gene, a gene involved in cell-cycle regulation. Mutations in the *p53* gene are concentrated in four regions ("hot spots"): in codons 129–146, 171–179, 234–260, and 270–287. These hot spots correspond to regions conserved in evolution, underlining their functional importance.

B. Mutations of the *p53* gene in different tumors

Mutations frequently occur (in 12–13% of all tumors) in codons 175, 248, and 273. These contain CG-rich sequences, which can be readily methylated to 5-methyl-cytosine. This in turn leads to the replacement of the cytosine by thymine after the next replication (after Levine et al. 1991).

C. Familial multiple tumors due to mutation of the *p53* gene

The Li–Fraumeni syndrome (McKusick no.*114480) is an autosomal dominant hereditary disorder in which affected individuals develop one or several tumors (1). In the family shown here, six individuals (I-1, II-2, II-3, III-1, III-2, and III-5) have a CGG (Arg) to TGG (Trp) mutation in codon 248 of the *p53* gene. They have an increased risk of developing tumors, whereas III-3 and III-4 do not. Tumors in patients with Li–Fraumeni syndrome show a characteristic spectrum (2) (after Malkin et al. 1990). In the pedigree shown, only four of the six persons at risk have, in fact, developed tumors.

D. Model of function of the *p53* gene

Normally, the *p53* gene is inactive (1). Mice homozygous for a deletion and humans with a germ line mutation of the *p53* gene develop normally (Donehower et al. 1992). However, *p53* plays an important role in regulating growth in damaged cells (2). DNA damage in cells leads to increased expression of *p53* and interruption of the cell cycle in G1. If DNA repair is successful, the cell can continue its cycle. If repair is not successful, the cell dies (cell death, apoptosis). Damaged cells with p53 protein that is mutant are not arrested in G1. There is not enough time for repair, and in the next S phase the damage will be doubled (after DP Lane. Nature. 1992; 358: 15–16).

References

Donehower LA, et al. Mice deficient for p53 are developmentally normal, but susceptible to spontaneous tumors. Nature. 1992; 356: 215.

Frebourg T, et al. Germ-line p53 mutations in 15 families with Li–Fraumeni syndrome. Am J Hum Genet. 1995; 56: 608–615.

Harris CC, Hollstein M. Clinical implications of the p53 tumor-suppressor gene. N Engl J Med. 1993; 18: 1318–1327.

Levine AJ, Momand J, Finlay CA. The p53 tumor suppressor gene. Nature. 1991; 351: 453.

Malkin D, et al. Germ line p53 mutations in a familial syndrome of breast cancer, sarcomas, and other neoplasms. Science. 1990; 250: 1233–1238.

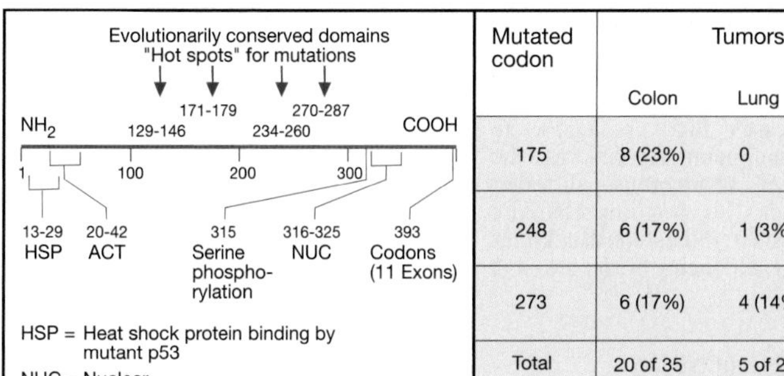

HSP = Heat shock protein binding by mutant p53

NUC = Nuclear localization signal

ACT = Transcription activating domain

A. P53 protein

Mutated codon	Tumors		
	Colon	Lung	Total
175	8 (23%)	0	11 (12%)
248	6 (17%)	1 (3%)	11 (12%)
273	6 (17%)	4 (14%)	12 (13%)
Total	20 of 35 (57%)	5 of 29 (17%)	34 of 94 (36%)

B. Mutations in the p53 gene in different tumors

I
II
III

mt mt mt N mt mt N N mt

Mutation in codon 248: CGG ⟶ TGG (Arg) (Trp)

mt= Mutation present

N= Mutation not present

1. Autosomal dominant inheritance

Other tumors possible: lung, prostate, pancreas, colon, lymphoma, melanoma

Brain 12%

Soft-tissue sarcoma 12%

Breast cancer 25%

Adrenal cortex 1%

Bone 6% (Osteosarcoma)

Bone marrow 6% (Leukemia)

2. Distribution of tumors

C. Mutations of the p53 gene in familial multiple tumors (Li-Fraumeni syndrome)

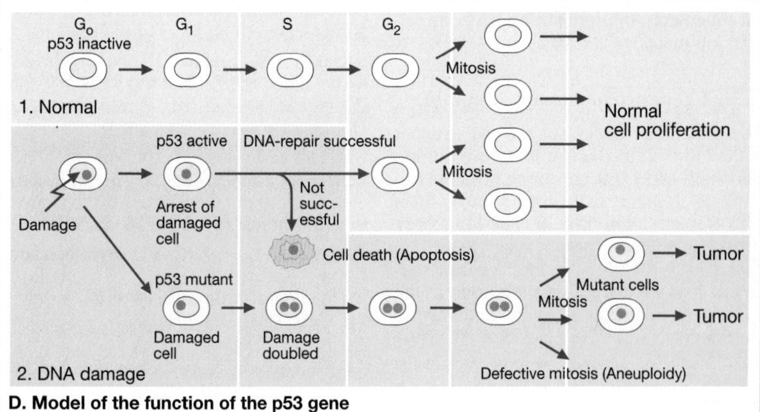

D. Model of the function of the p53 gene

Neurofibromatosis 1 and 2

The neurofibromatoses are clinically and genetically different autosomal dominant hereditary diseases that predispose to benign and malignant tumors of the nervous system. Numerous different forms are known. The most important are neurofibromatosis 1 (NF1, von Recklinghausen disease) and neurofibromatosis 2 (NF2).

A. The main signs of NF1

NF1 is very variable. Lisch nodules of the iris (1) in more than 90% of patients, café-au-lait spots (2) (more than five spots of more than 2-cm diameter are considered diagnostic) in more than 95%, and multiple neurofibromas (3) in more than 90% of patients are the most important signs. In addition, various skeletal anomalies (scoliosis in about 30%, pseudoarthrosis in about 3%) may be present. Malignant tumors of the nervous system occur in about 3–5%. About 50% of affected persons have a new mutation.

B. *NF1* gene

The *NF1* gene lies on the long arm of chromosome 17. It consists of at least 56 exons of different sizes. It spans about 350 kb and encodes an 11–13-kb transcript. Digestion with *Eco*RI (E) produces different fragments that constitute a useful restriction map.

C. NF1 gene product (neurofibromin)

The *NF1* gene codes for a gene product with 2485 amino acids. Between amino acids 840 and 1200, this large protein contains a domain that corresponds to a GTPase-activating protein. The homology includes a gene product in yeast (*S. cerevisiae*), IRA1 (inhibitor of ras mutants). It is assumed that mutations at the *NF1* locus interrupt a signal pathway to the *ras* genes (after Xu G, et al. Cell. 1990; 62: 599–608).

D. Candidate gene for NF2

NF2 is much less frequent than NF1 and is essentially limited to the area of the acoustic nerve (eighth cranial nerve), from which an acoustic neuroma arises (usually bilaterally). However, other tumors of the brain may occur, especially schwannomas and meningiomas. *NF2* codes for a 4.5-kb transcript (schwannomin, SCH). The *NF2* locus lies on chromosome 22 (22q12). A region of contig II (contiguous sequences, see p. 356) is shown of 28 overlapping cosmids in steplike arrangement. A systematic analysis of over 100 *Eco*RI fragments identified three CpG islands. With their help, genes with transcripts of 2.5 kb (EWS, Ewing sarcoma. Delattre O, et al. Nature. 1992; 359: 162–165), of 4.0 kb (unknown gene), and of 3.9 kb (NEFH, neurofilament heavy chain) could be identified. In addition, a fourth, previously unknown gene with a 4.5-kb RNA transcript was identified. This is the candidate gene for neurofibromatosis 2. It has been named *schwannomin (SCH)*, after the Schwann cells of the nervous system in which NF2 tumors originate, and *merlin* by other authors (*m*eosin-*e*zrin-*r*adixin-*l*ike-prote*in*, a member of the 4.1 cytoskeleton protein family, see p. 319).

References

MacCollin M, et al. Mutational analysis of patients with neurofibromatosis 2. Am J Hum Genet. 1994; 55: 314–320.

Riccardi VM, Eichner JE. Neurofibromatosis. Phenotype, Natural History and Pathogenesis. 2nd ed. Baltimore: Johns Hopkins University Press; 1992.

Rouleau GA, et al. Alteration in a new gene encoding a putative membrane-organizing protein causes neurofibromatosis type 2. Nature. 1993; 363: 515–521.

Trofatter JA, et al. A novel Moesin-, Ezrin-, Radixin-like gene is a candidate for the neurofibromatosis 2 tumor suppressor. Cell. 1993; 72: 791–800.

Xu G, et al. The neurofibromatosis type 1 gene encodes a protein related to GAP. Cell. 1990; 62: 599–608.

Neurofibromatosis 1 (NF1)
(von Recklinghausen disease)

Autosomal dominant
Frequency 1 in 3000
Gene locus on 17q
Café-au-lait spots
Lisch nodules in the iris
Multiple neurofibromas
Skeletal anomalies
Predisposition to tumors of the nervous system
50% new mutations

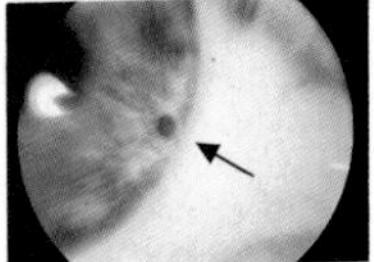

1. Lisch nodule

2. Café-au-lait spot

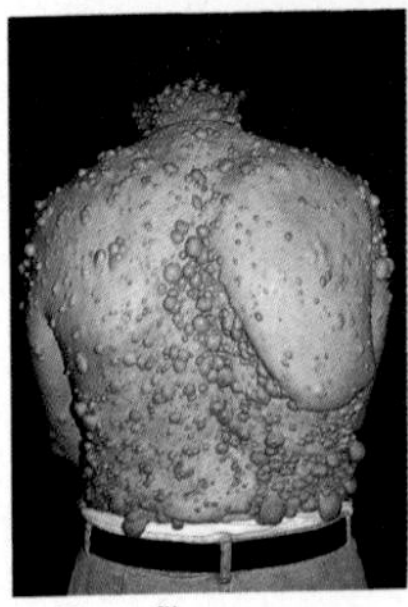

3. Neurofibromas

A. Main manifestations of neurofibromatosis 1

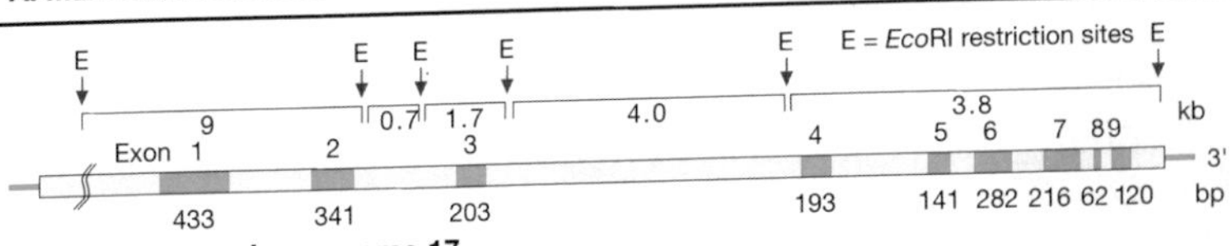

B. NF1 gene on chromosome 17

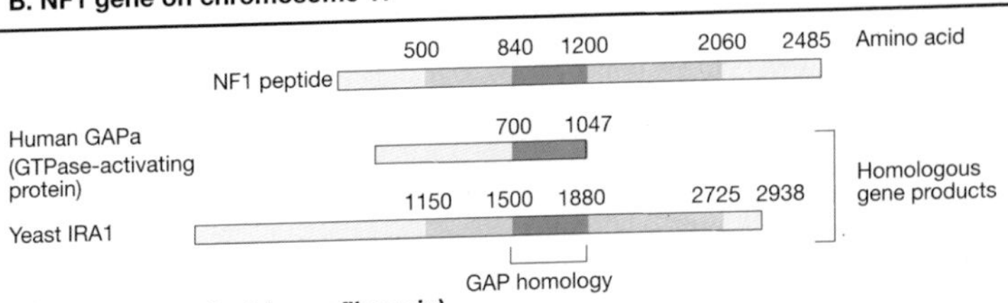

C. NF1 gene product (neurofibromin)

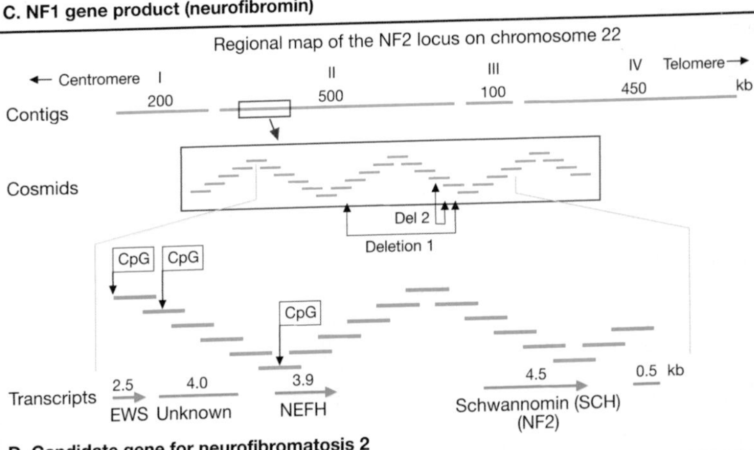

D. Candidate gene for neurofibromatosis 2

Hemoglobin

Hemoglobin and myoglobin are the oxygen-transporting proteins in vertebrates. Hemoglobin is found in red blood cells, myoglobin in muscle. Hemoglobin arose from myoglobin during the course of evolution. Hemoglobin has four, myoglobin one oxygen-binding site. Their genes and their three-dimensional protein structures are completely known in atomic detail. Different types of hemoglobin, which are optimally adapted to pre- and postnatal life, have evolved in mammals from an ancestral gene.

A. Types of hemoglobin

Hemoglobin consists of four subunits (globin chains), two pairs of identical polypeptides, each polypeptide being attached to a heme group with an oxygen-binding site. The different kinds of hemoglobins, which are characteristic of different stages of development (embryonic, fetal, and postnatal), differ in the composition of their chains: The hemoglobin of adults (HbA) contains two α and two β chains ($\alpha_2\beta_2$). A small proportion of adult hemoglobin has two α and two δ chains (HbA_2: $\alpha_2\delta_2$). Hemoglobin formed during the fetal period (HbF) contains two α and two γ chains ($\alpha_2\gamma_2$). In the embryonic stage, ζ chains are joined to ϵ or γ chains (Hb Gower 1: $\zeta_2\epsilon_2$, and Hb Portland: $\zeta_2\gamma_2$); two α and two ϵ chains form Hb Gower 2 ($\alpha_2\epsilon_2$).

B. Hemoglobins in thalassemia

The thalassemias are a group of genetically determined disorders of hemoglobin synthesis. Thalassemia occurs due to the absence or reduced synthesis of a globin chain, which results in unstable hemoglobin. It affects either the β chain (β-thalassemia) or the α chain (α-thalassemia). Since the α chain is a component of both fetal and adult hemoglobins, the α-thalassemias (α-thal) are especially severe. Hemoglobins with four identical globin chains are completely unstable and incompatible with life (HbH with four β chains, Hb Bart's with four γ chains). Theoretically, a homotetramer of four δ chains could be formed, but because of the normally slow rate of synthesis of δ chains, these are clinically insignificant. The α-thalassemias occur in very different grades of severity because two neighboring genes code for the α chain (see hemoglobin genes, p. 280).

C. Evolution of hemoglobin

Since hemoglobin has four polypeptide chains, it is a much more efficient oxygen carrier than the single-chained myoglobin molecule. Furthermore, the existence of different globin chains confers a selective advantage in evolution because their slight functional differences result in optimal adaption to the differences in oxygen concentration before and after birth. The genes for individual hemoglobin chains arose from myoglobin by a series of gene duplications during evolution. The evolutionary age of the individual Hb chains can be estimated from their differences in relation to the rate of mutation. When mammals began to evolve about 100 million years ago, the genes for α and β chains were present, whereas β-like chains (ϵ, γ, δ) evolved later.

D. Globin formation in ontogeny

Different types of globin chains are formed at different developmental stages: embryonic hemoglobin during the early embryonic period (to about the 12th week), fetal hemoglobin from about the 12th week until birth, and adult hemoglobin thereafter. They differ in oxygen-binding affinity. Thus, oxygen delivery is optimized for different phases of development. The site of synthesis also differs. During the fetal phase, globin chains are synthesized mainly in the liver, whereas after birth they are synthesized in red blood cell precursors in the bone marrow.

References

see p. 373.

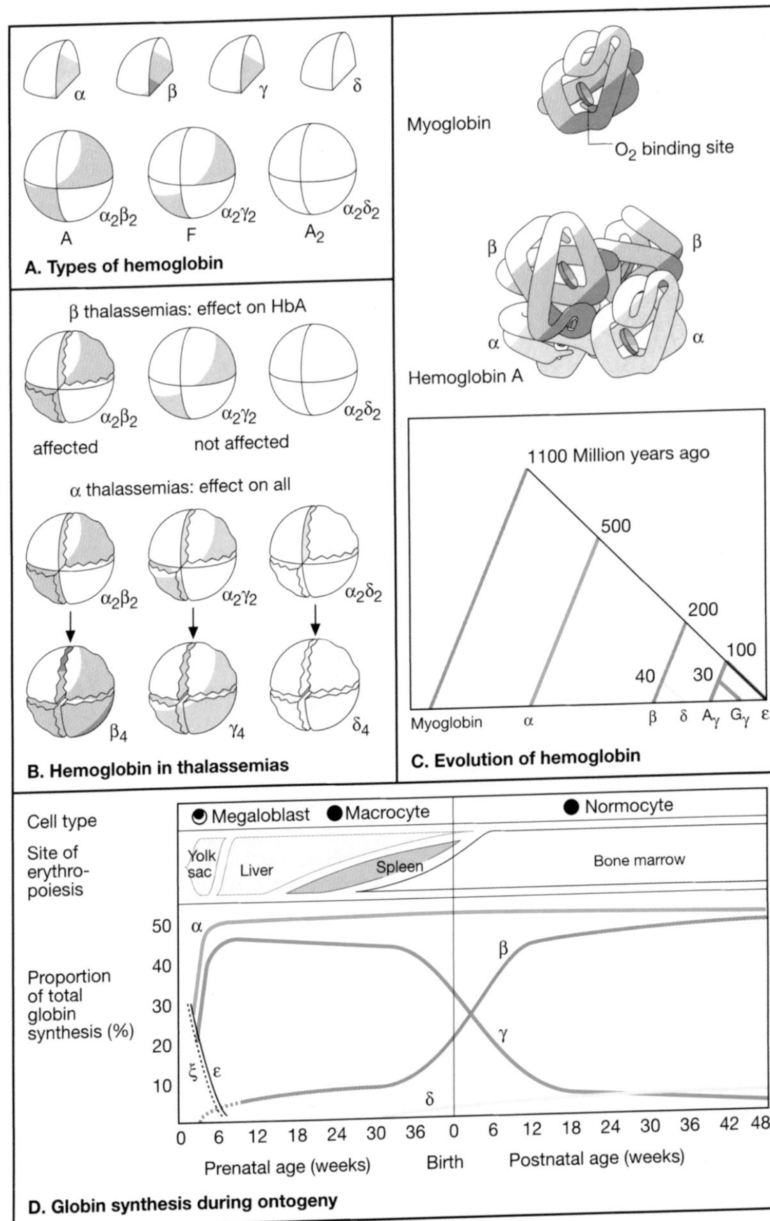

A. Types of hemoglobin

B. Hemoglobin in thalassemias

C. Evolution of hemoglobin

D. Globin synthesis during ontogeny

Hemoglobin Genes

Each of the globin polypeptide chains is coded for by a gene. In humans and in other mammals, the β-like genes (β, γ, δ) are located together in a cluster on one chromosome, while the α genes are located on another. They are arranged in the order of their activation during ontogeny.

A. The β-globin and α-globin genes

The β-globin-like genes (ϵ, Gγ, Aγ, δ, β) of man are located on the short arm of chromosome 11 in region 1, band 5.5 (11p15.5). They span about 60 000 base pairs (bp), or 60 kb (kilobases) of DNA. There are two γ genes, Aγ and Gγ, which differ only in codon 136. Codon 136 of Aγ is alanine, and of Gγ, glycine. A pseudogene ($\psi\beta_1$) is located between the Aγ gene and the δ gene. It is similar to the β gene, but has been permanently altered by deletion and an internal stop codon, so that it cannot code for a functional polypeptide. A region that jointly regulates these genes is located upstream (in the 5' direction) from the β genes (LCR, long-range control region).
In humans, two α-globin genes are located on the short arm of chromosome 16 (16p13.11 to 16p13.33) on a DNA segment of about 30 kb. A ζ gene, which is active only during the embryonic period, lies in the 5' direction. Three pseudogenes: ψζ, $\psi\alpha_2$, and $\psi\alpha_1$, are located in between. A further gene, θ, with unknown function, has been identified in this region.

B. Structure of the β-globin and α-globin genes

As a result of their origin from a common ancestral gene, all globin genes have a similar structure. Their coding sequences are arranged in three exons. Each globin transcription unit includes nontranslated sequences at the 5' and the 3' ends (see section on the structure of eukaryotic genes, p. 50). The lengths of the β- and α-globin exons are similar (e.g., exon 1 of the β gene has 30 codons; exon 1 of the α gene, 31 codons), whereas the lengths of the introns differ.

C. Tertiary structure of the β-globin chain

The three-dimensional structures of myoglobin and of the hemoglobin α and β chains are very similar, although their amino acid sequences correspond in only 24 of 141 positions. The β chain, with 146 amino acids, is somewhat longer than the α chain, with 141 amino acids. The structural similarity is functionally significant: The oxygen-binding region lies inside the molecule, where it is protected, and oxygen uptake from the aqueous surroundings is reversible.

D. Domains of the β-chain

Three functional and structural domains can be distinguished in all globin chains. They correspond to the three exons of the gene. Two domains, consisting of amino acids 1–30 and 105–146 (coded for by exons 1 and 3), are located on the outside. They are mainly formed of hydrophilic amino acids. A third domain, lying inside the molecule (coded for by exon 2), contains the oxygen-binding site and consists mainly of nonpolar hydrophobic amino acids.
The amino acid sequences of the hemoglobins of more than 60 investigated species are identical in nine positions. These invariant positions are especially important for the function of the molecule. Changes (mutations) in the invariant positions affect function so severely that they are not tolerated.

References

see p. 373.

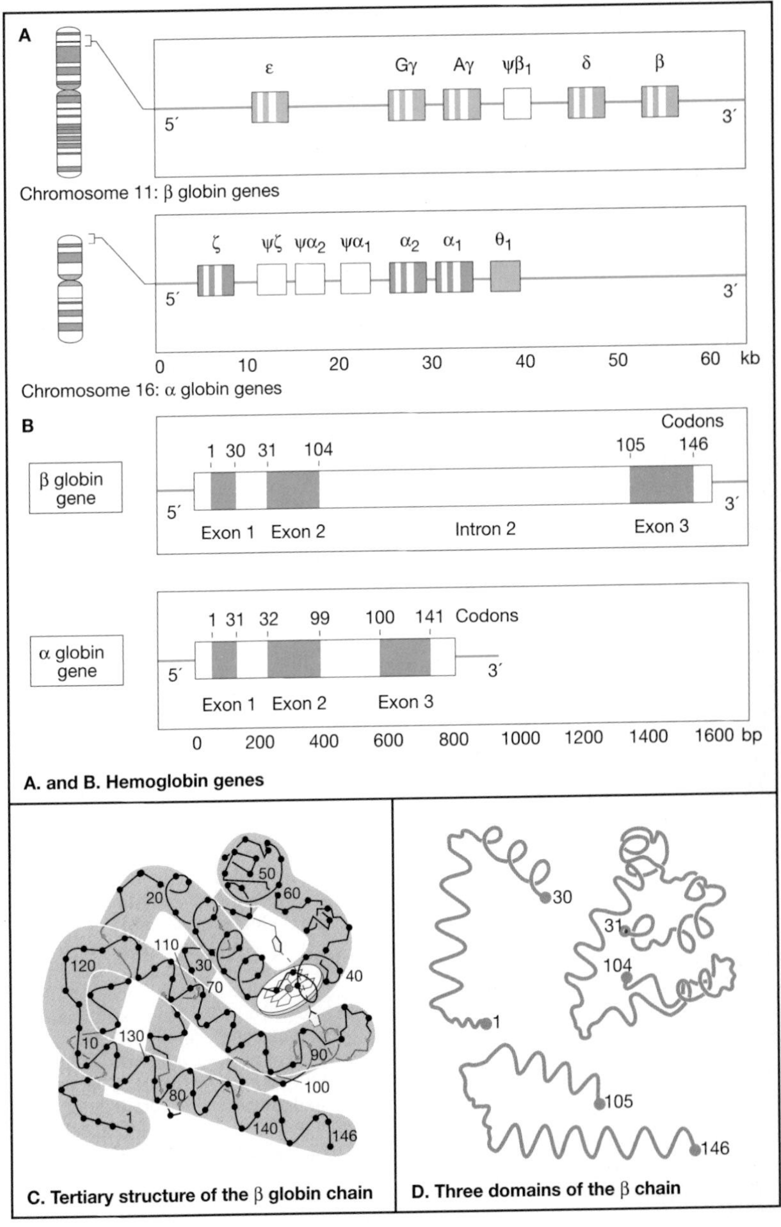

A. and B. Hemoglobin genes

C. Tertiary structure of the β globin chain

D. Three domains of the β chain

Sickle Cell Anemia

Sickle cell anemia is a severe progressive disease resulting from homozygosity for a mutation in the β-globin gene. It is especially frequent in Africa and in the black population of North America. With a frequency of 1 in 500, it is an important cause of morbidity and mortality in these regions. The disease is transmitted by autosomal recessive inheritance. Heterozygous carriers can be readily identified (see p. 68).

A. Sickle cells: erythrocytes deformed due to hemoglobin S

In a normal blood smear under the light microscope (1), erythrocytes (red blood cells) appear as regular round disks of about 7 μm diameter. Since a normal red blood cell is non-nucleated and biconcave, the center appears paler than the periphery. The erythrocytes of affected persons are deformed and resemble sickles (2). However, even the non-sickle-shaped red cells display unusual sizes and shapes. In the course of the disease, acute crises called sickle crises (3) occur, during which sicklelike cells are greatly increased and completely dominate the blood picture. Heterozygotes show occasional sickle cells but do not suffer from sickle crises, and at the most have only very mild signs and symptoms. (Figure from H. Lehmann & R.H. Huntsman, 1974).

B. Result of a mutation: sickle cell anemia

All manifestations of sickle cell anemia are due to the substitution of a single nucleotide base in the β-globin gene. The sickle cell mutation is the transversion of the second nucleotide base of codon 6, adenine (A), to thymine (T). This changes the codon GAG, for glutamic acid, to GTG, the codon for valine. During the 1950s, Vernon M. Ingram determined the amino acid sequence of hemoglobin and found that the only difference between sickle cell hemoglobin (HbS) and normal adult hemoglobin (HbA) was this exchange in the β chain. This has far-reaching pathophysiological consequences and explains all manifestations of the disease. Sickle cell hemoglobin (HbS) is less soluble than normal hemoglobin and does not allow normal erythrocyte distortion. It crystallizes in the deoxy state and forms small rods. Thus, the erythrocytes become firm and deform into sickle cells. Unlike normal erythrocytes, sickle cells are unable to pass through small arteries and capillaries. These become clogged and cause local oxygen deficiency in the tissues, followed by infection. As a rule, learning disability due to frequent illness occurs. Defective erythrocytes are destroyed (hemolysis). Chronic anemia and its numerous sequelae such as heart failure, liver damage, and infection are the result.

C. Selective advantage for HbS heterozygotes in malarial regions

Heterozygotes for the sickle cell mutation are relatively resistant to malarial infection. Erythrocytes of heterozygotes for the sickle cell mutation are a less favorable environment for the malaria parasite than those of normal homozygotes. Thus, heterozygotes develop malaria in a much milder form or not at all. However, this protection is at the expense of the affected homozygotes (HbS/HbS): Although they do not contract malaria, they suffer from the severe hemoglobin disorder. The protection against malaria conferred by sickle cell heterozygosity is an advantage in regions where malaria is common. With reduced morbidity and mortality, heterozygotes have a higher probability of survival and of being able to reproduce (selective advantage). This explains the high frequency of the sickle cell gene observed there (see p. 160). The sickle cell mutation has arisen independently in at least four or five different malaria-infested regions and has subsequently spread out in the respective populations. Sickle cell anemia is the best example in humans of a selective advantage in heterozygotes for a mutant allele that leads to severe illness in the homozygous state.

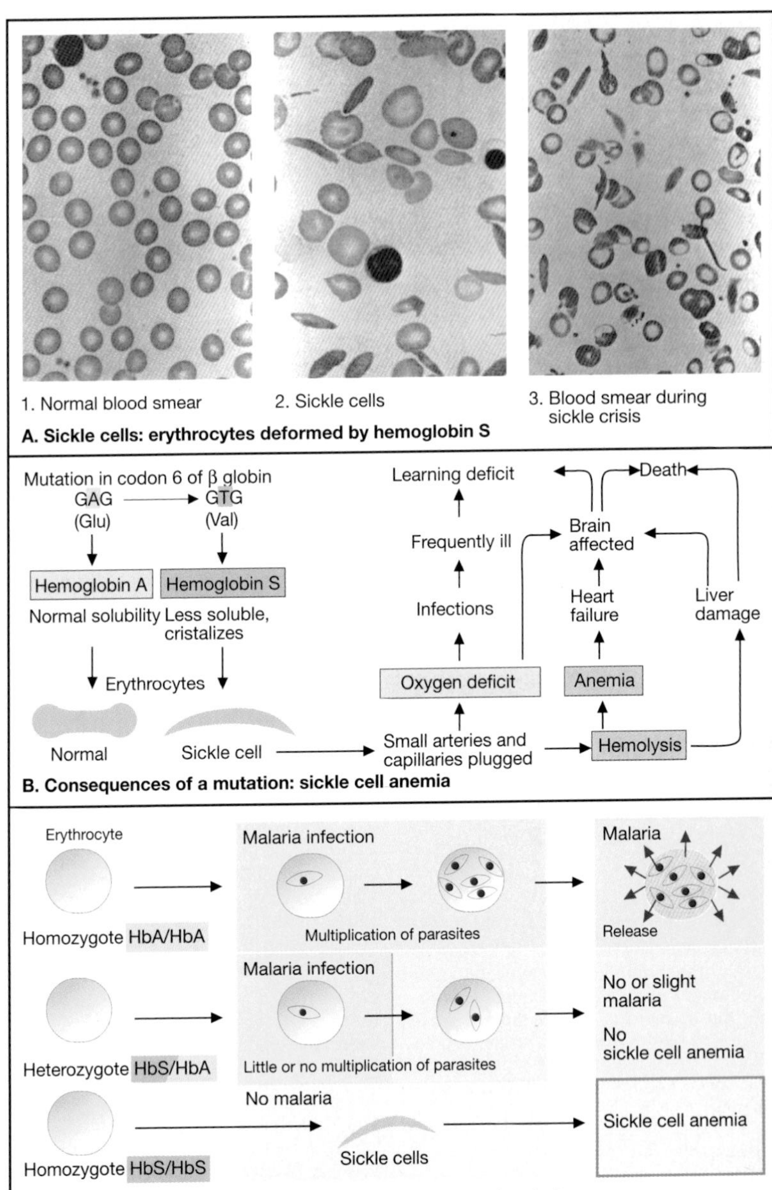

1. Normal blood smear
2. Sickle cells
3. Blood smear during sickle crisis
A. Sickle cells: erythrocytes deformed by hemoglobin S
Mutation in codon 6 of β globin
GAG
GTG
(Glu)
(Val)
Hemoglobin A
Hemoglobin S
Normal solubility
Less soluble, cristalizes
Erythrocytes
Normal
Sickle cell
Learning deficit
Death
Brain affected
Frequently ill
Infections
Heart failure
Liver damage
Oxygen deficit
Anemia
Small arteries and capillaries plugged
Hemolysis
B. Consequences of a mutation: sickle cell anemia
Erythrocyte
Malaria infection
Malaria
Multiplication of parasites
Release
Homozygote HbA/HbA
Malaria infection
No or slight malaria
No sickle cell anemia
Heterozygote HbS/HbA
Little or no multiplication of parasites
No malaria
Sickle cell anemia
Sickle cells
Homozygote HbS/HbS
C. Selective advantage of HbS heterozygotes in areas of malaria

Mutations of the Globin Genes

All types of mutations have been demonstrated in the globin genes. The most frequent are point mutations in a single codon. The functional consequences vary, depending on the electrical charge and size of the substituted amino acid and its position in the polypeptide. If one of the hydrophilic amino acids at the surface is replaced by a hydrophobic amino acid (e.g., valine for glutamic acid in the sickle cell mutation), profound physicochemical changes will result. Mutations may decrease elasticity of the molecule, alter oxygen affinity, or cause instability.

A. Point mutations of the β-globin gene

Over 300 point mutations in the β-globin gene and over 100 in one of the α-globin genes have been documented. Two clinically important mutations affect codon 6: the sickle cell mutation, 6 Glu → Val (sickle cell hemoglobin, HbS, resulting in the incorporation of valine instead of glutamic acid) and 6 Glu → Lys (hemoglobin C, HbC, incorporating lysine instead of glutamic acid in codon 6). Compound heterozygotes with the HbS mutation on one chromosome and the HbC on the other (HbSC) are not rare. The marked methemoglobin formation in Hb Zürich and Hb Saskatoon results from substitutions for histidine (His) in codon 63, which alter the oxygen-binding region of the hemoglobin molecule.

B. Deletion due to unequal crossing-over within a gene

Marked sequence homology of certain regions of the globin genes may lead to nonhomologous pairing and unequal crossing-over during meiosis, e.g., in the regions of codons 90–94 of one DNA strand and codons 95–98 of the other. This explains the deletion of codons 91–95 in hemoglobin Gun Hill.

C. Unequal crossing-over between similar genes

The sequence homology of the β-globin-like genes (explained by their common evolution) may lead to unequal crossover between regions of the two γ-globin genes (γA and γG), the δ-globin gene, or the β-globin gene. The best known example is partial deletion of the δ and of the β loci (δ–β fusion) in hemoglobin Lepore. The corresponding duplication of the δ/β sequence results in hemoglobin anti-Lepore.

D. Unstable hemoglobin due to chain elongation

If one of the globin chains is too long, it will destabilize the tetrameric hemoglobin molecule. Hemoglobin Cranston (Hb_{Cr}) (1) arises from the insertion of two nucleotide bases (adenine and guanine) into positions 1 and 2 of codon 145 (tyrosine) of the β chain, which leads to a shift of the reading frame. This changes the normal stop codon UAA into AGU, the RNA codon for threonine (Thr). As a result, the normally nontranslated sequences that follow the stop codon are now translated, and a polypeptide is formed that is 11 amino acids too long, extending to position 157. With hemoglobin Constant Spring (2), the α chain is lengthened by mutation of the stop codon UAA to CAA, which codes for glutamine (Gln) (transversion from thymine to cytosine). The sequences that normally follow the stop codon now become translated, and a peptide that is 31 amino acids too long is formed. A number of other chain-elongation mutations due to similar mechanisms, such as with hemoglobin Ikaria (an α chain with 172 amino acid residues), have been described.

Reference

Weatherall DJ, Clegg JB, Higgs DR, Wood WG. The hemoglobinopathies. In: The Metabolic and Molecular Bases of Inherited Disease. Scriver CR, Beaudet AL, Sly WS, Valle D, eds. 7th ed. New York: McGraw-Hill; 1995: 3417–3484.

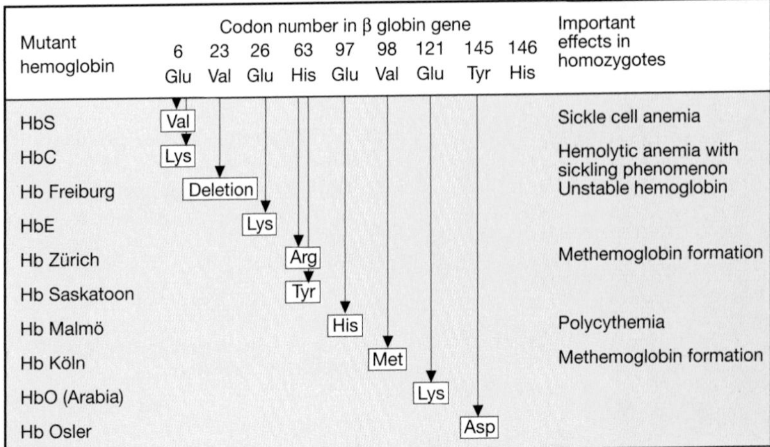

A. Examples of point mutations in the β-globin gene (10 of 310)

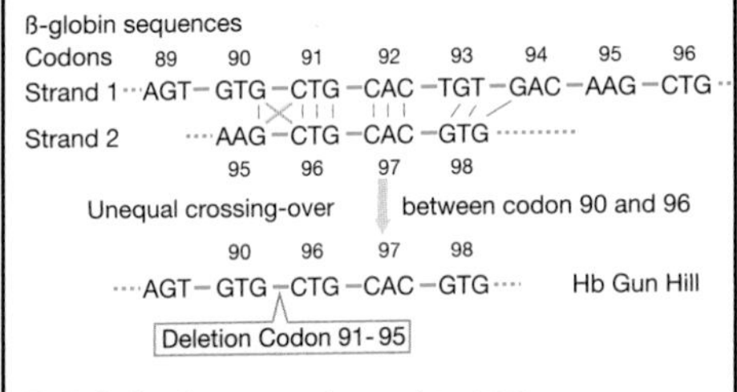

B. Deletion by unequal crossing-over within a gene

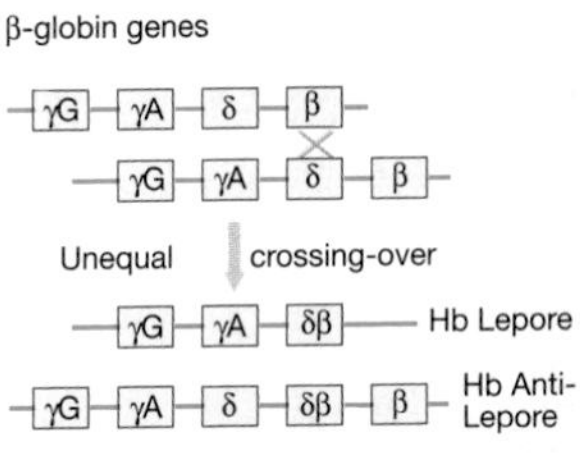

C. Unequal crossing-over between similar genes

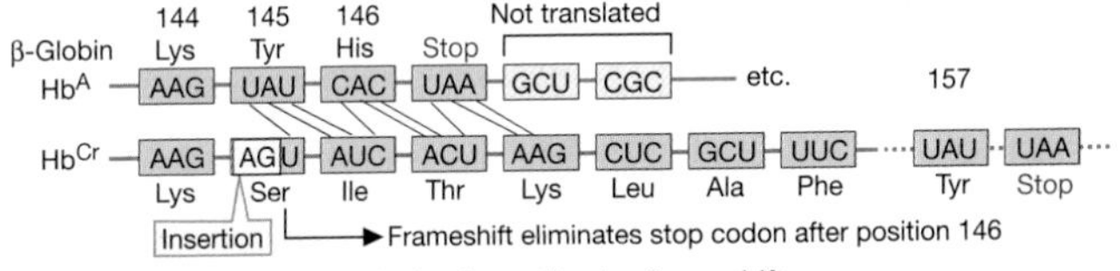

1. Hemoglobin Cranston: chain elongation by frameshift

141 Arg · Stop · Not translated

α-Globin HbA — CGU — UAA — CGU — GGA — GCC ······ GUC — UUU — GAA — UAA — AGU — CUG ··· Poly

HbConstant Spring — CGU — CAA — GCU — GGA — GCC ······ GUC — UUU — GAA — UAA — AGU — CUG ··· Poly A

Arg · Gln 142 · Ala 143 · Gly 144 · Ala 145 · Val 170 · Phe 171 · Glu 172 · Stop

Mutation T→C

2. Hemoglobin Constant Spring: chain elongation by mutation in the stop codon

D. Unstable hemoglobin due to chain elongation

The Thalassemias

The thalassemias are a heterogeneous group of diseases caused by decreased or absent formation of a globin chain. The thalassemias occur predominantly in the Mediterranean region, parts of Africa, and Southeast Asia (the word thalassemia is derived from *thalassa*, the Greek word for sea). In these regions, they are a significant cause of morbidity and mortality, but they are frequent because heterozygotes are protected from severe malaria infection (see pp. 160, 279).

A. Thalassemia, a chronic anemia

Depending on which globin chain is not formed in sufficient amounts, α-thalassemia, β-thalassemia, or δβ-thalassemia results. This leads to chronic anemia, which causes the various manifestations of thalassemia. Oxygen deficiency in the peripheral tissues leads to increased extramedullary (outside the bone marrow) blood formation. A tendency toward infection, undernourishment, and other signs characterize the severe clinical picture (photos from D.J. Weatherall & J.B. Clegg, The Thalassemia Syndromes, 3rd ed. Oxford; 1981).

B. β-Thalassemia and α-thalassemia

The thalassemias have a wide spectrum of different genotypes and phenotypes (disease manifestations and course). In the β-thalassemias (1), complete absence (β^0) is distinguished from decreased formation (β^+) of the β chain. With the α-thalassemias (2), one, two, three, or all four loci for α-globin may be affected. Altogether, there are 12 principal genotypes. In individuals with two mutations at the α-loci (α-thalassemia), the two can lie either on the same chromosome (thal-1) or on different chromosomes (thal-2). Thal-1 occurs mainly in Southeast Asia; thal-2, mainly in Africa. Each α gene is located within a 4-kb region of homology, interrupted by small, nonhomologous regions. The most frequent mechanism for the origin of a chromosome with only one α-globin gene is nonhomologous crossing-over between two α-globin gene loci after mispairing of the homologous chromosomes during meiosis.

C. β-Thalassemia due to different mutations

Many mutations in the β-globin gene region can lead to β-thalassemia. The mutations may also occur in noncoding sequences (5' to exon 1 and within introns).

D. Haplotypes resulting from polymorphic restriction sites in the β-globin gene cluster

From the presence or absence of recognition sites of a number of restriction enzymes (restriction fragment length polymorphism, RFLP), different haplotypes can be distinguished in the β-globin-like gene region (β-globin gene cluster). Each haplotype is characterized by the presence or absence of several polymorphic restriction sites. By establishing the haplotypes of the affected and unaffected within a family, the mutation-carrying haplotype can be identified (indirect genotype analysis). Different mutations have occurred on the background of different haplotypes. Frequently, a particular mutation is linked to a distinct haplotype (linkage disequilibrium). This reflects the time elapsed since the mutation first occurred in the population, where it has been maintained by selection (data after S.E. Antonarakis et al. Hum Genet. 1985; 69: 1–14).

α-Thalassemia may be associated with mental retardation. Two different syndromes can be distinguished: One occurs in patients with a large (1–2 megabase) deletion on the tip of chromosome 16 including the α-globin gene cluster (ATR-16 syndrome). The other is an X-linked disorder with a remarkably uniform phenotype and a mild form of HbH disease without α-globin deletion. A trans-acting regulatory factor appears to be encoded on the X chromosome.

References see p. 373.

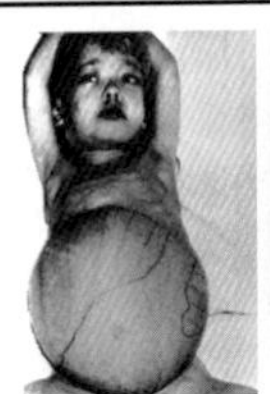

Different forms of thalassemia:

α: Decreased synthesis of α globin
β: Decreased synthesis of β globin
δβ: Decreased synthesis of δ and β globin

→ Unstable hemoglobin → Chronic anemia

A. Thalassemia, a chronic anemia

1. β-Thalassemias

Genotype		Phenotype
+ / −	β° heterozygote	Thalassemia minor (asymptomatic)
(+) / (+)	β⁺ heterozygote	
(+) / (+)	β⁺ homozygote	Thalassemia intermedia (not transfusion dependent)
+ / −	β° heterozygote	
− / −	β° homozygote (β° Thalassemia)	Thalassemia major (transfusion dependent)
oder		
(+) / −	β⁺/β° homozygote (β⁺ Thalassemia)	

2. α-Thalassemias

Genotype	Phenotype
α α / α α	Normal
α − / α α	"Silent carrier" (normal)
− − / α α (thal-1)	Thalassemia
α − / α − (thal-2)	Thalassemia
α − / − −	HbH Disease (HbH = β_4)
− − / − −	Hydrops fetalis

B. β-thalassemia and α-thalassemia

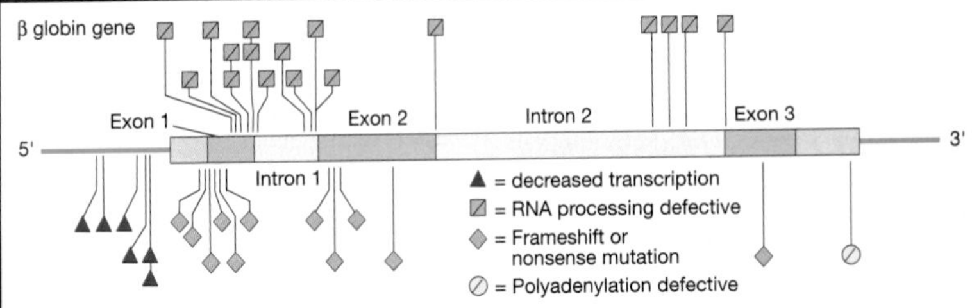

C. β-Thalassemia due to different mutations

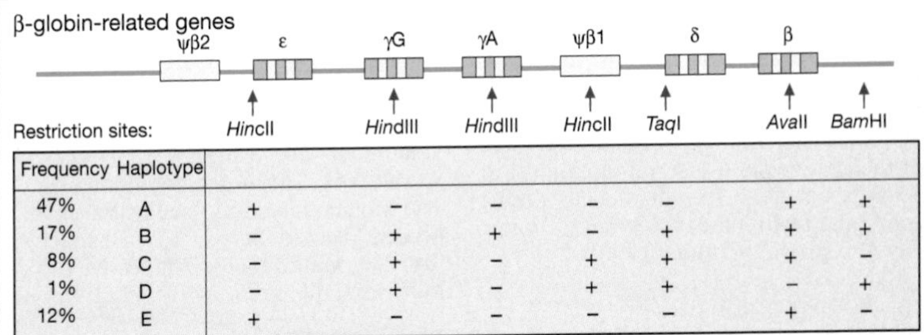

Frequency	Haplotype	HincII	HindIII	HindIII	HincII	TaqI	AvaII	BamHI
47%	A	+	−	−	−	−	+	+
17%	B	−	+	+	−	+	+	+
8%	C	−	+	−	+	+	+	−
1%	D	−	+	−	+	+	−	+
12%	E	+	−	−	−	−	+	−

D. Haplotypes resulting from polymorphic restriction sites in the β-globin gene cluster

Hereditary Persistence of Fetal Hemoglobin (HPFH)

Hereditary persistence of fetal hemoglobin (HPFH) refers to a genetically heterogeneous group of diseases in which the temporal expression of the β-globin genes during development has been altered. Individuals with HPFH produce increased amounts of fetal hemoglobin (HbF). Under some conditions, HbF may be the only β-globin-like gene product formed. Clinically, HPFH is relatively benign, although HbF is not optimally adapted to postnatal conditions. Analysis of HPFH has yielded insight into the control of globin gene transcription and the effect of mutations in noncoding sequences.

A. Large deletions in the β-globin gene cluster

A number of very large deletions in the β-globin gene cluster region are known, especially in the 3' direction. The deletions show different distributions in different ethnic populations, reflecting that they originated at different points in time. δβ-Thalassemia and failure of β-globin production have been the result in some cases.

B. Mutations in noncoding sequences of the promoter region

Mutations in the noncoding sequences of the promoter region at the 5' end of the β-globin cluster (on the 5' side of the γ-globin genes) can also lead to hereditary persistence of fetal hemoglobin. Although the highly conserved sequences CACCC, CCAAT, or ATAAA are not affected, the number of observed mutations substantiates the significance of the remaining noncoding sequences (long-range transcription control). They are probably required for the changes in transcription control of the different gene loci that occur during embryonic and fetal development. (Figure after Gelehrter & Collins, 1990).

C. Frequent mutations of β-thalassemia in different populations

β-Thalassemia mutants occur in different ethnic populations with different frequencies. Since a few mutations are quite frequent in certain populations, preventative diagnostic programs to determine the risk of disease are possible. (Data after S.E. Antonarakis et al. Hum Genet. 1985; 69: 1–14).

According to estimates of the WHO (Bull World Health Org. 1983; 61: 63–80) about 142 million persons are heterozygotes for hemoglobin diseases worldwide. Substantial numbers are due to the β-thalassemias in Asia (over 60 million), α_0-thalassemia in Asia (30 million), HbE/β-thalassemia in Asia (84 million), and sickle cell heterozygosity in Africa (50 million), India, the Caribbean, and the USA (about 50 million). At least 200 000 severely affected homozygotes are born annually, about 50% due to sickle cell anemia and 50% to thalassemia (Weatherall 1991).

References

Antonarakis SE, Kazazian HH jr, Orkin SH. DNA polymorphism and molecular pathology of the human globin gene clusters. Hum Genet 1985; 69: 1–14.

Gelehrter TD, Collins F. Principles of Medical Genetics. Baltimore: Williams & Wilkins; 1990.

Orkin SH, Kazazian HH. The mutation and polymorphism of the human β-globin gene and its surrounding DNA. Ann Rev Genet. 1984; 8: 131–171.

Stamatoyannopoulos G, Niehuis AW, Leder P, Majerus PW, eds. The Molecular Basis of Blood Diseases. Philadelphia: WB Saunders; 1987.

Weatherall DJ, Clegg JB, Higgs DR, Wood WG. The hemoglobinopathies. In: The Metabolic and Molecular Bases of Inherited Disease. Scriver CR, Beaudet AL, Sly WS, Valle D, eds. 7th ed. New York: McGraw-Hill; 1995: 3417–3484.

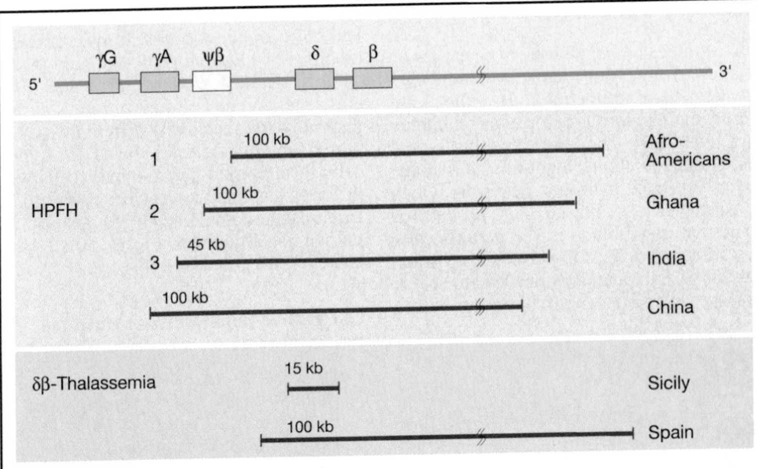

A. Large deletions in the β-globin cluster

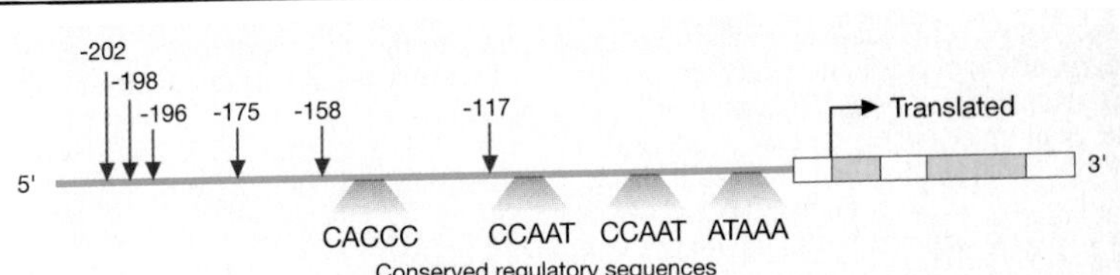

B. Mutations in noncoding sequences in the promoter of γ-globin genes cause hereditary persistence of fetal hemoglobin (HPFH)

β-thal mutation	Frequency	Ethnic group	Type
Intron 1 (110 G → A)	35%	Mediterranian	β^+
Codon 39 (C → T)	27%	Mediterranian	β°
TATA-Box (-29 A → G)	39%	Afro-Americans	β^+
Poly A (T → C)	26%	Afro-Americans	β^+
Intron 1 (5 G → C)	36%	India	β^+
Partial deletion (619 nt)	36%	India	β°
Codon 71-72 frameshift	49%	China	β°
Intron 2 (654 C → T)	38%	China	β°

C. Frequent mutations in β-thalassemia in different populations

Peroxisomal Diseases

Peroxisomes are small round organelles of about 0.5–1.0 μm diameter (somewhat smaller than mitochondria). They are found mainly in the cytoplasm of kidney and liver cells. They are the site of some important metabolic functions. The name is derived from hydrogen peroxide, which is formed as an intermediary product of oxidative metabolism in the peroxisomes. A number of defects in peroxisome formation or peroxisome enzymes lead to severe diseases in humans (peroxisomal diseases).

A. Biochemical reactions in peroxisomes

The electron micrograph (1) shows peroxisomes in a section of rat liver. The dark striated structures within the organelles consist of urates (peroxisomes contain an enzyme that oxidizes uric acid). Peroxisomes have both catabolic (substances are degraded) and anabolic (substances are synthesized) functions (2). Two biochemical reactions are especially important: a peroxisomal respiratory chain and the β-oxidation of very long-chain fatty acids. In the peroxisomal respiratory chain (3), certain oxidases and catalases act together. Specific substrates of the oxidases are organic metabolites of intermediary metabolism. Very long-chain fatty acids are broken down by β-oxidation (4) in a cycle with four enzymatic reactions. Energy production in peroxisomes is relatively inefficient compared with that of mitochondria. While free energy in mitochondria is mainly preserved in the form of ATP (adenosine triphosphate), in peroxisomes it is mostly converted into heat. Peroxisomes are probably a very early adaption of living organisms to oxygen. (Photo from C. de Duve: Die Zelle. Expedition in die Grundstruktur des Lebens. Heidelberg: Spektrum der Wissenschaft; 1986).

B. Peroxisomal diseases

Several peroxisomal diseases are known in man; the six most important are listed. All are autosomal recessive hereditary disorders. Patients with neonatal adrenoleukodystrophy do not form sufficient amounts of plasmalogens and cannot adequately degrade phytanic acid and pipecolic acid. When cultured fibroblasts from patients with genetically different types of peroxisomal diseases are fused, the hybrid cells form normal peroxisomes (cells with different defects can correct each other). Eight such complementation groups are known (N. Shimozawa et al. Am J Hum Genet. 1993; 52: 834–844).

C. Cerebro-hepato-renal syndrome type Zellweger

Patients with this autosomal recessive hereditary disease have a characteristic facial appearance (1–4), extreme muscle weakness (5), and a number of accompanying manifestations such as calcified stippling of the joints on X-ray (6), renal cysts (7, 8), and clouding of the lens and cornea. The severe form of the disease (type Zellweger) usually leads to death before age one year (Photos 1–5 from: E. Passarge, A.J. McAdams: J Pediatr. 1967; 70: 691–702, 6–8 unpublished).

References

Folz SJ, Trobe JD. The peroxisome and the eye. Survey of Ophthalmology. 1991; 35: 353–368.

Lazarow PB, Moser HW. Disorders of peroxisome biogenesis. In: The Metabolic Bases of Inherited Disease. 7th ed. Scriver CR, Beaudet AL, Sly WS, Valle D., eds. New York: McGraw–Hill; 1995: 2287–2324.

Shimozawa N, Tsukamoto T, Suzuki Y, et al. A human gene responsible for Zellweger syndrome that affects peroxisome assembly. Science. 1992; 255: 1132–1255.

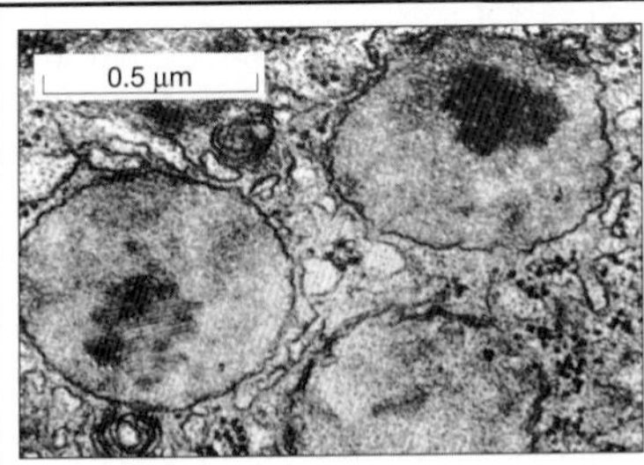

1. Peroxisomes in a rat liver cell

a) Catabolic

H_2O_2-involving cellular respiration
β Oxidation of long-chain fatty acids,
Prostaglandins, cholesterol side chains and others
Purines, urates
Pipecolic acid, dicarboxy acids
Ethanol, methanol

b) Anabolic

Phospholipids (Plasmalogen)
Cholesterol, bile acids
Gluconeogenesis
Glyoxalate transamination

2. Function of peroxisomes

O_2 —Oxidases→ H_2O_2 —Catalase→ 2 H_2O and heat (→ O_2)

R-H_2 → R; R'-H_2 → R'

R: D- and L-Amino acids, hydroxy acids, purines, urates, oxalate polyamines, fatty acid derivatives

R': Ethanol, Methanol, Nitrites, Quinones, Formates

3. Peroxisomal respiratory chain

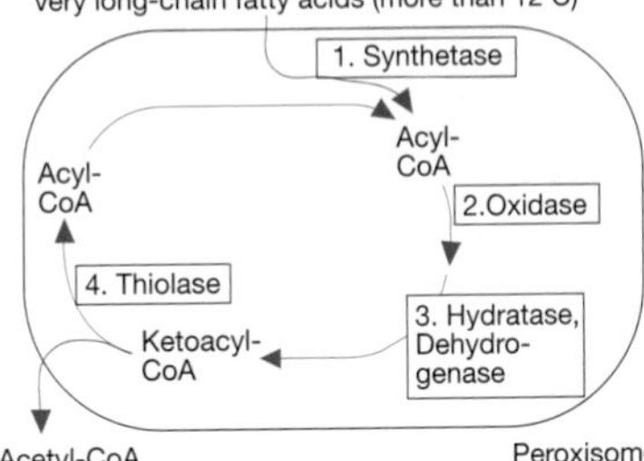

4. β Oxidation

A. Biochemical reactions in peroxisomes

214100 Cerebro-hepato-renal syndrome Zellweger
202370 Neonatal Adrenoleukodystrophy
266510 Infantile Refsum disease
239400 Hyperpipecolic acidemia
215100 Rhizomelic chondrodysplasia punctata
259900 Primary hyperoxaluria type I
and others

B. Peroxisomal diseases

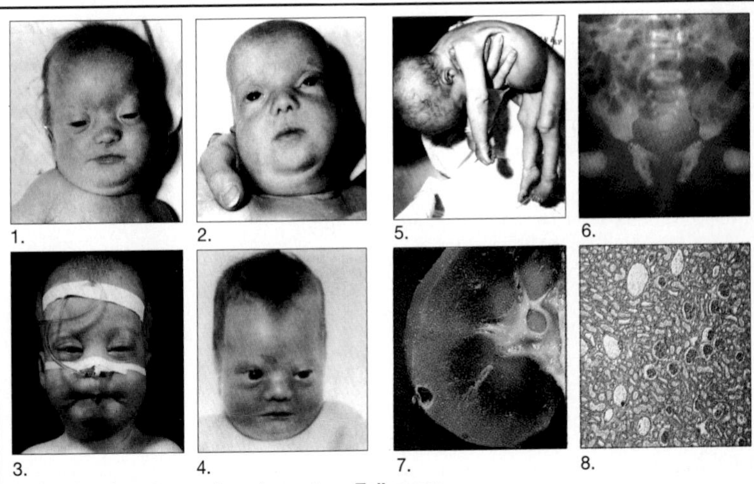

C. Cerebro-hepato-renal syndrome type Zellweger

Lysosomes and Endocytosis

Lysosomes are membrane-enclosed cytoplasmic organelles with a diameter of 0.05–0.5 µm. They contain a wide variety of active hydrolytic enzymes (hydrolases) such as glycosidases, sulfatases, phosphatases, lipases, phospholipases, proteases, and nucleases (lysosomal enzymes) in an acid milieu (pH about 5). Lysosomal enzymes enter a lysosome by means of a recognition signal (mannose-6-phosphate) and a corresponding receptor.

A. Receptor-mediated endocytosis and lysosome formation

Extracellular molecules to be degraded are taken into the cell by endocytosis. First, the molecules are bound to specific cell surface receptors (receptor-mediated endocytosis). The loaded receptors are concentrated in an invagination of the plasma membrane (coated pit). This separates from the plasma membrane and forms a membrane-enclosed cytoplasmic compartment (coated vesicle). Hormones, growth factors, energy-delivering proteins, and numerous viruses and toxins also enter cells by receptor-mediated endocytosis (see p. 88). The cytoplasmic lining of the vesicle consists of a network of a trimeric protein, clathrin. The clathrin coat is quickly lost within the cell, and an endosome forms, which fuses with membrane vesicles from the Golgi apparatus to form larger endosomal compartments. Here, the receptors are separated from the ligands and are returned to the cell surface in membrane vesicles (receptor recycling). Parts of the membrane are also reused. The ligands are now within a multivesicular body (endolysosomes). Hydrolases (lysosomal enzymes) are transported from the Golgi apparatus to an endolysosome in clathrin-enclosed vesicles after they become equipped with a recognition signal (mannose-6-phosphate receptor), required for uptake into the endolysosome and for normal functioning of the lysosome. There are different classes of endolysosomes, defined according to relative acidity, receptor content, biochemical composition, morphological appearance, and other characteristics. The acid milieu in the lysosomes is maintained by a hydrogen pump in the membrane, which hydrolyzes ATP and uses the energy produced to move H^+ ions into the lysosome. Some of the mannose-6-P receptors are transported back to the Golgi apparatus.

B. Mannose-6-phosphate receptors

There are two types of mannose-6-P receptor molecules, which differ in their binding properties and their cation dependence. They consist of either 2 or 16 extracellular domains with different numbers of amino acids. The cDNA of Ci-MPR (cation-independent mannose-6-P receptor) is identical with insulinlike growth factor II (IGF-2). Thus, Ci-MPR is a multifunctional binding protein.

C. Biosynthesis of the recognition signal

Two enzymes are essential for the formation of mannose-6-phosphate recognition signals: a phosphate transferase and a phosphoglycosidase. The phosphate is delivered by uridine-diphosphate-N-acetylglucosamine (UDP-GlcNAc) (uridine-5'-diphosphate-N-acetylglucosamine-glycoprotein-N-acetylglucosaminyl-phosphotransferase). A second enzyme, (N-acetylglucosamine-1-phosphodiester-N-acetylglucosaminidase) cleaves off the N-acetylglucosamine, leaving the phosphate residue at position 6 of the mannose (figure after D.D. Sabatini and M.B. Adesnik, 1989; and C. de Duve, 1986; the relative sizes of the individual structures are not to scale).

References

see p. 372.

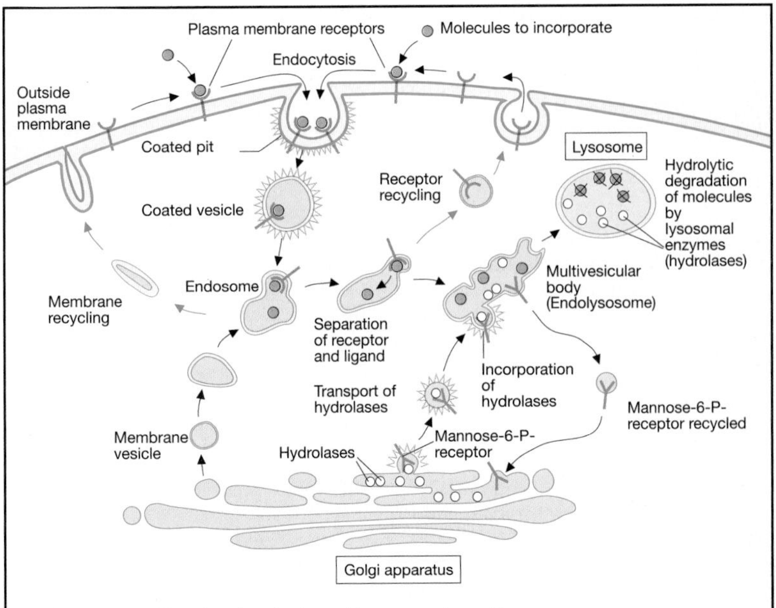

Plasma membrane receptors
Molecules to incorporate
Endocytosis
Outside plasma membrane
Coated pit
Coated vesicle
Receptor recycling
Lysosome
Hydrolytic degradation of molecules by lysosomal enzymes (hydrolases)
Membrane recycling
Endosome
Separation of receptor and ligand
Multivesicular body (Endolysosome)
Incorporation of hydrolases
Transport of hydrolases
Mannose-6-P-receptor recycled
Membrane vesicle
Hydrolases
Mannose-6-P-receptor
Golgi apparatus

A. Receptor-mediated endocytosis and lysosome formation

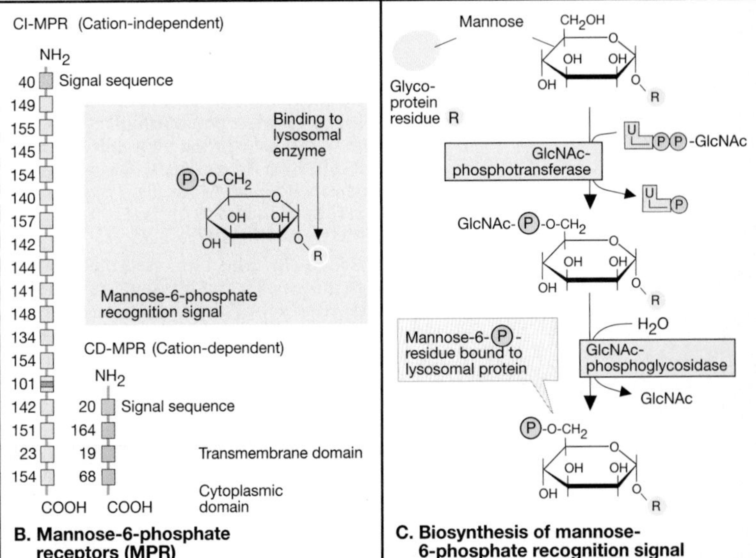

B. Mannose-6-phosphate receptors (MPR)

C. Biosynthesis of mannose-6-phosphate recognition signal

Diseases Due to Lysosomal Enzyme Defects

Mutations in genes for enzymes that degrade complex macromolecules in lysosomes (lysosomal enzymes) lead to disease. Clinical signs and biochemical and cellular manifestations depend on the altered enzyme's specificity in lysosomal metabolism. With such an enzyme defect, macromolecules that normally would be degraded are stored (storage disease). This occurs at different rates, so that each disease has its own characteristic course. Twelve groups of diseases due to genetically determined disorders of specific lysosomal function are known, each with about three to ten individually defined diseases.

A. Defective uptake of enzymes into lysosomes: I-cell disease (mucolipidosis type II)

Due to a mutation of the gene on chromosome 4 for the phosphotransferase needed to form the mannose-6-phosphate recognition signal (see p. 292), hydrolases cannot be taken up into lysosomes. Unlike normal cultured fibroblasts (1), those of patients (2) contain numerous densely packed cytoplasmic *i*nclusion bodies (thus, the name I-cell disease). The vesicular inclusions consist of hydrolases that cannot enter the lysosomes because the mannose-6-phosphate recognition signal is absent. Numerous enzymes are missing from the lysosomes, while their concentration in other parts of the cells and in body fluids is increased. Patients (3) show a severe progressive clinical picture, with the first signs usually apparent in the first half year of life.

B. Degradation of heparan sulfate by eight lysosomal enzymes

Heparan sulfate is an example of a macromolecule that is degraded stepwise by different lysosomal enzymes. Lysosomal enzymes are bond-specific, not substrate-specific. Thus, they also degrade other glycosaminoglycans, such as dermatan sulfate, keratan sulfate, and chondroitin sulfate (mucopolysaccharides). Specific enzyme defects cause the mucopolysaccharide storage diseases.

The first step in mucopolysaccharide degradation is the removal of sulfate from the terminal iduronate group by an iduronate sulfatase. A defect in the gene that codes for this enzyme leads to mucopolysaccharide storage disease type II (Hunter). The gene is located on the X chromosome, so that Hunter disease is transmitted by X-chromosomal inheritance. All other mucopolysaccharidoses are autosomal recessive. In the next step (2), the terminal iduronate is split off by an α-L-iduronidase. A mutation of the gene coding for this enzyme in the homozygous state leads to mucopolysaccharidosis (MPS) type I (Hurler/Scheie). In the next three steps a mutation (in the homozygous state) of a gene coding for one of the enzymes causes mucopolysaccharidosis type III (Sanfilippo). The four genetically and enzymatically different types (III-A to III-D) cannot be distinguished clinically. MPS type VII (Sly), due to a defect of β-glucuronidase, has a further characteristic clinical picture.

References

Kornfeld S, Sly WS. I-cell disease and Pseudo-Hurler polydystrophy: Disorders of lysosomal enzyme phosphorylation and localization. In: Scriver CR, et al. eds. The Metabolic and Molecular Bases of Inherited Disease. 7th ed. New York: McGraw–Hill; 1995: 2495–2508.

Neufeld EF, Muenzer J. The mucopolysaccharidoses. In: The Metabolic and Molecular Bases of Inherited Disease. 7th ed. Scriver CR, Beaudet AL, Sly WS, Valle D, eds. New York: McGraw–Hill; 1995: 2465–2494.

Sabatini DD, Adesnik MB. The biogenesis of membranes and organelles, In: The Metabolic and Molecular Bases of Inherited Disease. 7th ed. Scriver CR, Beaudet AL, Sly WS, Valle D, eds. New York: McGraw–Hill; 1995: 459–553.

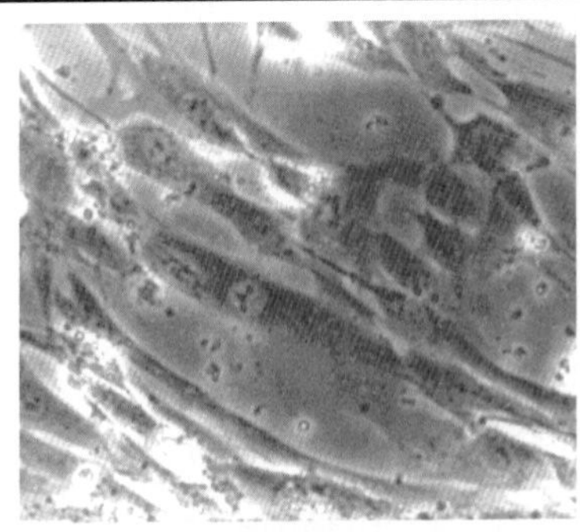

1. Normal fibroblast culture

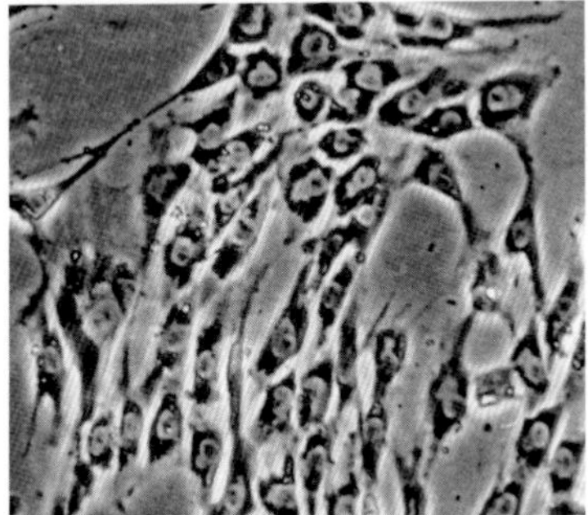

2. Fibroblast culture in I-cell disease

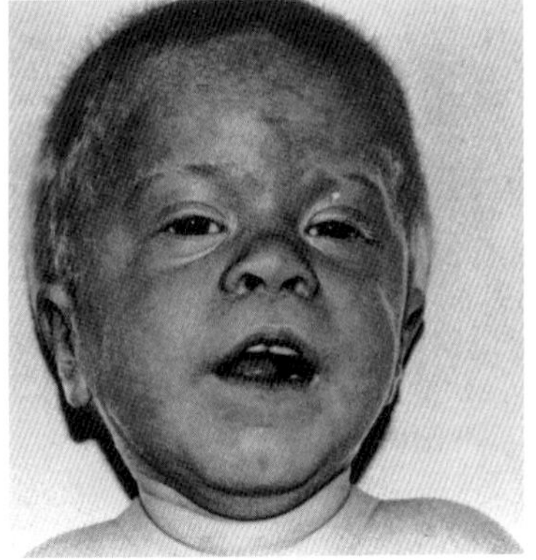

3. Patient with I-cell disease

A. Defective uptake of enzymes in lysosomes: I-cell disease

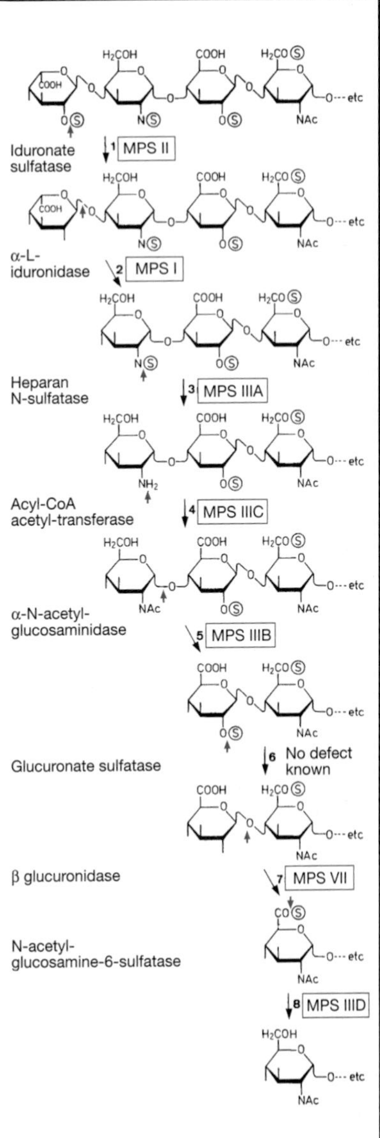

B. Degradation of heparan sulfate by eight lysosomal enzymes

Mucopolysaccharide Storage Diseases

The mucopolysaccharide storage diseases (the mucopolysaccharidoses) are a clinically and genetically heterogeneous group of lysosomal storage diseases caused by defects in different enzymes for mucopolysaccharide degradation (glycosaminoglycans). Except for mucopolysaccharide storage disease type II (Hunter), all are transmitted by autosomal recessive inheritance.

A. Mucopolysaccharide storage disease type I (Hurler)

At first nearly inapparent, the early signs of the disease occur at about 1–2 years of age, with increasing coarsening of the facial features, retarded mental development, limited joint mobility, enlarged liver, umbilical hernia, and other signs. X-rays show coarsening of skeletal structures (dysostosis multiplex). The photographs show the same patient at different ages (own data).

B. Mucopolysaccharide storage disease type II (Hunter)

This type of mucopolysaccharidosis is transmitted by X-chromosomal inheritance. Four cousins from one pedigree are shown. Clinically, the disease is similar to, but less rapidly progressive than, MPS type I. (Photos from E. Passarge et al., Dtsch Med Wschr. 1974; 99: 144–158).

Classification of the mucopolysaccharide storage diseases (MPS)

MPS Type	Enzyme defect	Important manifestations
IH (Hurler)	α-L-iduronidase	Dysostosis multiplex, severe developmental disorder, corneal clouding, a.o.
IS (Scheie)	α-L-iduronidase	Stiff joints, corneal clouding, normal mental development
II (Hunter) (X-chromosomal)	Iduronate sulfatase	Dysostosis multiplex, no corneal clouding
III (Sanfilippo)		
A	Heparan *N*-sulfatase	Severe psychomotor retardation beginning about age 6–8 years, relatively mild somatic signs.
B	α-*N*-acetylglucosaminidase	
C	Acetyl-CoA: α-glucosaminide *N*-acetyltransferase	
D	*N*-acetylglucosamine-6-sulfate sulfatase	
IV (Morquio)		
A	Galactose-6-sulfatase	Corneal clouding, severe skeletal changes, short stature, odontoid hypoplasia, normal mental development
B	β-galactosidase	
VI (Maroteaux–Lamy)	N-acetylgalactosamine-4-sulfatase (aryl-sulfatase B)	Dysostosis multiplex, corneal clouding, normal mental development
VII (Sly)	β-glucuronidase	Dysostosis multiplex, corneal clouding

(After V.A. McKusick: Mendelian Inheritance in Man. 11th ed., 1994)

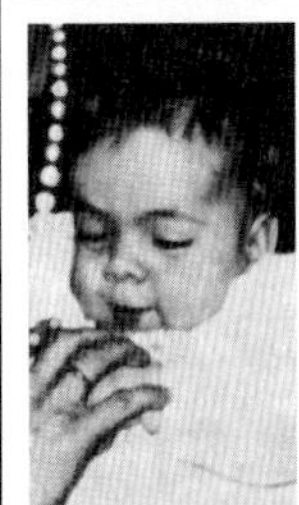

8 weeks

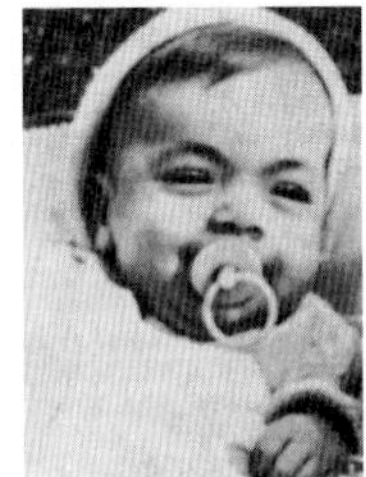
7 months

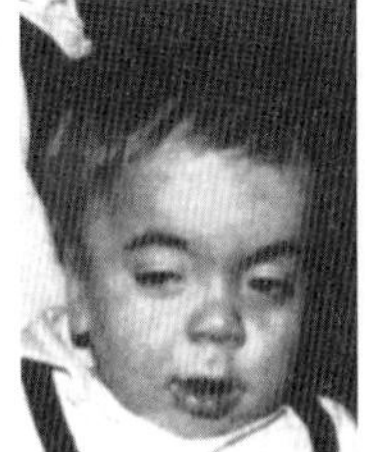
2 1/4 years

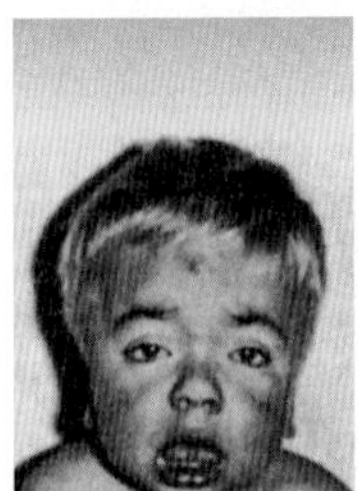
3 3/4 years

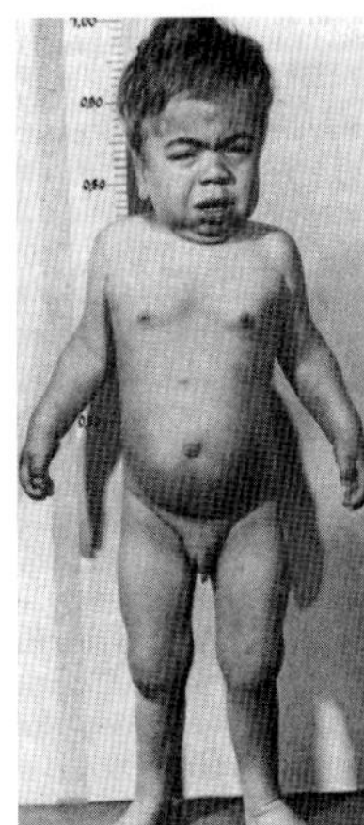
5 years

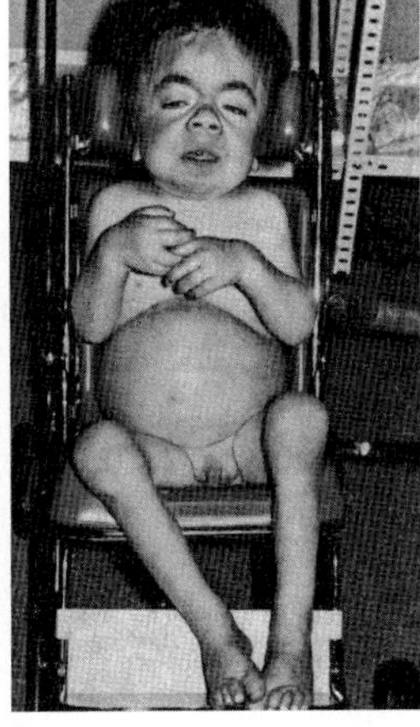
8 years

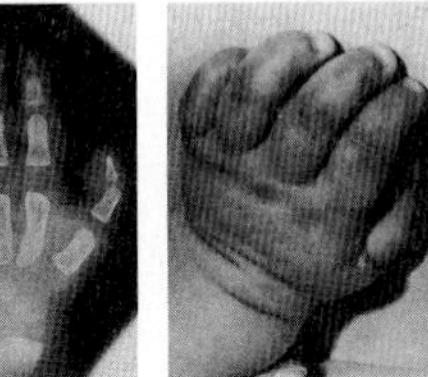
Dysostosis multiplex

Joint contractures

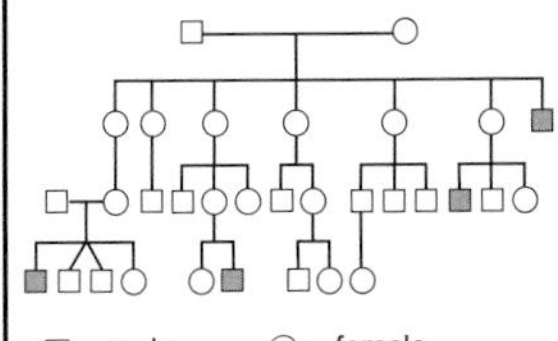

X-Chromosomal inheritance

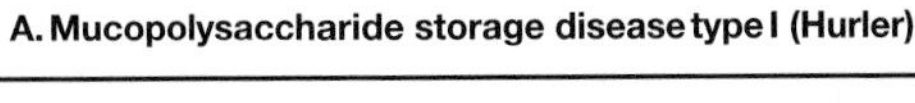
A. Mucopolysaccharide storage disease type I (Hurler)

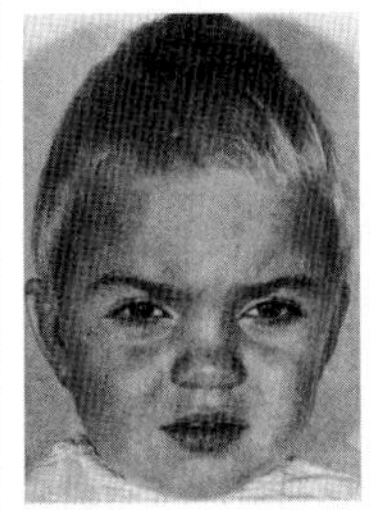
4 1/2 years

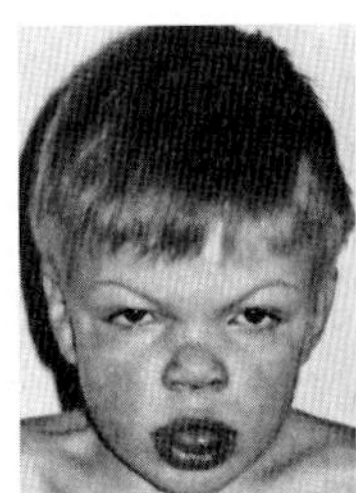
10 years

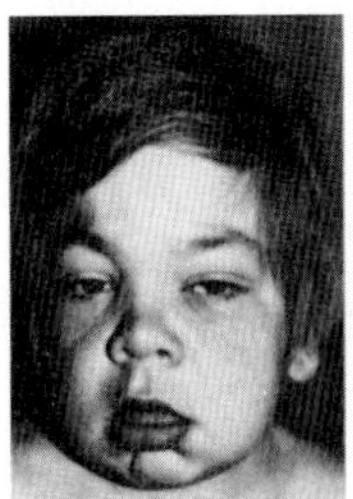
13 years

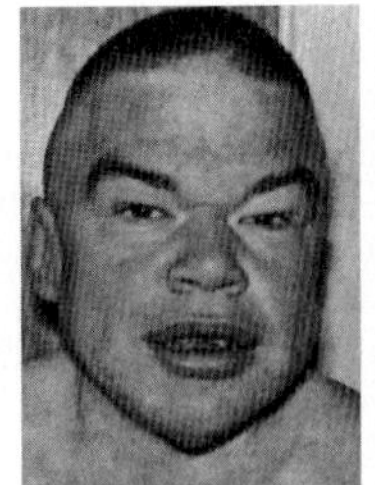
21 years

B. Mucopolysaccharide storage disease type II (Hunter)

Familial Hypercholesterolemia

Familial hypercholesterolemia (FH) is an autosomal dominant hereditary disease affecting about 1 in 500 persons. It is caused by mutations affecting the structure and function of a cell surface receptor for a cholesterol-ester-binding protein (low-density lipoprotein, LDL). The disease leads to highly elevated concentrations of LDL-bound cholesterol in plasma and increased deposits of cholesterol in blood vessels, skin, and tendons.

A. Familial hypercholesterolemia

The common signs of the disease (1) differ in severity depending on whether the mutant allele is present in the heterozygous state (frequency 1:500) or in the homozygous state (approx. 1:250 000). In homozygotes, early atherosclerosis (deposits of cholesterol plaques in the arteries, atheromas) leads to cardiac infarcts and early death, between 5 and 30 years of age. In heterozygotes, the average life expectancy is reduced to about 50 years. A prominent sign is cholesterol-containing deposits (xanthomas) in the skin, especially at the metacarpophalangeal joints (2), over the elbows and knees, on the buttocks, and under the eyelids. In the anterior chamber of the eye, on the lower edge of the iris, small opaque rims from fat deposits (arcus lipoides, 3) are visible. Affected individuals have very high concentrations of cholesterol in the blood (hypercholesterolemia, 4). LDL degradation is decreased as a result of decreased numbers of cell surface LDL receptors (5). (Photo 3 from H. Schuster, Berlin).

B. The LDL receptor

The LDL receptor is a membrane-bound protein composed of 839 amino acids. It consists of five domains, three extracellular, one within the membrane (transmembrane domain), and one with the carboxyl end in the cytoplasm. The ligand-binding domain consists of seven cysteine-rich units, each with 40 amino acids. In humans the receptor gene consists of 18 exons, which span about 45 kb on chromosome 19 (19p13.1–13.3). A signal sequence, coded for by exon 1, serves to direct (targeting) the receptor protein during endocytosis. (Illustration redrawn from a figure of H. Schuster, Berlin).

C. LDL receptor-mediated endocytosis

The LDL receptor mediates the endocytosis of LDL. The receptors loaded with LDL accumulate in a coated pit (1), which separates from the plasma membrane and forms an endocytotic vesicle (2). This transports LDL molecules to a lysosome. (Photo from R.G.W. Anderson, M.S. Brown, J.L. Goldstein, Cell. 1977; 10: 351).

D. Homology with other proteins

The central domain of the LDL receptor shows homology with the EGF growth factor receptor (*e*pidermal *g*rowth *f*actor, see p. 262). The regions of the LDL receptor lying within and close to the membrane are similar to three plasma proteins for blood clotting. The LDL receptor gene contains components of genes of very different functions. Thus, the LDL receptor gene is a member of two supergene families, namely, of EGF growth factors and of plasma proteins. (Figure after Goldstein & Brown, 1989).

References

see p. 374.

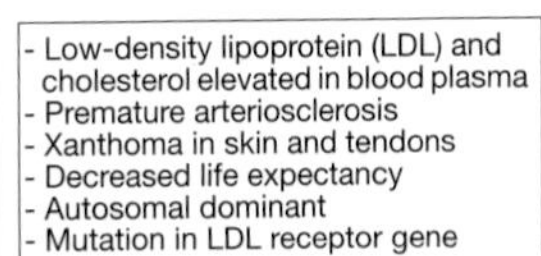

1. General features

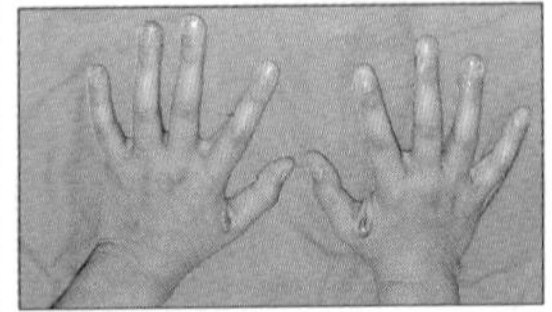

2. Xanthoma formation

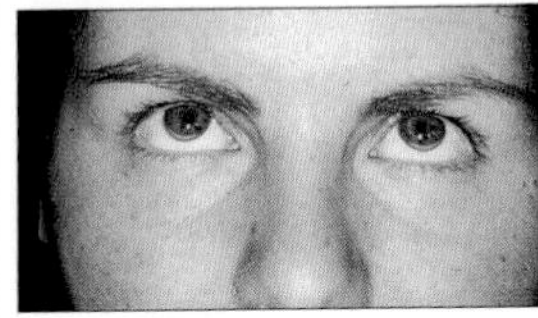

3. Arcus lipoides

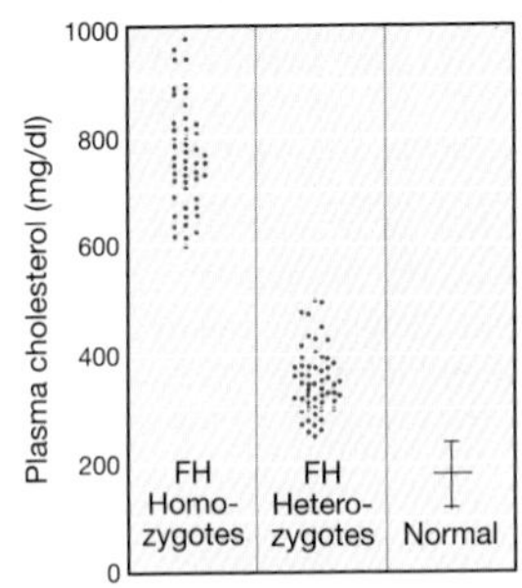

4. Hypercholesterolemia

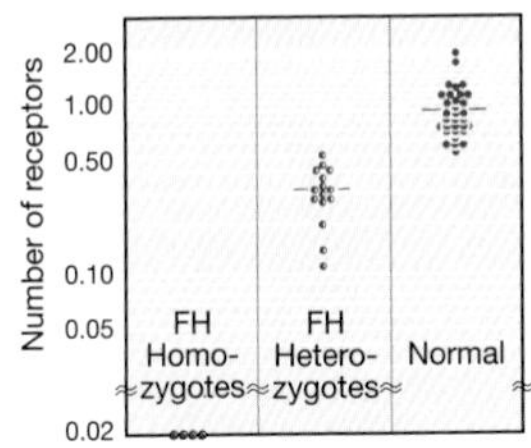

5. LDL receptors decreased

A. Familial hypercholesterolemia

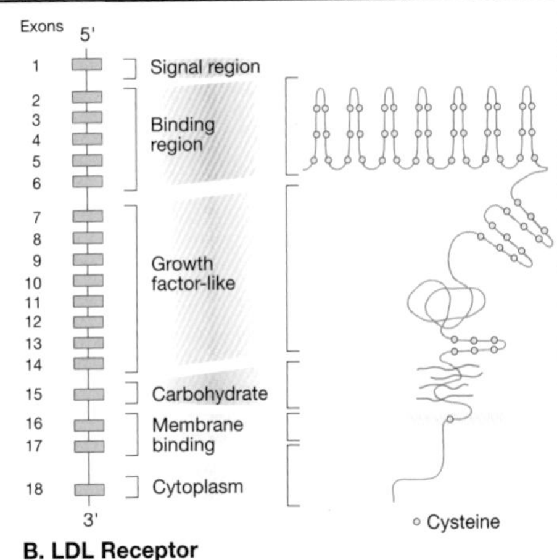

B. LDL Receptor

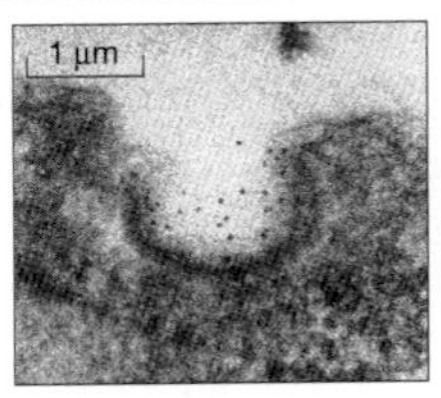

a Coated pit

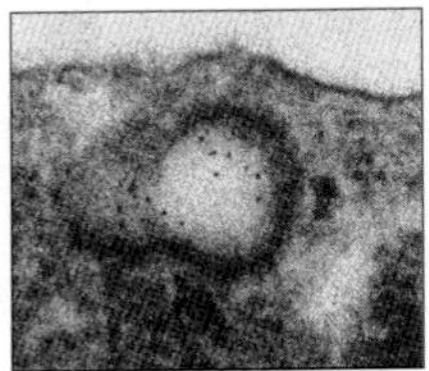

b Endocytotic vesicle

Electron micrographs of fibroblasts in culture that have taken up LDL molecules (black dots, made visible by binding to ferritin).

C. Receptor-mediated endocytosis of LDL

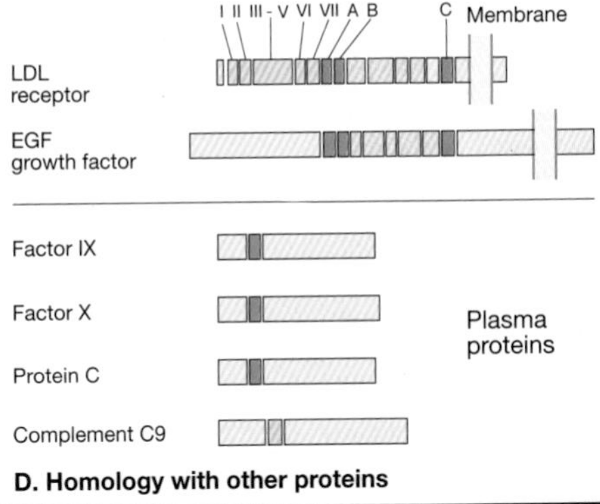

D. Homology with other proteins

Mutations in the LDL Receptor Gene

Different mutations of the LDL receptor gene are known. Their effects depend on their location within the gene. Mutant forms of the LDL receptor and mutant forms of the ligand (apoprotein B-100) both lead to familial hypercholesterolemia (Schuster et al. 1990).

A. Intracellular LDL receptor metabolism and classes of mutation

The metabolism of the LDL receptor corresponds to that of a lysosomal system. Low density lipoprotein (LDL), consisting of cholesterol and the apoprotein B-100, are taken up into the cell in a "coated pit" by receptor-mediated endocytosis and transported to an endosome in a "coated vesicle." In the lysosome, the ligand (LDL) and the receptor are separated by the action of hydrolases. The receptor is recycled and returned to the cell surface. The LDL protein is broken down into amino acids, and cholesterol is released in the process. The cholesterol derived from LDL regulates intracellular cholesterol metabolism. Normally, only as much cholesterol is formed as is needed for the synthesis of cell membranes, steroid hormones, and bile acids, and the uptake of exogenous cholesterol is decreased by reduced biosynthesis of LDL receptors. This is regulated by decreased activity of the enzyme 3-hydroxy-3-methylglutaryl-CoA reductase (HMG-CoA reductase). Cholesterol activates acyl-CoA cholesterol transferase (ACAT); so within certain limits, excess cholesterol can be stored as cholesterol ester. Normal LDL receptor metabolism can be disrupted by mutations at several stages. Five principal classes of mutations are distinguished: (1) no formation of the receptor protein (R^o mutations), (2) defective intracellular transport, (3) defective LDL binding, (4) defective endocytosis (R^+ mutations), and (5) failure to release the LDL inside the endosome (recycling-defective alleles). (Modified after Goldstein & Brown).

B. Different mutations in the LDL receptor gene and their functional consequences

Deletions and point mutations have been described. Some insertions and duplications have arisen by unequal crossing-over in Alu-repetitive sequences.
A total of 127 LDL receptor mutations have been described (Goldstein et al. 1995). They affect the signal sequence (3), the ligand-binding region (48), the epidermal growth factor homology domain (44), the O-linked sugar and transmembrane domain (3), the cytoplasmic tail (4), and the promoter (3).

C. Point mutation in the LDL receptor gene

Direct sequencing demonstrates a mutation in exon 9. First, exon 9 is amplified by PCR (P1 and P2 = primers 1 and 2). The mutation in codon 408, GTG (valine) to GTA (methionine), produces a recognition site (N) for *Nla*III (GATC) that is not normally present. This results in two fragments of 126 and 96 base pairs (bp) instead of the usual 222-bp fragment. Thus, affected individuals (1 and 3 in the pedigree) have two smaller fragments of 126 and 96 kb (2) in addition to the 222–kb fragment. Sequence analysis of the patient (individual 1 in the pedigree) demonstrates the mutation by the presence of an additional adenine (A) next to the normal guanine (3). With this knowledge about the mutation, the latter can be indirectly demonstrated within a family by the additional recognition site for a restriction enzyme (Data from H. Schuster, Berlin).

References

see p. 374.

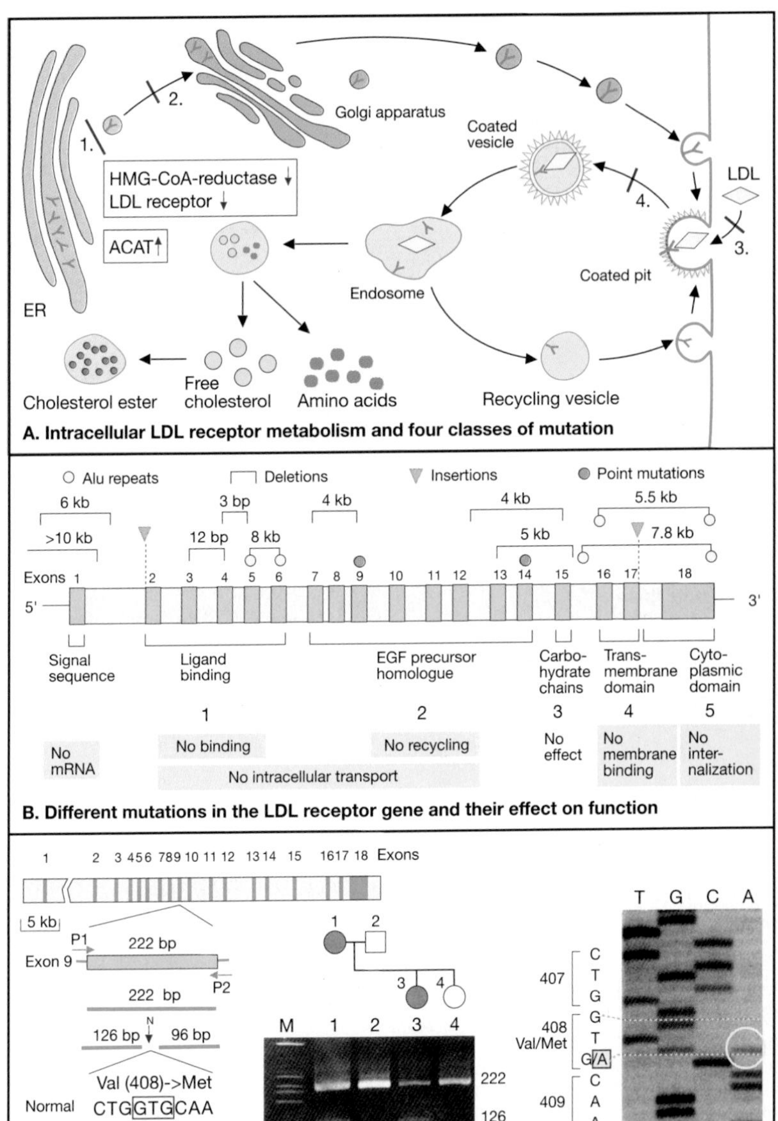

A. Intracellular LDL receptor metabolism and four classes of mutation

B. Different mutations in the LDL receptor gene and their effect on function

C. Point mutations in the LDL receptor gene

Protease Inhibitor α_1-Antitrypsin

α_1-Antitrypsin (α_1-AT) is an essential protease inhibitor in blood plasma. It binds to a wide range of proteases, such as elastase, trypsin, chemotrypsin, thrombin, and bacterial proteases. Its most important physiological effect is the inhibition of leukocyte elastase, a protease that breaks down the elastin of the pulmonary alveolar walls. Deficiency of α_1-antitrypsin leads to increasing destruction of the pulmonary alveoli, obstructive emphysema of the lungs, and in newborns, a form of hepatitis.

A. α_1-Antitrypsin

α_1-Antitrypsin (1) in humans is a glycoprotein composed of 394 amino acids and 12% carbohydrate. It is coded for by a 10.2-kb gene with five exons on chromosome 14 (14q32.1).

B. α_1-Antitrypsin deficiency

The uninhibited action of leukocyte elastase on the elastin of the pulmonary alveoli leads to chronic obstructive pulmonary emphysema (1) (X-ray from N. Konietzko, Essen). The most frequent deficiency allele is Pi(Z). The plasma concentration of α_1-AT with genotype PiZZ (homozygote) is usually about 12–15% of normal (with the normal allele M). MZ heterozygotes have 64%, and MS heterozygotes 86% of MM homozygote activity. α_1-Antitrypsin deficiency in the lung can be corrected by intravenous administration of α_1-antitrypsin. (After R.H. Ingram).

C. α_1-Antitrypsin: protein, gene, and important mutations

The α_1-antitrypsin protein has three oligosaccharide side chains at positions 46, 83, and 247. The protein is highly polymorphic because of differences in the amino acid sequence and in carbohydrate side chains. The reactive site is located at position 358/359 (methionine/serine). Clinically, the most important mutations affect codons 213 (PiZ), 256 (PiP), 264 (PiS), 342 (PiZ), and 357 (Pi[Pittsburgh]). The gene contains variant restriction enzyme sites, which can be used for reliable diagnosis. Today the diagnosis is often made using the PCR reaction.

D. Synthesis of α_1-antitrypsin

The α_1-AT gene is expressed in liver cells (hepatocytes). The gene product is channeled through the Golgi apparatus and released from the cell (secreted). The Z mutation leads to aggregation of the enzyme in the liver cells, with too little of it being secreted. The S mutation leads to premature degradation. About 2–4% of the population in Central and Northern Europe are MZ heterozygotes.

E. Reactive center of protease inhibitors

α_1-Antitrypsin is one member of a family of protease inhibitors that show marked homology, especially at their reactive centers. Oxidizing substances have an inhibitory effect and inactivate the molecule. Smokers have a much more rapid course of α_1-AT deficiency disease (onset of dyspnea at 35 years of age instead of 45–50).

References

Cox DW. α_1-Antitrypsin deficiency. In: Scriver CR, Beaudet AL, Sly WS, Valle D, eds. The Metabolic and Molecular Bases of Inherited Disease. 7th ed. New York: McGraw–Hill; 1995: 4125–4158.

Lancet editorial: α_1-Antitrypsin Z and the liver. Lancet. 1992; 340: 402–403.

Lomas DA, et al. The mechanism of Z α_1-antitrypsin accumulation in the liver. Nature. 1992; 357: 605–607.

Owen MC, et al. Mutation of antitrypsin to antithrombin: α_1-antitrypsin Pittsburgh (358 Met → Arg), a fatal bleeding disorder. N Engl J Med. 1983; 309: 694–698.

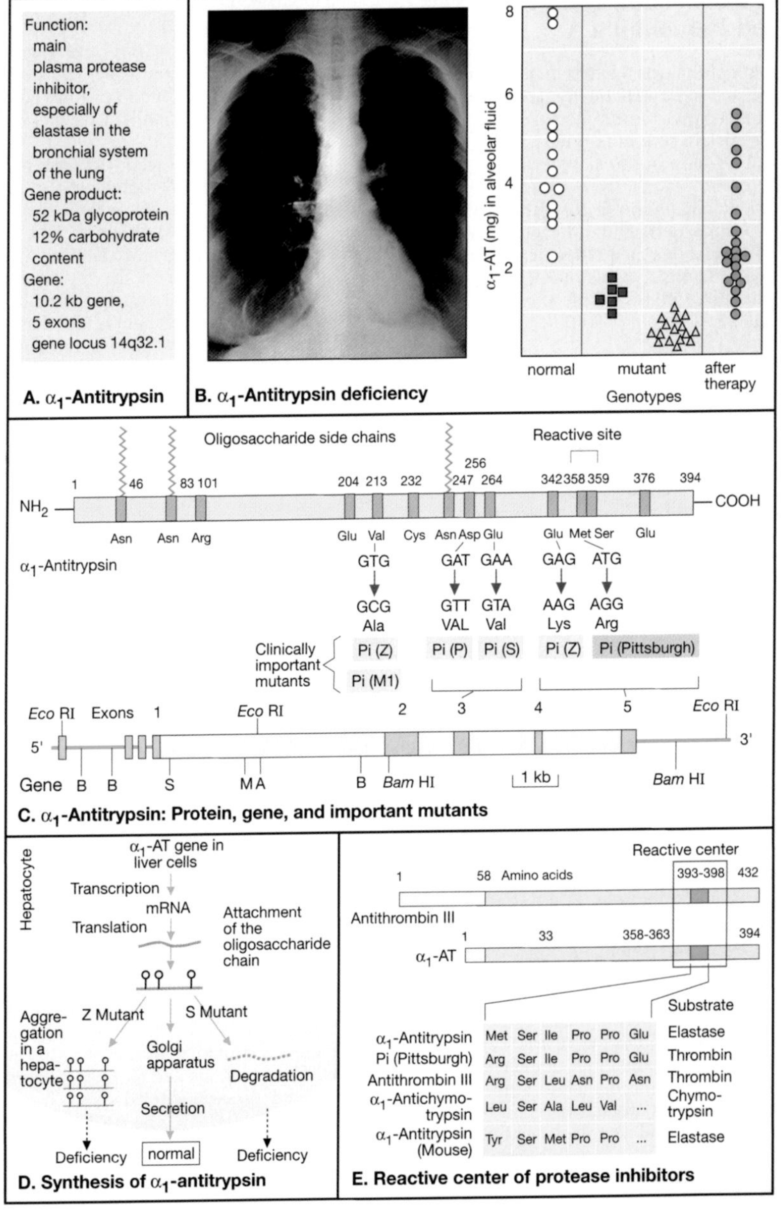

Function:
main plasma protease inhibitor, especially of elastase in the bronchial system of the lung
Gene product:
52 kDa glycoprotein
12% carbohydrate content
Gene:
10.2 kb gene,
5 exons
gene locus 14q32.1
A. α1-Antitrypsin
B. α1-Antitrypsin deficiency
α1-AT (mg) in alveolar fluid
8
6
4
2
normal
mutant
after therapy
Genotypes
Oligosaccharide side chains
Reactive site
1
46
83 101
204 213 232
256
247 264
342 358 359
376
394
NH2
COOH
Asn Asn Arg
Glu Val Cys Asn Asp Glu
Glu Met Ser Glu
α1-Antitrypsin
GTG
GAT GAA
GAG ATG
GCG Ala
GTT VAL
GTA Val
AAG Lys
AGG Arg
Clinically important mutants
Pi (Z)
Pi (P)
Pi (S)
Pi (Z)
Pi (Pittsburgh)
Pi (M1)
Eco RI
Exons
1
Eco RI
2
3
4
5
Eco RI
5'
3'
Gene
B
B
S
M A
B
Bam HI
1 kb
Bam HI
C. α1-Antitrypsin: Protein, gene, and important mutants
Hepatocyte
α1-AT gene in liver cells
Transcription
mRNA
Translation
Attachment of the oligosaccharide chain
Z Mutant
S Mutant
Aggregation in a hepatocyte
Golgi apparatus
Degradation
Secretion
Deficiency
normal
Deficiency
D. Synthesis of α1-antitrypsin
Reactive center
1
58 Amino acids
393-398
432
Antithrombin III
1
33
358-363
394
α1-AT
Substrate
α1-Antitrypsin Met Ser Ile Pro Pro Glu Elastase
Pi (Pittsburgh) Arg Ser Ile Pro Pro Glu Thrombin
Antithrombin III Arg Ser Leu Asn Pro Asn Thrombin
α1-Antichymotrypsin Leu Ser Ala Leu Val ... Chymotrypsin
α1-Antitrypsin (Mouse) Tyr Ser Met Pro Pro ... Elastase
E. Reactive center of protease inhibitors

Blood Coagulation Factor VIII and Hemophilia A

Hemophilia was the first major disease recognized to be genetically determined. The Talmud refers to its increased occurrence in males in certain families, corresponding with X-chromosomal inheritance. Hemophilia A results from the deficiency of blood coagulation factor VIII; hemophilia B, from a deficiency of factor IX. Factor VIII functions as a cofactor in the activation of factor X to factor Xa during the intermediate phase of the coagulation cascade.

A. X-chromosomal inheritance of hemophilia A

A classic example of the X-chromosomal inheritance of hemophilia A was seen in some royal families in Europe in the last century and first half of this century.

B. Blood coagulation factor VIII

When activated by thrombin, factor VIII protein consists of five subunits (A1, A2, A3, C1, C2) held together by calcium ions (1). The inactive factor VIII protein (2) contains three domains (A, B, C). Domain A occurs in three homologous copies (A1, A2, A3), domain C in two (C1, C2), and domain B in one copy. In humans, the gene for factor VIII (3) maps to the distal long arm of the X chromosome in region 2, band 8 (Xq28). It consists of 26 exons and spans 186 000 base pairs (186 kb), corresponding to about 0.1% of the whole X chromosome. Noteworthy in this gene are the large exon 14 (3106 base pairs), which codes for the B domain, and a large intron of 32 000 base pairs between exons 22 and 23. Most point mutations occur in DNA sequences involving TCGA, the recognition sequence for the restriction enzyme *Taq*I. It contains the dinucleotide CG, which is easily mutated. Since the cytosine of this dinucleotide is frequently methylated, and deamination of methyl cytosine leads to a C to T transition, mutations in CG dinucleotide regions are frequent (see p. 41). Mutation of TCGA to TTGA creates a stop codon (TGA), resulting in a truncated factor VIII protein. Even a stop codon at position 2307 leads to severe hemophilia, although only the last 26 amino acids are missing (J. Gitschier et al. Nature. 1985; 315: 427).
Polymorphic restriction sites (RFLPs, restriction fragment length polymorphisms) can be utilized for molecular genetic diagnosis of hemophilia A (4). When present, a variant recognition sequence (B*) for the restriction enzyme *Bcl*I in the region of exons 17 and 18 produces a fragment of 879 base pairs and a fragment of 286 base pairs; when absent, a single fragment of 1165 base pairs. This can be used in RFLP diagnosis (5): The index patient (II-1) with hemophilia A carries the 879-bp fragment. This fragment indicates the mutation. His sister (II-2) has an affected son (III-2) who also carries the 879-bp fragment, inherited from his mother. A brother (III-1) does not carry this fragment and thus is not at risk for the disease. (Note that II-3 is not affected and carries a fragment of 1165 bp, not 879 bp as shown).
In addition to point mutations, factor VIII gene rearrangements involving the long intron 22 are frequent.

C. Severity and factor VIII activity

Hemophilia occurs with a frequency of about 1 in 10 000 male newborns. Severity and frequency of bleeding are dependent on the degree of residual factor VIII activity.

References

Goodeve AC, et al. Factor VIII rearrangements in patients with severe haemophilia A. Lancet. 1994; 343: 329–330.

Kazazian HH Jr, et al. Hemophilia A and parahemophilia: Deficiencies of coagulation factors VIII and V. In: Scriver CR, et al. eds. The metabolic and molecular bases of inherited disease. 7th ed. New York: McGraw–Hill; 1995: 3241–3267.

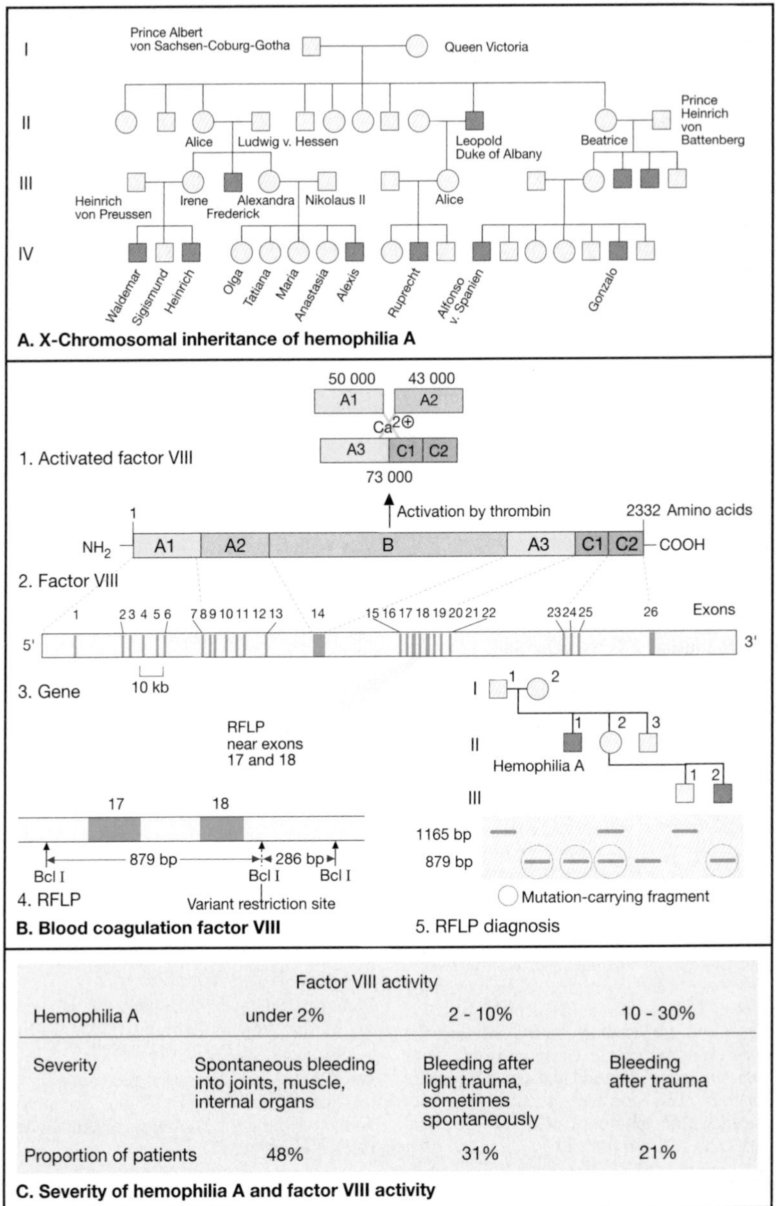

	Factor VIII activity		
Hemophilia A	under 2%	2 - 10%	10 - 30%
Severity	Spontaneous bleeding into joints, muscle, internal organs	Bleeding after light trauma, sometimes spontaneously	Bleeding after trauma
Proportion of patients	48%	31%	21%

C. Severity of hemophilia A and factor VIII activity

Von Willebrand Factor

Von Willebrand factor (vWF) is a complex multimeric protein found in plasma, platelets, and subendothelial connective tissue. It has two basic biological functions: It binds to specific receptors on the surface of platelets and subendothelial connective tissue, and it forms bridges between platelets and damaged regions of a vessel. Furthermore, it binds to clotting factor VIII and stabilizes it. Deficiency of vWF leads to decreased or absent platelet adhesion and to secondary deficiency of factor VIII (von Willebrand disease or von Willebrand–Jürgens syndrome). Hereditary deficiency of vWF is the most common bleeding disorder in man, with a frequency of about 1:250 for all forms, including the mild ones, and about 1:8000 for severe forms.

A. Von Willebrand cDNA and prepropeptide

Von Willebrand factor is formed in endothelial cells, in megakaryocytes, and possibly in some other tissues and is coded for by a large (178 kb) gene with 52 exons of various sizes on chromosome 12 (12p12–pter). The cDNA of vWF is about 8.7 kb long. The corresponding mRNA codes for a primary peptide (prepro–vWF) of 2813 amino acids, including a signal peptide of 22 amino acids, a segment of 741 amino acids (vW antigen II), and a subunit of four different domains (A–D), which together make up more than 90% of the sequence. The three A domains (A1–A3) in the middle contain binding sites for collagen, heparin, and thrombocytes. Three small B domains are on the carboxy side of the D4 domain, before the two C domains. vWF contains 8.3% cysteine (234 of 2813 amino acids), concentrated at the amino and carboxy ends, whereas the three A domains are cysteine-poor. After posttranslational modification, the mature plasma vWF contains twelve oligosaccharide side chains (19% of the total weight is carbohydrate).

B. Biosynthesis of the von Willebrand factor

vWF is first formed as a prepropeptide. After the signal peptide is removed, two pro-vWF units attach to each other at their carboxy ends by means of numerous disulfide bridges to form a dimer. The dimers represent the repetitive units, or protomers, of mature vWF. The pro-vWF dimers are transported to the Golgi apparatus, where the pro-vWF (vW antigen II or vWagII) is removed. The function of the propeptide (vWagII) is not known. Mature vWF and vWagII are stored in Weibel–Palade bodies in epithelial cells. The mature subunits and vWagII contain binding sites for factor VIII, heparin, collagen, ristocetin + platelets, and thrombin-activated platelets.

C. Classification of von Willebrand diseases

Von Willebrand disease is a heterogeneous group of disorders divided into several subtypes. In types I and III, the defect is quantitative; in type II, qualitative. Dominant and recessive phenotypes with vWF deficiency often cannot be readily distinguished because heterozygosity may not be manifest and can only be determined by laboratory tests. Type I with subtypes A and B is the most frequent group (70% of all patients). vWF deficiency may simulate platelet dysfunction or hemophilia.

References

Sadler JE. Von Willebrand disease. In: Scriver CR, Beaudet AL, Sly WS, Valle D, eds. The Metabolic and Molecular Bases of Inherited Disease. 7th ed. New York: McGraw–Hill; 1995: 3269–3287.

Manusco DJ, et al. Structure of the gene for human von Willebrand factor. J Biol Chem. 1989; 264: 19514–19527.

Wise RJ, et al. Autosomal recessive transmission of hemophilia A due to a von Willebrand factor mutation. Hum Genet. 1993; 91: 367–372.

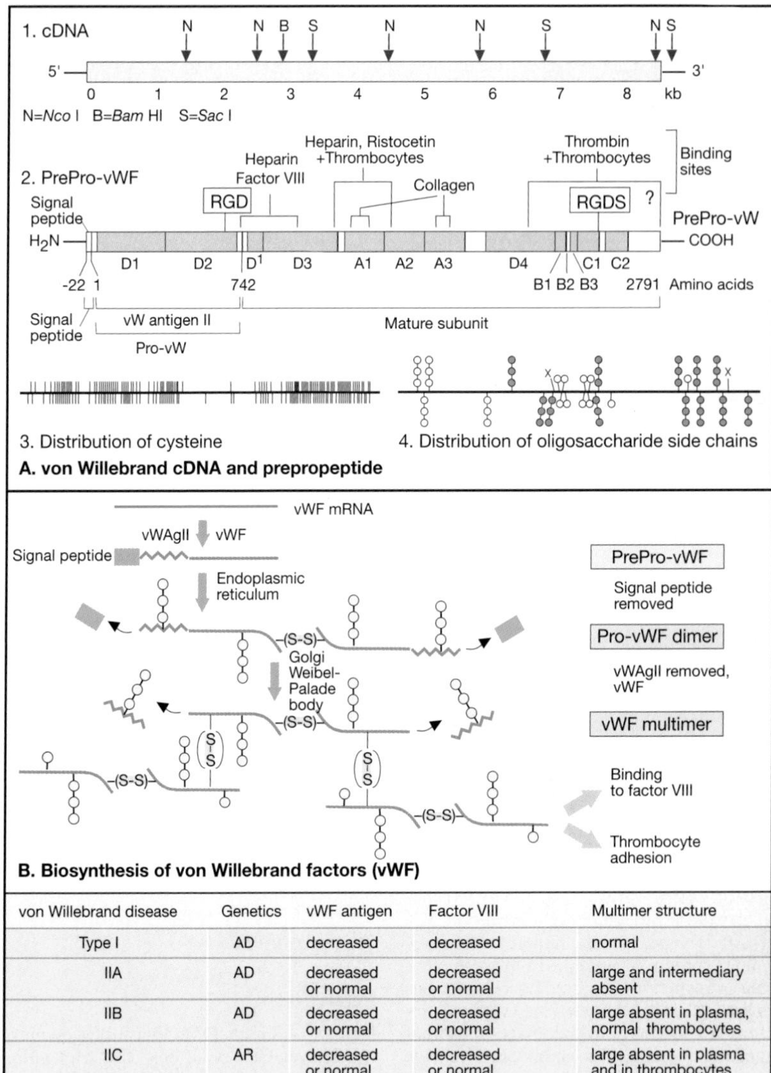

A. von Willebrand cDNA and prepropeptide

B. Biosynthesis of von Willebrand factors (vWF)

von Willebrand disease	Genetics	vWF antigen	Factor VIII	Multimer structure
Type I	AD	decreased	decreased	normal
IIA	AD	decreased or normal	decreased or normal	large and intermediary absent
IIB	AD	decreased or normal	decreased or normal	large absent in plasma, normal thrombocytes
IIC	AR	decreased or normal	decreased or normal	large absent in plasma and in thrombocytes
IID	AD	normal	normal	large absent
IIE	AD	decreased	normal	large absent
III	AR	absent	greatly decreased	absent

C. Classification of von Willebrand diseases

Pharmacogenetics

Many medications are degraded at different rates in different individuals. This has a genetic basis. Enzymes coded for by genes with different alleles may have different catabolic rates, which in turn can result in genetically determined differences in the reaction to drugs (pharmacogenetics).

A. Malignant hyperthermia due to abnormal regulation of a calcium channel in muscle cells

Malignant hyperthermia is a severe, life–threatening complication of anesthesia that may occur in persons with extreme hypersensitivity to halothane and similar agents used in general anesthesia. Normally, a nerve impulse depolarizes the plasma membrane of a nerve ending at the nerve–muscle endplate (1) (motor endplate), and the volt-gated calcium channel in the plasma membrane of the nerve ending is temporarily opened. The massive influx of calcium into the cell (the extracellular Ca^{++} concentration is about 1000 times higher than the intracellular) triggers the release of acetylcholine. Binding of the latter to the acetylcholine receptor of the muscle cell temporarily opens the receptor-controlled cation (Na^{+}) channels. This opens calcium channels located in the sarcoplasmic reticulum of the muscle cell. The resulting rapid increase in Ca^{++} concentration in the cytosol causes the myofibrils in the muscle cell to contract. The calcium channels in the sarcoplasmic reticulum are regulated by a receptor (ryanodin receptor) (2). Ryanodin (an alkaloid) binds to the calcium channel. The ryanodin receptor is a protein with four transmembrane domains. Mutations in the ryanodin receptor lead to greatly increased sensitivity to halothane and other anesthetic agents (3), which cause muscle spasm, drastic elevation of temperature (hyperthermia), acidosis, and cardiac arrest (4). Malignant hyperthermia is inherited as an autosomal dominant trait (5). One gene in man lies on chromosome 19 at 19q13.1 (MacLennan, D.H., Phillips, M.S.: Malignant hyperthermia. Science. 1992; 256: 789–794). Additional loci are on 7q, 17q, and 3q13.1 (Subrak R, et al. Am J Hum Genet. 1995; 56: 684–691). The mutant haplotype of a given family can be determined by segregation analysis. A ryanodin receptor mutation has been demonstrated in porcine malignant hyperthermia (J. Fujii, et al. Science. 1991; 253: 448–451).

B. Serum pseudocholinesterase deficiency (Butyrylcholinesterase)

About 1 in 200 individuals reacts to muscle relaxants, such as suxamethonium (succinylcholine), with prolonged muscle relaxation and respiratory arrest. In such persons, serum pseudocholinesterase activity is decreased. Persons at risk cannot be identified by determining their pseudocholinesterase activity alone (1), but determining dibucaine inhibition of their enzyme activity. Whereas homozygous normal persons show 80% enzyme activity after dibucaine administration, persons at risk show only 20%. Individuals with intermediate values of 60% are regarded as heterozygotes (2). A number of different alleles can lead to different degrees of reduced enzyme activity. (Figure after H. Harris: The Principles of Human Biochemical Genetics, 2nd ed. Amsterdam: North–Holland Publishing Co; 1975). This enzyme is now referred to as butyrylcholinesterase because it hydrolyzes butyrylcholine more readily than acetylcholine.

C. Examples of genetically determined adverse reactions to drugs

See opposite table.

Reference

Kalow W, Grant DM. Pharmacogenetics. In: Scriver CR, et al. eds. The Metabolic and Molecular Bases of Inherited Disease. 7th ed. New York: McGraw–Hill; 1995: 293–326.

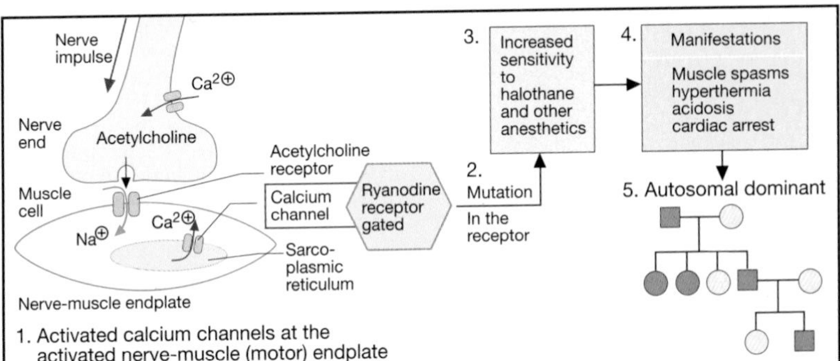

A. Malignant hyperthermia due to a calcium channel disorder in muscle cells

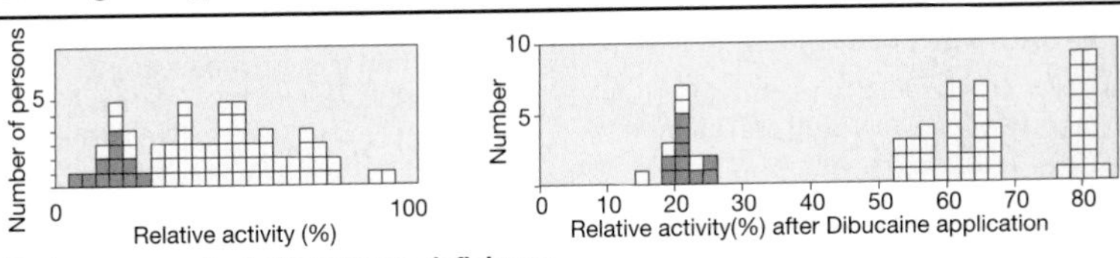

B. Serum pseudocholinesterase deficiency

Defect	Relevant chemical substance	Clinical consequence	Frequency	Pathogenesis	Genetics
Coumarin resistance	Coumarin (warfarin)	Ineffective anticoagulation therapy	Rarer than 1:80 000	Increased vitamin K affinity due to enzyme or receptor defect	Autosomal dominant
Increased sensitivity to isoniazid	Isoniazid, Sulfamethazine, Phenelzine, Hydralazine, etc.	Polyneuritis, lupus-like reaction	In about 50%	Decreased activity of liver Isoniazid acetylase	Autosomal recessive
Isoniazid ineffective	Isoniazid, Sulfamethazine, Phenelzine, Hydralazine	Reduced antituberculous effect		Increased INH excretion	Autosomal dominant
Glucose-6-phosphate dehydrogenase (G6PD)-deficiency	Sulfonamides, antimalaria drugs, Nitrofurantoin, Vicia faba	Hemolysis	Rare in Europeans, frequent in Africa	G6PD deficiency in erythrocytes	X-chromosomal (many mutant forms)
Hemoglobin Zürich	Sulfonamides	Hemolysis	Rare	Unstable hemoglobin due to point mutation in β-globin (arginine instead of histidine in position 63)	Autosomal dominant
Hemoglobin H	Sulfonamides	Hemolysis	Rare	Unstable hemoglobin of 4 β chains due to deletion of the α loci	Autosomal dominant
Glaucoma in adults (some forms)	Corticoids	Glaucoma	Frequent	Unknown	Possibly autosomal dominant

C. Examples of genetically determined reactions to pharmaceuticals

Cytochrome P450 Genes

Complex chemical substances, such as drugs or plant toxins, are degraded by an oxidation system (monooxygenases) in the endoplasmic reticulum of liver cells. These enzymes (collectively referred to as cytochrome P450) absorb light maximally at 450 nm after binding to CO. Cytochrome P450 is the last enzyme in the essential electron-transporting chain in microsomes of the liver and mitochondria of the adrenal cortex. A large system of evolutionarily related genes code for the different P450 proteins in mammals.

A. Cytochrome P450 system

The cytochrome P450 system (1) consists of oxidizing enzymes (mixed mono-oxygenases). They represent the first phase of detoxification: A substrate (RH) is oxidized to ROH utilizing atmospheric oxygen (O_2), with water (H_2O) being formed as a byproduct. A reductase delivers hydrogen ions (H^+) either from NADPH or NADH. A characteristic feature of P450 enzymes (2) is that a single chemical substrate frequently can be degraded by several P450 enzymes and that a single P450 protein can oxidize a number of structurally different chemical substances. The capacity to metabolize and detoxify a wide range of chemical substances is considerable. However, the enzyme activities of phase I and phase II must be well coordinated, since toxic intermediates with undesirable side effects occasionally arise in the initial stages of phase II.

B. Debrisoquin metabolism

Debrisoquin is an isoquinoline-carboxamidine. It was used to treat high blood pressure until it was found to cause severe side effects in 5–10% of the population. These persons have reduced activity of a degrading enzyme, debrisoquin-4-hydroxylase. A number of other medications, including β-adrenergic blockers, antiarrhythmics, and antidepressives, are also degraded by this enzyme and may also cause untoward reactions in persons with low activity. Individuals with a slow rate of degradation show an increased ratio of debrisoquin/4-hydrodebrisoquin (1). The enzyme is coded for by the *450-db1* gene, a member of the cytochrome P450-IID family (CYP2D). Mutations may cause aberrant splicing and produce a variant pre-mRNA containing an additional intron (F.J. Gonzalez et al., 1988).

C. CYP gene superfamily (cytochrome P450 genes)

The cytochrome P450 genes in mammals are designated CYP genes. They make up a superfamily of genes that resemble each other in exon/intron structure and that code for similar gene products. An evolutionary pedigree has been derived based on comparisons of their cDNA sequences. According to this pedigree, the CYP gene family arose during the last 1500–2000 million years. It is assumed that especially the CYP-2 family developed in response to toxic substances in plants that had to be detoxified by animal organisms. At least 30 gene duplications and gene conversions have led to an unusually diverse repertoire of CYP genes.

References

Ayesh R, et al. Metabolic oxidation phenotypes as markers for susceptibility to lung cancer. Nature. 1984; 312: 169–170.

Nerbert DW, Gonzalez FJ. P450 genes. Structure, evolution, and regulation. Ann Rev Biochem. 1987; 56: 945–994.

Gonzalez FJ, et al. Characterization of the common genetic defect in humans deficient in debrisoquine metabolism. Nature. 1988; 331: 442–446.

Gonzalez FJ, Nebert DW. Evolution of the P450 gene superfamily: animal–plant "warfare," molecular drive, and human genetic differences in drug oxidation. Trends Genet. 1990; 6: 182–186.

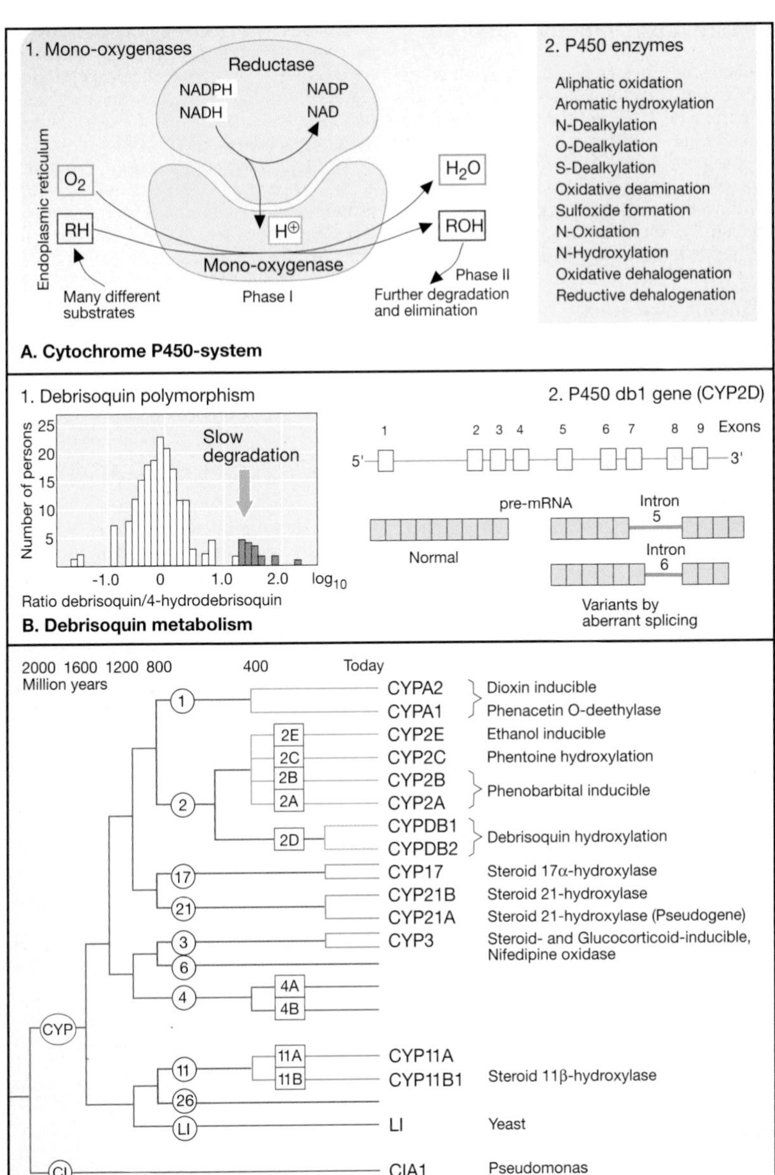

A. Cytochrome P450-system

B. Debrisoquin metabolism

C. CYP gene superfamily (cytochrome P450 genes)

Insulin and Diabetes Mellitus

Diabetes mellitus is one of the most common diseases of the Western world, occurring in about 1–2% of the population. The blood sugar is abnormally elevated due to a variety of causes, including genetic factors. With time, this leads to numerous complications, such as myocardial infarction, stroke, renal failure, vascular damage leading to amputation, and blindness.

A. Insulin formation

In humans, the gene for insulin is located on the short arm of chromosome 11 in region 1, band 5.5. With 1430 base pairs, it is a small gene. It consists of a signal sequence (L, leader) and two exons. The gene is expressed exclusively in β cells of the islands of Langerhans of the pancreas. A β-cell-specific enhancer is located at the 5' end of the gene, and a variable number of tandemlike repeats (VNTR) are located further upstream. The primary transcript is spliced to produce the mRNA template for preproinsulin. The signal peptide (24 amino acids) is removed, and the B chain and A chain are joined by two disulfide bridges. Proper binding and three-dimensional structuring require the presence of a connecting peptide (C peptide). The complete insulin molecule consists of an A chain of 21 amino acids and a B chain of 30 amino acids. The signal peptide of the insulin molecule is required for secretion.

B. Insulin receptor

Insulin initiates its physiological effect by binding to a receptor (insulin receptor). When bound to insulin, the insulin receptor functions as an enzyme and phosphorylates tyrosine in the target proteins. This is the intracellular signal for the metabolic processes induced by insulin.

C. Diabetes mellitus (simplified model)

Diabetes mellitus is classified into two basic types: type I (insulin-dependent diabetes mellitus, IDDM) and type II (non-insulin-dependent, NIDDM). The majority of diabetes type I cases are caused by external factors, such as certain viral infections, on a background of genetic susceptibility. Diabetes type II is mainly due to genetic factors, but also in part to overnourishment. Apart from an autosomal dominant hereditary form with onset in young adults, it is not a monogenic disorder. Monozygotic twins are concordant for type II in about 40–50% of cases and for type I in about 25%, as opposed to a risk of less than 10% for type I in first-degree relatives (about 2–7% according to family relationship and age at onset of disease). Diabetes mellitus is a secondary manifestation of a number of genetically determined diseases, e.g., insulin receptor defect (insulin-resistance syndrome).

D. Influence of genes of the HLA-D region

Genetic susceptibility to diabetes type I is especially influenced by certain alleles of class I MHC genes (see p. 252). The presence of alleles DR3 and DR4, especially in DR3/DR4 heterozygotes, is associated with susceptibility to diabetes type I. DR2 confers relative resistance to diabetes. Genes conferring susceptibility to diabetes have been located at several sites of the genome.

References

Bennett ST, et al. Susceptibility to human type 1 diabetes at *IDDM2* is determined by tandem repeat variation at the insulin gene minisatellite locus. Nature Genet. 1995; 9: 284–292.

Davies JL, et al. A genome-wide search for human type 1 diabetes susceptibility genes. Nature. 1994; 371: 130–136.

Taylor SI. Diabetes mellitus. In: Scriver CR, et al. eds. The Metabolic and Molecular Bases of Inherited Diseases. 7th ed. New York: McGraw–Hill; 1995: 843–896.

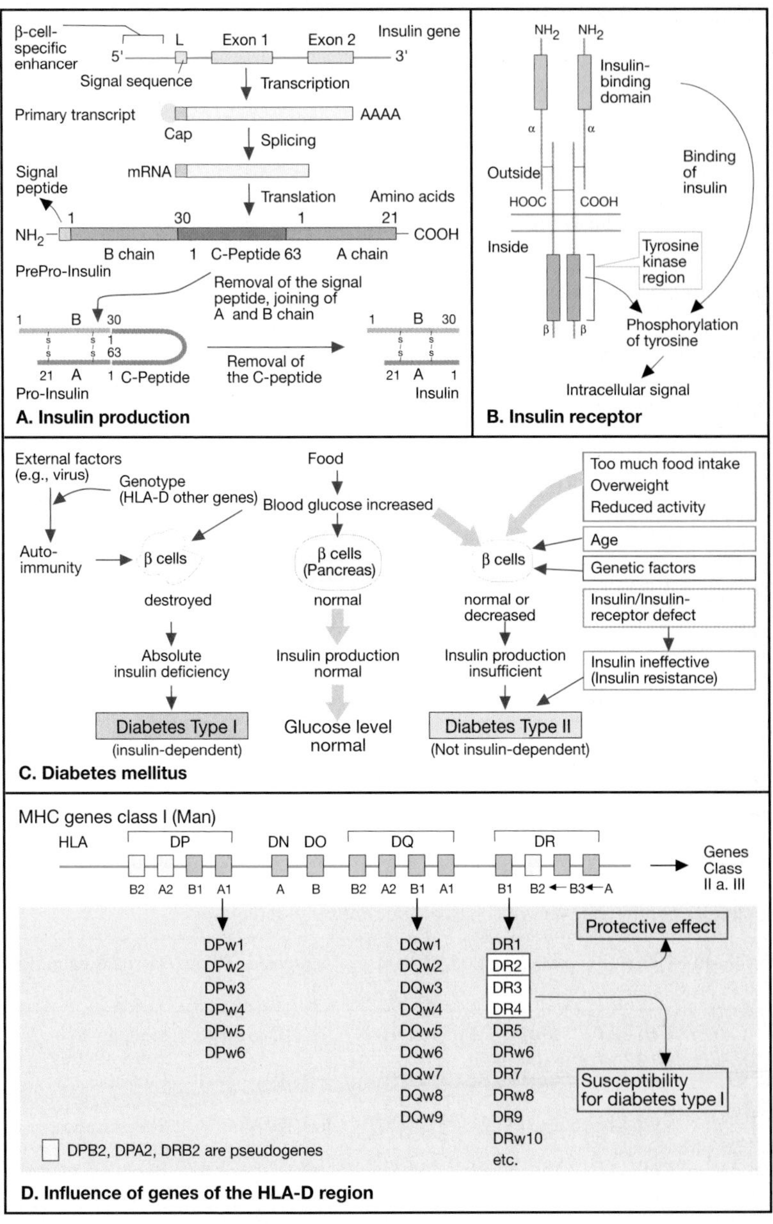
β-cell-specific enhancer
L
Exon 1
Exon 2
Insulin gene
5'
3'
Signal sequence
Transcription
Primary transcript
AAAA
Cap
Splicing
Signal peptide
mRNA
Translation
Amino acids
1
30
1
21
NH2
COOH
B chain
1 C-Peptide 63
A chain
PrePro-Insulin
Removal of the signal peptide, joining of A and B chain
Removal of the C-peptide
Pro-Insulin
Insulin
A. Insulin production
NH2 NH2
Insulin-binding domain
Binding of insulin
Outside
HOOC
COOH
Inside
Tyrosine kinase region
Phosphorylation of tyrosine
Intracellular signal
B. Insulin receptor
External factors (e.g., virus)
Genotype (HLA-D other genes)
Food
Blood glucose increased
Too much food intake
Overweight
Reduced activity
Auto-immunity
β cells
β cells (Pancreas)
β cells
Age
Genetic factors
destroyed
normal
normal or decreased
Insulin/Insulin-receptor defect
Absolute insulin deficiency
Insulin production normal
Insulin production insufficient
Insulin ineffective (Insulin resistance)
Diabetes Type I
(insulin-dependent)
Glucose level normal
Diabetes Type II
(Not insulin-dependent)
C. Diabetes mellitus
MHC genes class I (Man)
HLA
DP
DN
DO
DQ
DR
B2 A2 B1 A1
A B
B2 A2 B1 A1
B1 B2 B3 A
Genes Class II a. III
DPw1
DPw2
DPw3
DPw4
DPw5
DPw6
DQw1
DQw2
DQw3
DQw4
DQw5
DQw6
DQw7
DQw8
DQw9
DR1
DR2
DR3
DR4
DR5
DRw6
DR7
DRw8
DR9
DRw10
etc.
Protective effect
Susceptibility for diabetes type I
DPB2, DPA2, DRB2 are pseudogenes
D. Influence of genes of the HLA-D region

Genes for Collagen Molecules

Collagen, the most abundant protein in mammals, constitutes about one quarter of the total body protein. There are more than 20 different types, which occur in skin, bones, tendons, cartilage, blood vessels, teeth, basement membranes, the corneas and vitreous bodies, and supporting tissues of the internal organs. Collagen forms interlinked, insoluble threads (fibrils) of unusual strength. A fiber of 1 mm diameter can hold a weight of almost 10 kg.

A. Collagen structure

The amino acid sequence of collagen is simple and periodic (1). Every third amino acid is glycine (Gly). Other amino acids alternate between the glycines. The general structural motif is $(Gly–X–Y)_n$. X is either proline or hydroxyproline; Y is either lysine or hydroxylysine. Three chains of collagen form a triple helix (3). In collagen type I, the helix is composed of two identical α1 chains and an α2 chain. It is first formed as a precursor molecule, procollagen (4). Procollagen peptidases remove peptides at the N-terminal and C-terminal ends to form tropocollagen (5). Tropocollagen molecules are connected by the numerous hydroxylated proline and lysine residues to form collagen fibrils (6). Each fibril consists of staggered, parallel rows of end-to-end tropocollagen molecules, separated by gaps (7). Collagen fibrils are visible as transverse stripes under the electron microscope (8). (Photo from L. Stryer, Biochemistry, 3rd ed., 1988).

B. Prototype of a gene *(COL2A1)* for procollagen type II (α1[II])

A procollagen molecule is coded for by a gene consisting of 52 exons. The translated part of exon 1 (85 base pairs) codes for a signal peptide necessary for secretion. Exon sizes differ, with one exon coding for 5, 6, 11, 12, or 18 periodic Gly–X–Y units. The genes for procollagen types I, II, and III differ in that some exons are fused, but otherwise they are similar, especially for the three main fibrillar collagen types (I, II, III). Collagens form a multigene family with more than 28 members, their genes being located on 12 different chromosomes.

C. Gene structure and procollagen type a1(I)

The 52 exons of the *COL1A1* gene correspond to the different domains (A to G) of procollagen α1(I). The *COL1A2* gene for procollagen α2(I) is about twice as large (ca. 40 kb) as the *COL1A1* gene because the introns between the exons are on average twice as long as in *COL1A1*. Unlike Proα1(I), Proα2(I) does not contain domain B (globular portion). More than ten distinct human diseases are caused by mutations in one of the ten genes encoding collagen.

References see p. 316.

Examples of different types of collagen (four of twenty-three)

Collagen Type	Gene	Localization	Chain	Molecule	Tissue distribution
I	*COL1A1* *COL1A2*	17q21–22 7q21–22	α1(I)	$[\alpha1(I)]_2\ \alpha2(I)$	Skin, tendons, bone, arteries
II	*COL2A1*	12q13–14	α1(II)	$[\alpha1(II)]_3$	Cartilage, vitreous body
III	*COL3A1*	2q31–32	α1(III)	$[\alpha1(III)]_3$	Skin, arteries, uterus
IV	*COL4A1* *COL4A2* (6 α chains)	13q33–34	α1(IV) α2(IV)	$[\alpha1(IV)]_2\ \alpha2(IV)$ $[\alpha1(IV)]_3$ $[\alpha2(IV)]_3$	Basement membranes

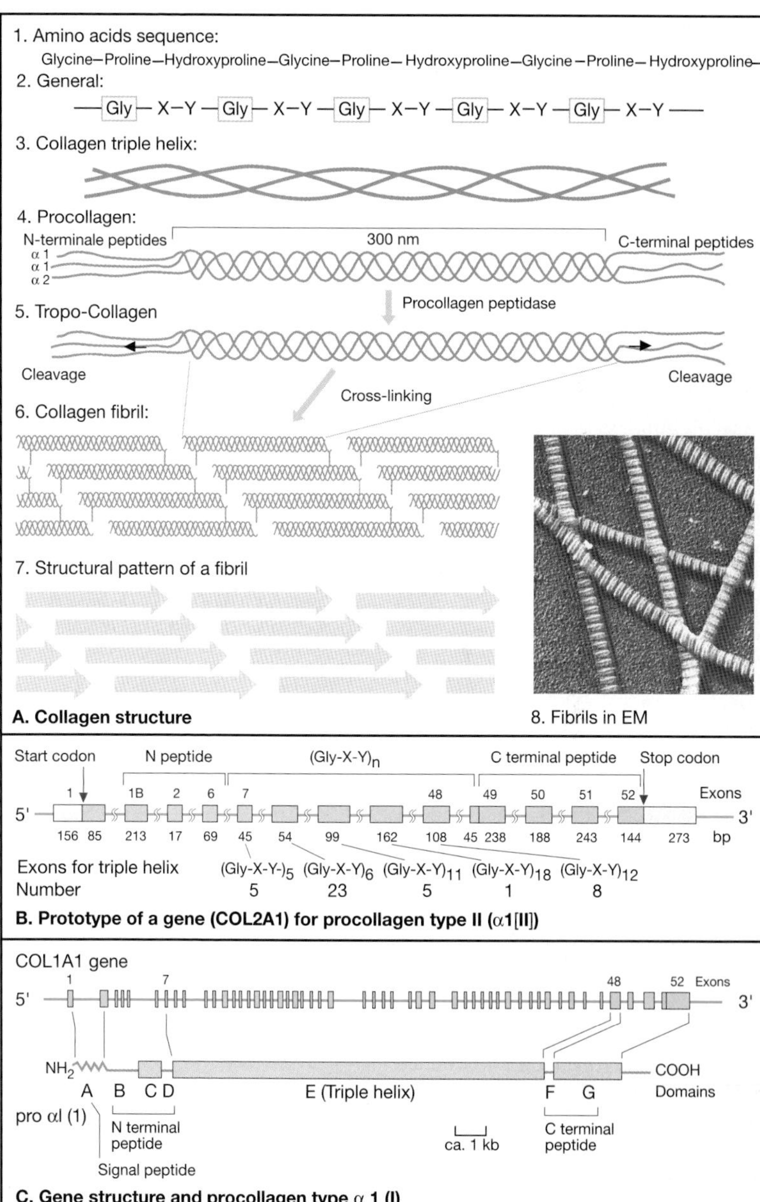

A. Collagen structure

B. Prototype of a gene (COL2A1) for procollagen type II (α1[II])

C. Gene structure and procollagen type α 1 (I)

Osteogenesis Imperfecta

Osteogenesis imperfecta (OI) ("brittle bone disease") is a heterogeneous group of clinically and genetically different types of diseases with a total frequency of at least 1 in 10 000 individuals. Spontaneously occurring bone fractures, bone deformity, small stature, defective dentition (dentinogenesis imperfecta), hearing impairment due to faulty formation of the auditory ossicles, and blue sclerae (the fully developed conjunctiva of the eye is thinner than normal, so that refracted light is shifted toward blue) occur in the various forms to different extents and with different grades of severity, depending on the type of mutation. OI has been demonstrated in a 7th-century Egyptian mummy (cited from Byers,1993).

A. Molecular mechanisms in osteogenesis imperfecta

Some types of mutation may lead to reduced production of proα1(I) (1 and 2), e.g., deletion of a *COL1A1* allele, a transcription or splicing defect, or faulty formation of collagen fibrils. The relative excess of proα2(I) molecules becomes degraded. Thus, less procollagen than normal is formed, but it is not defective. Numerous other types of mutations can lead to defective procollagen (3) (figure after R.J. Wenstrup et al., Hum Genet. 1990; 46: 975–982). Mutations in the proα1(I) gene are more severe than mutations in the proα2(I) gene because a greater amount of defective collagen is formed.

B. Mutations and phenotype

The location of a mutation in the gene influences the phenotype. Generally, mutations in the 3' region are more serious than mutations in the 5' region (position effect). Mutations of the proα1(I) chain are more severe than those in the proα2(I) chain (chain effect). The substitution of a larger amino acid for glycine, which is indispensible for the formation of the triple helix, leads to severe disorders (size effect). Different types of mutations may occur, such as deletions, mutations in the promoter or enhancer, and splicing mutations. The codons (AAG, AAA) for the amino acid lysine, which occurs frequently in collagen, are readily transformed into a stop codon by substitution of the first adenine by a thymine (TAG or TAA), so that a short, unstable procollagen is formed. Splicing mutations may lead to the loss of exons (exon-skipping). (Figure from P.H. Byers, Trends Genet. 1990; 6: 293–300).

C. Different forms of osteogenesis imperfecta

Osteogenesis imperfecta may be classified according to severity into four basic phenotypes (Sillence classification). Although the classification does not correspond to the types of mutation, it has in general proved clinically useful. OI types I and IV are less severe than type II (lethal in infancy) and type III. Three X-rays show a relatively mild (but for the patient nevertheless very disabling) deformity of the tibia and fibula in OI type IV (1); severe deformities in the tibia and fibula in OI type III (2); and the distinctly thickened and shortened long bones in the lethal OI type II (3). Mutations in OI are autosomal dominant, the severe forms being due to de novo mutations. Germline mosaicism has been shown to account for rare instances of affected sibs being born to unaffected parents.

References

Byers PH. Osteogenesis imperfecta. In: Connective Tissue and Its Heritable Disorders. Broyce PM, Steinmann B, eds. New York: Wiley–Liss; 1993: 137–350.

Byers PH. Disorders of collagen synthesis and structure. In: Scriver CR, Beaudet AL, Sly WS, Valle D, eds. The Metabolic and Molecular Bases of Inherited Disease. 7th ed. New York: McGraw–Hill; 1995: 4029–4077.

Chu M-L, Prockop DJ. Collagen gene structure. In: Connective Tissue and Its Heritable Disorders. Broyce PM, Steinmann B, eds. New York: Wiley–Liss; 1993: 149–165.

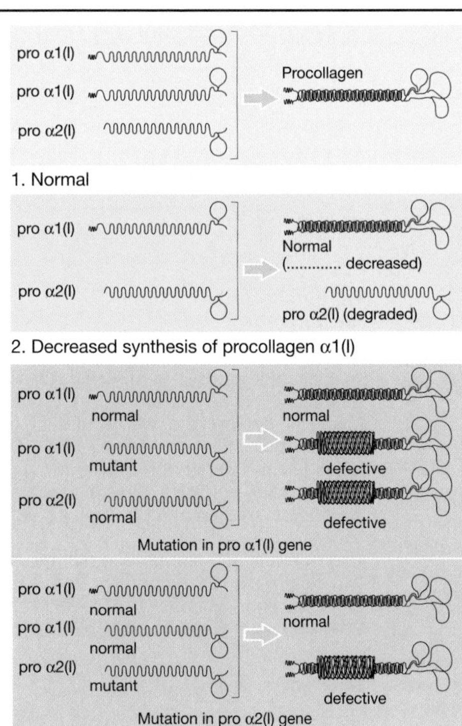

A. Molecular mechanisms in osteogenesis

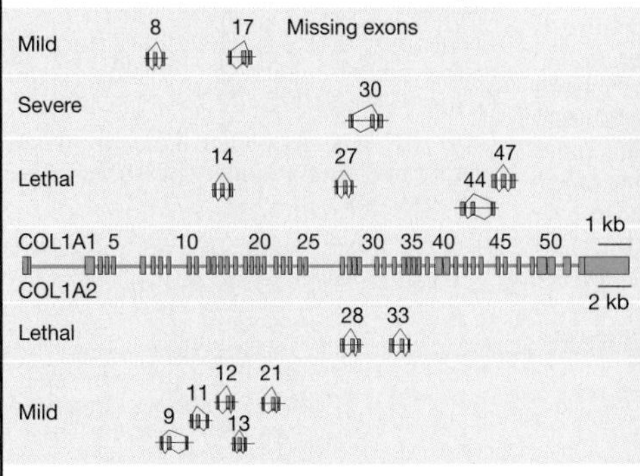

B. Mutations and phenotype

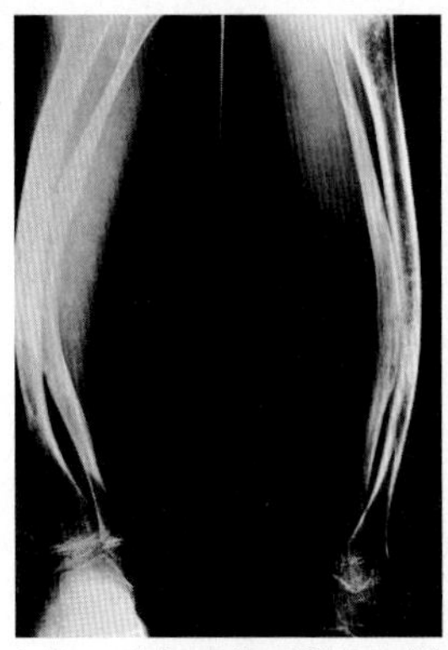

1. Bone deformation (OI type IV)

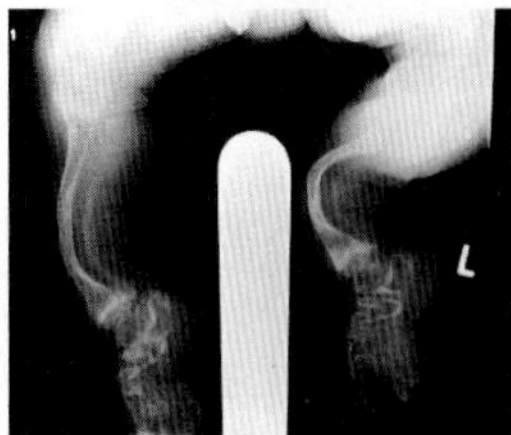

2. Severe deformation (OI type III)

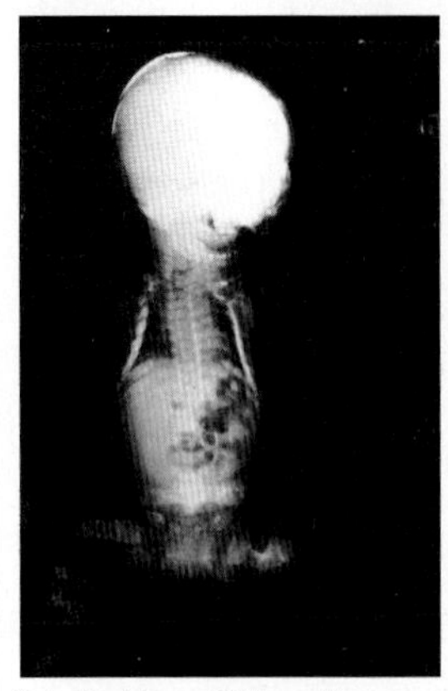

3. Lethal form (OI type II)

C. Different forms of Osteogenesis imperfecta

Cytoskeletal Proteins in Erythrocytes

Eukaryotic cells contain supportive proteins that confer stability while allowing flexibility (cytoskeletal proteins). The membrane skeleton is a network of structural proteins underlying the plasma membrane and partly associated with it. Erythrocytes must meet extreme requirements: About a half a million times during a 4-month lifespan, they traverse small capillaries with diameters less than that of the erythrocytes themselves. Membrane flexibility is also essential for muscle cell function. Thus, it is not surprising that the cytoskeletal proteins of erythrocytes and muscle cells are similar.

A. Erythrocytes

A normal erythrocyte is maintained in a characteristic biconcave discoid form by the cytoskeletal proteins. Genetic defects in different cytoskeletal proteins lead to characteristic erythrocyte deformations: as ellipses (elliptocytes), as spheres (spherocytes), or as cells with a mouthlike area (stomatocytes) or thornlike projections (acanthocytes). The various forms are the result of defects of different proteins. (Scanning electron micrograph from Davies & Lux, 1989).

B. Skeletal proteins in erythrocytes

SDS polyacrylamide gel electrophoresis differentiates numerous membrane-associated erythrocyte proteins. Each band of the gel is numbered, and the individual proteins are assigned to them. The main proteins include α- and β-spectrin, ankyrin, an anion-channel protein (band-3 protein), proteins 4.1 and 4.2, actin, and others. The chromosomal localization of their genes and associated diseases due to mutations are known for man and mouse.

C. Alpha- and β-spectrin

The main component of cytoskeletal proteins is spectrin, a long protein composed of a 260-kdal α chain and a 225-kdal β chain. The chains consist of 20 (α chain) and 18 (β chain) subunits, each with 106 amino acids. Each subunit is composed of three α-helical protein strands running counter to one another. Subunit 10 and subunit 20 of the α chain consist of five, instead of three, parallel chains. The individual subunits are assigned to different domains (I–V in the α chain and I–IV in the β chain).

D. Proteins of the erythrocyte membrane

The rod-shaped spectrin proteins, which run parallel to the erythrocyte plasma membrane (1), are attached to the anion channels by ankyrin and to the glycophorin molecules by protein 4.1. The anion channels in erythrocytes are important for CO_2 transport. Glycophorins (A, B, C) are transmembrane proteins with several carbohydrate units. Actin is the main protein for muscle contraction and cell flexibility. (Diagram after E. J. Luna & A. L. Hitt, Science. 1992; 258: 955–964).

References

Becker PS, Lux SE. Hereditary spherocytosis and hereditary elliptocytosis. In: Scriver CR, Beaudet AL, Sly WS, Valle D. eds. The Metabolic and Molecular Bases of Inherited Diseases. 7th ed. New York: McGraw–Hill; 1995: 3513–3560.

Davies KA, Lux SE. Hereditary disorders of the red cell membrane skeleton. Trends Genet. 1989; 5: 222–227.

Luna EJ, Hit, AL. Cytoskeleton plasma membrane interactions. Science. 1992; 258: 955–964.

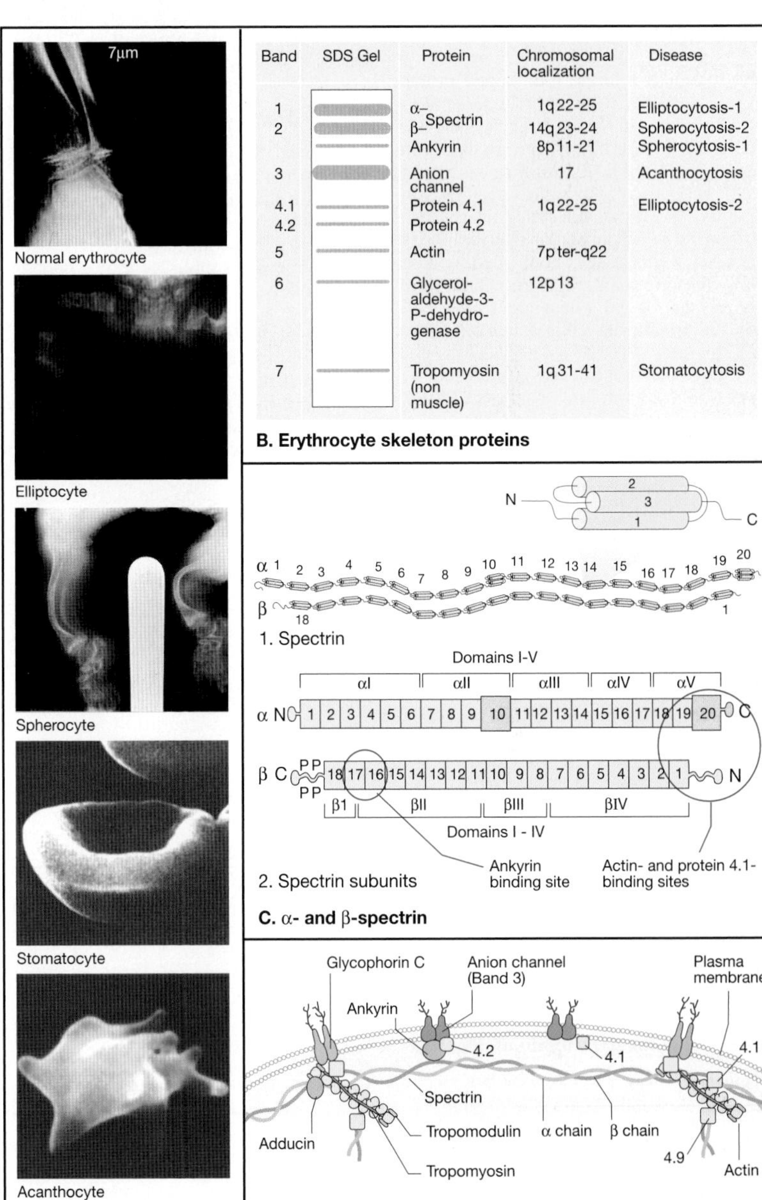

A. Erythrocytes

Band	SDS Gel	Protein	Chromosomal localization	Disease
1		α–Spectrin	1q22-25	Elliptocytosis-1
2		β–Spectrin	14q23-24	Spherocytosis-2
		Ankyrin	8p11-21	Spherocytosis-1
3		Anion channel	17	Acanthocytosis
4.1		Protein 4.1	1q22-25	Elliptocytosis-2
4.2		Protein 4.2		
5		Actin	7pter-q22	
6		Glycerol-aldehyde-3-P-dehydrogenase	12p13	
7		Tropomyosin (non muscle)	1q31-41	Stomatocytosis

B. Erythrocyte skeleton proteins

C. α- and β-spectrin

D. Proteins in the erythrocyte membrane

Dystrophin

Dystrophin is a large, rod-shaped protein in muscle cells. It has structural similarities with spectrin (see p. 318). It is the product of the largest known gene in man, consisting of 2400 kb (2.4 million base pairs) on the short arm of the X chromosome. Mutations of the dystrophin gene cause two important forms of muscular dystrophy. Expression of this gene is controlled by complex transcriptional regulation and splicing events; it is independent and cell-specific in muscle cells (skeletal, heart, and smooth muscle cells), nerve cells (glia, neurons, peripheral nerves), and other organs (kidney, liver, lung, pancreas, testis).

A. Dystrophin–glycoprotein complex

Dystrophin is a cytoskeletal protein of muscle cells. Its C terminus forms a complex with six other membrane-associated proteins in the sarcolemmal membrane (dystrophin-associated glycoprotein complex, DAGC). The N-terminus of dystrophin is bound to actin. The largest of the dystrophin-associated proteins, dystroglycan (156 kdal), is bound extracellularly to laminin. Proteins 25, 50, 35, and 43 lie in the sarcolemma; protein 59 lies intracellularly. Dystrophin is required to stabilize the whole protein complex (Luna & Hitt 1992; Hilton-Johns & Squire 1993). Dystrophin has three different presumed functions: (1) stabilization of the muscle cell membrane, (2) binding to contractile forces within and outside of the cell, (3) other special membrane functions in noncontractile cells, especially of the nervous system (Ahu & Kunkel 1993). (Diagram from J. M. Ervasti and K. P. Campbell, Cell. 1991; 66: 1121–1131).

B. Model of the dystrophin molecule

Dystrophin, the largest member of the spectrin superfamily, is composed of 3685 amino acids (mol. wt. 427 kD), which form four characteristic domains: (1) actin-binding domains, (2) triple-helix segments as in spectrin, (3) cysteine-rich domains, and (4) C-terminal domains. (Diagram from M. Koenig et al., Cell. 1988; 53: 219–228).

C. The dystrophin gene

In humans, the dystrophin gene lies on the short arm of the X chromosome in region 2 band 1 (Xp21). This region consists of three subbands and contains other important genes that lead to disease when mutated. The dystrophin gene (*DMD* gene, Duchenne muscular dystrophy gene) codes for a transcript of about 14 kb. A number of genomic clones are available as probes for the gene and can be used to detect structural rearrangements. cDNA clones from expressed DNA regions (exons) are also available. Owing to its size, this gene cannot be examined in one piece. By using restriction enzymes, large fragments of the gene can be isolated and assigned to different exon positions.

D. cDNA of the dystrophin gene

The most frequent *DMD* gene deletions are located in the 5' region (upper third) (in about 60% of patients) and in the middle of a large intron between exons 44 and 45 (40% of patients). About 5% of patients show duplications of one or more exons. In addition, an increasing number of point muations are being identified.
The components of the dystrophy-associated protein complex have recently been renamed and their function has been elucidated to a greater extent than referred to in section A (Tinsley et al. 1994). The 156-kD extracellular protein is α-dystrophin, and the transmembrane 43-kD protein is β-dystrophin. The 59-kD membrane-bound protein (59 DAP or A1) is a syntrophin, a member of a complex family of proteins. The 50-kD glycoprotein is adhalin. α-Dystrophin binds to laminin and merosin situated in the extracellular matrix. Deficiency in adhalin is observed in patients with severe childhood autosomal recessive muscular dystrophy, which has been mapped to 13q12.

References see p. 375.

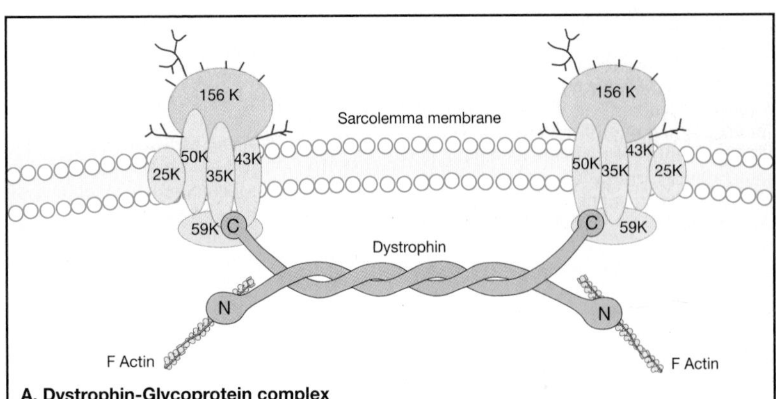

A. Dystrophin-Glycoprotein complex

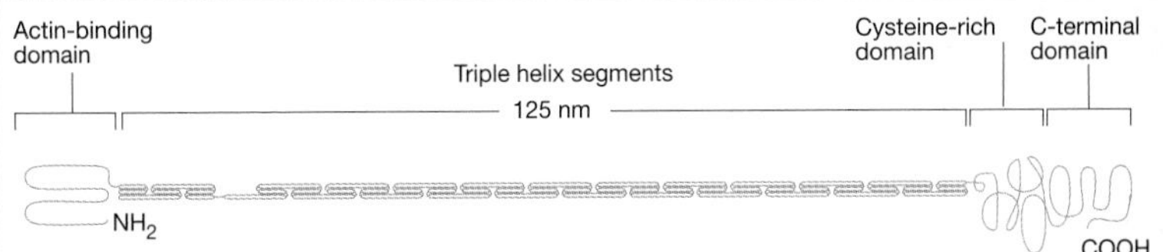

B. Model of the dystrophin molecule

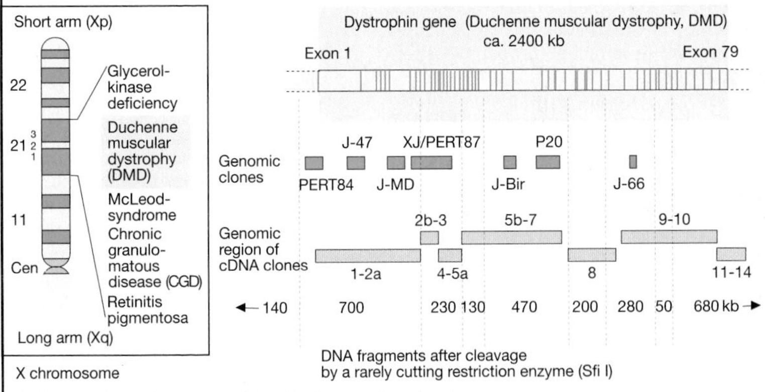

C. The dystrophin gene on the short arm of the X chromosome

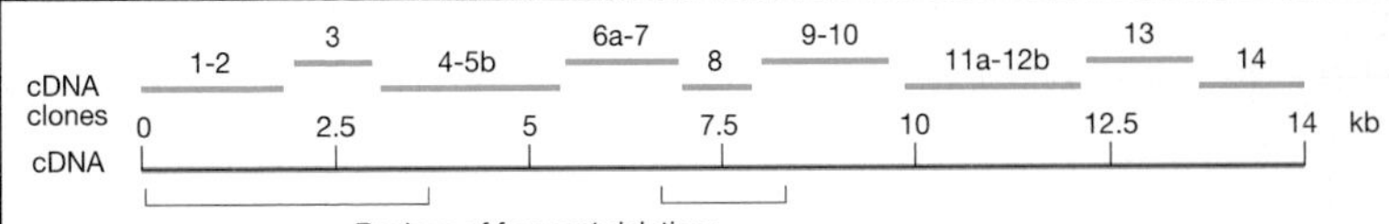

D. cDNA of the dystrophin gene

Duchenne Muscular Dystrophy

Muscular dystrophy of the Duchenne type (DMD) is an X-chromosomal hereditary disease occurring in about 1 in 3500 males. The mutation rate is high, with about one third of the patients having new mutations. Germ cell mosaicism can occur, so that in some cases there may be uncertainty about the genetic risk. Small deletions are quite common. In Duchenne muscular dystrophy, they lead to a shift of the reading frame. Thus, either no dystrophin or a structurally markedly altered molecule is formed. In contrast, in the clinically milder form muscular dystrophy type Becker (BMD), the reading frame is not altered, and a truncated but still somewhat functional dystophin molecule is formed.

A. Clinical signs

This disease, first described by the French physician Duchenne in 1861, begins in the 2nd or 3rd year of life with gait disturbance and increasing muscle weakness: first in the legs, then in the arms. The calves are enlarged without accompanying increase in strength (pseudohypertrophy), and the anterior curvature of the lumbar spine is abnormally increased (lordosis). Most patients require a wheelchair by the age of 10 years. The muscle weakness is progressive, and death occurs by the end of the second decade of life. (Drawing from Duchenne, 1861, and Gowers, 1879).

B. Dystrophin analysis

Dystrophin, normally present in the plasma membrane of muscle cells and demonstrable by special stains (1), is absent in patients (2). Female heterozygotes show a patchy distribution of muscle cells with normal and defective dystrophin (3) due to the effects of X-inactivation (p. 190). (Photographs from R. Gold, Dept. of Neurology, University of Würzburg).
Dystrophin is not confined to muscle. Slightly different isoforms are expressed in the brain (cerebral cortex, hippocampus, cerebellar Purkinje cells) and the retina.

C. Analysis using DNA markers

Alterations of the *DMD* gene can be demonstrated either directly or indirectly by polymorphic DNA markers. The pedigree shows three affected male individuals (II-3, who could not be investigated, III-1, and III-2). The two patients have allele 1 in common in their Southern blot analysis (*Taq*I with the probe L1.28 at locus *DXS7*), inherited from their mothers (II-1 and II-2, respectively). II-4 is not affected and does not have allele 1. Thus, allele 1 must carry the mutation. II-5 is a heterozygote (2–1), but since her two sons (III-3 and III-4) are not affected even though they have inherited allele 1, recombination must have occurred between the disease locus (*DMD*) and the marker locus (*DXS7*). Since these two loci lie about 15 cM apart, recombination will occur in about 15% of cases. Individual II-6 is homozygous for allele 2 (2–2), so that the probability for heterozygosity for the mutant allele is 15%. Today, markers from the gene itself are being used to avoid uncertainty due to recombination. (Data from C.R. Müller-Reible, Institute of Human Genetics, University of Würzburg).

D. Important forms of muscular dystrophy in man

Several other forms of genetically determined muscular dystrophy are known in man. Course, diagnosis, and molecular genetic analysis depend on the basic disorder.

References

see p. 373.

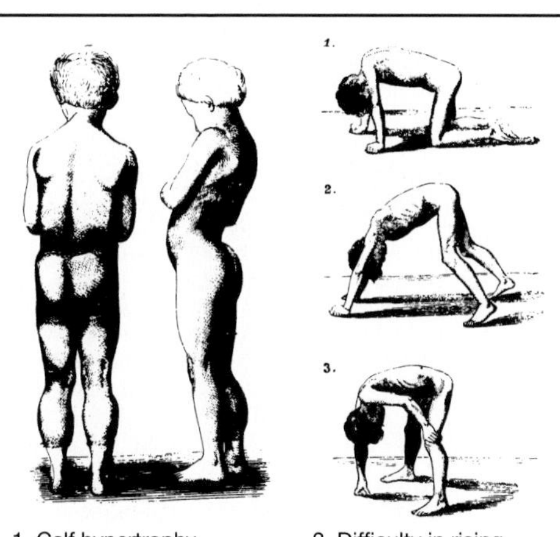

1. Calf hypertrophy and lordosis

2. Difficulty in rising (Gower's sign)

A. Clinical signs of Duchenne muscular dystrophy

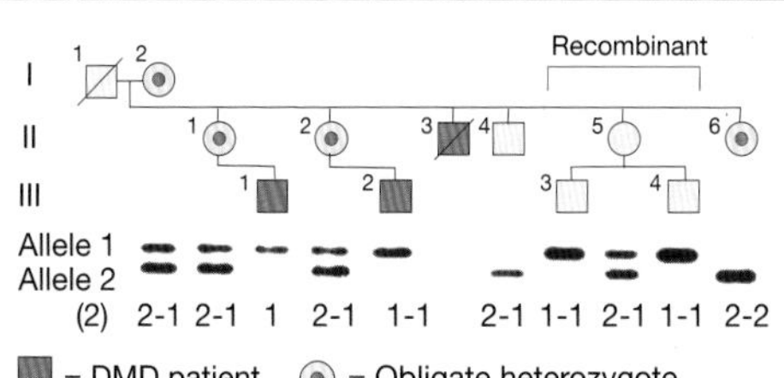

■ = DMD patient ◉ = Obligate heterozygote

C. Investigation of a family with DMD by DNA marker

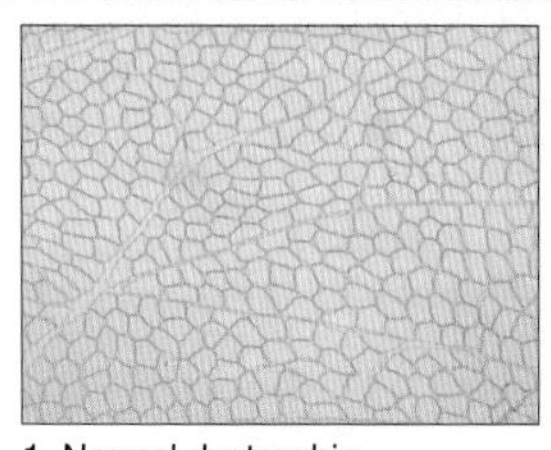

1. Normal dystrophin

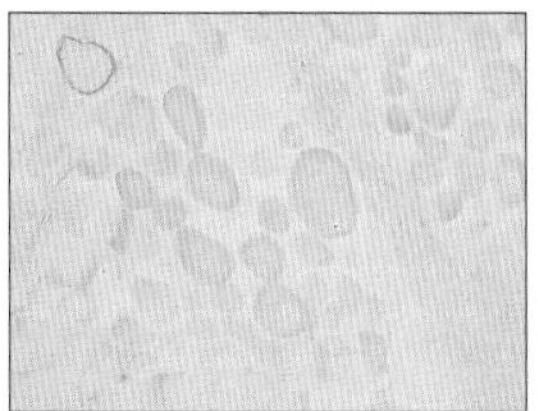

2. Dystrophin absent

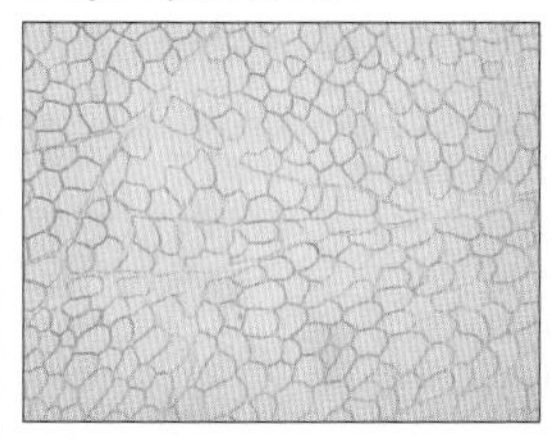

3. Areas lacking dystrophin in heterozygotes

B. Dystrophin analysis in muscle cells

Disease	Chromosomal location	McKusick Nr.
X-chromosomal:		
Muscular dystrophy Duchenne	Xp21.2	310200
Muscular dystrophy Becker (allelic with DMD)	Xp21.2	310200
Muscular dystrophy Emery-Dreifuss	Xq28	310300
Autosomal dominant:		
Myotonic dystrophy	19q13	160900
Facioscapulo-humeral dystrophy	4q35-qter	158900
Oculo-pharyngeal muscular dystrophy	Unknown	164300
Autosomal recessive:		
Duchenne-like muscular dystrophy	13q12-13	253700
Congenital muscular dystrophy–type Fukuyama	9q31-33	253800
Limb-girdle muscular dystrophy	15q15-q22, other loci	253600

D. Important forms of hereditary muscular dystrophy in man

Sex Determination

In the early 1940's, the French embryologist Alfred Jost observed that when the undifferentiated gonads were removed from a male rabbit fetus before male development had begun, it developed as a female. In 1959, chromosomal analysis of two disorders in man, Turner syndrome and Klinefelter syndrome, yielded the first evidence that genetic factors on the Y chromosomes of mammals are important in determining male sex. In contrast, sex in drosophila is determined by the ratio of the number of autosomes to the number of X chromosomes (2A:2X = female; 2A:1X = male). Today it is firmly established that a specific gene on the mammalian Y chromosome (*SRY*, sex-related Y) is responsible for the induction of male sex development during embryogenesis (sex determination).

A. Determination of male phenotype by the Y chromosome

Individuals with Turner syndrome have only one X chromosome (no Y chromosome) and a female phenotype, although incompletely developed and usually accompanied by malformations. Individuals with Klinefelter syndrome have two X chromosomes, a Y chromosome, and a male phenotype, although also incompletely developed (p. 344). Even individuals with four X chromosomes and one Y chromosome (49,XXXXY) have a male phenotype.

B. Sex-determining region SRY on the Y chromosome (in man)

Not all of the Y chromosome (1) — just a small region (about 35 kb in man) on the short arm — is required for male sex determination. The relevant region in man lies in the distal short arm of the Y chromosome at Yp11.32. The short arm and the proximal half of the long arm of the Y chromosome have been divided into seven intervals (here, interval 1, subdivided into 1A1–1C, is shown) (2). The most distal region of the short arm is the pseudoautosomal region (PAR). This region is homologous to the distal segment of the short arm of the X chromosome. Homologous pairing occurs here with crossing-over during meiosis (thus, the term pseudoautosomal). The physical map of the pseudoautosomal region and the proximal half of interval 1 (1A1–1B) span somewhat more than 2500 kb in man (3). Intervals 2–7 contain no genes for male sex determination. The crucial portion of the Y chromosome for male sex determination in man is about 35 kb (in the mouse about 14 kb) of a region designated SRY (sex-related Y chromosomal region) in the interval 1A1 proximal of the pseudoautosomal region. (After Wolf U, Schempp W, Scherer G. Molecular Biology of the Human Y Chromosome. Rev Physiol Biochem Pharmacol. 1992; 121: 148–213).

C. Male development of a chromosomally female transgenic mouse with the *Sry* gene

Clinical observations and experimental evidence indicate that the presence of *SRY* induces male development, irrespective of the presence of the remainder of the Y chromosome. A chromosomally female transgenic mouse (XX) showed normal male development after the 14-kb DNA fragment carrying the Sry region of a mouse Y chromosome was implanted into its blastocyst. (Figure from P. Koopman et al., Nature. 1991; 351: 117–121).

D. *Sry* expression during embryonic gonadal development of the mouse

During embryonic development of an XY mouse, *Sry* (a segment of about 14 kb on the murine Y chromosome homologous with the SRY segment in man, but written Sry for the mouse) is expressed only between days 10.5 and 12.5. (Figure from P. Koopman & J. Gubbay: The biology of Sry. Seminars Develop Biol. 1991; 2: 259–264). The subsequent events leading to male development are initiated during this short time of expression.

References see p. 375.

No Y chromosome present		Y chromosome present	
45, XO One X chromosome only	46, XX	46, XY	47, XXY Additional X chromosome
Turner syndrome female	Normal female	Normal male	Klinefelter syndrome male

A. Determination of the male phenotype by the Y chromosome

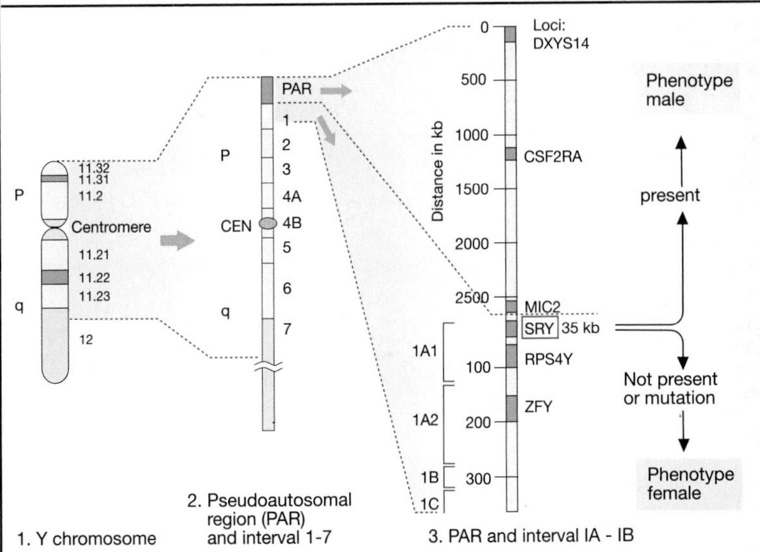

B. Sex-determining region SRY on the Y chromosome

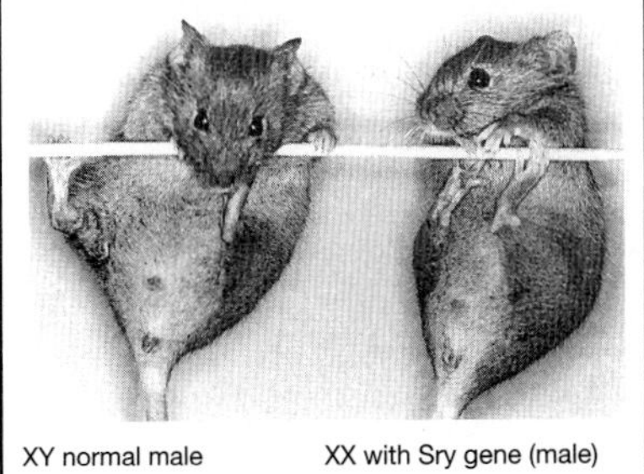

XY normal male XX with Sry gene (male)

C. Male development of a transgenic XX mouse with Sry gene

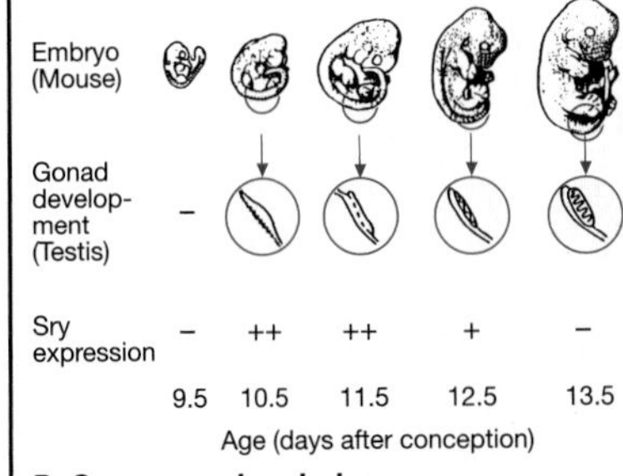

D. Sry expression during embryonal gonad development

Sex Differentiation

Sex differentiation (development of a given sex) consists of many genetically regulated, hierarchical developmental steps. In mammals, the development of male structures requires induction by appropriate genes.

A. Indifferent anlagen of sex differentiation

The gonads (1), the efferent ducts (mesonephric and paramesonephric) (2), and the external genitalia (3) all develop from an indifferent stage. At about the end of the sixth week of pregnancy in humans, after the primordial germ cells of the embryo have migrated into the initially undifferentiated gonads, an inner portion (medulla) and an outer portion (cortex) of the gonads can be distinguished. When a normal Y chromosome is present, early embryonic testes develop at about the 10th week of pregnancy under the influence of a testis-determining factor (TDF). If a normal Y or TDF (SRY) is not present, ovaries develop. The wolffian ducts, the precursors of the male efferent ducts (vas deferens, seminal vesicles, and prostate), develop under the influence of testosterone, a male steroid hormone formed in the fetal testis. At the same time, the müllerian ducts—precursors of the fallopian tubes, the uterus, and the upper vagina—are suppressed by a hormone, the müllerian inhibition factor (MIF).

When testosterone is absent or ineffective, the wolffian ducts degenerate. The muellerian ducts develop under the influence of estradiol, a hormone produced by the fetal ovaries. The external genitalia (3) in humans do not develop until relatively late, starting in the 15th to 16th week. Full development of male external genitalia depends on a derivative of male-inducing testosterone, 5-dihydrotestosterone, a metabolite of testosterone produced by the enzymatic action of 5α-reductase.

B. Time of events in sex differentiation

Sex differentiation proceeds in a cascade-like manner, with a series of temporally regulated successive steps at different levels of differentiation. After the primordial germ cells migrate into the undifferentiated gonads, early embryonic testes develop under the influence of testis-determining factor (TDF) if a Y chromosome is present. TDF is identical with the Y-specific sequences of the SRY region (see p. 324).

During normal male differentiation, the further development of the muellerian ducts is suppressed by a testis-derived hormone, the *m*uellerian *i*nhibition *f*actor (MIF). Testosterone can exert its effect only in the presence of an appropriate intracellular receptor (androgen receptor TFM, see p. 329).

When a Y chromosome is not present or when the SRY region is missing or altered by mutation, testes are not formed. In this case the wolffian ducts cease to develop. In the absence of testes, ovaries develop from the undifferentiated gonads; the wolffian ducts degenerate; and the müllerian ducts differentiate into uterine tubes, uterus, and the upper vagina.

Testosterone also has an effect on the central nervous system ("brain imprinting"). It is assumed that this is required for the psychosexual orientation apparent later in life. When testosterone is absent or ineffective due to a receptor defect, gender orientation is female.

In the majority of genetically determined disorders of sexual differentiation, the levels of differentiation do not correspond (pseudohermaphroditism). In true hermaphroditism (e.g., when both types of cells, those with XY and those with XX chromosome complements, are present) male and female structures exist side by side.

References

see p. 375.

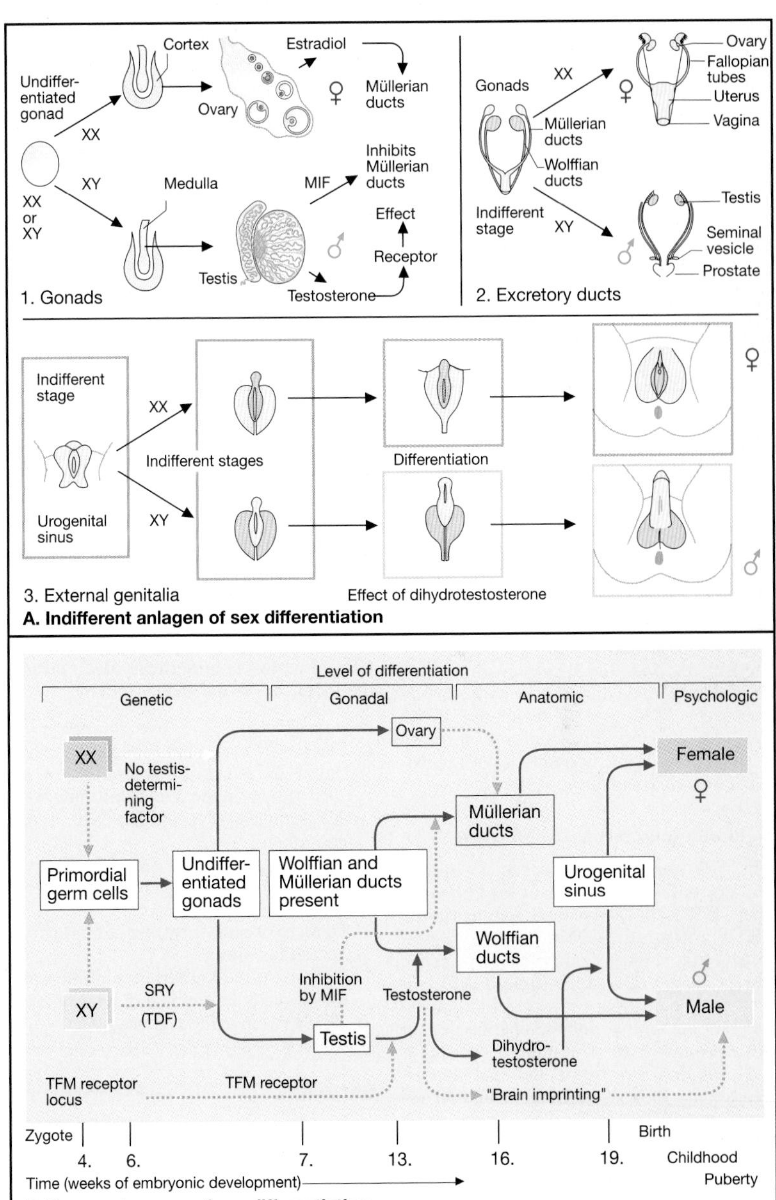

A. Indifferent anlagen of sex differentiation

B. Temporal course of sex differentiation

Genetically Determined Disorders of Sexual Development

Normal sexual development is the result of numerous genes. Mutation or chromosomal rearrangements of any of these genes causes partial or total failure of sex differentiation. The classification of genetically determined disorders of sexual development takes the different developmental processes into account. Pinpointing the basic defect is a prerequisite for diagnosis and treatment.

A. Male-determining region *SRY* on the Y chromosome

Normally, the male-determining Y-specific DNA sequences on the Y chromosome (*SRY*) remain on the Y chromosome and are not transferred to the X chromosome during the homologous pairing and crossing-over of meiosis. However, since the male-determining region *SRY* is located very close to the pseudoautosomal region (PAR), crossing-over in the PAR border region may result in a transfer of the *SRY* region to the X chromosome. This results in a male individual with an XX karyotype (XX male). Conversely, if the *SRY* region is missing from a Y chromosome, a female phenotype with XY chromosomes (XY female) results.

B. Point mutations in the *SRY* gene

The *SRY* gene in humans has no introns and codes for a transcript of 1.1 kb (Su & Lau, 1993). In the critical region of the SRY protein, the DNA-binding region (HMG box, high mobility group protein), the exchange of a single amino acid can lead to sex reversal. Affected individuals have an incomplete female phenotype and XY chromosomes. The majority of mutations are de novo (figure and data after U. Wolf et al., 1992). Mutations in an *SRY*-related gene *SOX9* on human chromosome 17q24 cause sex reversal and campomelic dysplasia (Wagner T, et al. Cell. 1994; 70: 1111–1120; Foster JW, et al. Nature. 1994; 372: 525–530).

C. Androgen receptor and TFM

The testis-determining factor (TDF), formed under the influence of the *SRY* gene, induces the development of the fetal testis from an undifferentiated gonad. The fetal testis produces testosterone, the hormone that induces male sexual differentiation. Testosterone is taken up by cells of the target tissues (wolffian ducts and urogenital sinus) (1). In the urogenital sinus, testosterone is converted into dihydrotestosterone (DHT) by the enzyme 5α-reductase. Both testosterone and dihydrotestosterone bind to an intracellular receptor (androgen receptor). The activated hormone–receptor complex (TR* or DR*) acts as a transcription factor for genes that regulate the differentiation of the wolffian ducts and the urogenital sinus. Thus, normal male fetal development is dependent on normal biosynthesis of testosterone and normal receptors. Androgen receptor mutations lead to disorders of sexual development (2) with X-chromosomal inherited complete or incomplete androgen resistance (testicular feminization, TFM).

Classification of genetically determined disorders of sexual development

1. Defects of sex determination due to mutation or structural aberration of the *SRY* region on the Y chromosome (e.g., XY gonadal dysgenesis, XX males, and others)
2. Defects of androgen biosynthesis (e.g., adrenogenital syndrome due to 21-hydroxylase deficiency)
3. Defects of androgen receptors (testicular feminization)
4. Defects of the müllerian inhibition factor (so-called hernia uteri syndrome)
5. XO/XY gonadal dysgenesis

References

see p. 375.

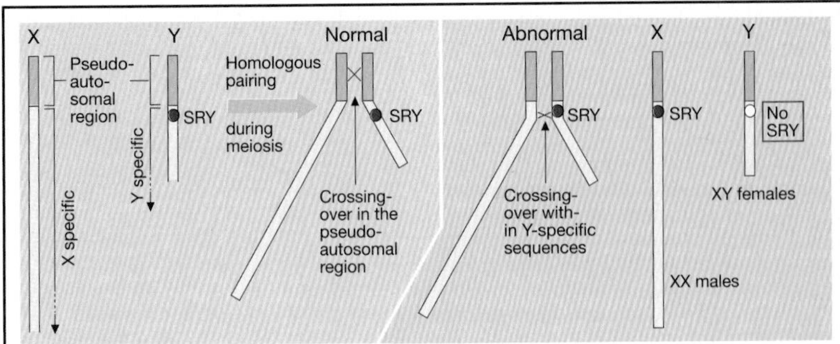

1. SRY remains on the Y chromosome 2. Transfer of SRY to the X chromosome

A. Male-determining region SRY on the Y chromosome

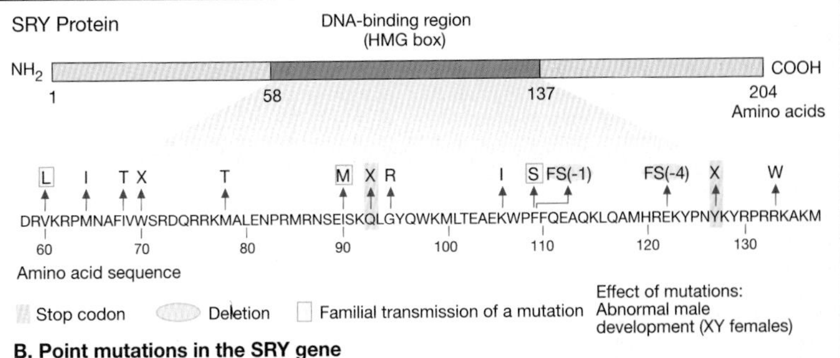

B. Point mutations in the SRY gene

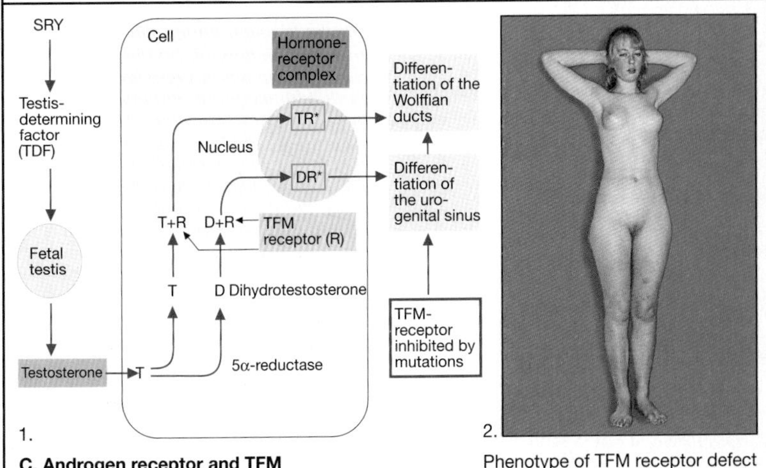

C. Androgen receptor and TFM

Phenotype of TFM receptor defect

Congenital Adrenal Hyperplasia

This disorder, also called adrenogenital syndrome (AGS), is caused by a genetically determined deficiency of cortisol, a steroid hormone produced in the fetal adrenal cortex. A compensatory increase in adrenocortical hormone (ACTH) excretion leads to secondary enlargement (hyperplasia) of the adrenal cortex (congenital adrenal hyperplasia), increased production of prenatal steroids and their metabolites with androgenic effects, and incomplete female sex differentiation.

A Clinical phenotype and genetics

Girls are born with ambiguous or virilized genitalia (1). The adrenal cortex is enlarged (2). Increased production of androgenic metabolites causes masculinization. The cortisol deficiency (3) leads to life-threatening crises due to loss of sodium chloride (salt-wasting) that require prompt treatment. AGS is an autosomal recessive heritable disorder (4). Untreated girls develop a male physical appearance (5). In boys, the early signs are limited to salt-wasting. Initially, skeletal maturation is accelerated and the children are tall for their age; however, they stop growing prematurely and eventually are too short. Besides the classic form of the disorder with a frequency of 1:5000, there are other forms with less pronounced masculinization due to different mutations (genetic heterogeneity).

B. Biochemical defect

The enzymatic conversion of progesterone to deoxycortisol (DOC) by hydroxylation at position 21 (steroid-21-hydroxylase) is decreased. As a result, the concentration of 17-hydroxy-progesterone (17-OH-P) is increased.

C. Gene locus and gene structure

21-Hydroxylase belongs to the family of cytochrome P450 (CYP) oxidases (see p. 310). Two gene loci relevant for AGS, *21–OH A* and *21–OH B* (new designations: *CYP21A* and *CYP21B*), are located on the short arm of chromosome 6 in the HLA system region (see p. 252). Thus, the AGS-mutation-carrying chromosome can be indirectly identified by HLA typing of the affected child and its parents. The *CYP21B* gene consists of 10 exons of various sizes that span almost 60 kb of genomic DNA. The *CYP21A* gene is not transcribed because it contains deletions and stop codons (pseudogene). Genes for the complement factors 4A (*C4A*) and 4B (*C4B*) are located 5' to *CYP21A* and *CYP21B*, respectively.

D. Molecular genetic analysis

Different mutations are known, and most are readily detectable by DNA analysis of the region. With Southern blot analysis after *Taq*I digestion and use of a cDNA probe, the normal *21–OHB* gene (*CYP21B*) is represented by a 3.7-kb fragment and the *21–OHA* gene (*CYP21A*) by a 3.2-kb fragment (1). Panel (2) shows the individual patterns of the two genes: a band of 3.7 kb and a band of 3.2 kb. When the *CYP21A* gene is deleted (no clinical effect), the 3.2.-kb band is absent (3). In the classic form of AGS, the normal 21-OH-B (*CYP21B*) gene is frequently deleted, so that the 3.7-kb fragment is absent (4). Heterozygotes for deletion of the *CYP21B* gene (5) may show decreased intensity of the 3.7-kb band, since one gene is absent. Patients with mild signs without salt loss often have partial or total duplication of the *CYP21A* gene (6). Deletions and duplications are relatively frequent because unequal crossing-over occurs due to the similarity of the *21-OHA* and *21-OHB* genes. (Diagram after New et al., 1989).

Reference

New M, et al. The adrenal hyperplasias. In: The Metabolic Basis of Inherited Disease. 6th ed. Scriver CR, Beaudet AL, Sly WS, Valle D, eds. New York: McGraw–Hill; 1989: 1881–1917.

additional references p. 375.

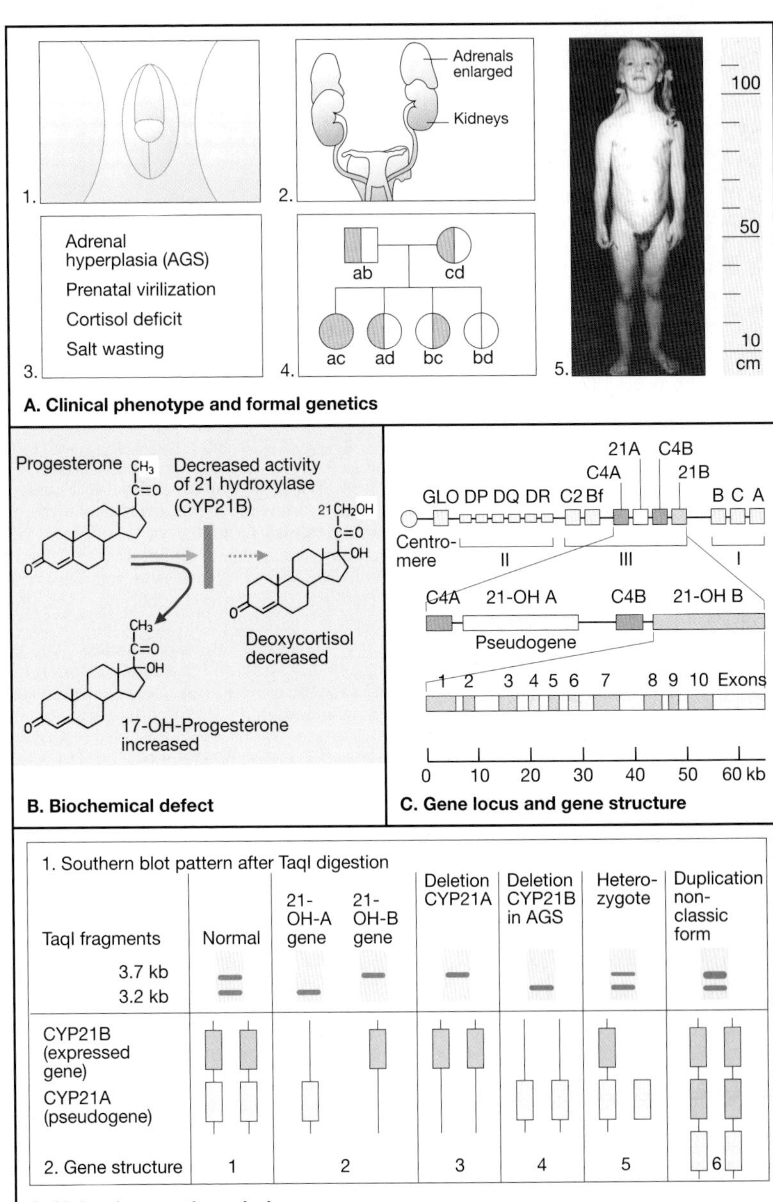

A. Clinical phenotype and formal genetics

B. Biochemical defect

C. Gene locus and gene structure

D. Molecular genetic analysis

Cystic Fibrosis Due to Chloride Ion Channel Defects

Cystic fibrosis (mucoviscidosis) is the most frequent autosomal recessive hereditary disease in populations of European origin (1: 2000, heterozygote frequency 1:22). The isolation of the gene in 1989, a milestone of medical genetics, has contributed to markedly improved diagnosis, recognition of heterozygous gene carriers, and understanding of the pathophysiology. The high frequency of heterozygotes is thought to be the result of selection because heterozygotes presumably are less liable to develop severe cholera than normal homozygotes.

A. Cystic fibrosis, the most frequent autosomal recessive hereditary disease in Europe

The disease is severe and progressive. The bronchial system is especially affected, with viscous mucous formation and frequent infections. Gastrointestinal signs may also occur due to disorders of the pancreas. Pancreatic insufficiency occurs in about 85% of patients. In addition, there are relatively mild forms of the disease due to different mutations. About 5–10% of newborns with cystic fibrosis develop severe intestinal obstruction (meconium ileus). In about 2–5%, the liver is involved. Males are almost always, and females are frequently infertile. The average life expectancy is about 20–30 years. Many of the infertile men with absence of the vas deferens are heterozygous for a CF mutation.

B. Positional cloning of the *CF* gene

The gene for cystic fibrosis was isolated on the basis of its chromosomal location (positional cloning). After the gene was mapped to the long arm of chromosome 7 (7q31), a DNA segment of about 1.5 Mb (megabase = about 1500 kb) from this region was isolated. A combination of chromosome jumping and chromosome walking, starting from a series of flanking polymorphic marker loci (e.g., locus *D7S340* proximally and *D7S424* distally) was used to isolate a segment of about 250 kb with the candidate genes. These were tested for *CF*. Identification of the *CF* gene was based on the demonstration of mutations in patients and their absence in normal controls, on searches for evolutionarily conserved coding sequences, on sequencing, on elucidation of the intron/exon structure, and on demonstration of expression in epithelial cells, especially of the pancreas, salivary glands, sweat glands, gastrointestinal tract, and reproductive tract.

C. The *CF* gene and CFTR protein

The *CF* gene is very large (over 230 kb); its 24 exons code for a 6.5-kb transcript, from which a protein with 1480 amino acids is translated. The protein is a membrane-bound regulator of a chloride ion channel designated CFTR (cytic fibrosis transmembrane conduction regulator). It contains two transmembrane domains with six transmembrane subunits each, two nucleotide-binding domains, and a regulatory domain. It forms a phosphorylation-regulated chloride ion channel in the apical membrane of epithelial cells.
In 70% of the patients, three base pairs in exon 10 (codon 508) are deleted. As a result, the amino acid phenylalanine (F) is missing (thus the designation ΔF508). This mutation accounts for about 70% of those in central and western Europe. Five other mutations account for a further 15% of cases. Some mutations lead to mild forms of the disease with a quantitative decrease of function of the chloride ion channel (Sheppard et al., 1993).
In a study of 102 males with congenital bilateral absence of the vas deferens, 28 different mutations in the CF gene were found in 73 men. Of these, 54 had a mutation in one copy, 19 in both copies (M. Chillón et al., New Engl. J. Med. 1995; 332: 1475–1480).

References

see p. 376.

Cystic fibrosis (Mucoviscidosis)

Severe progressive disease of the bronchial system and gastro-intestinal tract

Disturbed function of a chloride ion channel by mutations of one gene

Autosomal recessive

Gene locus 7q31.3

Disease incidence approx. 1:2000

Heterozygote frequency approx. 1:22

Mutation Delta F508 in approx. 70%

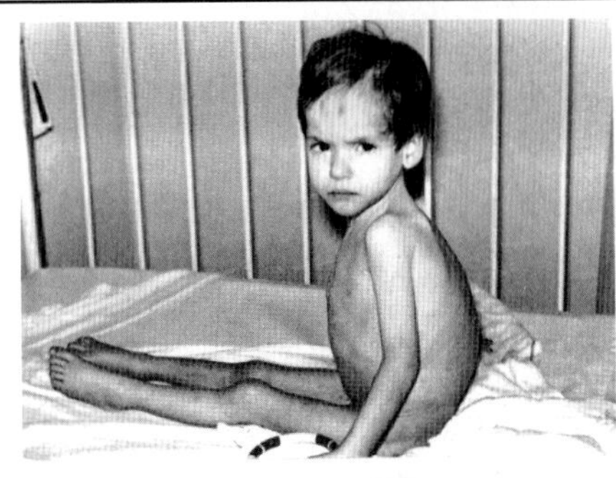

A. Cystic fibrosis, the most frequent recessive disease in Europe and N.America

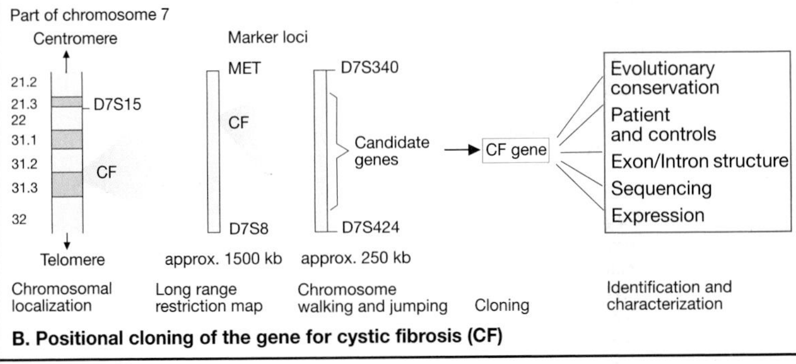

B. Positional cloning of the gene for cystic fibrosis (CF)

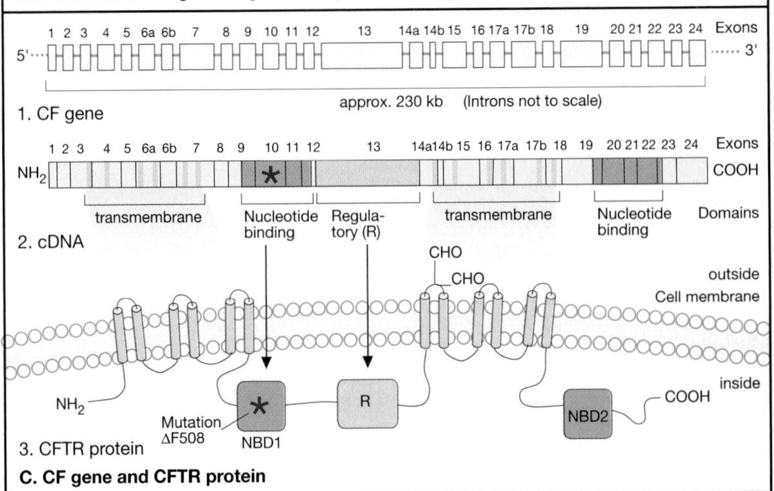

C. CF gene and CFTR protein

Unstable Number of Trinucleotide Repeats

Heritable changes in the number of repeated groups of three nucleotides each (trinucleotide or triplet repeat) represent a new class of mutations in man, for which there is no parallel in other organisms. They occur either within the gene and are translated or outside the gene in an untranslated region, and they are unstable during transmission through the germline. Unaffected persons may carry a premutation, which may be converted to a full mutation when passed through the germline to the next generation. Therefore, the effects of the mutation may be of varied severity in affected members within the same family. Occasionally, there is regression and a generation is skipped.

A. Genetic diseases with increased numbers of trinucleotides

Some important genetically determined diseases are based on a greater than normal number of trinucleotides: Huntington disease, fragile X syndrome, myotonic dystrophy, spinobulbar muscular atrophy type Kennedy, and spinocerebellar ataxia type 1 (H.T. Orr et al. Nature Genet. 1993; 4: 221–226). As can be expected, further diseases resulting from unstable DNA segments have been found.

B. Huntington disease

Huntington disease is a progressive disease of the brain. Within 5 to 10 years, it leads to complete loss of motor control and intellectual abilities (1). It usually begins around age 40–50 with uncoordinated movements (chorea, St. Vitus' dance), excitation, hallucinations, and psychological changes. The disease is transmitted by autosomal dominant inheritance and shows complete penetrance. It presents an affected family with two difficult problems: (I) due to its late onset, carriers of the mutation usually have completed their family planning before the disease is manifest, and (II) children of affected persons first learn as young adults that they are at a 50% risk of developing the disease later in life. Thus, the recent introduction of a direct predictive DNA diagnostic procedure is very important. However, before such a genetic test is carried out, it must be established through genetic counseling that the persons at risk have decided for themselves whether they want to have the test performed. The gene, precisely localized in March 1993 (The Huntington's Disease Collaborative Research Group. Cell. 1993; 72: 971–983), is located between the markers D4S127 and D4S125 on the distal short arm of chromosome 4 (2). It spans 210 kb and codes for a protein (called "huntingtin") of unknown function. The 5' end of the gene contains numerous copies of a trinucleotide sequence consisting of cytosine, adenine, and guanine (CAG), a codon for the amino acid glutamine. Normally, the gene has 10–34 CAG repeats; in patients there are 42–100. The new diagnostic test (3) demonstrates that affected individuals (here, individuals 1, 2, and 4) have enlarged DNA fragments due to expanded CAG repeats. (Findings of the Institut für Humangenetik of the Universität Göttingen with kind permission; C. Zühlke et al., Hum Mol Genet. 1993; 2: 1467–1469).

C. Myotonic dystrophy (MDY1)

Myotonic dystrophy is an autosomal dominant hereditary disease that predominantly affects the central nervous and muscular systems (1). The myotonia causes a mask-like facies (2). The disease is very variable and in many families shows increasing severity in consecutive generations (anticipation). An increased number of CTG repeats, more than 50 copies compared with 5–35 in normal individuals (3), is found immediately beyond the 3' end of the gene in affected persons. This is demonstrated in a Southern blot as an enlarged DNA fragment (4). (Schematic representation of a Southern blot at the gene locus D19S95, probe pBBO.7 after DNA cleavage with *Eco*RI. After H. G. Harley et al., Lancet. 1992; 339: 1125–1128).

References

see p. 376.

Disease (Examples)	Gene	Frequency	Trinucleotide	Normal Number	Mutant Allele	Chromosome
Huntington's chorea	CH	1:10 000	$(CAG)_n$	11-34	42-100	4p16.3
Fragile X syndrome	FMR1	1:2 000	$(CGG)_n$	10-50	52-500	Xq27.3
Myotonic dystrophy	MDY1	1:8 000	$(CTG)_n$	5-35	50-200	19q13.3
Spino-bulbar muscular atrophy (Kennedy)	SBMA	<1:50 000	$(CTG)_n$	11-31	40-65	Xq21.3-22

A. Genetic diseases due to increased number of trinucleotides

Severe progressive disease of the central nervous system

Loss of motor and intellectual control

Onset age 25 - 60

Autosomal dominant

CAG repeat size increased

Predictive diagnosis possible, but problematic

1. Main manifestations

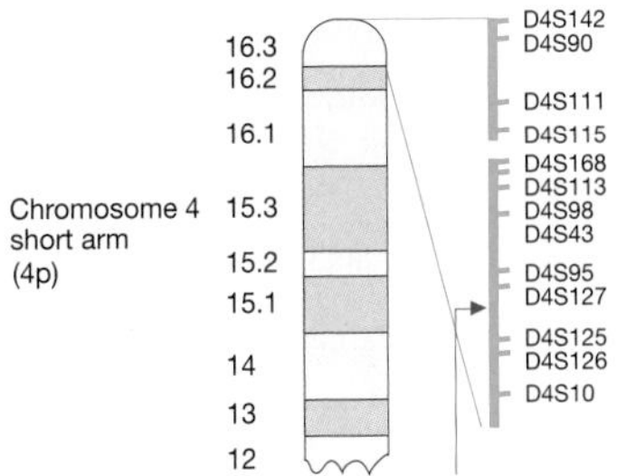

2. Localization of the gene

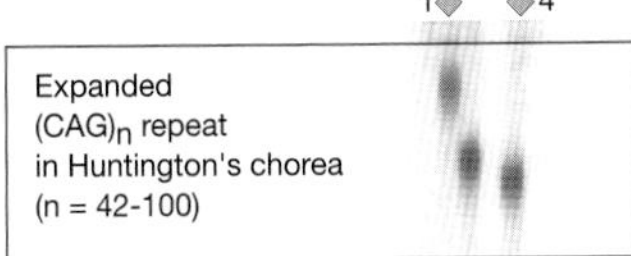

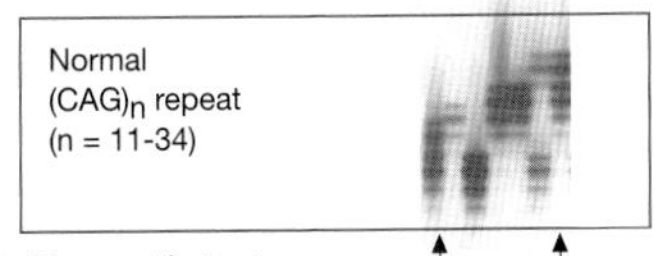

3. Diagnostic test

B. Huntington's chorea

Muscle weakness,

Myotonia, mask-like face,

Cataract, alopecia,

Variable expression,

Autosomal dominant,

CTG repeat increased

1. Main manifestations

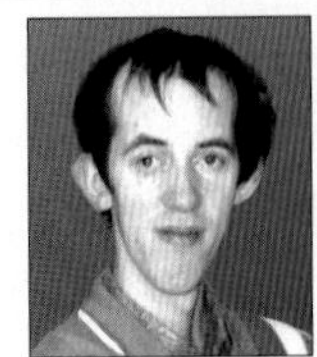

2. Phenotype

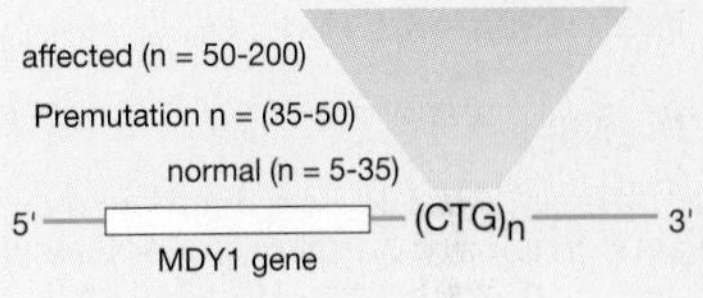

3. Expanded CTG repeat in Myotonic Dystrophy

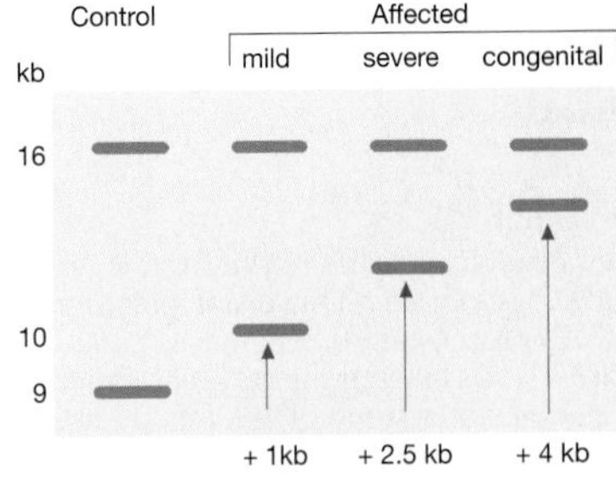

Diagram of a Southern blot at gene locus D19S95 (probe pBB0.7)

4. Correlation with degree of severity

C. Myotonic Dystrophy (MDY1)

Fragile X Syndrome (*FMR1* Gene)

The fragile X syndrome (McKusick *309550; other designations: fraX syndrome, X-chromosomal mental retardation with fragile site on the X chromosome, Martin–Bell syndrome) is the most frequent form of hereditary mental retardation in males, with a frequency of about 1 : 1000–1500 individuals. The responsible mutation usually consists of an increased number of unstable nucleotide repeats. Unlike with classic X-chromosomal inheritance, there are males without manifestations, and a large proportion of female carriers show partial manifestations.
The unstable expansion of a trinucleotide repeat (CGG) is located in the 5'-untranslated region of the *FMR1* gene. Recent findings indicate that an increase beyond 200 repeats impedes the migration of the 40S ribosomal subunit. This causes translational suppression (Y. Feng et al., Science. 1995; 268: 731–734).

A. Phenotype

The phenotype is very variable. The mental retardation varies; there is no distinct neurological dysfunction. In adult males, the testes are enlarged (macroorchidism). Affected individuals can usually be well integrated into the family and learn to function in a familiar environment (photographs from E. Passarge & A. Schmidt, Issues and Reviews in Teratology. 1985; 3: 95–133).

B. Fragile site Xq27.3

The gene locus (FRAXA) for the gene (*FMR1*) is located on the distal long arm of the X chromosome in region 2, band 7.3 (Xq27.3). In this region the great majority of patients and some of the female heterozygotes show a constriction ("fragile site," because frequent breaks occur here) in the affected X chromosome in about 2–25% of metaphases. The constriction must be induced by folic acid deficiency in the culture medium, and it must be differentiated from other fragile sites in this region.

C. Expanded CGG repeats in the fragile X syndrome

The heritable unstable sequences explain two unusual characteristics of the fraX syndrome: (I) the transition from a premutaton (about 60–200 CGG repeats) without clinical manifestation into a full mutation (more than 200 CGG repeats) during transmission through the germline and (II) differences in the FRAXA locus within a given family (1). Intellectually and cytogenetically normal males (proportion 20%) may carry and transmit the mutation (normal male transmitter). The mothers of all affected males are heterozygotes, and all mutations must pass through the female germline before a phenotypic effect is manifest. Phenotypically affected females (proportion 40–50%) always receive the fraX mutation from their mothers and not from their fathers. Daughters of normal male carriers are not phenotypically affected.
Daughters of normal male transmitters have a higher risk of having affected sons (76% penetrance) than other heterozygotes (18% penetrance). The size of the CGG repeats may increase from one generation to the next. In the untranslated 5' region of the *FMR1* gene, affected individuals carry about 200–500 copies of trinucleotide repeats [$(CGG)_9$ $AGG(CGG)_9AGG(CGG)_{10}$] instead of the normal 10–50 copies [abbreviated: $p(CGG)_n$]. This corresponds to about 30 codons for arginine.
The different numbers of CGG repeats can be demonstrated in Southern blots as DNA fragments of different sizes (3). The normal gene is represented by a small DNA fragment (S). A premutation leads to slightly enlarged fragments (*S). The full mutation is characterized by large fragments (L). With this procedure, a reliable diagnosis of the genotype is possible. (Photograph of a Southern blot: *Hin*dIII digestion and hybridization with probe Ox1.1; P. Steinbach, Ulm).

References

see p. 377.

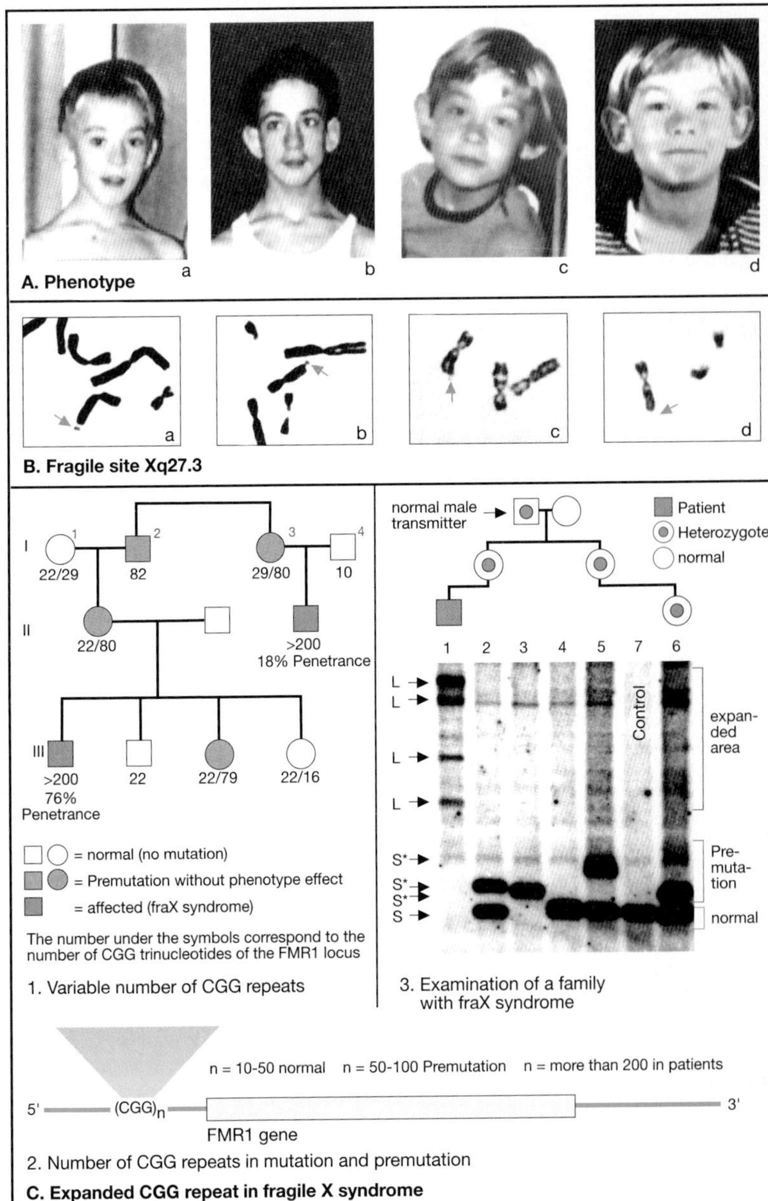

A. Phenotype
a
b
c
d
B. Fragile site Xq27.3
I
II
III
22/29
82
29/80
10
22/80
>200
18% Penetrance
>200
76%
Penetrance
22
22/79
22/16
= normal (no mutation)
= Premutation without phenotype effect
= affected (fraX syndrome)
The number under the symbols correspond to the number of CGG trinucleotides of the FMR1 locus
1. Variable number of CGG repeats
normal male transmitter
Patient
Heterozygote
normal
1 2 3 4 5 7 6
Control
L
S*
S
expanded area
Premutation
normal
3. Examination of a family with fraX syndrome
n = 10-50 normal
n = 50-100 Premutation
n = more than 200 in patients
5'
(CGG)n
3'
FMR1 gene
2. Number of CGG repeats in mutation and premutation
C. Expanded CGG repeat in fragile X syndrome

Increased Chromosomal Instability (Bloom Syndrome, Fanconi Anemia, Ataxia Telangiectasia)

A number of autosomal recessive hereditary diseases show chromosomal instability, visible in metaphase chromosomes. The most important examples are Bloom syndrome (BS), Fanconi anemia (FA), and ataxia telangiectasia (AT). The features these diseases have in common are disturbances of growth and development, defects of the immune system and bone marrow function, and a predisposition to develop malignant tumors.

A. Bloom syndrome (BS)

In Bloom syndrome (1), prenatal and postnatal growth deficiency is pronounced (birth weight 2000 g, birth length ca. 40 cm, adult height around 150 cm). The phenotype (2) includes a narrow face. Usually, but not always, a sunlight-induced erythema develops on the cheeks, eyelids, mouth, ears, and back of the hands (a and b). The photo on the right (c) shows a boy with Bloom syndrome and acute leukemia. Metaphase cells show about a tenfold increase in the rate of sister chromatid exchanges (SCE), approx. 60 instead of about 6 per metaphase in normal cells (3). (Sister chromatid exchanges are explained in the glossary, p. 392). Metaphases of patients contain increased breaks in one or both chromatids and exchanges between homologous chromosomes in about 1–2% of cells.

In Bloom syndrome, malignancies occur at a much younger age than usual. In a series of 147 patients between 0 and 4 years of age, 57 children (38%) developed a total of 77 malignant tumors (German 1993). Several patients had more than one primary tumor. The proportion of acute leukemias, lymphomas, and adenocarcinomas is especially high. Chemotherapy is extremely poorly tolerated. The predisposition of patients with Bloom syndrome to malignant tumors is due to an increased rate of mutations and recombination in somatic cells. Thus, patients become homozygous at many gene loci, which may result in loss of function of both alleles of a tumor suppressor gene.

B. Fanconi anemia (FA)

Fanconi anemia (hereditary pancytopenia) is an autosomal recessive hereditary malformation syndrome (1) with variable clinical expression. The main hematologic signs, aplastic anemia, thrombocytopenia, and leukopenia, occur in the first year of life. Pre- and postnatal growth deficiency (2), hypoplastic or absent thumbs (3), and short or absent radii are characteristic physical signs. The cytogenetic manifestations are breaks in one or both chromatids of metaphase chromosomes in ca. 5–15% of cells and occasional exchanges between nonhomologous chromosomes. Fanconi anemia is genetically heterogeneous. At least four different complementation groups with impaired DNA stability have been identified (C.A. Stratheie et al., Nature. 1992; 356: 763–767).

C. Ataxia telangiectasia (AT)

Ataxia telangiectasia (Louis–Bar syndrome) (1) is a group of autosomal recessive hereditary diseases with neurological and immunological defects. Characteristic telangiectases (small vascular lesions) develop in the conjunctiva of the eyes during the first year of life (2). Cerebellar ataxia also begins in the first year. It is initially progressive, but eventually becomes stationary. Immune deficiency leads to severe infections of the lungs and the bronchial tree. Lymphoma and leukemia are relatively frequent. Metaphase cells reveal clonal rearrangements with preferred involvement of chromosomes 7 and 14 (3). Radiation-induced repair processes are impaired. Thus, patients are extremely sensitive to roentgen rays. Even diagnostic X-ray examinations should be done only when absolutely necessary.

References

see p. 377.

Bloom syndrome

Extreme intrauterine and postnatal growth retardation

Chromosomal instability

Predisposition to leukemias, lymphomas, and other tumors

Immune defects

Sunlight-induced erythema of the face

Hypo- and hyper-pigmented skin areas

Autosomal recessive

Gene locus on chromosome 15

1. Main features

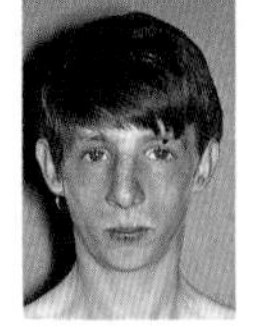

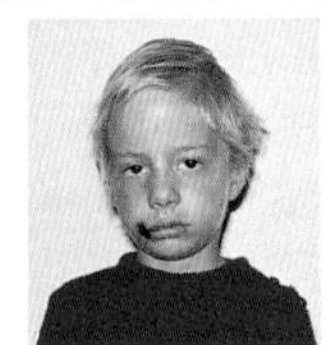

2. Phenotype a b c

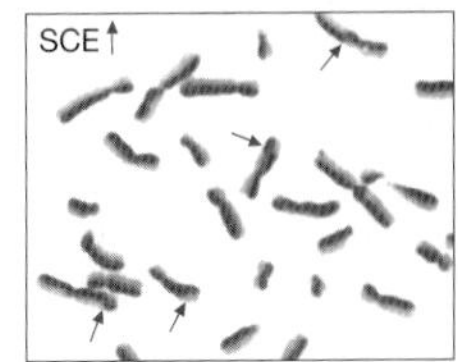

Bloom syndrome Normal control

3. Increased rate of sister chromatid exchanges (SCE)

A. Bloom syndrome (BS)

Fanconi Anemia

Growth retardation
Skeletal defects (e.g., radius and thumb)

Bone marrow failure

Skeletal and kidney malformation

Localized pigment changes

Autosomal recessive

Several gene loci

1. Main features

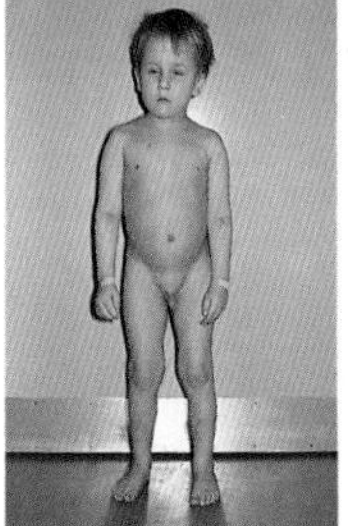

2. Phenotype

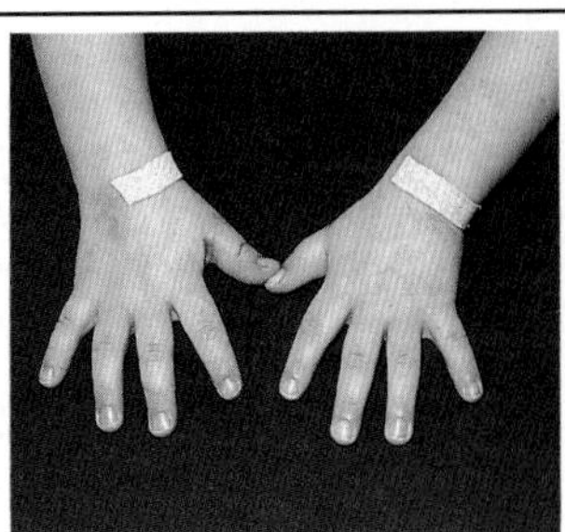

3. Thumb hypoplasia

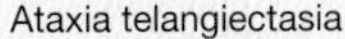

B. Fanconi Anemia (FA)

Ataxia telangiectasia

Cerebellar ataxia

Immune defects

Telangiectases of the conjunctivae

Predisposition to tumors (lymphoma, leukemia)

Extreme radiation sensitivity

Autosomal recessive

Several gene loci

1. Main features

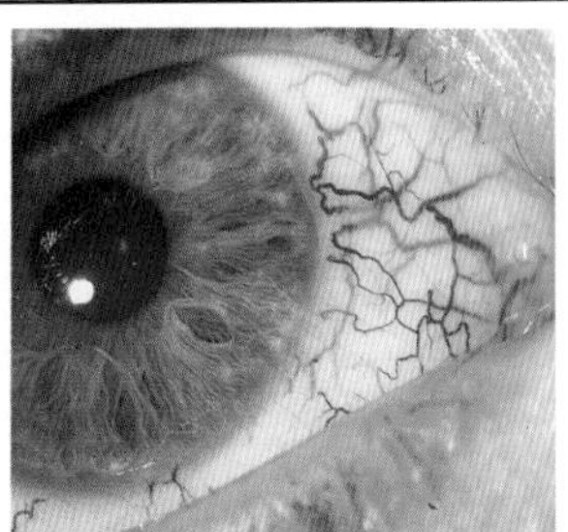

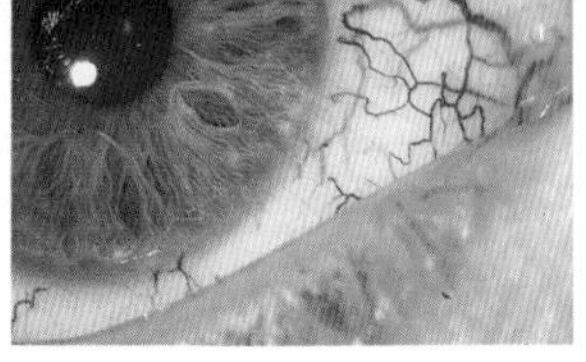

2. Telangiectases

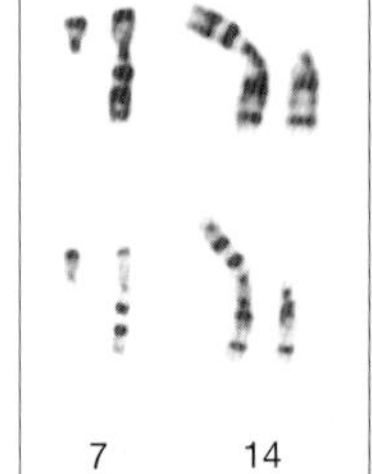

3. Translocation 7q;14p

C. Ataxia telangiectasia (AT)

Parent-Specific Loss of Function of a Chromosomal Region

Prader–Willi (PWS) and Angelman syndromes (AS) are clinically different diseases with developmental dysfunction and unusual behavior. They are caused by loss of function of two chromosomal regions located close together on the proximal long arm of chromosome 15. These two disorders are distinguished by the functional loss of either a paternal or a maternal allele.

A. Two syndromes of the same chromosomal region

An interstitial deletion of the proximal long arm of a chromosome 15 (15q11–13) is found in about 70% of patients with PWS or AS. If the deletion affects the paternal chromosome, Prader–Willi syndrome results; if it affects the maternal chromosome, Angelman syndrome results.
In Prader–Willi syndrome, the deletion 15q11–13 is of paternal origin. Neonatal hypotonia, developmental delay, hyperphagia/obesity, short stature, small hands and feet, hypogonadism, and hypopigmentation characterize this disease.
In Angelman syndrome, the deletion 15q11–13 is of maternal origin. Mental retardation, especially with poor speech development, friendly disposition with laughing for no apparent reason, hyperactive behavior, seizures, typical EEG changes, and hypopigmentation are characteristic signs.

B. Parental origin of the deletion

DNA markers from the affected region are used to determine the origin of the deletion by Southern blot analysis. In Prader–Willi syndrome, the paternal allele (pat) is missing; in Angelman syndrome, the maternal allele (mat).

C. Uniparental disomy (UPD)

In about a third of patients with PWS and about 20–30% with AS, no deletion can be demonstrated. Instead, uniparental disomy (UPD) may be found in about 30% of PWS and 2% of AS cases. In the (schematic) Southern blot, it can be determined whether the same allele (allele 1) is present twice, in this case of maternal origin (isodisomy) (1), or whether two different alleles (alleles 1 and 2) have come from the same parent, here the mother (heterodisomy) (2). In 1–2% of patients with PWS and 20% of patients with AS, neither a deletion nor uniparental disomy can be demonstrated. In these cases, the disease may show familial occurrence.

D. Chromosomal region

The Prader–Willi and Angelman chromosomal regions span 1000–1500 kb. The critical chromosomal region for PWS is about 400 kb and includes the two loci D15S63 (probe PW71) and SNRPN (small nuclear riboprotein N, a gene). SNRPN represents the candidate gene for PWS. The critical region for AS, about 100 kb further distally, is defined by the locus D15S10 (probes 3–21).

E. A model for genomic imprinting

Gene loci in the PWS/AS chromosomal region are subject to genomic imprinting. The PWS region is active on the paternal chromosome and inactive on the maternal chromosome. The AS chromosomal region is active on maternal chromosome 15, but inactive on the paternal. The different parental origins can be demonstrated by analysis of the methylation pattern at locus D15S63 (probe PW71) because DNA at the inactive regions is hypermethylated (B. Dittrich et al., 1992).
Imprinting in 15q11-13 affects multiple genes and appears to be controlled by single imprinting center in this region. Mutations of the imprinting center can be transmitted silently through the germ line of one sex, but block the resetting of the imprint in the opposite sex (Buiting et al., 1995). Depending on whether they occur on a grandmaternal or grandpaternal chromosome, they cause Prader-Willi or Angelman syndrome.

References see p. 377.

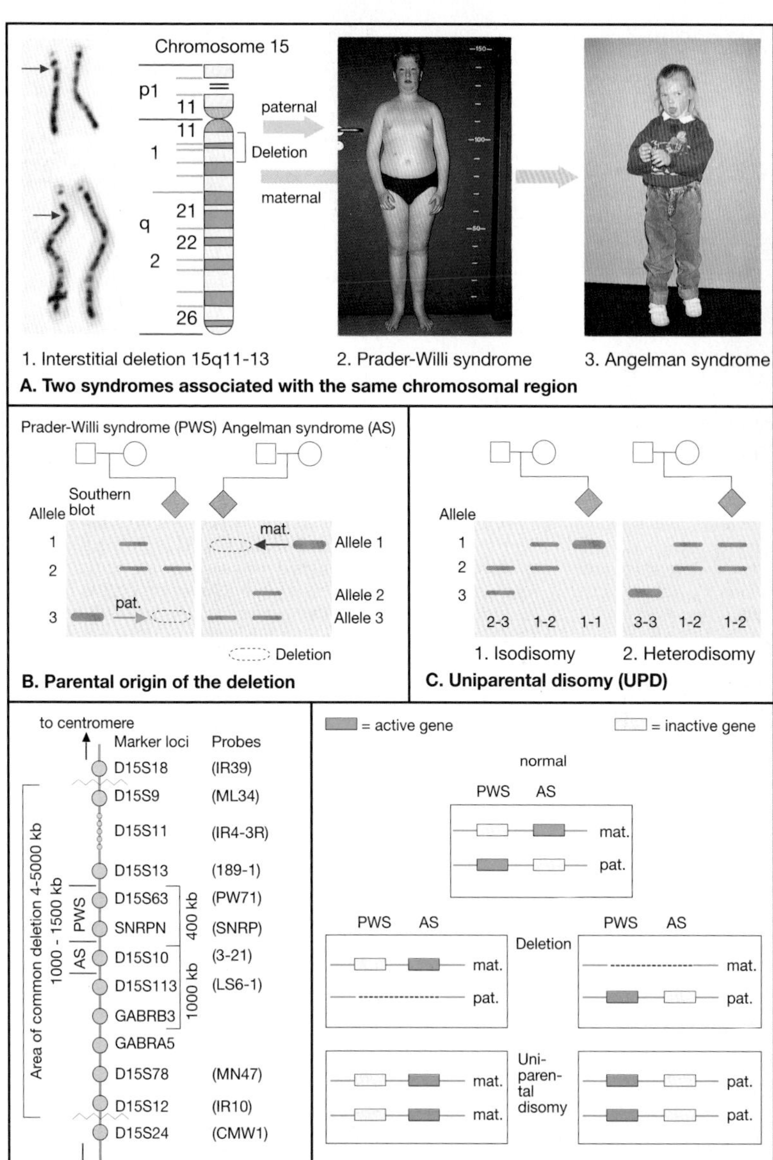
Chromosome 15
p1
11
11
1
Deletion
paternal
maternal
q
21
22
2
26
1. Interstitial deletion 15q11-13
2. Prader-Willi syndrome
3. Angelman syndrome
A. Two syndromes associated with the same chromosomal region
Prader-Willi syndrome (PWS) Angelman syndrome (AS)
Southern blot
Allele
1
2
3
pat.
mat.
Allele 1
Allele 2
Allele 3
Deletion
B. Parental origin of the deletion
Allele
1
2
3
2-3 1-2 1-1
3-3 1-2 1-2
1. Isodisomy
2. Heterodisomy
C. Uniparental disomy (UPD)
to centromere
Marker loci
Probes
D15S18 (IR39)
D15S9 (ML34)
D15S11 (IR4-3R)
D15S13 (189-1)
D15S63 (PW71)
SNRPN (SNRP)
D15S10 (3-21)
D15S113 (LS6-1)
GABRB3
GABRA5
D15S78 (MN47)
D15S12 (IR10)
D15S24 (CMW1)
Area of common deletion 4-5000 kb
1000 - 1500 kb
PWS
AS
400 kb
1000 kb
to telomere
chromosomal breakpoints
D. Chromosomal region
= active gene
= inactive gene
normal
PWS AS
mat.
pat.
Deletion
Uni-parental disomy
Prader-Willi syndrome
Angelman syndrome
E. Model for genomic imprinting

Autosomal Trisomies

A trisomy (the presence of three homologous chromosomes instead of the usual two) arises prezygotically during meiosis due to faulty distribution (nondisjunction) of a chromosome pair. It may also arise after fertilization (postzygotic) during somatic cell division (mitosis). In this case, trisomy is present in a certain proportion of cells (chromosomal mosaicism). Trisomy leads to a phenotype characteristic for the particular chromosome, although in humans most trisomies are lethal in early embryonic development.

A. Trisomy in jimsonweed *(Datura stramonium)*

In 1922, Blakeslee observed that triploid and tetraploid jimsonweed plants *(Datura stramonium)* differ little in phenotype. However, when plants contained three copies of only one of the twelve chromosomes (trisomy), and two each of the others, a characteristic appearance resulted for each of the trisomies (from Blakeslee AF. Variation in Datura due to changes in chromosome number. Amer Naturalist. 1922; 56: 16–31).

B. Trisomies in mouse

During the 1970s, A. Gropp and co-workers investigated the effect of trisomies on the development of the mouse. Trisomic mice, resulting from the segregation of translocations, had a developmental profile and certain morphological changes characteristic for each trisomy (1). Embryos with a chromosome missing (monosomies) died very early in gestation (figure from A. Gropp, 1982). A mouse embryo with trisomy 12 shows an open skull cap and other malformations on the 14th day of development (2), unlike other embryos of the same age (H. Winking, Lübeck, In: W. Traut: Chromosomen. Klassische und molekulare Cytogenetik, Heidelberg: Springer; 1991; Boué et al., 1985). Only trisomy 19 is compatible with survival until birth (day 21), but the brain is too small (3). These animals die shortly after birth.

C. Autosomal trisomies in man

Of the 22 autosomes in man, only 3 occur regularly as trisomies in live-born infants: trisomy 21, trisomy 18, and trisomy 13. They differ in phenotype and course of disease. Other trisomies are not observed in live-born infants because they are lethal in early embryonic life, and not compatible with life at birth (see p. 344). Trisomy 21 causes the clinical picture of Down syndrome (formerly called mongolism).

D. Nondisjunction as the cause of trisomy

Especially in trisomy 21, the frequency of nondisjunction depends on the age of the mother at the time of conception (1). The age of the father has very little or no influence.

Nondisjunction may occur during the first or the second maturation division (meiosis I or meiosis II) (2). The difference can be established by appropriate chromosomal markers. If nondisjunction occurs in meiosis I, the three chromosomes will be different (1 + 1 + 1), whereas if nondisjunction occurs during meiosis II, two of the three chromosomes will be identical (2 + 1). In humans, about 70% of nondisjunctions occur in meiosis I, and 30% in meiosis II.

References

Boué A, Gropp A, Boué J. Cytogenetics of pregnancy wastage. Adv Hum Genet. 1985; 14: 1–57.

Epstein CJ. Down syndrome (trisomy 21). In: Scriver CR, et al., eds. The Metabolic and Molecular Bases of Inherited Disease. 7th ed. New York: McGraw–Hill, 1995: 749–794.

Therman E, Susman M. Human Chromosomes. Structure, Behavior, Effects. 3rd ed. Heidelberg: Springer Verlag; 1993.

Traut W. Chromosomen. Klassische und Molekulare Cytogenetik. Heidelberg: Springer Verlag; 1991.

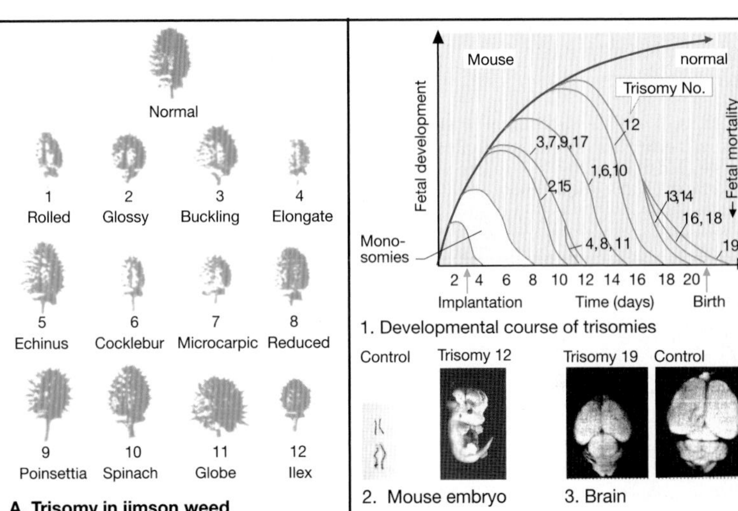

A. Trisomy in jimson weed (*Datura stramonium*)

1. Developmental course of trisomies

2. Mouse embryo

3. Brain

B. Trisomies in the mouse

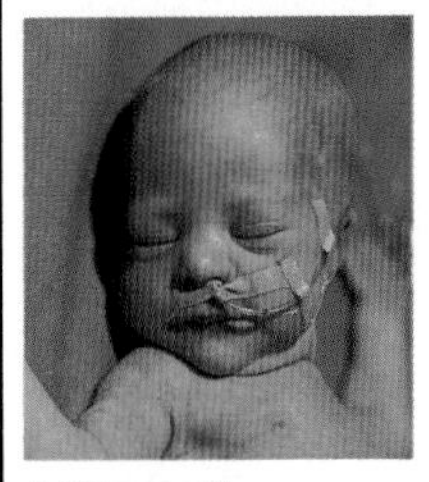

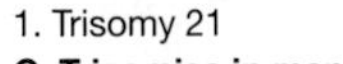

1. Trisomy 21

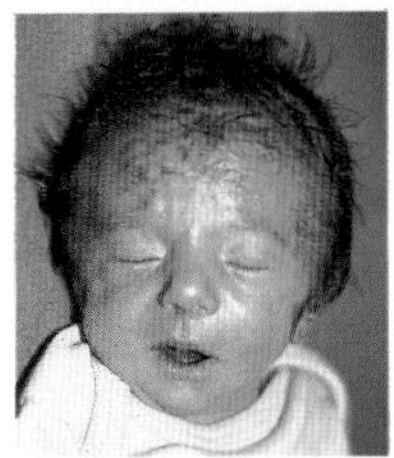

2. Trisomy 18

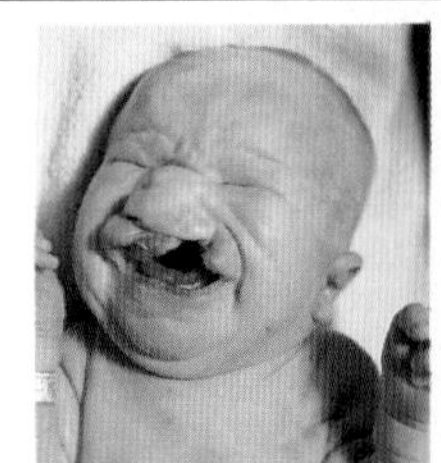

3. Trisomy 13

C. Trisomies in man

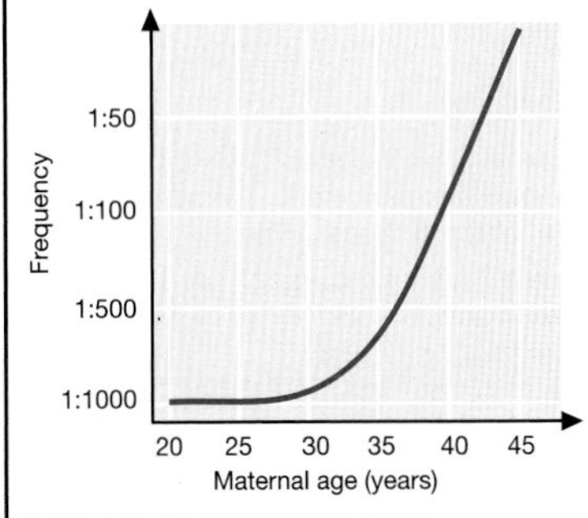

1. Trisomy 21 in liveborn infants

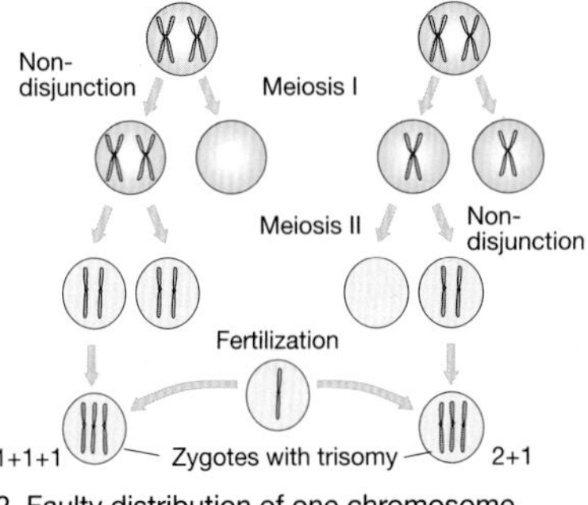

2. Faulty distribution of one chromosome

D. Nondisjunction as cause of trisomy

Other Deviations from the Normal Number of Chromosomes

In addition to the autosomal trisomies, there are other conditions associated with an abnormal number of chromosomes. They involve either the entire set of chromosomes (triploidy or tetraploidy) or the X chromosome or Y chromosome. Deviations from the normal number of X or Y chromosomes comprise about half of all chromosomal aberrations in man (total frequency about 1:400).

A. Triploidy

Triploidy is one of the most frequent chromosomal aberrations in man (1). Possible causes include a diploid spermatocyte, a diploid oocyte, or fertilization of an egg cell by two spermatozoa (dispermy). Triploidy usually leads to spontaneous miscarriage within the first four months of pregnancy. The fetus shows numerous severe malformations (2), such as cardiac defects, cleft lip and palate, skeletal defects, and others. The additional chromosome set may be of either maternal or paternal origin, with different clinical consequences.

B. Monosomy X (Turner syndrome)

Monosomy X (karyotype 45,XO) is a frequent chromosomal aberration, representing about 5% of those in humans at conception. However, of 40 zygotes with monosomy X, only 1 will develop to birth. The phenotypic spectrum is very wide. During the fetal stage, (1) lymphedema of the head and neck result in cystic hygroma, large multilocular thin-walled lymphatic cysts. Congenital cardiovascular defects, especially involving the aorta, and kidney malformations are frequent. An important component of the disease is the absence of ovaries, which develop only as connective tissue (streak gonads). Small stature is always a feature (average adult height about 150 cm). In newborns, webbing of the neck (pterygium colli) may be present as a residual of the lymphedema (clinical picture of Ullrich–Turner syndrome). On the other hand, the manifestations may be mild (2). Very frequently, pure monosomy is not present, but rather chromosomal mosaicism with normal cells (45,XO/46,XX) or a structurally altered X chromosome (deletion of the short arm, isochromosome of the long or short arm, ring chromosome, or other).

C. Additional X or Y chromosomes

An additional X chromosome in males (47,XXY) leads to the clinical picture of Klinefelter syndrome after puberty when untreated (1). This includes tall stature, absent or decreased development of male secondary sex characteristics, and infertility due to absent spermatogenesis. With an additional Y chromosome (47,XYY) no unusual phenotype results (2). Girls with three X chromosomes (47,XXX) are also physically unremarkable (3). However, learning disorders and delayed speech development have been observed in some of these children.

D. Wide spectrum of chromosomal aberrations in human fetuses

The relative proportions of the various trisomies observed in fetuses after spontaneous abortion differ. The most frequent is trisomy 16, which accounts for about 5% of all autosomal trisomies. Autosomal monosomies lead to death of the embryo within the first days or weeks (data after Lauritsen, 1982).

References

DeGrouchy J, Turleau C. Clinical Atlas of Human Chromosomes. 2nd ed. New York: John Wiley & Sons; 1984.

Schinzel A. Catalogue of Unbalanced Chromosome Aberrations in Man. Berlin: W de Gruyter; 1984.

Lauritsen JG. The cytogenetics of spontaneous abortion. Res Reproduct. 1982; 14: 3–4.

1.
Triploidy

- Most frequent chromosomal aberration (15%) in fetuses following spontaneous abortion
- Severe growth retardation, early lethality
- Occasional liveborn infant with severe malformation
- Dispermia a frequent cause

2.
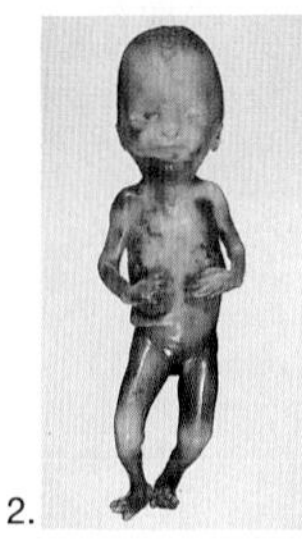

3.
1 2 3 4 5 X
6 7 8 9 10 11 12
13 14 15 16 17 18
19 20 21 22 Y

A. Triploidy

1.

2.
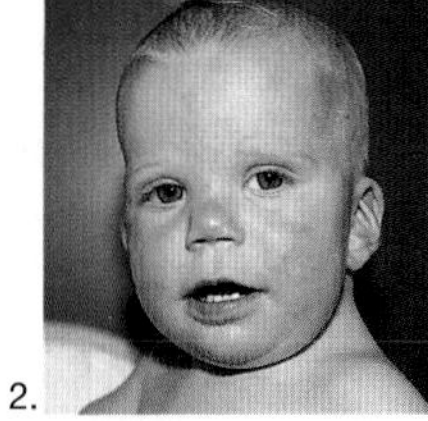

3.
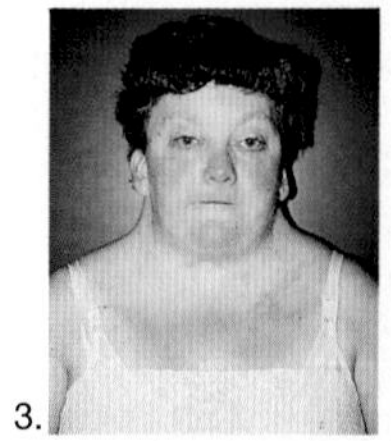

B. Monosomy X (Turner syndrome; 45, XO)

1. XXY
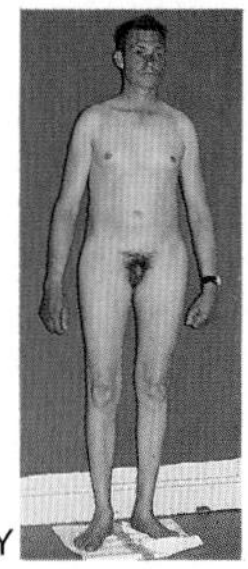

2. XYY
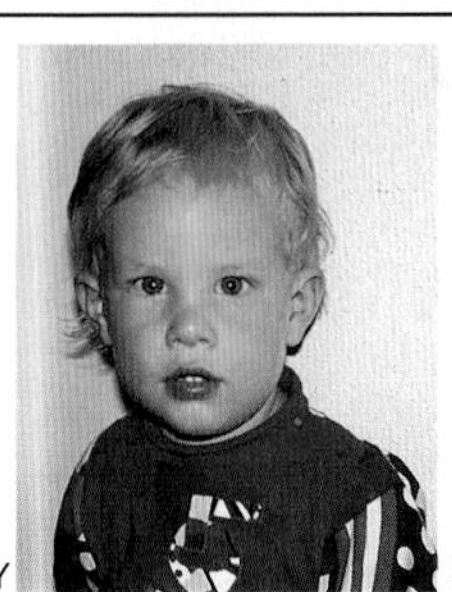

3. XXX
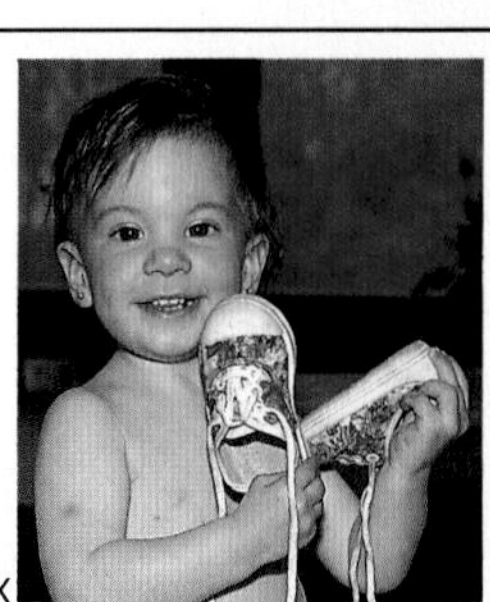

C. Additional X or Y chromosome

Proportion of autosomal trisomies in 669 trisomic fetuses following spontaneous abortion

Trisomic chromosomes	1	2	3	4	5	6	7	8	9	10	11	12	13	14	15	16	17	18	19	20	21	22
Proportion (%)	0.0	4.9	0.6	2.5	0.2	0.5	4.0	3.9	2.7	2.0	0.3	1.0	4.6	4.6	7.7	32.3	0.6	5.1	0.2	2.7	9.4	10.2

D. Wide spectrum of chromosomal aberrations in human fetuses

Deletion and Duplication

Several dozen diseases due to deletion or duplication of defined chromosomal segments and recognizable by their clinical features (phenotype) are known in man. Specific chromosomal aberrations are associated with certain phenotypes because chromosome-specific gene loci are involved. Three examples will be presented.

A. Cri du chat syndrome (5p–)

In 1963, Lejeune and his co-workers in Paris described a clinical picture in children with a partial deletion of the short arm of a chromosome 5 (5p–) and retarded mental and physical development. About 15% of the patients show a translocation of chromosome 5. In these cases, the risk of recurrence of the disorder is increased. Affected children come to attention during the newborn period and early infancy because of their prolonged, high-pitched crying resembling that of a kitten (cri du chat, cat cry). The catlike crying usually disappears during the first two years of life. The face is roundish; the eyes are relatively widely spaced (hypertelorism). The phenotype changes with age.

B. Wolf–Hirschhorn syndrome (4p–)

In 1964, U. Wolf and co-workers in Freiburg and K. Hirschhorn and co-workers in New York independently described a clinical picture associated with partial deletion of the short arm of a chromosome 4 (4p–). Patients showed marked physical and mental retardation in addition to a characteristic phenotype (1, 2) and malformations such as facial clefts, scalp defects, hypospadias, and congenital heart defects. The deletion is usually de novo (3). In some patients, the deletion can only be demonstrated by molecular genetic methods.

C. Duplication 5q

In addition to deletions, duplications are an important cause of morbidity. This example shows the phenotype of a chromosomal aberration in children of different ages. The patients shown are sisters with partial duplication of the distal region of a chromosome 5 (5q33 to the telomere), about 10% of the distal long arm of a chromosome 5. The chromosomal imbalance was the result of the segregation of a reciprocal translocation between a chromosome 5 (5q33–qter) and a chromosome 8 (8p23–pter) in the father. Three of the children had a partial duplication of 5q. Down-turned lateral corners of the eyes, protruding forehead, and a small mouth are apparent in the affected fetus (1) and at ages 5 months (2) and 8 years (3) (from Bartsch-Sandhoff & Liersch, 1977; Passarge et al., 1982).

References

Altrogge HC. Deficienz des kurzen Arms von Chromosomen der B-Gruppe (4p–; 5p–). Ein Beitrag zur Korrelation von Phänotyp und Karyotyp. Med Dissertation. Universität Hamburg. 1971.

Bartsch-Sandhoff M, Liersch R. Partial duplication 5q syndrome: Phenotypic similarity in two sisters with identical karyoype, partial duplication 5q33–5qter and partial deficiency 8p23–pter. Ann Génét. 1977; 20: 281–284.

Passarge E, Bartsch-Sandhoff M, Rehder H. Fetal manifestation of a chromosomal disorder: partial duplication of the long arm of chromosome 5 (5q33–qter). Teratology. 1982; 25: 221–225.

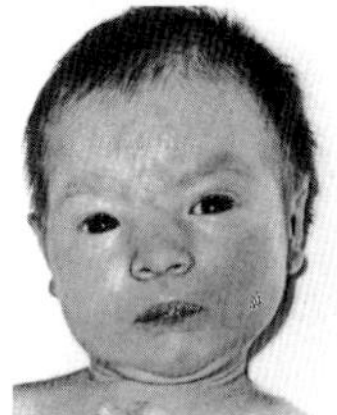
7 days

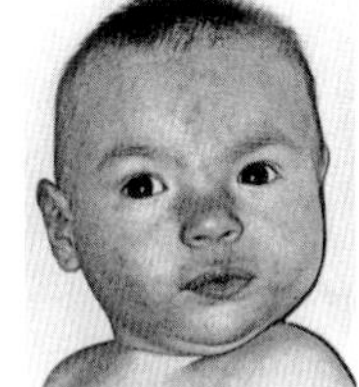
5 months

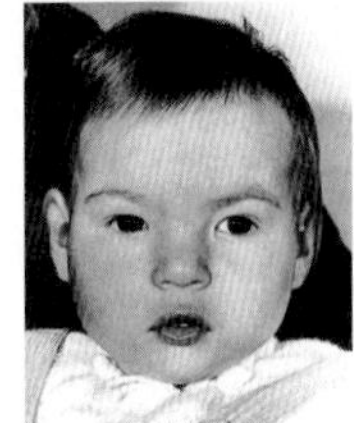
9 months

(same patient)

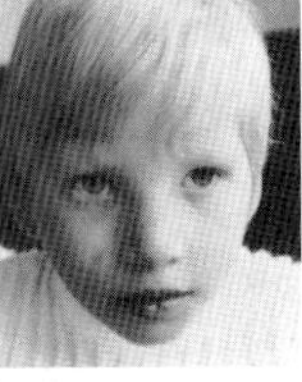
3 years

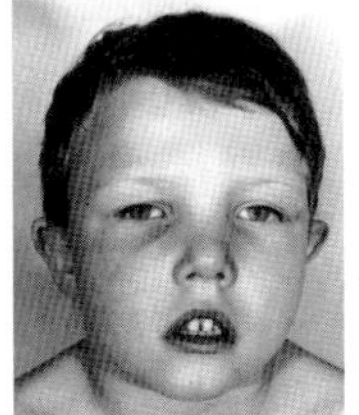
6 years

8 1/2 years

(different patients)

A. Cri-du-chat syndrome (Deletion 5p-)

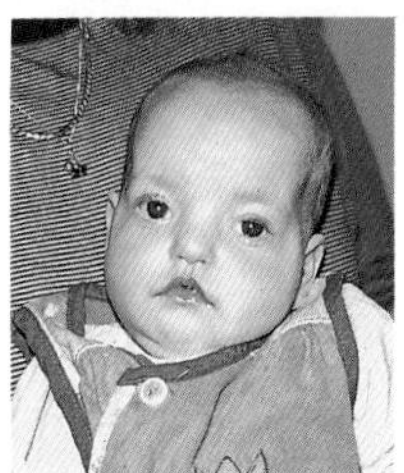
1. Age: 1 1/4 years

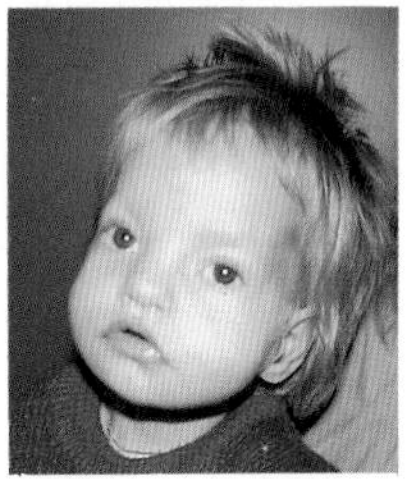
2. Age: 2 1/2 years

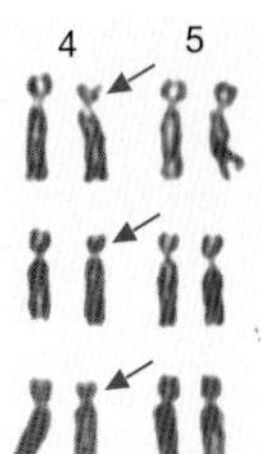

3. Deletion 4p-

B. Wolf-Hirschhorn syndrome (Deletion 4p-)

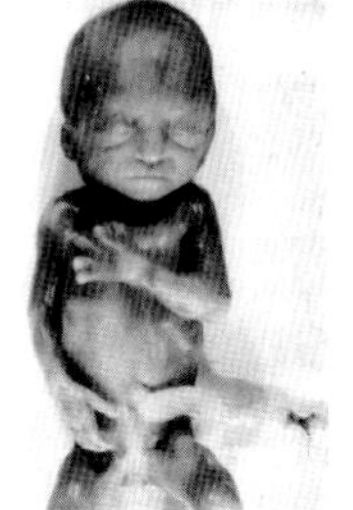
1. Fetus: 22nd week

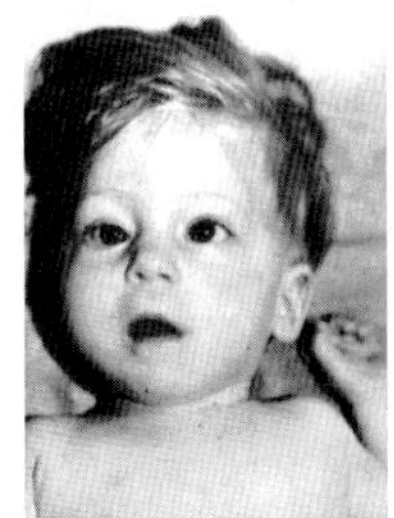
2. 5 months

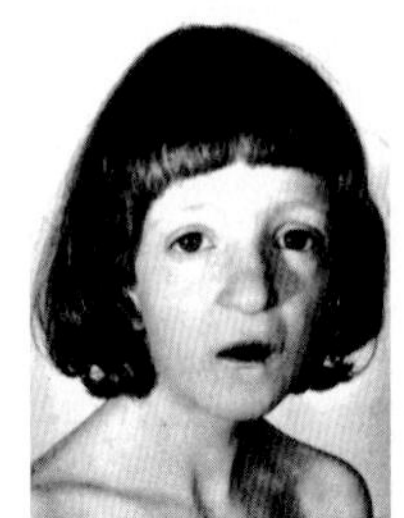
3. 8 years

C. Duplication 5q

Genome Organization

Genomes of Bacteriophages and Bacteria (Prokaryotic Genomes)

Bacteriophages are viruses that multiply in bacteria. The genome of a bacteriophage usually consists of double-stranded DNA, although some phage genomes consist of single-stranded DNA or of RNA. The phage genome size ranges from 1.6 kb to over 150 kb, representing anywhere from a few to over 200 genes. The bacterial genome is mainly contained in a circular DNA molecule incorporating 99% of a bacterium's genes. In addition, bacteria often contain small independent DNA molecules known as plasmids.

A. Genome of the bacteriophage ϕX174

One of the smallest phages is ϕX174. The coat (capsid) of ϕX174 is shaped like an icosahedron (20 plane surfaces), and the genome consists of 5386 nucleotides in the form of single-stranded DNA. F. Sanger and co-workers sequenced the genome completely in 1977 and found that several of the ten genes of ϕX147 overlap. After ϕX147 enters a host cell, a complementary DNA strand is formed (replicative DNA). Conventional semiconservative replication and viral reproduction can then proceed. The genes are designated by the letters A–J; their basic functions or gene products are indicated in the figure.

B. Overlapping genes in ϕX174

In ϕX174 phages, the genes A and B, B and C, and D and E partially overlap. In these overlapping regions of the genome, the reading frame is used by different genes. Gene E begins with the start codon ATG, the first two nucleotides of which (AT) are part of the codon TAT for tyrosine in gene D. Similarly, the stop codon TGA for the E gene is part of codon GTG (valine) and ATG (methionine) of the D gene.

C. Genome of *Escherichia coli*

The genome of *Escherichia coli* is contained in the DNA double helix of a ring-shaped chromosome. This has been divided into sections 0 to 100. It consists of about 4 x 10^6 base pairs (ca. 4700 kb) and includes approx. 2800 genes, around 1600 of which have been mapped. About a fourth of the genes are organized into 75 different operons, whereas the other, independently occurring genes appear to be scattered over the genome. An operon is a series of functionally related, juxtaposed genes that are regulated together. As an example, a simplified gene map of the operon for the synthesis of lactose, galactose, tryptophan, and histidine is shown. The direction of transcription may be clockwise (histidine operon) or counterclockwise. Most genes of the E. coli genome are present as a single copy; only the genes for ribosomal RNA (rRNA) are present in multiple copies. Most strains of E. coli contain seven genes for rRNA. Therefore, they can double their protein content every 20 minutes during cell division. Originally, the E. coli genome was analyzed by observing the order in which genes were transferred during conjugation (see p. 80). In 1955, E. Wollman and F. Jacob mapped the first genes in E. coli by a technique called interrupted conjugation. The experiment consisted in mixing Hfr and F^- cells and taking samples after culturing for different lengths of time. The genes were transferred in a defined sequence, corresponding to their position. Since the transfer of all E. coli DNA requires about 100 minutes, the circular map was divided into units from 0 to100. Since conjugation mapping is not very precise (2 minutes correspond to about 80 kb DNA), other mapping methods were also used. Today, techniques using recombinant DNA have replaced the classical methods.

References

Brown TA. Genetics. A Molecular Approach. 2nd ed. London: Chapman & Hall; 1992.

Knippers R, Phillipsen P, Schäfer KP, Fanning E. Molekulare Genetik, 5th ed. Stuttgart: Thieme; 1990.

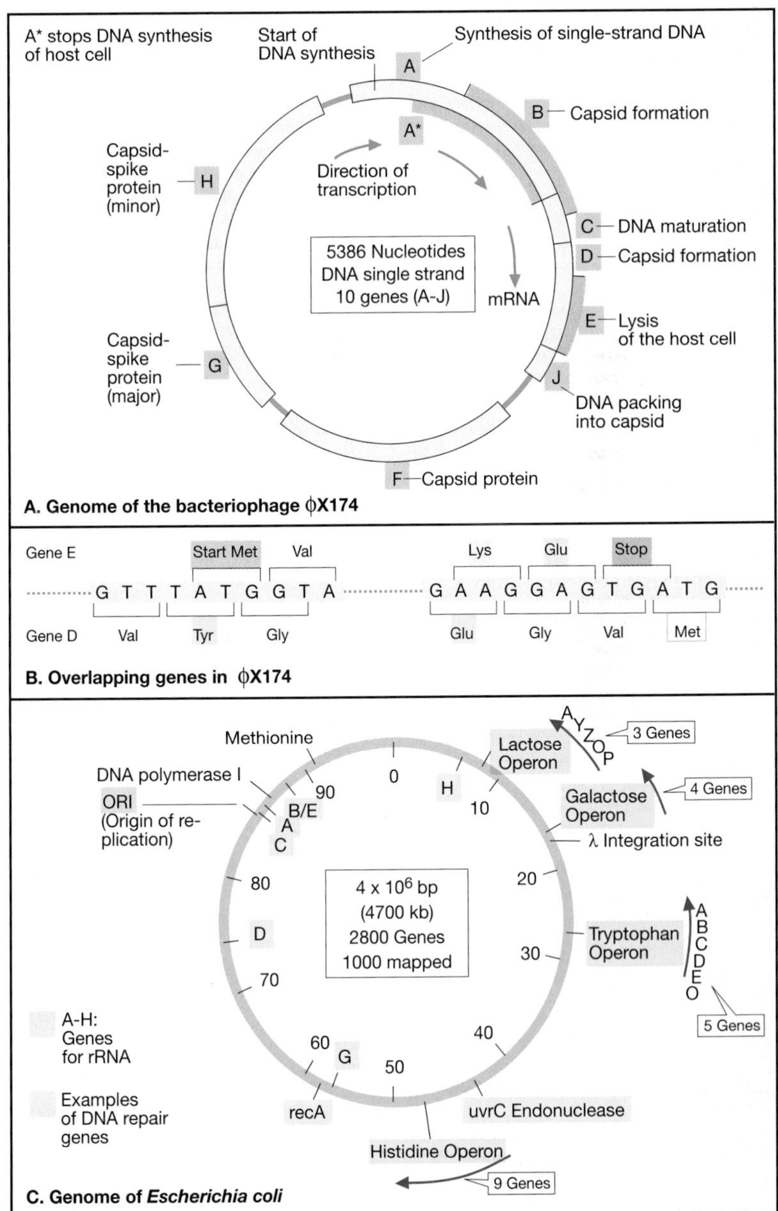

A. Genome of the bacteriophage ϕX174

B. Overlapping genes in ϕX174

C. Genome of *Escherichia coli*

Structural Elements of the Eukaryotic Genome

The genome of man contains 3 billion base pairs (3 x 10^9 bp, equal to 3 million kb or 3000 Mb [megabases]). It consists of very different types of sequences, which represent different structural elements. The number of structural human genes is estimated to be about 60000.

A. Satellite DNA

When DNA is centrifuged in a cesium chloride density gradient, the main band is formed at 1.701 g · cm^{-3}. This represents the main portion of DNA (1). In addition, there are three additional bands (satellites), which are less dense because their CG content is low. One distinguishes classic satellite DNA (2) made up of repeats of 100–6500 bp, minisatellites (3) of 10–20 bp repeats, and microsatellites (4) of 2–5 bp repeats. AT-rich and CG-rich segments can be recognized. Microsatellites are the most frequent form of repetitive DNA. Their general structure is $(CA)_n$, where n equals about 2–10. The human genome contains 50000–100000 polymorphic $(CA)_n$ blocks. Since the particular number of these blocks at a given site is stable, they represent an important genetic marker system. To investigate them, only two steps are needed: amplification by the polymerase chain reaction (PCR) and electrophoresis.

B. Long interspersed repeat sequences (LINE)

Long interspersed repeat sequences (LINE) contain repetitive sequences up to 6500 bp long that are adenine-rich at their 3' ends. They may contain one or two open reading frames (ORF), although they are usually shorter and contain no ORF. LINE elements are thought to have arisen by transposition (see p. 354). The genomes of mammals contain 20000–60000 copies of LINE sequences. The major LINE is the L1 sequence, a segment that spans up to 6.4 kb.

C. Short interspersed repeat sequences (SINE)

Short interspersed repeat sequences (SINE) consist of numerous families of midsize repetitive segments of similar nucleotide sequences with an average of 300 bp. Their basic structure is a tandem duplication of CG-rich segments separated by adenine-rich segments. The most frequent SINE sequences in humans are the Alu family (Alu sequences). With about 500000 copies, they make up about 3–6% of the total genome of man (about one copy occurs every 6 kb). An Alu sequence consists of two 130-bp tandem duplications with A-rich sections between them. The 3' side ("right side") contains an insertion of 31 bp. They are named Alu sequences because many contain the target sequence AGCT of the restriction enzyme *Alu*I.

D. Repetitive and singular sequences

About 70–80% of the genome of man consists of extragenic DNA (highly repetitive and moderately repetitive nucleotide base sequences) (1), and about 20–30%, of DNA within the genes or gene region (singular gene sequences) (2). Less than 10% consists of coding DNA in exons.

E. Gene clusters and multigene families

Functionally related genes may lie on the same chromosome (gene cluster) (1), but more frequently they lie on different chromosomes and make up multigene families (2).

F. Pseudogenes

Pseudogenes are genelike sequences without function. A conventional pseudogene has lost its function as a result of a nonsense mutation. A retropseudogene (processed pseudogene) is derived from the mRNA of a gene from which single-stranded DNA was formed by reverse transcription. After being processed into double-stranded DNA, it was integrated into the genome. A retropseudogene contains no introns and has no promoter.

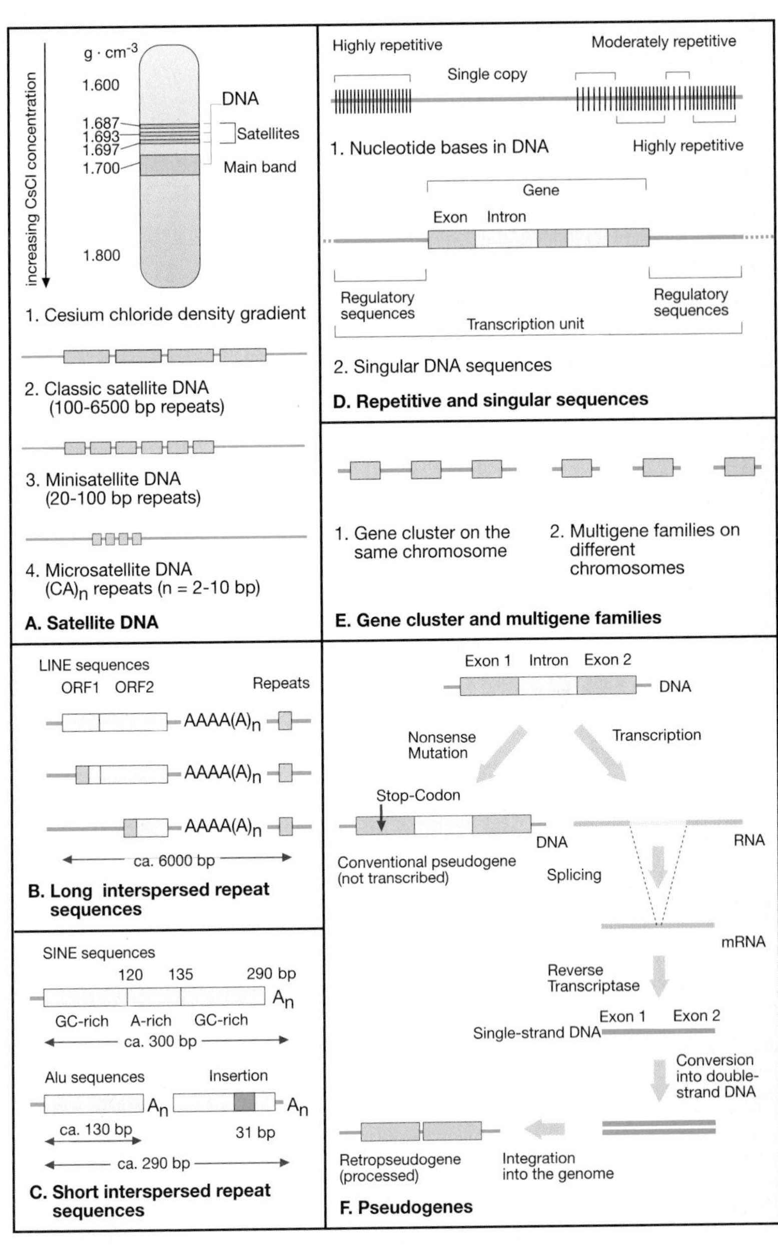

A. Satellite DNA

B. Long interspersed repeat sequences

C. Short interspersed repeat sequences

D. Repetitive and singular sequences

E. Gene cluster and multigene families

F. Pseudogenes

The Dynamic Genome: Mobile Genetic Elements

In the 1940s Barbara McClintock observed an unusual phenomenon during her genetic investigations of Indian corn (maize, *Zea mays*), namely, "jumping genes" (mobile genetic elements). A mobile element causes a break in a chromosome at the site of an insertion and causes a gene locus to move to a different location in the chromosome (transposition). During the last 20 years, mobile elements were found in every organism in which they have been sought: in bacteria, drosophila, the nematode *C. elegans*, and in mammals, including man. These observations have resulted in the concept of a dynamic genome that is by no means fixed and unchangeable. Mutations due to mobile elements inserted into a gene have also been demonstrated in man.

A. Stable and unstable mutations in Indian corn

McClintock observed not only stable mutations (e.g., violet corn kernels) but also fine or somewhat coarser pigment spots (variegation) in some kernels due to unstable mutations.

B. Effect of mutation and transposition

A gene at the C locus produces a violet pigment in cells of the aleurone of Indian corn. When a mobile element *(Ds)* inactivates this gene locus, the corn is colorless. If *Ds* is removed by transposition, C-locus function is restored and small pigmented spots appear.

C. Insertion and removal *(Ds)*

As defined by McClintock, an activator (*Ac* locus) is an element that can activate another locus, dissociation (*Ds*), and cause a break in the chromosome (1). While *Ac* can move independently (autonomous transposition), *Ds* can only move to another location of the chromosome under the influence of *Ac* (nonautonomous transposition). The C locus is inactivated by the insertion of *Ds* (2). Under the influence of *Ac*, *Ds* is then removed from some of the cells, and the C locus is returned to normal function. Since the cells of corn are of clonal origin, the time of transposition influences the phenotype. If transposition occurs early in development, the pigmented spots are relatively large; if it occurs late, the spots are small.

D. Transposons in bacteria

Mobile genetic elements are classified according to their effect and molecular structure: simple insertion sequences (IS) and complex transposons (Tn). A transposon contains additional genes, e.g., for antibiotic resistance in bacteria.
Transposition is a special type of recombination by which a DNA segment of about 750 bp to 10 kb is able to move from one position to another, either on the same or on another DNA molecule. The insertion occurs at an integration site (1) and requires a break (2) with subsequent integration (3). The sequences on either side of the integrated segment at the integration site are direct repeats. At both ends, each IS element or transposon carries inverted repeats whose lengths and base sequences are characteristic for different IS and Tn elements. The expression "direct" signifies that two copies of a sequence are oriented in the same direction (e.g., TTAG on each side of the integrated transposon). Direct and inverted repeats are evidence of the presence of a mobile genetic element. One E. coli cell contains on average about ten copies of such sequences. They have also been demonstrated in yeast, drosophila, and other eukaryotic cells.

References

see p. 375.

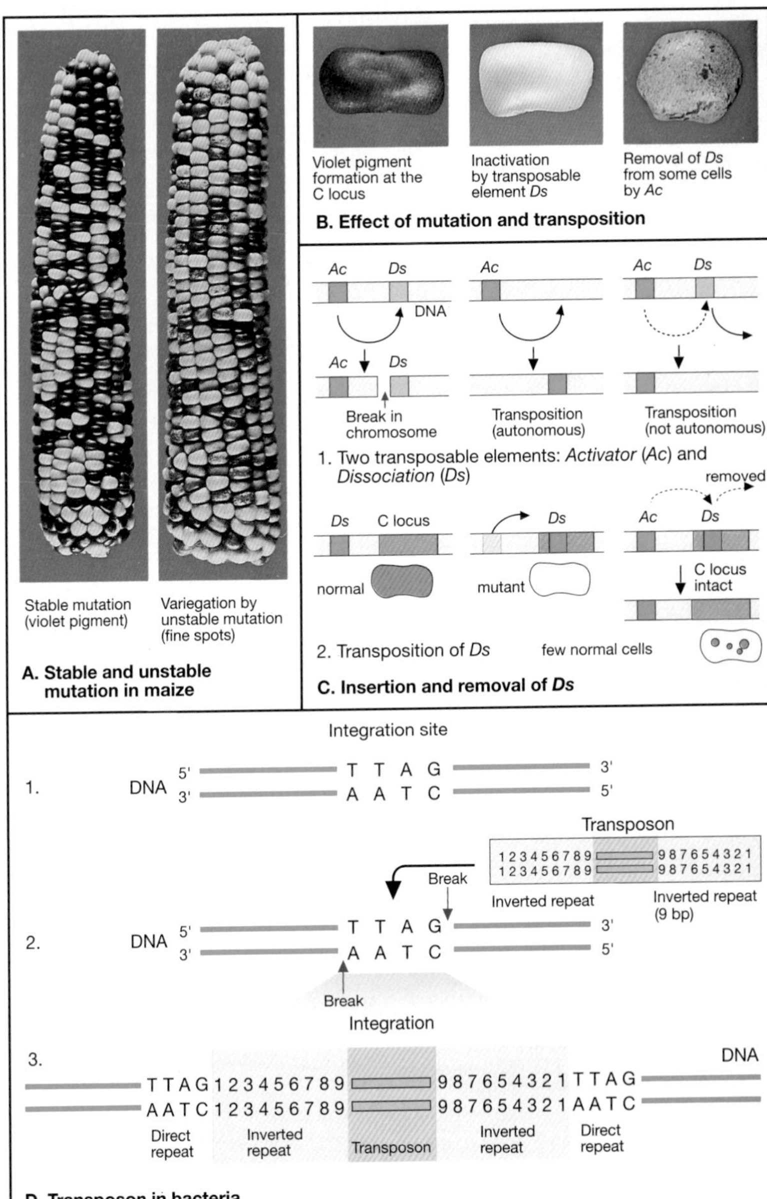
Violet pigment formation at the C locus
Inactivation by transposable element Ds
Removal of Ds from some cells by Ac
B. Effect of mutation and transposition
Ac
Ds
DNA
Break in chromosome
Transposition (autonomous)
Transposition (not autonomous)
1. Two transposable elements: Activator (Ac) and Dissociation (Ds)
removed
C locus
normal
mutant
C locus intact
2. Transposition of Ds
few normal cells
C. Insertion and removal of Ds
Stable mutation (violet pigment)
Variegation by unstable mutation (fine spots)
A. Stable and unstable mutation in maize
Integration site
1.
DNA
5'
3'
T T A G
A A T C
Transposon
123456789
987654321
Break
Inverted repeat
Inverted repeat (9 bp)
2.
Integration
3.
TTAG123456789
AATC123456789
987654321TTAG
987654321AATC
Direct repeat
Inverted repeat
Transposon
D. Transposon in bacteria

Mapping and Sequencing of the Genome

The genome is all of the nucleotides in the DNA of all chromosomes in a cell of an individual. Mapping (determination of the position of gene loci on the chromosomes) and sequencing (determination of the sequence of the nucleotide base pairs) are of immediate medical significance. Knowledge of the location of a gene locus is a prerequisite for DNA diagnosis of genetic disease. By sequencing, the structure and function of a gene and of its mutations leading to disorders can be elucidated. The Human Genome Project (a major international initiative created in 1989) aims to sequence the entire genome of man with its 3 billion base pairs by the year 2005. This enormous number written in the single-letter symbols (A, T, G, C) would fill 13 sets of the *Encyclopaedia Britannica* or 750 megabytes of computer disk space. There are similar projects for organisms with smaller genomes.

A. Physical and genetic gene maps

The physical map gives the position of a gene locus and its distance from other genes on the same chromosome in absolute values, expressed in base pairs and related to given positions along the chromosome. The genetic map gives the relative position of gene loci according to the frequency of recombination, expressed as recombination units, or centimorgans (cM). One centimorgan corresponds to a recombination frequency of 1%. Since recombination occurs almost twice as often in oocytes as in spermatocytes, the genetic map in females is about 40% longer. Each gene locus has an official designation with a defined abbreviation, using the letter D (for DNA), the number of the chromosome, and the number of the marker, e.g., D1S77. (Diagram after J.D. Watson et al., Recombinant DNA, 2nd ed., 1992).

B. Number of gene loci mapped in man

An autosomal gene was first mapped in humans in 1968. To date, almost 3000 human gene loci have been mapped, about half of these in the last 4 years (from V.A. McKusick, Mendelian Inheritance in Man, 11th ed., 1994).

C. STS content map

An STS (sequence tagged site) is a short DNA segment (usually about 500 bp) of known sequence. It can be amplified by PCR (see p. 72) and used as starting material for a map (STS content map). The presence or absence of different STSs (e.g., A–D) in different YAC clones (see p. 104) is determined (1). Their position or sequence can be established in overlapping fragments. This yields a segment of connected DNA sequences (contiguous sequences or contigs). For example, if locus A is present only in YAC 1 and 2, but not in 3 to 6, then YACs 1 and 2 must overlap. The presence of locus D in YACs 2, 3, and 4 indicates that these three overlap, etc. In this way, a series of overlapping fragments (contigs) with defined orientation can be characterized.

D. Segment of DNA sequenced

Automated procedures can be used to sequence large segments of DNA. Unlike the usual DNA-sequencing methods (p. 52), the automated precedures use fluorochrome-labeled bases. Each base is represented by a color peak specific for that base (A, green; T, red; C, blue; G, black). A short segment of 28 base pairs from the retinoblastoma gene (positions 45–72) as presented by an automatic sequencing apparatus is shown (findings of D. Lohmann and B. Horsthemke, Essen).

References

Olson MV. The human genome project. Proc Natl Acad Sci. 1993; 90: 4338–4344.

Green ED, Cox DR, Myers RM. The human genome project and its impact on the study of human disease. In: Scriver CR, et al. eds. The Metabolic and Molecular Bases of Inherited Disease. 7th ed. New York; McGraw–Hill; 1995: 401–436.

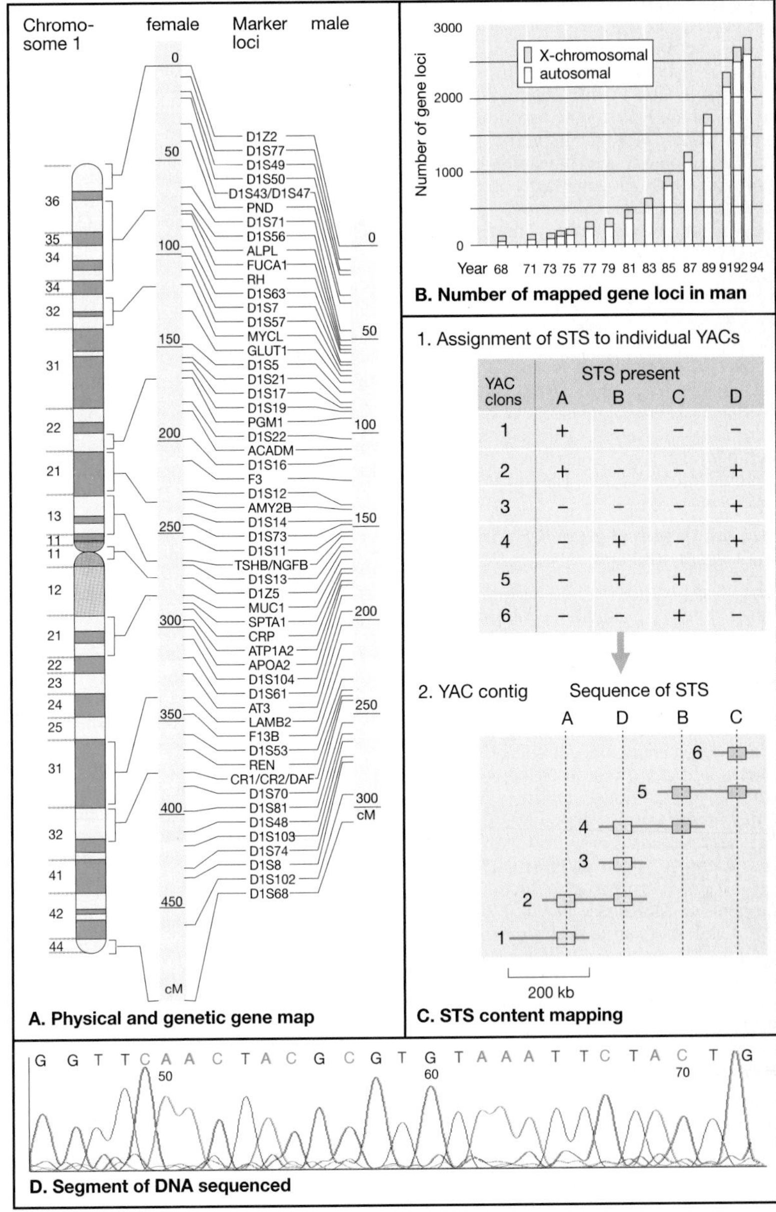

YAC clons	STS present A	B	C	D
1	+	–	–	–
2	+	–	–	+
3	–	–	–	+
4	–	+	–	+
5	–	+	+	–
6	–	–	+	–

A. Physical and genetic gene map

B. Number of mapped gene loci in man

C. STS content mapping

D. Segment of DNA sequenced

Comparative Gene Mapping

Gene mapping (determination of the positions of gene loci along a chromosome) has two goals: (1) to obtain the fundaments for analyzing gene function with respect to their biological significance, e.g., for development, for physiological function including in genetically determined diseases, for the aging process, for the origin of cancer, for immunological processes, for behavioral genetics, etc., and (2) comparison of the genomes and comparison of homologous genes in different organisms to help clarify evolutionary relationships. There are gene maps for about 30 mammalian species (O'Brien SJ ed. Genetic Maps. New York: Cold Spring Harbor; 1990: 1–1115. O'Brien SJ, Graves JAM. Report of the committee on comparative gene mapping. Cytogenet Cell Genet. 1991; 58: 1124–1151).

A. Comparison of gene loci in man, mouse, cat, and cow

The conservation of certain groups of genes during mammalian evolution can be demonstrated by comparing reference loci in four different orders of mammals. Many of the 23 loci (cloned genes) on chromosome 1 and many of the 19 loci on the X chromosome of humans (representing the order Primates) are found in the same order, although often on different chromosomes, in the house mouse (*Mus musculus*, representing Rodentia), the house cat (representing Carnivora), and the cow (representing Artiodactyla). Thus, the gene for PGD (6-phosphogluconate dehydrogenase, McKusick Nr. 172200) is on the distal short arm of chromosome 1 of man, on chromosome 4 of the mouse, C1 of the cat, and U1 of the cow. The homologous segment on chromosome C1 of the cat is unusually long. Numerous groups of genes are evolutionarily conserved. (After S.J. O'Brien et al., 1993).

B. The X chromosomes of man and mouse

About 60 chromosomal segments with the same sequences of two or more homologous loci are syntenic (on the same chromosome) in mouse and man (Davisson et al., 1990). Gene loci sequences on the X chromosomes of eutherian mammals are especially similar due to the conserving effect of X inactivation (S. Ohno, 1971). In the X chromosomes of humans and mice, there are five blocks of gene loci that have the same sequences, although they are in different regions and in part oriented in different directions; these have been retained for about 80 million years, since their divergence from a common ancestor. Three regions (white fields) on the X chromosome of humans and the X chromosome of mice do not have any of the loci that have been mapped in both species (after S.H. Lavale & Y. Boyd, 1993).

C. Sequence homologies of the X and the Y chromosomes of man

During the evolution of vertebrates, the X and the Y chromosomes were derived from an autosome. While the Y chromosome has become considerably smaller, the X chromosome has for the most part retained the gene loci of originally autosomal character. In five regions (I–V), the X chromosome and Y chromosome of man contain sequence homologies. The homologous sequences on the extreme distal ends of both chromosomes (I and II) correspond to a pseudoautosomal region. In region I, homologous pairing and crossing-over occur regularly during meiosis. (After U. Wolf et al., 1992).

References

see p. 378.

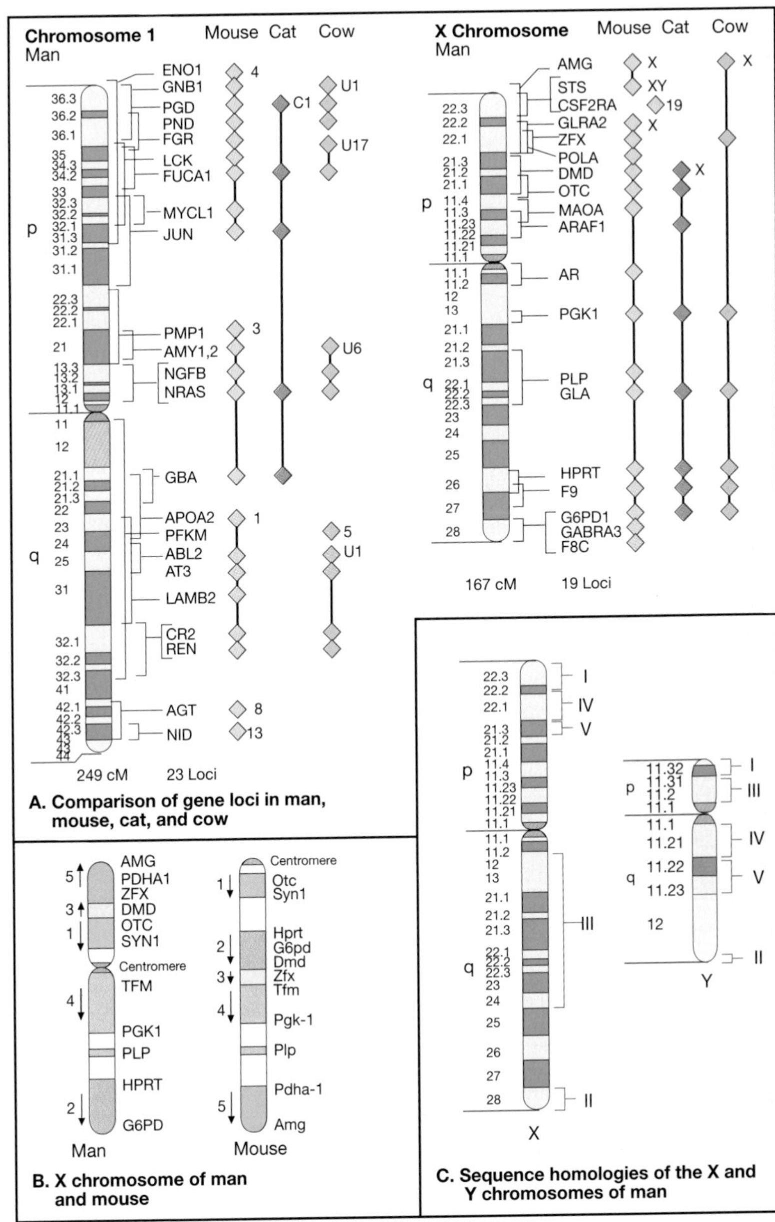

A. Comparison of gene loci in man, mouse, cat, and cow

B. X chromosome of man and mouse

C. Sequence homologies of the X and Y chromosomes of man

Chromosomal Location of Human Genetic Diseases

At about 800 of the 2800 mapped gene loci, mutations are known to cause defined diseases. A selection of gene loci for diseases mapped to the genome of man are presented in the following five tables. The presentation is based on a similar map published by V. A. McKusick in many editions of "The Morbid Anatomy of the Human Genome" (McKusick, 1994; McKusick & Amberger, 1993). For reasons of clarity, only a portion of the gene loci actually localized are included, and diseases of special significance are emphasized. However, this does not mean that the diseases not included are less important.

Most genetically determined diseases are rare, but the occurrence of a particular disease represents an urgent problem for the affected family. Genetically determined diseases are not a marginal group in medicine, but together represent a considerable proportion of all disease processes in medicine. This fact is obscured because all organ systems and all age groups are affected. Since individual physicians as a rule become acquainted with only a limited sector of genetically determined diseases, they may be led to underestimate their full significance. Thus, dealing with so many different disorders represents a considerable challenge for the physician.

At present, about 2800 expressed genes have been mapped to autosomes in man, and about 230, to the X chromosome. Mutations in these genes lead to monogenic genetic diseases. For more than 800 clinical disorders the intragenic lesion is known. The mapping methods are many and varied, and they complement each other. McKusick relates the following figures: 1683 genes mapped by hybrid-cell analysis, 629 by linkage studies in families, 361 by RFLP analysis, 193 by gene-dose determination, 179 by determination of homologies of syntenic loci, 174 through chromosomal aberrations, 36 by radiation-induced gene segregation in vitro, and 171 by other means (V.A. McKusick, personal communication, May 19, 1993).

Of the several methods employed to isolate and identify disease genes, i.e., loci at which mutations cause a disease, the following have been used most often: (1) positional cloning based on knowing the chromosomal position, (2) candidate gene approach, where a previously known gene with a role in a normal function is shown to contain mutations, and (3) mapping to a small region of deletion or to the breakpoint of a chromosome translocation.

A complete list of localized diseases with gene symbol, McKusick number, and reports of chromosomal localization can be found in the McKusick catalog, 11th ed., 1994. This catalog contains reports about gene loci, their alleles, and phenotypes. The first edition appeared in 1966. Since then, each new edition has regularly increased in size with the expanding knowledge in this field. The map is also available on a computer on-line system.

Numerous diseases are clinically similar (similar phenotype), but have different causes (principle of genetic heterogeneity). Allelic heterogeneity (due to different mutant alleles at one gene locus) and hetero-allelic heterogeneity (due to mutations at different gene loci) can be differentiated. In practice, this is important for genetic diagnosis and counseling, since the mode of inheritance and course of disease may differ.

References

Collins FS, Galas D. A new five-year plan for the U.S. Human Genome Project. Science. 1993; 262: 43–46.

Collins FS. Positional cloning moves from perditional to traditional. Nature Genet. 1995; 9: 347–350.

McKusick VA. Mendelian Inheritance in Man. A Catalog of Human Genes and Genetic Disorders. 11th ed. Baltimore: Johns Hopkins University Press; 1994.

McKusick VA, Amberger JS. The morbid anatomy of the human genome: chromosomal location of mutations causing disease. J Med Genet. 1993; 30: 1–26.

Strachan T. The Human Genome. Oxford: Bios Scientific Publishers; 1992.

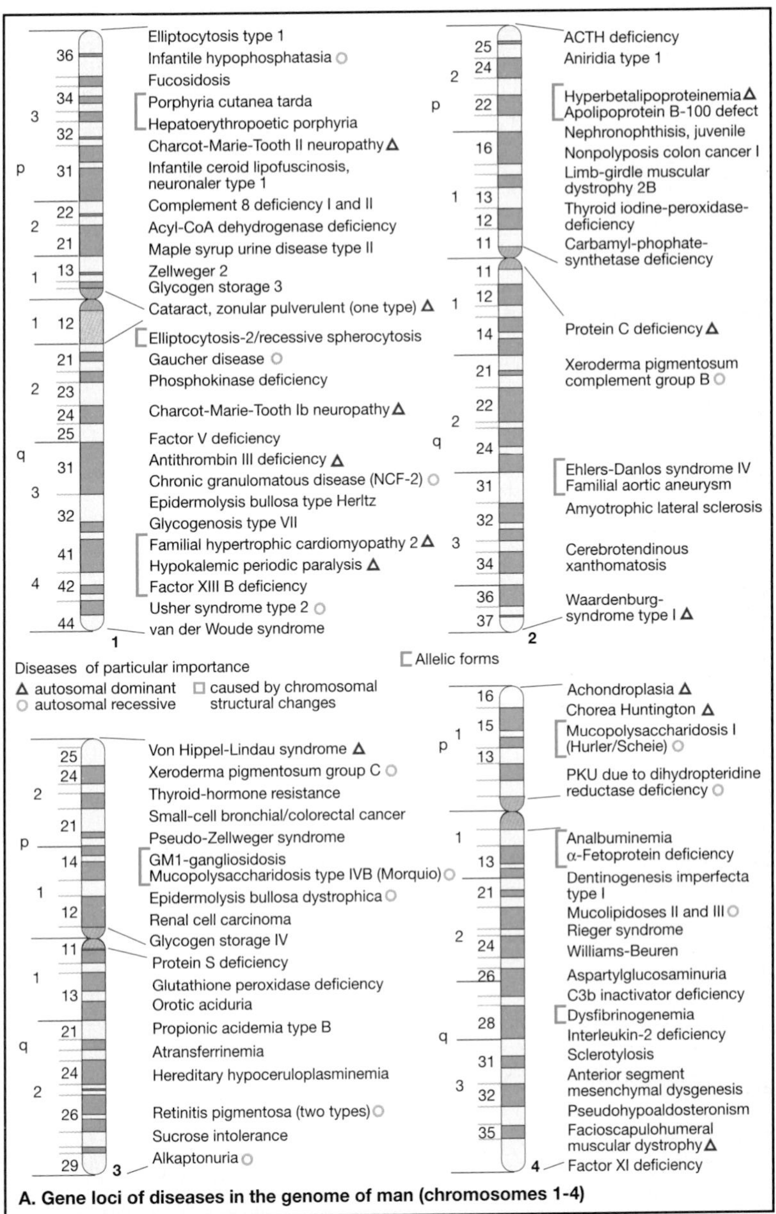

A. Gene loci of diseases in the genome of man (chromosomes 1-4)

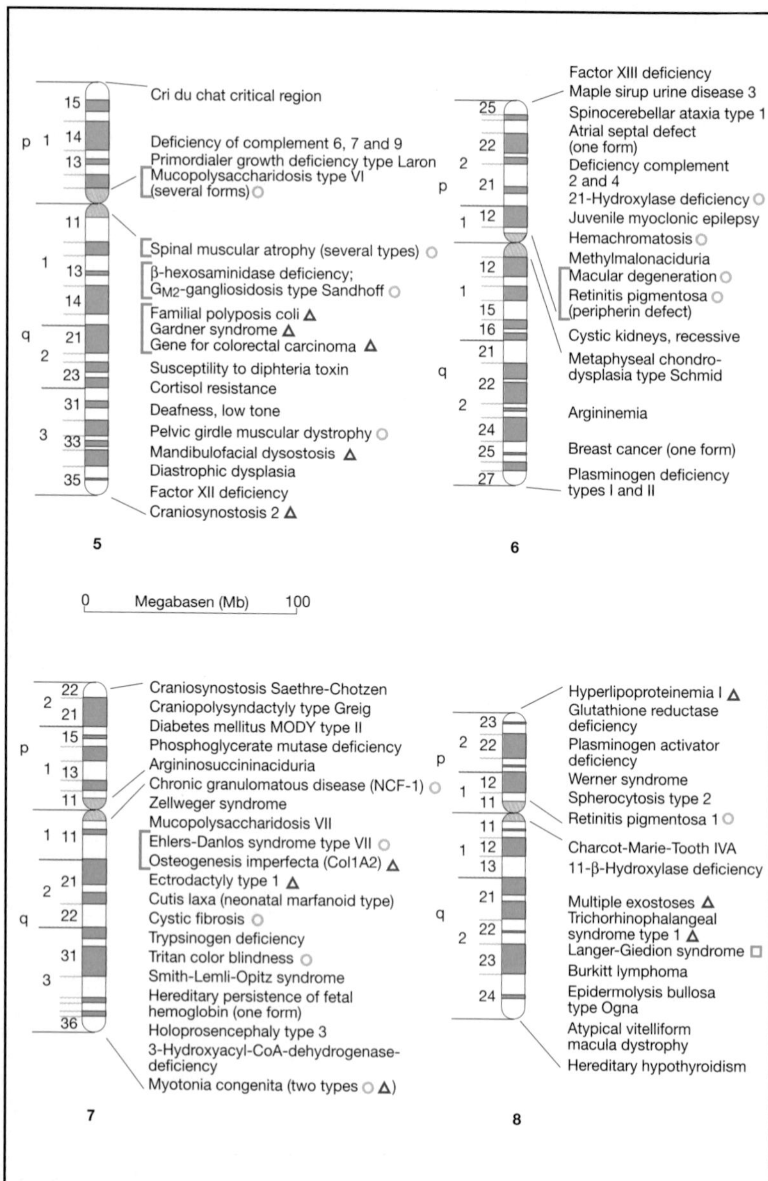

A. Gene loci of diseases on the human genome (chromosomes 5-8)

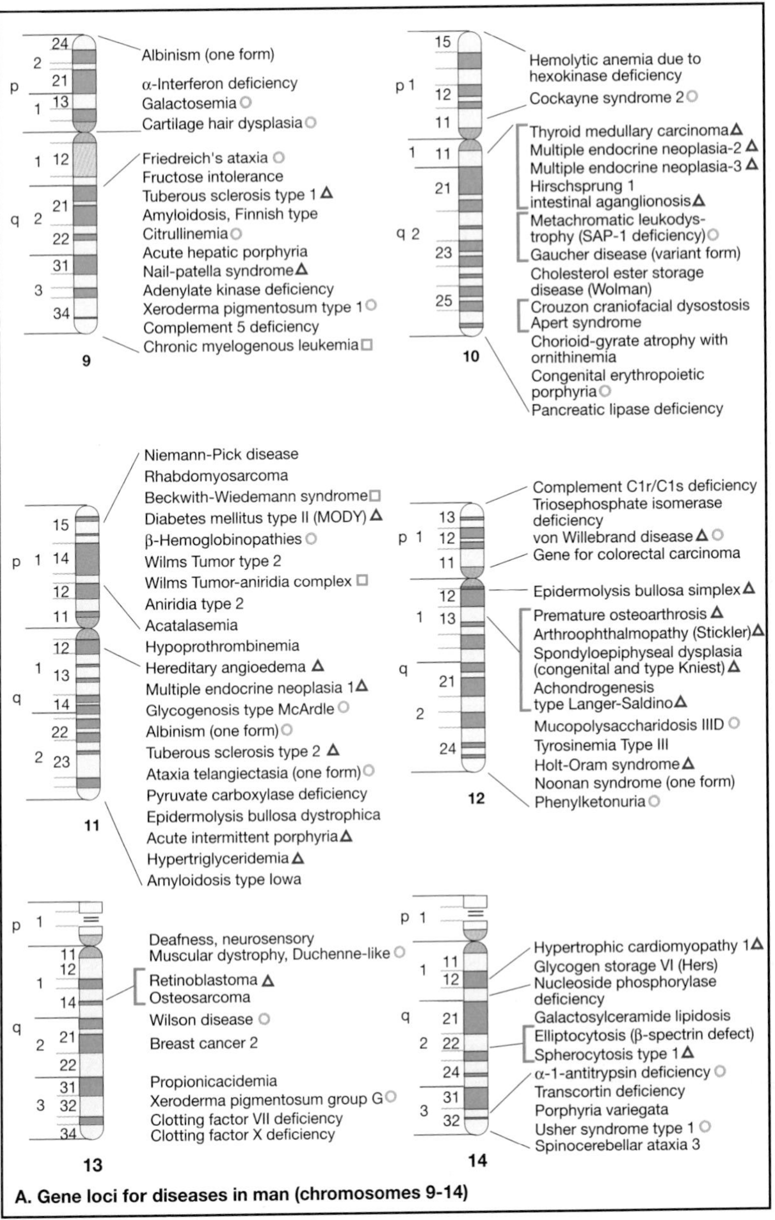

A. Gene loci for diseases in man (chromosomes 9-14)

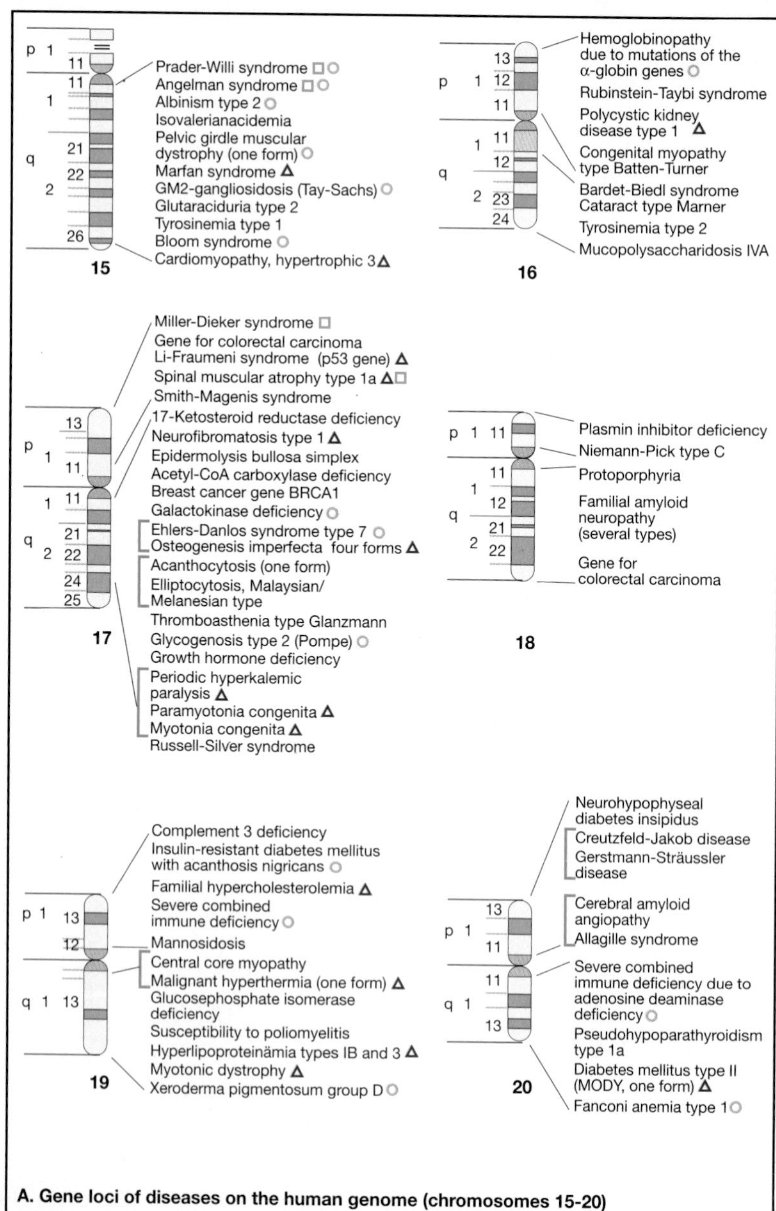

A. Gene loci of diseases on the human genome (chromosomes 15-20)

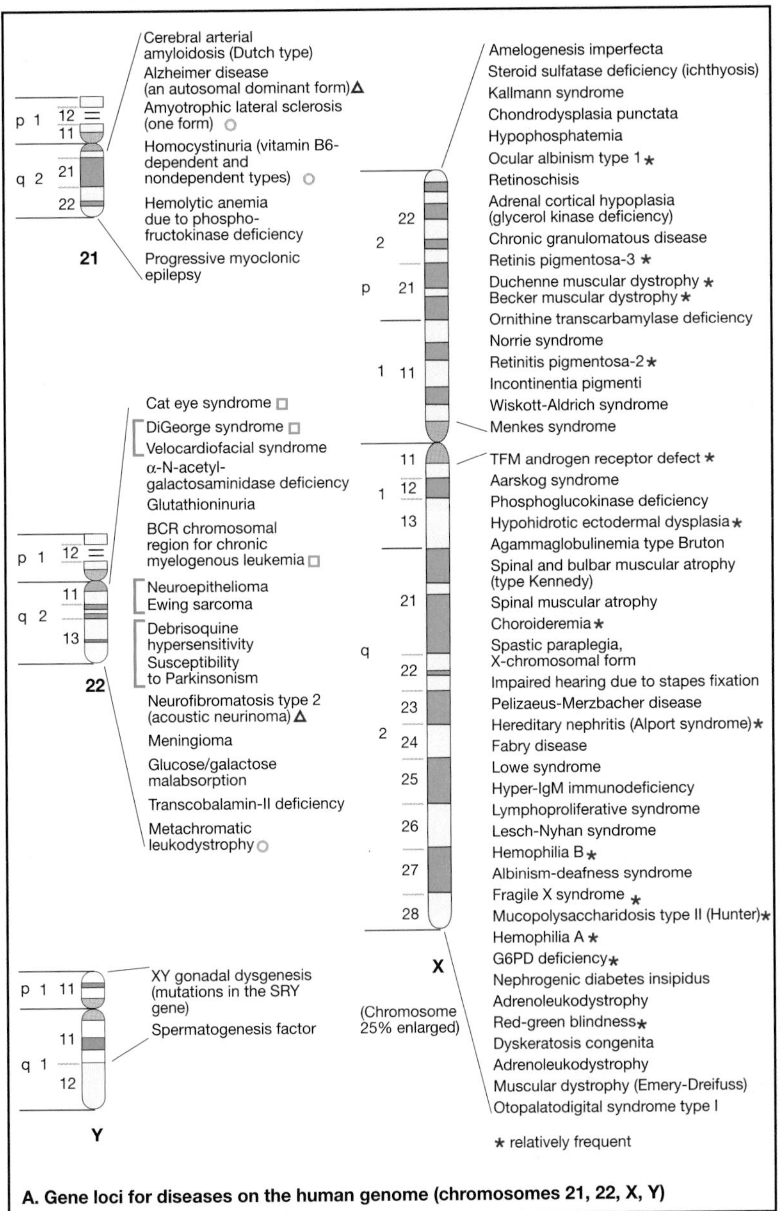

A. Gene loci for diseases on the human genome (chromosomes 21, 22, X, Y)

Chromosomal Location of Human Genetic Diseases

(Alphabetic list to the maps on pp. 361–365)

Aarskog syndrome (X ch.)
Acanthocytosis (one form) (ch. 17)
Acatalasemia (ch. 11)
Acetyl-CoA carboxylase deficiency (ch.17)
Achondroplasia (ch. 4)
ACTH deficiency (ch. 2)
Acute hepatic porphyria (ch. 9)
Acute intermittent porphyria (ch. 11)
Acyl-CoA dehydrogenase deficiency (ch. 1)
Adenylate kinase deficiency (ch. 9)
Adrenal cortical hypoplasia with glycerol kinase deficiency (X ch.)
Adrenoleukodystrophy (X ch.)
α-Interferon deficiency (ch. 9)
Agammaglobulinemia (X ch.)
Alagille syndrome (ch. 20)
Albinism (one form) (ch. 9)
Albinism (one form) (ch. 11)
Albinism type 2 (ch. 15)
Albinism–deafness syndrome (X ch.)
Alkaptonuria (ch. 3)
Alpha-N-acetylgalactosamindase deficiency (ch. 22)
Alpha-1-antitrypsin deficiency (ch. 14)
Alpha-fetoprotein deficiency (ch. 4)
Alzheimer disease (one form) (ch. 21)
Amelogenesis imperfecta (X ch.)
Amyloidosis, Finnish type (ch. 9)
Amyloidosis type Iowa (ch. 11)
Amyotrophic lateral sclerosis (one form) (ch. 21)
Amyotrophic lateral sclerosis, juvenile (ch. 2)
α-N-acetylgalactosamindase deficiency (ch. 22)
Analbuminemia (ch. 4)
Androgen receptor defect (X ch.)
Angelman syndrome (ch. 15)
Aniridia type 1 (ch. 2)
Aniridia type 2 (ch. 11)
Anterior segmental mesenchymal dysgenesis (ch. 4)
Antithrombin III deficiency (ch. 1)
Apert syndrome (ch. 10)
Apolipoprotein B-100 defect (ch. 2)
Argininemia (ch. 6)
Argininosuccinicaciduria (ch. 7)
Arthroophthalmopathy (Stickler syndrome) (ch. 12)
Aspartylglucosaminuria (ch. 4)
Ataxia telangiectasia (one form) (ch. 11)
Atransferrinemia (ch. 3)
Atrial septal defect (one form) (ch. 6)
Atypical vitelliform macular dystrophy (ch. 8)
Bardet–Biedl syndrome (ch. 16)
BCR chromosomal region for chronic myelogenous leukemia (ch. 22)
Becker muscular dystrophy (X ch.)
Beckwith–Wiedemann syndrome (ch. 11)
Beta-hemoglobinopathies (ch. 11)
Beta-hexosaminidase deficiency; GM^2–gangliosidosis type Sandhoff (ch. 5)
11-Beta-hydroxylase deficiency (ch. 8)
Bloom syndrome (ch. 15)
Breast cancer gene BRCA1 (ch. 17)
Breast cancer gene BRCA 2 (ch. 13)
Breast cancer (one form) (ch. 6)
Burkitt lymphoma (ch. 8)
Carbamylphosphate synthetase–I deficiency (ch. 4)
Cardiomyopathy, familial hypertrophic type 3 (ch. 15)
Cartilage–hair dysplasia (ch. 9)
Cat eye syndrome (ch. 22)
C3b-inactivator deficiency (ch. 4)
Central core myopathy (ch. 19)
Cerebral amyloid angiopathy (ch. 20)
Cerebral arterial amyloidosis (Dutch type) (ch. 21)
Cerebrotendinosis xanthomatosis (ch. 2)
Charcot–Marie–Tooth neuropathy type1b (ch. 1)
Charcot–Marie–Tooth neuropathy type II (ch. 1)
Charcot–Marie–Tooth neuropathy type IVa (ch. 8)
Cholesteryl ester storage disease (Wolman) (ch. 10)
Chondrodysplasia punctata (X ch.)
Choroid gyrate atrophy with ornithinemia (ch. 10)
Choroideremia (X ch.)
Chronic granulomatous disease (NCF-1) (ch. 7)
Chronic granulomatous disease (NCF-2 deficiency) (ch. 1)
Chronic granulomatous disease (X ch.)

Chronic myelogenous leukemia (ch. 9)
Citrullinemia (ch. 9)
Clotting factor VII deficiency (ch. 13)
Clotting factor X deficiency (ch. 13)
Cockayne syndrome 2 (ch. 10)
Colon cancer, familial nonpolyposis type 1 (ch. 2)
Colorectal adenocarcinoma (ch. 12)
Colorectal carcinoma (ch. 5)
Colorectal carcinoma/Li–Fraumeni syndrome (ch. 17)
Complement 2 and 4 deficiency (ch. 6)
Complement 3 deficiency (ch. 19)
Complement 5 deficiency (ch. 9)
Complement 6, 7, and 9 deficiency (ch. 5)
Complement 8 deficiency 1 and 2 (ch. 1)
Complement C1r/C1s deficiency (ch. 12)
Congenital erythropoietic porphyria (ch. 10)
Congenital myopathy type Batten–Turner (ch. 16)
Cortisol resistance (ch. 5)
Craniopolysyndactyly type Greig (ch. 7)
Craniosynostosis Saethre–Chotzen (ch. 7)
Craniosynostosis type 2 (ch. 5)
Creutzfeldt–Jakob disease (ch. 20)
Cri du chat critical region (ch. 5)
Crigler–Najjar syndrome (ch. 1)
Crouzon craniofacial dysostosis (ch. 10)
Cutis laxia (neonatal marfanoid type) (ch. 7)
Cystic fibrosis (ch. 7)
Deafness, low-tone (ch. 5)
Deafness, neurosensory (ch. 13)
Debrisoquin hypersensitivity (ch. 22)
Dentinogenesis imperfecta type I (ch. 4)
Diabetes mellitus type MODY (ch. 11)
Diabetes mellitus type MODY (one form) (ch. 20)
Diabetes mellitus, MODY type II (ch. 7)
Diastrophic dysplasia (ch. 5)
DiGeorge syndrome (ch. 22)
Duchenne muscular dystrophy (X ch)
Dysfibrinogenemia (ch. 4)
Dyskeratosis congenita (X ch.)
Ectrodactyly type 1 (ch. 7)
Ehlers–Danlos syndrome type 4 (ch. 2)
Ehlers–Danlos syndrome type 7 (ch. 7)
Ehlers–Danlos syndrome type 7A1 (ch. 17)
Elliptocytosis (β-spectrin defect) (ch. 14)
Elliptocytosis, Malaysian/Melanesian type (ch. 17)
Elliptocytosis-2/recessive spherocytosis (ch. 1)
Epidermolysis bullosa dystrophica (ch. 3)
Epidermolysis bullosa simplex (ch. 12)
Epidermolysis bullosa simplex (ch. 17)
Epidermolysis bullosa type Herlitz (ch.1)
Epidermolysis bullosa type Ogna (ch. 8)
Ewing sarcoma (ch. 22)
Fabry disease (X ch.)
Facioscapulohumeral muscular dystrophy (ch. 4)
Factor V deficiency (ch. 1)
Factor XI deficiency (ch. 4)
Factor XII deficiency (ch. 5)
Factor XIIIa deficiency (ch. 6)
Factor XIII B deficiency (ch. 1)
Familial amyloid neuropathy (several types) (ch. 18)
Familial aortic aneurysm (ch. 2)
Familial hypercholesterolemia (ch. 19)
Familial hypertrophic cardiomyopathy (ch. 1)
Familial polyposis coli (ch. 5)
Fanconi anemia type 1 (ch. 20)
Fragile X syndrome (X ch.)
Friedreich's ataxia (ch. 9)
Fructose intolerance (ch. 9)
Fucosidosis (ch. 1)
G6PD deficiency (X ch.)
Galactokinase deficiency (ch. 17)
Galactose epimerase deficiency (ch. 1)
Galactosemia (ch. 9)
Galactosylceramide lipidosis (ch. 14)
Gardner syndrome (ch. 5)
Gaucher disease (ch. 1)
Gaucher disease (variant form) (ch. 10)
Gerstmann–Sträussler disease (ch. 20)
Glucose/galactose malabsorption (ch. 22)
Glucosephosphate isomerase deficiency (ch. 19)
Glutaricaciduria type 2 (ch. 15)
Glutathione peroxidase deficiency (ch. 3)
Glutathione reductase deficiency (ch. 8)
Glutathionuria (ch. 22)
Glycogenosis type VII (ch. 1)
Glycogenosis type McArdle (ch. 11)
Glycogenosis type 2 (Pompe) (ch. 17)
Glycogen storage type 3 (ch. 1)
Glycogen storage type 4 (ch. 3)
Glycogen storage VI (Hers) (ch. 14)
GM1-gangliosidosis (ch. 3)
GM2-gangliosidosis (Tay–Sachs) (ch. 15)
GM2-gangliosidosis type Sandhoff (ch. 5)

Growth hormone deficiency (ch. 17)
Hemochromatosis (ch. 6)
Hemoglobinopathies due to mutations of the α-globin genes (ch. 16)
Hemolytic anemia due to hexokinase deficiency (ch. 10)
Hemolytic anemia due to phosphofructokinase deficiency (ch. 21)
Hemophilia A (X ch.)
Hemophilia B (X ch.)
Hepatoerythropoietic porphyria (ch. 1)
Hereditary angioedema (ch. 11)
Hereditary congenital hypothyroidism (ch. 8)
Hereditary hypoceruloplasminemia (ch. 3)
Hereditary nephritis (Alport syndrome) (X ch.)
Hereditary persistence of fetal hemoglobin (one form) (ch. 7)
Hirschsprung disease (chs. 10 and 13)
Holoprosencephaly type 3 (ch. 7)
Holt–Oram syndrome (ch. 12)
Homocystinuria (B6-responsive and B6-nonresponsive forms) (ch. 21)
Huntington chorea (ch. 4)
3-Hydroxyacyl-CoA dehydrogenase deficiency (ch. 7)
21-Hydroxylase deficiency (ch. 6)
Hyperbetalipoproteinemia (ch. 2)
Hyper-IgM immune deficiency (X ch.)
Hyperlipoproteinemia type 1 (ch. 8)
Hyperlipoproteinemia type 1b (ch. 19)
Hyperlipoproteinemia type 3 (ch. 19)
Hypertriglyceridemia (ch. 11)
Hypertrophic cardiomyopathy (ch. 14)
Hypochondroplasia (ch. 4)
Hypohidrotic ectodermal dysplasia (X ch.)
Hypophosphatemia (X ch.)
Hypoprothrombinemia (ch. 11)
Immune deficiency, severe combined (ch. 19)
Immunodeficiency due to ADA deficiency (ch. 20)
Impaired hearing (lower frequencies) (ch. 5)
Impaired hearing due to stapes fixation (X ch.)
Infantile ceroid lipofuscinosis, neuronal type (ch. 1)
Infantile hypophosphatasia (ch. 1)
Insulin-resistant diabetes mellitus with acanthosis nigricans (ch. 19)
Interleukin 2 deficiency (ch. 4)
Intestinal aganglionosis (Hirschsprung) (ch. 10 and 13)
Isovalericacidemia (ch. 15)
Juvenile myoclonic epilepsy (ch. 6)
Kallmann syndrome (X ch.)
17-Ketosteroid reductase deficiency (ch. 17)
Lamellar cataract (one type) (ch. 1)
Langer–Giedion syndrome (ch. 8)
Lesch–Nyhan syndrome (X ch.)
Li–Fraumeni syndrome (ch. 17)
Limb–girdle muscular dystrophy 2b (ch. 2)
Lowe syndrome (X ch.)
Lymphoproliferative syndrome (X ch.)
Macular degeneration (ch. 6)
Malignant hyperthermia (ch. 19, others)
Mandibulofacial dysostosis (Franceschetti–Klein syndrome) (ch. 5)
Mannosidosis (ch. 19)
Maple syrup urine disease type 2 (ch. 1)
Maple syrup urine disease type 3 (ch. 6)
Marfan syndrome (ch. 15)
Meningioma (ch. 22)
Menkes syndrome (X ch.)
Metachromatic leukodystrophy (ch. 22)
Metachromatic leukodystrophy (SAP–1 deficiency) (ch. 10)
Metaphyseal chondrodysplasia type Schmid (ch. 6)
Methylmalonicaciduria (ch. 6)
Miller–Dieker syndrome (ch. 17)
Morquio syndrome B (ch. 3)
Mucolipidosis types II and III (ch. 4)
Mucopolysaccharidosis type I (Hurler/Scheie) (ch. 4)
Mucopolysaccharidosis type II (X ch.)
Mucopolysaccharidosis type IVa (ch. 16)
Mucopolysaccharidosis type IVb (ch. 3)
Mucopolysaccharidosis type VI (Maroteaux–Lamy) (ch. 5)
Mucopolysaccharidosis type VII (ch. 7)
Multiple endocrine neoplasia type 1 (ch. 11)
Multiple endocrine neoplasia type 2 (ch. 10)
Multiple endocrine neoplasia type 3 (ch. 10)
Multiple exostoses (ch. 8)
Muscular dystrophy, Duchenne-like (ch. 13)
Muscular dystrophy type Becker (X ch.)

Muscular dystrophy type Duchenne (X ch.)
Muscular dystrophy type Emery–Dreifuss (X ch.)
Myotonia congenita (ch. 17)
Myotonia congenita (two types) (ch. 7)
Myotonic dystrophy (ch. 19)
Myotubular myopathy (X ch.)
Nail–patella syndrome (ch. 9)
Nephrogenic diabetes insipidus (X ch.)
Nephronophthisis, juvenile (ch. 2)
Neuroepithelioma (ch. 22)
Neurofibromatosis type 1 (ch. 17)
Neurofibromatosis type 2 (acusticus neurinoma) (ch. 22)
Neurohypophyseal diabetes insipidus (ch. 20)
Niemann–Pick disease (ch. 11)
Niemann–Pick type C (ch. 18)
Noonan syndrome (one locus) (ch. 12)
Norrie syndrome (X ch.)
Nucleoside phosphorylase deficiency (ch. 14)
Ocular albinism (X ch.)
Ornithine transcarbamylase deficiency (X ch.)
Oroticacidemia (ch. 3)
Osteogenesis imperfecta (ch. 17)
Osteogenesis imperfecta (COL1A2) (ch. 7)
Osteosarcoma (ch. 13)
Otopalatodigital syndrome type 1 (X ch.)
Pancreatic lipase deficiency (ch. 10)
Paramyotonia congenita (ch. 17)
Pelizaeus–Merzbacher disease (X ch.)
Pelvic girdle muscular dystrophy (ch. 5 and ch. 15)
Periodic hyperkalemic paralysis (ch. 17)
Phenylketonuria (PKU) (ch. 12)
Phosphoglucokinase deficiency (X ch.)
Phosphoglycerate mutase deficiency (ch. 7)
Phosphokinase deficiency (ch. 1)
PKU due to dihydropteridine reductase deficiency (ch. 4)
Plasmin inhibitor deficiency (ch. 18)
Plasminogen activator deficiency (ch. 8)
Plasminogen deficiency, types I and II (ch. 6)
Polycystic kidney disease (one form) (ch. 16)
Porphyria cutanea tarda (ch. 1)
Porphyria variegata (ch. 14)
Prader–Willi syndrome (ch. 15)
Primordial growth deficiency type Laron (ch. 5)
Progressive myoclonic epilepsy (ch. 21)
Propionicacidemia type A (ch. 13)
Propionicacidemia type B (ch. 3)
Protein C deficiency (ch. 2)
Protein S deficiency (ch. 3)
Protoporphyria (ch. 18)
Pseudoaldosteronism (ch. 4)
Pseudohypoparathyroidism type 1a (ch. 20)
Pseudo-Zellweger syndrome (ch. 3)
Red–green blindness (X ch.)
Renal cell carcinoma (ch. 3)
Retinitis pigmentosa (perpherin defect) (ch. 6)
Retinitis pigmentosa type 1 (ch. 8)
Retinitis pigmentosa type 2 (X ch.)
Retinitis pigmentosa type 3 (X ch.)
Retinitis pigmentosa type 5 (ch. 3)
Retinoblastoma (ch. 13)
Retinoschisis (X ch.)
Rhabdomyosarcoma (ch. 11)
Rieger syndrome (ch. 4)
Rubinstein–Taybi syndrome (ch. 16)
Russell–Silver syndrome (ch. 17)
Sclerotylosis (ch. 4)
Sex reversal (XY females due to mutation in the SRY gene) (Y ch.)
Small cell bronchial carcinoma/colorectal cancer (ch. 3)
Smith–Lemli–Opitz syndrome (ch. 7)
Smith–Magenis syndrome (ch. 17)
Spastic paraplegia (X chromosomal form) (X ch.)
Spermatogenesis factor (Y ch.)
Spherocytosis type 1 (ch. 14)
Spherocytosis type 2 (ch. 8)
Spinal muscular atrophy (X ch.)
Spinal muscular atrophy Ia (ch. 17)
Spinal muscular atrophy IVa (ch. 8)
Spinal muscular atrophy Werdnig–Hoffmann and other types (ch. 5)
Spinocerebellar ataxia type 1 (ch. 6)
Spinocerebellar ataxia type 3 (ch. 14)
Spondyloepiphyseal dysplasia (congenital type) (ch. 12)
Spondyloepiphyseal dysplasia (type Kniest) (ch. 12)
Steroid sulfatase deficiency (ichthyosis) (X ch.)
Sucrose intolerance (ch. 3)

Susceptibility to diphtheria toxin (ch. 5)
Susceptibility to Parkinsonism (ch. 22)
Susceptibility to poliomyelitis (ch. 19)
T cell leukemia/lymphoma (ch. 14)
TFM androgen receptor defect (X ch.)
Thrombasthenia type Glanzmann (ch. 17)
Thyroid hormone resistance (ch. 3)
Thyroid iodine peroxidase deficiency (ch. 2)
Thyroid medullary carcinoma (ch. 10)
Transcobalamin II deficiency (ch. 22)
Transcortin deficiency (ch. 14)
Trichorhinophalangeal syndrome type 1 (ch. 8)
Triosephosphate isomerase deficiency (ch. 12)
Tritan color blindness (ch. 7)
Trypsinogen deficiency (ch. 7)
Tuberous sclerosis type 1 (ch. 9)
Tuberous sclerosis type 2 (ch. 11)
Tuberous sclerosis type 3 (ch. 12)
Tyrosinemia type 1 (ch. 14)
Tyrosinemia type 2 (ch. 16)
Usher syndrome type 1 (ch. 14)
Usher syndrome type 2 (ch. 1)
Van der Woude syndrome (ch. 1)
Velocardiofacial syndrome (ch. 22)
Vitelline macular dystrophy (ch. 6)
Von Hippel–Lindau syndrome (ch. 3)
Von Willebrand disease (ch. 12)
Waardenburg syndrome type 1 (ch. 2)
Werner syndrome (ch. 8)
Williams–Beuren syndrome (ch. 4)
Wilms tumor–aniridia complex (ch. 11)
Wilms tumor type 2 (ch. 11)
Wilson's disease (ch. 13)
Xeroderma pigmentosum complementation group B (ch. 2)
Xeroderma pigmentosum group C (ch. 3)
Xeroderma pigmentosum group D (ch. 19)
Xeroderma pigmentosum group G (ch. 13)
Xeroderma pigmentosum type 1 (ch. 9)
XY gonadal dysgenesis (Y ch.)
Zellweger syndrome (ch. 7)
Zellweger syndrome type 2 (ch. 1)
Zonular cataract (ch. 1)

Caveat: There are numerous similar disorders caused by mutations of genes at other loci, sometimes with other modes of inheritance.
This list and the corresponding maps are not complete, but give examples. For a complete map and list, see McKusick's catalog, 11th edition, 1994.

References

A. General

Abbas AK, Lichtman AH, Pober JS. Cellular and Molecular Immunology. 2nd ed. Philadelphia: WB Saunders; 1994.

Alberts B, Bray D, Lewis J, Ruff M, Roberts K, Watson JD. Molecular Biology of the Cell. 3rd ed. New York: Garland; 1994.

Ayala FJ, Kieger JA Jr. Modern Genetics. 2nd ed. Menlo Park, California: Benjamin/ Cummings; 1984.

Bender K. The HLA System. 5th ed. Dreieich, Germany: Biotest Diagnostics; 1991.

Branden C, Tooze J. Introduction to Protein Structure. New York & London: Garland; 1991.

Brock TD, Madigan MT. Biology of Microorganisms. 6th ed. Englewood Cliffs, New Jersey: Prentice Hall; 1991.

Brown TA. Genetics: a molecular approach. 2nd ed. London: Chapman & Hall; 1992.

Camerino G, Goodfellow PN, eds. Genetics of disease. Current Opin Genet Develop. 1994; 4:357–497.

Campbell NA. Biology. 2nd ed. Redwood City, California: Benjamin Cummings; 1990.

Cavalli-Sforza LL, Menozzi P, Piazza A. The History and Geography of Human Genes. Princeton: Princeton University Press; 1994.

Childs B, Holtzman NA, Kazazian HH Jr, Valle DL, eds. Molecular Genetics in Medicine. Progress in Medical Genetics. New Series, vol. 7. Elsevier, Amsterdam, 1988.

Cold Spring Harbor Symposia on Quantitative Biology: Molecular Biology of Homo Sapiens vol. LI. Cold Spring Harbor Laboratory. 1986.

Darnell J, Lodish H, Baltimore D. Molecular Cell Biology. 2nd ed. New York: Scientific American Books, Freeman; 1990.

Davies KE, ed. Human Genetic Diseases: a practical approach. 2nd ed. Oxford: IRL Press; 1986.

Davies KE, Read AP. Molecular Basis of Inherited Disease, 2nd ed. Washington: IRL Press; 1992.

DeGrouchy J, Turleau C. Clinical Atlas of Human Chromosomes. 2nd ed. New York: John Wiley; 1984.

Duve C de. Die Zelle. Expedition in die Grundstruktur des Lebens. Heidelberg: Spektrum der Wissenschaft; 1984.

Emery AEH. Methodology in Medical Genetics: an introduction to statistical methods. 2nd ed. Edingburgh: Churchill Livingstone; 1986.

Emery AEH, Rimoin DL, Pyeritz RE, eds. Principles and Practice of Medical Genetics. 3rd ed. Edinburgh: Churchill Livingstone; (in press).

Fristrom JW, Clegg MT. Principles of Genetics. 2nd ed. San Francisco: WH Freeman; 1988.

Gardener EJ, Simmons MJ, Snustad DP. Principles of Genetics. 8th ed. New York: John Wiley; 1991.

Gelehrter TD, Collins FS. Principles of Medical Genetics. Baltimore: Williams & Wilkins; 1990.

Hentze MW, Kulozik AE, Bartram CR. Einführung in die medizinische Molekularbiologie: Grundlagen, Klinik, Perspektiven. Heidelberg: Springer; 1990.

Holtzman NA. Proceed with Caution: predicting genetic risks in the recombinant DNA era. Baltimore: Johns Hopkins University Press; 1989.

Kaplan JC, Delpeck M. Biologie moleculaire et Médicine. Médicine-Sciences. Paris: Flammarion; 1990.

King RA, Rotter JI, Motulsky AG, eds. The Genetic Basis of Common Diseases. Oxford: University Press; 1992.

Knippers R, Philippsen P, Schäfer KP, Fanning E. Molekulare Genetik. 5th ed. Stuttgart: Thieme; 1990.

Lawrence PH. The Making of a Fly: genetics of animal design. Oxford: Blackwell; 1992.

Leder P, Clayton DA, Rabenstein E. Introduction to Molecular Medicine. New York: Scientific American Books; 1994.

Lewin B. Genes V. Oxford: Oxford University Press; 1994.

Lindsten J, Pettersson U, eds. Etiology of Human Disease at the DNA Level. Nobel Symposium 80. New York: Raven Press; 1991.

Mange AP, Mange PJ. Genetics: human aspects. Sunderland, Massachusetts: Sinauer Associates; 1990.

McKusick VA. Mendelian Inheritance in Man: a catalog of human genes and genetic disorders. 11th ed. Baltimore: Johns Hopkins University Press; 1994.

Passarge E. Elemente der Klinischen Genetik: Grundlagen und Anwendung der Humangenetik in Studium und Praxis. Stuttgart: Fischer; 1979.

Passarge E. Genetik. In: Siegenthaler W, ed. Klinische Pathophysiologie. 7th ed. Stuttgart: Thieme; 1994:1–51.

Rees AR, Sternberg JE. From Cells to Atoms: an illustrated introduction to molecular biology. Oxford: Blackwell; 1984.

Rooney DE, Czepulkowski BH. Human Cytogenetics: a practical approach. 2nd ed. Oxford: IRL Press; 1992.

Rothwell NV. Understanding Genetics: a molecular approach. New York: Wiley-Liss; 1993.

Scriver CR, Beaudet CR, Sly WS, Valle D, eds. The Metabolic and Molecular Bases of Inherited Disease. 7th ed. New York: McGraw-Hill; 1995.

Singer M, Berg P. Genes and Genomes: a changing perspective. Oxford: Blackwell; 1991.

Stent GS, Calendar R. Molecular Genetics: an introductory narrative. 2nd ed. San Francisco: WH Freeman; 1978.

Stevenson RE, Hall JG, Goodman RM, eds. Human Malformations and Related Anomalies. Oxford: University Press; 1993.

Strachan T. The Human Genome. Oxford: Bios; 1992.

Stryer L. Biochemistry. 4th ed. New York: WH Freeman; 1995.

Suzuki DT, Griffiths AGF, Miller JH, Lewontin RC. An Introduction to Genetic Analysis. 3rd ed. New York: WH Freeman; 1986.

Therman E, Susman M. Human Chromosomes: structure, behavior, effects. 3rd ed. Heidelberg: Springer; 1993.

Traut W. Chromosomen: Klassische und molekulare Cytogenetik. Heidelberg: Springer; 1991.

Verma RS, Babu A. Human Chromosomes: manual of basic techniques. New York: Pergamon Press; 1989.

Vogel F, Motulsky AG. Human Genetics: problems and approaches. 2nd ed. Heidelberg: Springer; 1986.

Watson JD, Hopkins NH, Roberts JW, Steitz JA, Weiner AM. Molecular Biology of the Gene. 4th ed. Menlo Park, California: Bejamin/Cummings; 1987.

Watson J, Gilman M, Witkowski J, Zoller M. Recombinant DNA. 2nd ed. Scientific American Books, New York: WH Freeman; 1992.

Weatherall DJ. The New Genetics and Clinical Practice. 3rd ed. Oxford: Oxford University Press; 1991.

Weinberg RA, ed. Oncogenes and the Molecular Origins of Cancer. Cold Spring Harbor Laboratory Press; 1989.

Whitehouse HLK. Towards an Understanding of the Mechanisms of Heredity. 3rd ed. London: Edward Arnold; 1973.

Winnacker EL. Gene und Klone. Eine Einführung in die Gentechnologie. 2nd ed. Weinheim: Chemie; 1990.

B. References for Individual Tables

Odorant Receptor Family (p. 234)

Buck L, Axel R. A novel multigene family may encode odorant receptors: a molecular basis for odor recognition. Cell. 1991; 65:175–187.

Chess A, Femon I, Cedar H, Axel R. Allelic inactivation regulates olfactory receptor gene expression. Cell. 1994; 78:823–834.

Ngai J, et al. The family of genes encoding odorant receptors in the channel catfish. Cell. 1993; 72:657–666.

Ngai J, et al. Coding of olfactory information: topography of odorant receptor expression in the catfish olfactory epithelium. Cell. 1993; 72:667–680.

Parmentier M, et al. Expression of members of the putative olfactory receptor gene family in mammalian germ cells. Nature. 1992; 355:453–455.

Intestinal Tumors (p. 272)

Bronner CE, Baker SM, Morrisson PT, et al. Mutation in the DNA mismatch repair gene homologue *hMLH1* is associated with hereditary nonpolyposis colon cancer. Nature. 1994; 368:258–261.

Caspari W, et al. Familial adenomatous polyposis: mutations at codon 1309 and early onset of colon cancer. Lancet. 1994; 343:629–632.

Fearon ER, et al. Identification of a chromosome 18q gene that is altered in colorectal cancers. Science. 1990; 247:12–13.

Groden J, et al. Identification and characterization of the familial adenomatous polyposis coli gene. Cell. 1991; 66:589–600.

Ichii S, et al. Inactivation of both APC alleles in an early stage of colon adenomas in a patient with familial adenomatous polyposis (FAP). Hum Molec Genet. 1992; 6:387–390.

Joslyn G, et al. Identification of deletion mutations and three new genes at the familial polyposis locus. Cell. 1991; 66:601–613.

Kinzler KW, Vogelstein B. Colorectal tumors. In: Scriver CR, et al. eds. The Metabolic and Molecular Bases of Disease. 7th ed. New York; McGraw-Hill; 1995: 643–663.

Kinzler KW, et al. Identification of a gene located at chromosome 5q21 that is mutated in colorectal cancers. Science. 1991; 251:1366–1369.

Powell SM. APC mutations occur early during colorectal tumorigenesis. Nature. 1992; 359:235–237.

Papadopoulos N, Nicolaides NC, Wei YF, et al. Mutation of a mutL homolog in hereditary colon cancer. Science. 1994; 263:1625–1629.

Rustgi AK. Hereditary gastrointestinal polyposis and nonpolyposis syndromes. New Engl J Med. 1994; 331:1694–1702.

Hemoglobin and Thalassemia (pp. 278–288)

Antonarakis SE, Kazazian HH Jr, Orkin SH. DNA polymorphism and molecular pathology of the human globin gene clusters. Hum Genet. 1985; 69:1–14.

Gibbons RJ, et al. Mutations in a putative global transcriptional regulator cause X-linked mental retardation with α-thalassemia (ATR-X syndrome). Cell. 1995; 80:837–845.

Lehmann H, Huntsman RG. Man's Hemoglobins. Amsterdam: North-Holland; 1974.

Orkin SH, Kazazian HH. The mutation and polymorphism of the human beta-globin gene and its surrounding DNA. Ann Rev Genet. 1984; 8:131–171.

Stamatoyannopoulos G, Niehuis AW, Leder P, Majerus PW, eds. The Molecular

Basis of Blood Diseases. 2nd ed. Philadelphia: WB Saunders; 1994.

Wainscoat JS, et al. Evolutionary relationships of human populations from an analysis of nuclear DNA polymorphisms. Nature. 1986; 319:491–493.

Weatherall DJ, Clegg JB, Higgs DR, Wood WG. The hemoglobinopathies. In: Scriver CR, Beaudet AL, Sly WS, Valle D, eds. The Metabolic an Molecular Bases of Inherited Disease. New York: McGraw-Hill; 1995:3417–3484.

Lysosomes and Lysosomal Enzyme Defects (pp. 292–296)

Kornfeld S, Sly WS. I-cell disease and pseudo-Hurler-polydystrophy: Disorders of lysosomal enzyme phosphorylation and localisation. In: Scriver CR, Beaudet AL, Sly WS, Valle D, eds. The Metabolic and Molecular Bases of Inherited Disease. 7th ed. New York: McGraw-Hill; 1995:2495–2508.

Neufeld EF, Muenzer J. The mucopolysaccharidoses. In: Scriver CR, Beaudet AL, Sly WS, Valle D, eds. The Metabolic and Molecular Bases of Inherited Disease. New York: McGraw-Hill; 1995:2485–2494.

Passarge E, et al. Krankheiten infolge genetischer Defekte in lysosomalen Mucopolysaccharid-Speicherkrankheiten. Dtsch Med Wochenschr. 1974; 99:144–158.

Sabatini DD, Adesnik MB. The biogenesis of membranes and organelles. In: Scriver CR, Beaudet AL, Sly WS, Valle D, eds. The Metabolic and Molecular Bases of Inherited Disease. 7th ed. New York: McGraw-Hill; 1995:459–553.

Hypercholesterolemia and LDL Receptors (pp. 298–300)

Brown MS, Anderson ROW, Goldstein JL. Recycling receptors: the round trip itinerary of migrant membrane proteins. Cell. 1983; 32:663.

Goldstein JL, et al. Receptor-mediated endocytosis: concepts emerging from the LDL receptor system. Ann Rev Cell Biol. 1985; 1:1.

Goldstein JL, Brown MS. Familial hypercholesterolemia. In: Scriver CR, Beaudet AL, Sly WS, Valle D, eds. The Metabolic and Molecular Bases of Inherited Disease. 7th ed. New York: McGraw-Hill; 1995:1981–2030.

Schuster H, et al. Familial defective apolipoprotein B-100: comparison with familial hypercholesterolemia in 18 cases detected in Munich. Arteriosclerosis. 1990; 10:577–581.

Südhof TC, et al. The LDL receptor gene: a mosaic of exons shared with different proteins. Science. 1985; 228:815–822.

Protease Inhibitors (p. 302)

Carrell RW, et al. Structure and variation of human α_1-antitrypsin. Nature. 1982; 298:329–333.

Cox DW, Woo SLC, Mansfield T. DNA restriction fragments associated with α_1-antitrypsin indicate a single origin for deficiency allele PiZ. Nature. 1985; 316:79–81.

Cox DW. Alpha 1-antitrypsin deficiency. In: Scriver CR, Beaudet AL, Sly WS, Valle D, eds. The Metabolic and Molecular Bases of Inherited Disease. 7th ed. New York: McGraw-Hill; 1995:4125–4158.

Lancet Editorial: Alpha 1-antitrypsin, Z, and the liver. Lancet 1992; 340:402–403.

Lomas DA, et al. The mechanism of Z α_1-antitrypsin accumulation in the liver. Nature. 1992; 357:605–607.

Owen MC, et al. Mutation of antitrypsin to antithrombin. Alpha 1-antitrypsin Pittsburgh (358, MET → ARG), a fatal bleeding disorder. N Engl J Med. 1983; 309:694–698.

Blood Coagulation Factor VII and von Willebrand Factor (pp. 304–306)

Kazazian HH Jr, et al. Hemophilia A and parahemophilia: deficiencies of coagulation factors VIII and V. In: Scriver CR, et al. eds. The Metabolic and Molecular Bases of Inherited Disease. 7th ed. New York: McGraw-Hill; 1995:3241–3267.

Manusco DJ, et al. Structure of the gene for human von Willebrand factor. J Biol Chem. 1989; 264:19 514–19 527.

Sadler JE. Von Willebrand disease. In: Scriver CR, Beaudet AL, Sly WS, Valle D, eds. The Metabolic and Molecular Bases of Inherited Disease. 7th ed. New York: McGraw-Hill; 1995:3269–3287.

Vehar GA, et al. Factor VIII and factor V: biochemistry and pathophysiology. In: Scriver CR, Beaudet AL, Sly WS, Valle D, eds. The Metabolic Basis of Inherited Disease. New York: McGraw-Hill; 1989: 2155–2170.

Dystrophin and Duchenne Muscular Dystrophy (pp. 320–322)

Ahu AW, Kunkel LM. The structural and functional diversity of dystrophin. Nature Genet. 1993; 3:283–291.

Becker PS, Lux SE. Hereditary spherocytosis and hereditary elliptocytosis. In: Scriver CR, et al. eds. The Metabolic and Molecular Bases of Inherited Diseases. 7th ed. New York: McGraw-Hill; 1995: 3513–3560.

Davies KA, Lux SE. Hereditary disorders of the red cell membrane skeleton. Trends Genet. 1989; 5:222–227.

Ervasti JM, Campbell KP. Membrane organization of the dystrophin-glycoprotein complex. Cell. 1991; 66:1121–1131.

Harper PS. Myotonic dystrophy and other autosomal muscular dystrophies. In: Scriver CR, Beaudet AL, Sly WS, Valle D, eds. The Metabolic and Molecular Bases of Inherited Disease. 7th ed. New York: McGraw-Hill; 1995:4227–4251.

Hilton-Jones D, Squier MV. Muscular dystrophy. Lancet. 1993; 341:528–529.

Koenig M, Monaco AP, Kunkel LM. The complete sequence of dystrophin predicts a rod-shaped cytoskeletal protein. Cell. 1988; 53:219–228.

Luna EJ, Hitt AL. Cytoskeleton plasma membrane interactions. Science. 1992; 258:255–264.

Monaco AP, Kunkel LM. Cloning of the Duchenne/Becker muscular dystrophy locus. Adv Hum Genet. 1988; 17:61–98.

Mutsumura K, et al. Abnormal expression of dystrophin-associated proteins in Fukuyama-type congenital muscular dystrophy. Lancet. 1993; 341:521–522.

Tinsley JM, et al. Increasing complexity of the dystrophin-associated protein complex. Proc Natl Acad Sci. 1994; 91:8307–8313.

Tennyson CN, Klamut HJ, Worton RG. The human dystrophin gene requires 16 hours to be transcribed and is cotranscriptionally spliced. Nature Genet. 1995; 9:184–190.

Worton RG, Brooke MH. The X-linked muscular dystrophies. In: Scriver CR, et al. eds. The Metabolic and Molecular Bases of Inherited Disease. 7th ed. New York: McGraw-Hill; 1995:4195–4226.

Sex Determination and Sex Differentiation (pp. 324–330)

Affara N, Chalmers IJ, Ferguson-Smith MA. Analysis of the SRY gene in 22 sex-reversed XY females identifies four new mutations in the conservative DNA binding domain. Hum Molec Genet. 1993; 2:785–789.

Chandley AC, Cooke HJ. Human male fertility – Y-linked genes and spermatogenesis. Hum Mol Genet. 1994; 3:1449–1452.

Donahue PA, Parker K, Migeon CJ. Congenital adrenal hyperplasias. In: Scriver CR et al. eds. The Metabolic and Molecular Bases of Inherited Disease. 7th ed. New York: McGraw-Hill; 1995:2929–2966.

Ferguson-Smith MA, Goodfellow PN. SRY and primary sex-reversal syndromes. In: Scriver CR, et al. eds. The Metabolic and Molecular Bases of Inherited Disease. 7th ed. New York: McGraw-Hill; 1995: 739–748.

Griffin JE, McPhaul MJ, Russell DW, Wilson JD. The androgen resistance syndromes: Steroid 5α-reductase 2 deficiency, testicular feminization, and related disorders. In: Scriver CR, Beaudet AL, Sly WS, Valle D, eds. The Metabolic and Molecular Bases of Inherited Disease. 7th ed. New York: McGraw-Hill; 1995:2967–2998.

Hawkins JR. Sex determination. Hum Mol Genet. 1994; 3:1463–1467.

Kremer H, et al. Male pseudohermaphroditism due to a homozygous missense mutation of the luteinizing hormone receptor gene. Nature Genet. 1995; 9:160–164.

McElreavy K, et al. XY sex reversal associated with a deletion 5′ to the SRY "HMG box" in the testes-determining region. Proc Natl Acad Sci. 1992; 89:11 016–11 021.

Passarge E. Geschlechtsdifferenzierung. In: Bettendorf G, Breckwoldt M, eds. Reproduktionsmedizin. Stuttgart: Fischer; 1989:139–151.

Passarge E, Gillessen-Kaesbach G. Intersexualität. In: Stolecke H, ed. Endokrinologie des Kindes- und Jugendalters. 2nd ed. Heidelberg: Springer; 1992:711–734.

Su H, Lau YF. Identification of the transcriptional unit, structural organization, and promoter sequence of the human sex-determining region Y (SRY) gene, using a reverse genetic approach. Am J Hum Genet. 1993; 52:24–38.

Willard HF. The sex chromosomes and X chromosome inactivation. In: Scriver CR, et al. eds. The Metabolic and Molecular Bases of Inherited Disease. 7th ed. New York: McGraw-Hill; 1995:719–737.

Wolf U, Schempp W, Scherer G. Molecular biology of the human Y chromosome. Rev Physiol Biochem Pharmacol. 1992; 121:148–213.

Cystic Fibrosis (p. 332)

Collins FS. Cystic fibrosis: molecular biology and therapeutic implications. Science. 1992; 256:774–779.

Kerem B, et al. Identification of the cystic fibrosis gene: genetic analysis. Science. 1989; 245:1073–1080.

Miller C. Sickly channels in mild disease. Nature. 1993; 362:106.

Riordan JR, et al. Identification of the cystic fibrosis gene: cloning and characterization of complementary DNA. Science. 1989; 245:1066–1073.

Rommens JM, et al. Identification of the cystic fibrosis gene: chromosome walking and jumping. Science. 1989; 245:1059–1065.

Sheppard DN, et al. Mutations in CFTR-associated with mild-disease form Cl channels with altered pore properties. Nature. 1993; 362:160–164.

Tizzano EF, Buchwald M. Cystic fibrosis: beyond the gene to therapy. J Pediatr. 1992; 120:337–349.

Tsui LC. The spectrum of cystic fibrosis mutations. Trends Genet. 1992; 8:392–398.

Welsh MJ, et al. Cystic fibrosis. In: Scriver CR, et al. eds. The Metabolic and Molecular Bases of Inherited Disease. 7th ed. New York: McGraw-Hill; 1995:3799–3876.

Unstable Trinucleotide Repeats (p. 334)

Caskey CT, et al. Triplet repeat mutations in human disease. Science. 1992; 256: 784–789.

Harper PS. Myotonic dystrophy and other autosomal muscular dystrophies. In: Scriver CR, et al. eds. The Metabolic and Molecular Bases of Inherited Disease. 7th ed. New York: McGraw-Hill; 1995:4227–4251.

Hayden MR, Kremer B. Huntington disease. In: Scriver CR, et al. eds. The Metabolic and Molecular Bases of Inherited

Disease. 7th ed. New York: McGraw-Hill; 1995:4483–4510.

Kawaguchi Y, et al. CAG expansions in a novel gene for Machado-Joseph disease at chromosome 14q32.1. Nature Genet. 1994; 8:221–228.

Mandel JH. Questions of expansion. Nature Genet. 1993; 4:8–9.

Morell V. The puzzle of the triplet repeats. Science. 1993; 260:1422–1423.

Richards RI, Sutherland GR. Dynamic mutations causing human disease. Cell. 1992; 70:709–712.

Richards RI, Sutherland GR. Heritable unstable DNA sequences. Trends Genet. 1992; 8:249–253.

Riggins GJ, et al. Human genes containing polymorphic trinucleotide repeats. Nature Genet. 1992; 2:186–191.

Willems PJ. Dynamic mutations hit double figures. Nature Genet. 1994; 8:213–215.

Fragile X Syndrome (p. 336)

Eichler EE, et al. Fine structure of the FMR1 gene. Hum Molec Genet. 1993; 2:1147–1153.

Nussbaum RL, Ledbetter DH. The fragile X syndrome. In: Scriver CR, et al. eds. The Metabolic and Molecular Bases of Inherited Disease. 7th ed. New York: McGraw-Hill; 1995:795–810.

Oberle I, et al. Instability of a 550-base pair DNA segment and abnormal methylation in fragile X syndrome. Science. 1991; 252:1097–1102.

Parrish JE, et al. Isolation of a GCC repeat showing expansion in FRAXF, a fragile site distal to FRAXA and FRAXE. Nature Genet. 1994; 8:229–235.

Rousseau F, et al. A multicenter study on genotype, phenotype correlations in the fragile X syndrome, using direct diagnosis with probe StB12.3: the first 2,253 cases. Am J Hum Genet. 1994; 55:225–237.

Increased Chromosomal Instability (p. 338)

German J, Passarge E. Bloom's syndrome. XII. Report from the registry for 1987. Clin Genet. 1989; 35:57–69.

German J. Bloom's syndrome: a Mendelian prototype of somatic mutational disease. Medicine. 1993; 72:393–406.

Passarge E. Bloom's syndrome: the German experience. Ann Génét. 1991; 34:179–197.

Schroeder-Kurth TM, Auerbach AD, Obe G, eds. Fanconi Anemia. Heidelberg: Springer; 1989.

Stratheie CA, et al. A cloning of cDNAs for Fanconi's anemia by functional complementation. Nature. 1992; 356:763–767.

Prader-Willi and Angelman Syndromes (p. 340)

Buiting K, et al. Microdissection of the Prader-Willi syndrome chromosome region and identification of potential gene sequences. Genomics. 1990; 6:521–527.

Buiting K, et al. Inherited microdeletions in the Angelman and Prader-Willi syndromes define an imprinting centre on human chromosome 15. Nature Genet. 1995; 9:395–400.

Dittrich B, et al. Molecular diagnosis of the Prader-Willi and Angelman syndromes by detection of parent-of-origin specific DNA methylation in 15q11–13. Hum Genet. 1992; 90:313–315.

Nicholls RD. Genomic imprinting and uniparental disomy in Angelman and Prader-Willi syndromes: a review. Am J Med Genet. 1993; 46:16–25.

Reis A, et al. Imprinting mutations suggested by abnormal DNA methylation patterns in familial Angelman and Prader-Willi syndromes. Am J Hum Genet. 1994; 54:741–747.

Sapienza C, Hall JG. Genetic imprinting in human disease. In: Scriver CR, et al. eds. The Metabolic and Molecular Bases of Inherited Disease. 7th ed. New York: McGraw-Hill; 1995:437–458.

The Dynamic Genome (p. 354)

Federoff NV. Transposable genetic elements in maize. Scient Amer. 1984; 250:65–74.

Federoff NV. Maize transposable elements. Perspect Biol Med. 1991; 35:2–19.

Federoff NV, Botstein D, eds. The Dynamic Genome: Barbara McClintock's ideas in the century of genetics. New York: Cold Spring Harbor Laboratory Press; 1992.

Fox Keller E. A Feeling for the Organism: the life and work of Barbara McClintock. San Francisco: WH Freeman; 1983.

Schwartz RS. Jumping genes. N Engl J Med. 1995; 332:941–944.

Gene Mapping (p. 358–360)

Buetow KH, et al. Integrated human genome-wide maps constructed using the CEPH reference panel. Nature Genet. 1994; 6:391–393.

Davisson MT, Lalley PA, Peters J, Doolittle DP, Hillyard AL, Searle AG. Report of the comparative subcommittee for human and mouse homologies. Cytogenet Cell Genet. 1990; 55:434–456.

Fain PR, et al. A 2D crossover-based map of the human X chromosome as a model for map integration. Nature Genet. 1995; 9:261–266.

Green ED, Cox DR, Myers RM. The human genome project and its impact on the study of human disease. In: Scriver CR, et al. eds. The Metabolic and Molecular Bases of Inherited Disease. 7th ed. New York: McGraw-Hill; 1995:401–436.

Lavale SH, Boyd Y. Novel sequences conserved on the human and mouse X chromosomes. Genomics. 1993; 15:483–491.

O'Brien SJ, et al. Anchored reference loci for comparative genome mapping in mammals. Nature Genet. 1993; 3:103–112.

Tugendreich S, et al. Genes conserved in yeast and humans. Hum Mol Genet. 1994; 3:1509–1517.

Glossary

Definitions of Genetic Terms

Acentric – refers to a chromosome or chromatid without a centromere.

Acrocentric (White, 1945) – refers to a chromosome with a centromere that lies very close to one of the ends, dividing the chromosome into a long and a very short arm.

Allele (Johannsen, 1909) or **allelomorph** (Bateson and Saunders, 1902) – one of several alternative forms of a gene at a given gene locus.

Alu sequences – a family of related DNA sequences, each about 300 base pairs long and containing the restriction site for the Alu restriction enzyme; about 500 000 copies of Alu sequences are dispersed throughout the human genome.

Amber codon – the stop codon UAG.

Ames test – a mutagenicity test carried out with a mixture of rat liver and mutant bacteria.

Aminoacyl tRNA – a transfer RNA carrying an amino acid.

Amplification – production of additional copies of a given DNA sequence.

Anaphase (Strasburger, 1884) – a stage of mitosis and of meiosis I and II. Characterized by the movement of homologous chromosomes (or sister chromatids) toward opposite poles of the cell division spindle.

Aneuploid (Täckholm, 1922) – deviation from the normal number of chromosomes.

Aneusomy – deviation from the normal presence of homologous chromosomal segments. Aneusomy by recombination refers to the duplication/deficiency resulting from crossing-over within an inversion (inverted region).

Annealing – to cause complementary single strands of nucleic acid to pair and form double-stranded molecules (DNA with DNA, RNA with RNA, or DNA with RNA).

Antibody – a protein (immunoglobin) that recognizes and binds to an antigen as part of the immune response.

Antigen – a substance with a molecular surface structure that triggers an immune response, i.e., the production of antibodies, and/or that reacts with (its) specific antibodies (antigen–antibody reaction).

Antisense RNA – an RNA strand that is complementary to mRNA. Thus it cannot be used as a template for normal translation. The term antisense is generally used to refer to a sequence of DNA or RNA that is complementary to mRNA.

Attenuator – a DNA sequence regulating the termination of transcription, involved in controlling the expression of some operons in bacteria.

Australopithecus – the genus of fossil Hominidae from Eurasia. Walked erect; brain size between that of modern man and other modern primates; large and massive jaw. Lived about 4–5 million years ago.

Autoradiography (Lacassagne and Lattes, 1924) – photographic detection of a radioactive substance incorporated by metabolism into cells or tissue. The distribution of the radioactively labeled substance can be demonstrated, e.g., in tissue, cells, or metaphase chromosomes by placing a photographic film or photographic emulsion in contact with the preparation.

Autosome (Montgomery, 1906) – any chromosome except a sex chromosome (the latter usually designated X or Y). Autosomal refers to genes and chromosomal segments that are located on autosomes.

Auxotrophic (Ryan and Lederberg, 1946) – refers to cells or cell lines that cannot grow on minimal medium unless a certain nutritive substance is added (compare with Prototrophic).

Backcross – cross of a heterozygous animal with one of its homozygous parents. In a double backcross, two heterozygous gene loci are involved.

Bacteriophage – a virus that infects bacteria. Usually abbreviated as phage.

Banding pattern (Painter, 1939) – staining pattern of a chromosome consisting of alternating light and dark transverse bands. Each chromosomal segment of homologous chromosomes shows the same specific banding pattern, characterized by the distribution and size of the bands, which can be used to identify that segment. The term was introduced by Painter in 1939 for the linear pattern of strongly and weakly staining bands in polytene chromosomes of certain diptera (mosquitoes, flies). Each band is defined relative to its neighboring bands. The sections between bands are interbands.

Barr body – see X chromatin.

Base pair (bp) – in DNA, two bases—one a purine, the other a pyrimidine—lying opposite each other and joined by hydrogen bonds. Normal base pairs are A with T and C with G in a DNA double helix. Other pairs can be formed in RNA.

B cells – B lymphocytes.

Bimodal distribution – refers to a frequency distribution curve with two peaks. If the frequency distribution curve of a population trait is bimodal, it is frequently evidence of two different phenotypes distinguished on a quantitative basis.

Bivalent (Haecker, 1892) – pairing configuration of two homologous chromosomes during the first meiotic division. As a rule the number of bivalents corresponds to half the normal number of chromosomes in diploid somatic cells. Bivalents are the cytogenetic prerequisite for crossing-over of nonsister chromatids. During meiosis, a trisomic cell forms a trivalent of the trisomic chromosomes.

Breakpoint – site of a break in a chromosomal alteration, e.g., translocation, inversion, or deletion.

CAT (or CAAT) box – a regulatory DNA sequence in the 5' region of eukaryotic genes; transcription factors bind to this sequence.

cDNA – complementary DNA synthesized by the enzyme reverse transcriptase from RNA as the template.

Cell cycle – life cycle of an individual cell. In dividing cells, the following four phases can be distinguished: G_1 (interphase), S (DNA synthesis), G_2, and mitosis. Cells that do not divide are said to be in the G_0 phase.

Cell hybrid – in culture, a somatic cell generated by fusion of two cells of different species. It contains the complete or incomplete chromosome complements of the parental cells. Cell hybrids are an important tool in gene mapping.

Centimorgan – a unit of length on a linkage map [100 centimorgans (cM) = 1 Morgan]. The distance between two gene loci in centimorgans corresponds to their recombination frequency expressed as percentage, i.e., one cM corresponds to one percent recombination frequency. Named after Thomas H. Morgan (1866–1945), who initiated the classical genetic experiments in drosophila.

Centromere (Waldeyer, 1903) – chromosomal region to which the spindle fibers attach during mitosis or meiosis. It appears as a constriction at metaphase.

Chaperone – a protein needed to assemble or fold another protein correctly.

Chiasma (Janssens, 1909) – cytologically recognizable region of crossing-over in a bivalent. In some organisms the chiasmata move toward the end of the chromosomes (terminalization of the chiasmata) during late diplotene and diakinesis (see meiosis). The average number of chiasmata in autosomal bivalents is about 52 in human males, somewhat less in females. The number of chiasmata in man was first determined in 1956 in a paper that confirmed the normal number of chromosomes in man (C. E. Ford and J. L. Hamerton, Nature. 178: 1020, 1956).

Chimera (Winkler, 1907) – an individual or tissue that consists of cells of different genotypes of prezygotic origin.

Chromatid (McClung, 1900) – longitudinal subunit of a chromosome resulting from chromosome replication; two chromatids are held together by the centromere and are visible during early prophase and metaphase of mitosis and between diplotene and the second metaphase of meiosis. Sister chromatids arise from the same chromosome; nonsister chromatids are the chromatids of homologous chromosomes. After division of the centromere in anaphase, the sister chromatids are referred to as daughter chromosomes. A chromatid break or a chromosomal aberration of the chromatid type affects only one of the two sister chromatids. It arises after the DNA replication cycle in the S phase (see Cell cycle). A break that occurs before the S phase affects both chromatids and is called an isolocus aberration (isochromatid break).

Chromatin (Flemming, 1882) – the stained material that can be observed in interphase nuclei. It is composed of DNA, basic chromosomal proteins (histones), nonhistone chromosomal proteins, and small amounts of RNA.

Chromomere (Wilson, 1896) – each of the linearly arranged thickenings of the chromosome visible in meiotic and under some conditions also in mitotic prophase. Chromomeres are arranged in chromosome-specific patterns.

Chromosome (Waldeyer, 1888) – the gene-carrying structures, which are composed of chromatin and are visible during nuclear division as threadlike or rodlike bodies. *Polytene chromosomes* (Koltzhoff 1934; Bauer 1935) are a special form of chromosomes in the salivary glands of some diptera larvae (mosquitoes, flies).

Chromosome walking – sequential isolation of overlapping DNA sequences in order to find a gene on the chromosome studied.

Cis/trans (Haldane, 1941) – in analogy to chemical isomerism, refers to the position of genes of double heterozygotes (heterozygotes at two neighboring gene loci) on homologous chromosomes. When two genes lie on the same chromosome, they are in the *cis* position. If they lie on different chromosomes, they are in the *trans* position. The cis/trans test (Lewis, 1951; Benzer, 1957) uses genetic methods (*genetic complementation*) to determine whether two mutant genes are in the cis or in the trans position. With reference to genetic coupling, the expressions cis and trans are analogous to the terms coupling and repulsion, respectively.

Cistron (Benzer, 1957) – a functional unit of gene effect as represented by the cis/trans test. If the phenotype is mutant with alleles in the cis position and the alleles do not complement each other (genetic complementation), they are considered alleles of the same cistron. If they complement each other they are considered to be nonallelic. This definition of Benzer was later expanded (Fincham 1959): accordingly, a cistron now refers to a segment of DNA that codes for a unit of gene product. Within a cistron, mutations in the trans position do not complement each other. Functionally, the term cistron can be equated with the term gene.

Clone (Webber, 1903) – a population of molecules, cells, or organisms that have originated from a single cell or a single ancestor and are identical with it and with each other.

Cloning efficiency – a measure of the clonability of individual mammalian cells in culture.

Cloning vector – a plasmid, phage, or yeast artificial chromosome (YAC) used to carry a foreign DNA fragment for the purpose of cloning (producing multiple copies of the fragment).

Coding strand of DNA – the strand of DNA that bears the same sequence as the RNA strand (mRNA) that is used as a template for translation (sense RNA). The other strand of DNA, which directs synthesis of the mRNA, is the template strand (see Antisense RNA).

Codominant – see Dominant.

Codon (Crick, 1963) – a sequence of three nucleotides (a triplet) in DNA or RNA that codes for a certain amino acid or for the terminalization signal of an amino acid sequence.

Complementation, genetic (Fincham, 1966) – complementary effect of (restoration of normal function by) double mutants at different gene loci. An example, genetic complementation for xeroderma pigmentosum, is illustrated on page 42.

Concatemer – an association of DNA molecules with complementary ends linked head to tail and repeated in tandem. Formed during replication of some viral and phage genomes.

Concordance – the occurrence of a trait or a disease in both members of a pair of twins (mono- or dizygotic).

Conjugation – the transfer of DNA from one bacterium to another.

Consanguinity – blood relationship. Two or more individuals are referred to as consanguineous (related by blood) if they have one or more ancestors in common. A quantitative expression of parental consanguinity is the inbreeding coefficient.

Consensus sequence – a corresponding or identical DNA sequence in different genes or organisms.

Contig – a series of overlapping DNA fragments (contiguous sequences).

Cosmid – a plasmid carrying the cos site of λ phage in addition to sequences required for division. Serves as a cloning vector for DNA fragments up to 40 kb.

Cos site – a restriction site required of a small strand of DNA to be cleaved and packaged into the λ phage head.

Coupling (Bateson, Saunders, Punnett, 1905) – cis configuration of double heterozygotes.

Crossing-over (Morgan and Cattell, 1912) – the exchange of genetic information between two homologous chromosomes by chiasma formation in the diplotene stage of meiosis I; leads to genetic recombination of neighboring (linked) gene loci.
Crossing-over may also occur in somatic cells during mitosis (Stern, 1936). Unequal crossing-over (Sturtevant, 1925) results from mispairing of the homologous DNA segments at the recombination site. It results in structurally altered DNA segments or chromosomes, with a duplication in one and a deletion in the other.

Cyclin – a protein involved in cell cycle regulation.

Cytoplasmic inheritance – transmission of genetic information located in mitochondria. Since sperm cells do not contain mitochondria, the transmitted information is of maternal origin.

Cytoskeleton – network of stabilizing protein in the cytoplasm and cell membrane.

Dalton – a unit of atomic mass, equal to 1/12 the mass of the ^{12}C nuclide (1.657×10^{-24} gm).

Deficiency (Bridges, 1917) – loss of a chromosomal segment resulting from faulty crossing-over, e.g., by unequal crossing-over or by crossing-over within an inversion or within a ring chromosome. It arises at the same time as a complementary duplication. This is referred to as duplication/deficiency.

Deletion (Painter and Muller, 1929) – loss of part of or a whole chromosome or loss of DNA nucleotide bases.

Denaturation of DNA or RNA – separation of double-stranded nucleic acid molecules into single strands. Rejoining of the complementary single strands is referred to as renaturation.

Diakinesis (Haecker, 1897) – a stage during late prophase I of meiosis.

Dicentric (Darlington, 1937) – refers to a structurally altered chromosome with two centromeres.

Dictyotene – a stage of fetal oocyte development during which meiotic prophase is interrupted. In human females, oocytes attain the stage of dictyotene about 4 weeks before birth; further development of the oocytes is arrested until ovulation, at which time meiosis is continued.

Diploid (Strasburger, 1905) – cells or organisms that have two homologous sets of chromosomes, one from the father (paternal) and one from the mother (maternal).

Diplotene – a stage of prophase I of meiosis.

Discordance – the occurrence of a given trait or disease in only one member of a pair of twins.

Disomy, uniparental (UPD) – presence of two chromosomes of a pair from only one of the parents. One distinguishes UPD due to isodisomy, in which the chromosomes are identical, and heterodisomy, in which they are homologous.

Dispermy – the penetration of a single ovum by two spermatozoa.

Dizygotic – twins derived from two different zygotes (fraternal twins), as opposed to monozygotic (identical) twins, derived from the same zygote.

D loop – a region in mitochondrial DNA where one strand is paired with RNA, which displaces the other strand.

DNA (deoxyribonucleic acid) – the molecule containing the primary genetic information in the form of a linear sequence of nucleotides in groups of threes (triplets) (see codon).
Satellite DNA (sDNA) (Sueoka, 1961; Britten & Kohne, 1968) contains tandem repeats of nucleotide sequences of different lengths. sDNA can be separated from the main DNA by density gradient centrifugation in a cesium chloride solution, after which it appears as one or several bands (satellites) separate from that of the main body of DNA. In eukaryotes, light (AT-rich) and heavy (GC-rich) satellite DNA can be distinguished.
Microsatellites are small (2–5) tandem repeats of DNA nucleotides; minisatellites are tandem repeats of about 10–20 base pairs; classical satellite DNA consists of large repeats of 100–6500 bp (see p. 353).

DNA polymerase – a DNA-synthesizing enzyme. To begin synthesis, it requires a primer of RNA or a complementary strand of DNA.

DNase (deoxyribonuclease) – an enzyme that digests DNA.

Domain – a distinctive region of the tertiary structure of a protein or a particular region of a chromosome.

Dominant (Mendel, 1865) – refers to a genetic trait that can be observed in the heterozygous state. The terms "dominant" and "recessive" refer to the effects of the alleles at a given gene locus. The effects observed depend in part on the accuracy of observation. When the effects of each of two different alleles at a (heterozygous) locus can be observed, the alleles are said to be codominant. At the DNA level, allelic genes at two homologous loci are codominant.

Downstream – the 3' direction of a gene.

Drift, genetic (Wright, 1921) – random changes in gene frequency of a population. Especially relevant in small populations, where random differences in the reproductive frequency of a certain allele can change the frequency of the allele. Under some conditions an allele may disappear completely from a population (loss) or be present in all individuals of a population (fixation).

Duplication (Bridges, 1919) – addition of a chromosomal segment resulting from faulty crossing-over (see deficiency). It may also refer to additional DNA nucleotide base pairs. Duplication of genes (gene duplication) played an important role in the evolution of eukaryotes.

Electrophoresis (Tiselius, 1937) – separation of molecules by utilizing their different speeds of migration in an electrical field. As support medium, substances in gel form such as starch, agarose, acrylamide, etc., are used. Further molecular differences can be detected by modifications such as two-dimensional electrophoresis (electric field rotated 90° for the second migration) or cessation of migration at the isoelectric point (isoelectric focusing).

Elongation – addition of amino acids to a polypeptide chain.

Elongation factor – one of the proteins that associate with ribosomes while amino acids are added; EF in prokaryotes and eEF in eukaryotes.

Endocytosis – specific uptake of extracellular material at the cell surface. The material is surrounded by an invagination of the cell membrane, which pinches off to form a membrane-bound vesicle containing the material.

Endonuclease – one of a heterogeneous group of enzymes that cleave bonds between nucleotides of single- or double-stranded DNA or of RNA.

Endoplasmic reticulum – a complex system of membranes in the cytoplasm.

Endoreduplication (Levan and Hauschka, 1953) – chromosome replication during interphase without actual mitosis. Endoreduplicated chromosomes in metaphase consist of four chromatids lying next to each other, held together by two neighboring centromeres.

Enhancer – a cis-acting regulatory DNA segment that contains binding sites for transcription factors. An enhancer is located at various distances from the promoter. It causes an increase in the rate of transcription (about tenfold).

Enzyme (E. Büchner, 1897) – a protein that catalyzes a biochemical reaction. Enzymes consist of a protein part (apoenzyme), responsible for the specificity, and a nonprotein part (coenzyme), needed for activity. Enzymes bind to their substrates, which become metabolically altered or combined with other substances during the train of the reaction. Most of the enzymatically catalyzed chemical reactions can be classified into one of six groups:
(1) hydrolysis (cleavage with the addition of H_2O), by *hydrolases*;
(2) transfer of a molecular group from a donor to a receptor molecule, by *transferases*;
(3) oxidation and reduction, by *oxidases* and *reductases* (transfer of one or more electrons or hydrogen atoms from a molecule to be oxidized to another molecule that is to be reduced);
(4) isomerization, by *isomerases* (rearranging the position of an atom or functional group within a molecule);
(5) joining of two substrate molecules to form a new molecule, by *ligases* (*synthetases*);
(6) nonhydrolytic cleavage with formation of a double bond on one or both of the two molecules formed, by *lyases*.

Epigenetic influence – a factor that changes the phenotype without altering the genotype.

Episome (Jacob and Wollman, 1958) – a plasmid that can exist either independently in the cytoplasm or as an integrated part of the genome of its bacterial host.

Epistasis (Bateson, 1907) – interaction of genes at the same gene locus (allelic) or at different gene loci (nonallelic) to alter phenotypic expression.

Euchromatin (Heitz, 1928) – chromosome or chromosomal segment that stains less intensely than heterochromatin. Euchromatin corresponds to the genetically active part of chromatin that is not fully condensed in the interphase nucleus.

Eukaryote (Chatton, 1925) – plants and animals with cells that have a chromosome-containing nucleus, which divides during mitosis and meiosis, in contrast to prokaryotes.

Euploid (Täckholm, 1922) – refers to cells, tissues, or individuals with the complete normal chromosomal complement characteristic of that species (compare with Aneuploid, Heteroploid, Polyploid).

Excision repair – repair of bulk lesions in DNA in which a stretch of nucleotides (about 14 in prokaryotes and about 30 in eukaryotes) is excised from the affected strand and replaced by the normal sequence (resynthesis).

Exocytosis – specific process by which nondiffusable particles are transported through the cell membrane to be discharged into the cellular environment.

Exon (Gilbert, 1978) – a segment of DNA that is represented in the mature mRNA of eukaryotes (compare with Intron).

Exonuclease – an enzyme that cleaves nucleotide chains at their terminal bonds only (cf. Endonuclease).

Expression – the observable effects of an active gene.

Expression vector – a cloning vector containing DNA sequences that can be transcribed and translated.

Expressivity (Vogt, 1926) – refers to the degree of phenotypic expression of a gene or genotype. Absence of expressivity is also called nonpenetrance.

Fibroblast – type of connective tissue cell. Can be propagated in culture flasks containing suitable medium (*fibroblast cultures*).

Fingerprint – a characteristic pattern of fragments of DNA or peptides.

Fitness, biological – refers to the probability (between 0.0 and 1.0) that a gene will be passed on to the next generation. For a given genotype and a given environment, the biological (or reproductive) fitness is determined by survival rate and fertility.

Founder effect – presence of a particular allele in a population due to a mutation in a single ancestor.

Gamete (Strasburger, 1877) – a haploid germ cell, either a spermatozoon (male) or an ovum (female). In mammals, males are heterogametic (XY) and females homogametic (XX). In birds, females are heterogametic (ZW) and males homogametic (ZZ).

Gene (Johannsen, 1909) – a hereditary factor that constitutes a single unit of hereditary material. It corresponds to a segment of DNA that codes for the synthesis of a single polypeptide chain (cf. Cistron).

Gene amplification (Brown and David, 1968) – selective production of multiple copies of a given gene without proportional increases of other genes.

Gene bank – a collection of cloned DNA fragments that together represent the genome they are derived from (gene library).

Gene cluster (Demerec and Hartman, 1959) – a group of two or more neighboring genes of similar function, e.g., the HLA system or the immunoglobulin genes.

Gene conversion (Winkler, 1930; Lindgren, 1953) – interaction of alleles that leads to unequal genetic exchange during meiosis. Gene conversion refers to a process of nonreciprocal transfer of genetic information. One gene serves as a sequence donor, remaining unaffected, while the other gene receives sequences and undergoes variation.

Gene dosage – refers to the quantitative degree of expression of a gene.

Gene family – a set of evolutionarily related genes by virtue of identity or great similarity of some of their coding sequences.

Gene flow (Berdsell, 1950) – transfer of an allele from one population to another.

Gene frequency – the frequency of a given allele at a given locus in a population (allele frequency).

Gene locus (Morgan, Sturtevant, Muller, Bridges, 1915) – the position of a gene on a chromosome.

Gene map – the position of gene loci on chromosomes. A physical map refers to the absolute position of gene loci, their distance from each other being expressed by the number of base pairs between them. A genetic map expresses the distance of genetically linked loci by their frequency of recombination.

Gene product – the polypeptide or ribosomal RNA coded for by a gene.

Genetic code – the information contained in the triplets of DNA nucleotide bases used to incorporate a particular amino acid into a gene product.

Genetic marker – a polymorphic genetic property that can be used to distinguish the parental origin of alleles.

Genetics (Bateson, 1906) – the science of heredity and the hereditary basis of organisms; derived from Gk. *genesis* (origin).

Genome (Winkler, 1920) – all of the genetic material of a cell or of an individual.

Genotype (Johannsen, 1909) – all or a particular part of the genetic constitution of an individual or a cell (compare with Phenotype).

Germinal – refers to germ cells, as opposed to somatic cells.

G6PD – glucose-6-phosphate dehydrogenase.

Gyrase – a topoisomerase that unwinds DNA.

Haploid (Strasburger, 1905) – refers to cells or individuals with a single chromosome complement; gametes are haploid.

Haplotype (Ceppellini et al., 1967) – a combination of alleles at two or more closely linked gene loci on the same chromosome, e.g., in the HLA system.

Helicase – an enzyme that unwinds and separates the two strands of the DNA double helix by breaking the hydrogen bonds during transcription or repair.

Helix–loop–helix – a structural motif in DNA-binding proteins, such as some transcription factors.

Hemizygous – refers to genes and gene loci that are present in only one copy in an individual, e.g., on the single X chromosome in male cells (XY) or because the homologous locus has been lost.

Heritability (Lush, 1950; Falconer, 1960) – the ratio of additive genetic variance to the total phenotypic variance. Phenotypic variance is the result of the interaction of genetic and nongenetic factors in a population.

Heterochromatin (Heitz, 1928) – a chromosome or chromosomal segment that remains darkly stained in interphase, early prophase, and late telophase because it remains condensed, as all chromosomal material is in metaphase. This contrasts with euchromatin, which becomes invisible during interphase. Heterochromatin corresponds to chromosomes or chromosome segments showing little or no genetic activity. *Constitutive* and *facultative* heterochromatin can be distinguished (Brown, 1966). An example of facultative heterochromatin is the heterochromatic X chromosome resulting from inactivation of one X chromosome in somatic cells of female mammals. An example of constitutive heterochromatin is the centromeric heterochromatin that can be demonstrated as C bands.

Heterodisomy – presence of two homologous chromosomes from one parent only (compare with Isodisomy and UPD).

Heteroduplex – refers to a region of a double-stranded DNA molecule with noncomplementary strands that originated from different duplex DNA molecules.

Heterogametic (Wilson, 1910) – producing two different types of gametes, e.g., X and Y in (male) mammals or ZW in female birds.

Heterogeneity, genetic (H. Harris, 1953; F. C. Fraser, 1956) – an apparently uniform phenotype caused by two or more different genotypes.

Heterokaryon (Ephrussi & Weiss, 1965; Harris & Watkins, 1965; Okada & Murayama, 1965) – a cell having two or more nuclei with different genomes.

Heteroploid (Winkler, 1916) – refers to cells or individuals with an abnormal number of chromosomes.

Heterosis (Shull, 1911) – increased reproductive fitness of heterozygous genotypes compared with the parental homozygous genotypes, in plants and animals.

Heterozygous (Bateson and Saunders, 1902) – having two different alleles at a given gene locus (compare with Homozygous).

Hfr cell – a bacterium that possesses DNA sequences that lead to a high frequency of DNA transfer at conjugation.

HGPRT – hypoxanthine-guanine-phosphoribosyl transferase. An enzyme in purine metabolism that is inactive in Lesch–Nyhan syndrome.

Histocompatibility – tissue compatibility. Determined by the major histocompatibility complex MHC (see HLA).

Histone (Kossel, 1884) – chromosome-associated protein of the nucleosome. Histones H1–H4 form the core of a nucleosome.

HLA (Dausset, Terasaki, 1954) – human leukocyte antigen system A.

Hogness box – a nucleotide sequence that is part of the promoter in eukaryotic genes.

Homeobox – a highly conserved DNA segment in homeotic genes.

Homeotic gene – one of the developmental genes in drosophila that can lead to the replacement of one body part by another by mutation.

Homologous – refers to chromosomes or parts of chromosomes containing the same gene loci.

Homozygosity mapping – identification of chromosomal regions of homozygosity in individuals born of consanguineous parents.

Homozygous (Bateson and Saunders, 1902) – having identical alleles at a given gene locus.

Hox genes – clusters of mammalian genes containing homeobox sequences. They are important in embryonic development.

Hybridization – cross between two genotypically different plants or animals belonging to the same species. The term is often used in more narrow definitions: fusion of two single complementary DNA strands (DNA/DNA hybridization), fusion of complementary DNA and RNA strands (DNA/RNA hybridization), or the in vitro fusion of cultured cells of different species (cell hybridization).

Immunoglobulin – an antigen-binding molecule.

Imprinting, genomic – different expression of an allele or chromosomal segment depending on the parental origin.

Inbreeding coefficient (Wright, 1929) – measure of the probability that two alleles at a gene locus of an individual are identical by descent, i.e., that they are copies of a single allele of an ancestor common to both parents (IBD, identity-by-descent). Also, the proportion of loci at which the individual is homozygous.

Incidence – the rate of occurrence of a disease in a population. In contrast, prevalence is the percentage of a population that is affected with a particular disease at a particular time.

Inducer – a molecule that induces the expression of a gene.

Initiation factor – a protein that associates with the small subunit of a ribosome when protein synthesis begins (IF in prokaryotes, eIF in eukaryotes).

Insertion – insertion of chromosomal material of nonhomologous origin into a chromosome without reciprocal translocation.

Insertion sequence (IS) – a small bacterial transposon.

Intercalating agent – a chemical compound that can occupy a space between two adjacent base pairs in DNA.

Intron (Gilbert, 1978) – a segment of noncoding DNA within a gene (compare with Exon). It is transcribed, but removed from the primary RNA transcript before translation.

Inversion (Sturtevant, 1926) – structural alteration of a chromosome through a break at two sites with reversal of direction of the intermediate segment and reattachment. A *pericentric inversion* includes the centromere in the inverted segment. A *paracentric inversion* does not involve the centromere. An inversion per se does not cause clinical signs, but it represents a potential genetic risk because crossing-over may occur in the region of the inversion and lead to aneusomy in offspring (aneusomy by recombination). Chromosomal inversions played an important role in evolution.

Inverted repeat – two identical, oppositely oriented copies of the same DNA sequence. They are an important feature of retroviruses.

Isochromatid break – see Chromatid.

Isochromosome (Darlington, 1940) – a chromosome composed of two identical arms connected by the centromere, e.g., two long or two short arms of an X chromosome. Implies duplication of the doubled arm and deficiency of the absent arm. An isochromosome may have one or two centromeres.

Isodisomy – presence of two identical chromosomes from one of the parents (compare with Heterodisomy).

Isolate, genetic (Waklund, 1928) – a physically or socially isolated population that has not interbred with individuals outside of that population.

Isotype – closely related chains of immunoglobulins.

Isozyme or isoenzyme (Markert and Möller, 1959) – one of multiple distinguishable forms of enzymes of similar function in the same organism. Isoenzymes are a biochemical expression of genetic polymorphism.

Karyotype (Levitsky, 1924) – the chromosome complement of a cell, an individual, or a species.

Lagging strand of DNA – the new strand of DNA replicating in the 3' to 5' direction. It is synthesized in short fragments in the 5' to 3' direction (Okazaki fragments), which are subsequently joined together.

Lampbrush chromosome (Rückert, 1892) – a special type of chromosome found in the primary oocytes of many vertebrates and invertebrates during the diplotene stage of meiotic division and in drosophila spermatocytes. The chromosomes show numerous lateral loops of DNA that are accompanied by RNA and protein synthesis.

Lariat – an intermediate form of RNA during splicing when a circular structure with a tail is formed by a 5' to 2' bond.

Leader sequence – a short N-terminal sequence of a protein that is required for directing the protein to its target.

Leaky mutant – a mutation causing only partial loss of function.

Leptotene – a stage of meiosis.

Lethal equivalent (Morton, Crow, and Muller, 1956) – a gene or combination of genes that in the homozygous state is lethal to 100% of individuals. This may refer to a gene that is lethal in the homozygous state, to two different genes that each have 50% lethality, to three different genes each with 33% lethality, etc. It is assumed that each individual carries about 5–6 lethal equivalents.

Lethal factor (Bauer, 1908; Hadorn, 1959) – an abnormality of the genome that leads to death in utero, e.g., numerous chromosomal anomalies.

Ligand – a molecule that can bind to a receptor and thereby induce a signal in the cell, e.g., a hormone.

LINE (long interspersed nuclear element) – long interspersed repetitive DNA sequences (see p. 352).

Linkage disequilibrium (Kimura, 1956) – nonrandom association of alleles at closely linked gene loci that deviates from their individual frequency as predicted by the Hardy–Weinberg equilibrium. Linkage disequilibrium is usually due to the founder effect.

Linkage, genetic (Morgan, 1910) – localization of gene loci on the same chromosome close enough to cause deviation from independent segregation.

Linkage group – gene loci on the same chromosome, which are so close together that they usually are inherited together without recombination.

Linker DNA – a synthetic DNA double strand that carries the recognition site for a restriction enzyme and that can bind two DNA fragments. Also, the stretch of DNA between two nucleosomes.

Locus – see Gene locus.

Lymphocyte – cell of the immune system, of one of two general types: B lymphocytes from the bone marrow and thymus-derived T lymphocytes.

Lysosome (deDuve et al., 1955) – small cytoplasmic organelle containing hydrolytic enzymes.

Map distance – distance between gene loci, expressed either in physical terms (cM, centimorgan) or genetic terms (recombination frequency).

Marker, genetic – an allele used to recognize a particular genotype.

Megabase (Mb) – one million base pairs.

Meiosis (Farmer and Moore, 1905) – The special division of a germ cell nucleus that leads to reduction of the chromosome complement from the diploid to the haploid state. Prophase of the first meiotic division is especially important and consists of the following stages: leptotene, zygotene, pachytene, diplotene, diakinesis.

Mendelian inheritance (Castle, 1906) – inheritance according to the laws of Mendel, as opposed to extrachromosomal inheritance under the control of cytoplasmic hereditary factors (mitochondrial DNA).

Metabolic cooperation (Subak–Sharpe et al., 1969) – correction of a phenotype in cells in culture by contact with normal cells or cell products. An example of metabolic correction is the cross correction of cultured cells of different mucopolysaccharide storage diseases or correction of HGPRT-deficient cells by normal cells.

Metacentric – refers to chromosomes that are divided by the centromere into two arms of approximately the same length.

Metaphase (Strasburger, 1884) – stage of mitosis in which the contracted chromosomes are readily visible.

MHC (major histocompatibility complex) (Thorsby, 1974) – the principal histocompatiblity system, consisting of class I and class II antigen genes of the HLA system and of class III antigen genes.

Mitosis (Flemming, 1882) – nuclear division during the division of somatic cells, consisting of prophase, metaphase, anaphase, and telophase.

Mitosis index (Minot, 1908) – the proportion of cells present that are undergoing mitosis.

Mixoploidy (Nemec, 1910; Hamerton, 1971) – a tissue or individual having cells with different numbers of chromosomes (chromosomal mosaic).

MLC (mixed lymphocyte culture) (Bach and Hirschhorn, Bach and Lowenstein, 1964) – a test for differences in HLA–D phenotypes.

Modal number (White, 1945) – the number of chromosomes of an individual or a cell.

Monolayer (Abercrombie and Heaysman, 1957) – the single-layered sheet of cultured diploid cells on the bottom of a culture flask.

Monosomy (Blakeslee, 1921) – absence of a chromosome in an otherwise diploid chromosomal complement.

Monozygotic – pertaining to uniovular twins (compare with dizygotic).

Morphogen – a protein present in embryonic tissues in a concentration gradient that induces a developmental process.

Mosaic – tissue or individuals made up of genetically different cells, as a rule of the same zygotic origin (compare with Chimera).

mRNA (Brenner, Jacob, and Meselson, 1961; Jacob & Monod, 1961) – messenger RNA.

mtDNA – mitochondrial DNA.

Multigene family – a group of genes related by their common evolution.

Mutagen – a chemical or physical agent that can induce a mutation.

Mutation (de Vries, 1901) – permanent alteration of the genetic material. Different types include point mutations from exchange, loss, or insertion of base pairs within a gene and chromosomal mutation with alteration of the chromosome structure. A *missense mutation* is an alteration resulting in a gene product containing a substitution for a wrong amino acid. A *nonsense mutation* is an alteration that produces a stop codon in the midst of a genetic message so that a totally inadequate gene product is formed.

Mutation rate – the frequency of a mutation per locus per individual per generation.

Nondisjunction (Bridges, 1910) – faulty distribution of homologous chromosomes at meiosis. In mitotic nondisjunction, the distribution error occurs during mitosis.

Northern blot – transfer of RNA molecules to a membrane by a procedure similar to that of a Southern blot.

Nucleoside – compound of a purine or a pyrimidine base with a sugar (ribose or deoxyribose). Compare with Nucleotide.

Nucleosome (Navashin, 1912; Kornberg, 1974) – a subunit of chromatin consisting of DNA wound around histone proteins in a defined spatial configuration.

Nucleotide – Single monomeric building block of a polynucleotide chain that makes up nucleic acid. A nucleotide is a phosphate ester consisting of a purine or a pyrimidine base, a sugar (ribose or deoxyribose as a pentose), and a phosphate group (see p. 27).

Ochre codon – the stop codon UAA.

Okazaki fragment – a short nucleotide sequence that is synthesized on the lagging strand of DNA during replication (see p. 39).

Oncogene (Huebner and Todaro, 1969) – a DNA sequence of viral origin that can lead to malignant transformation of a eukaryotic cell after being integrated into the cell genome (see proto-oncogene).

Open reading frame (ORF) – a DNA sequence of variable length that does not contain stop codons and therefore can be translated (see p. 49).

Operator (Jacob and Monod, 1959) – the recognition site of an operon at which the negative control of genetic transcription takes place by binding to a repressor.

Operon (Jacob et al., 1960) – in prokaryotes, a group of functionally and structurally related genes that are regulated together.

Origin of replication (ORI) – site where DNA replication starts.

Pachytene (de Winiwarter, 1900) – stage of prophase meiosis I.

Palindrome (Wilson and Thomas, 1974) – adjacent inverted repetitive linear base sequences of DNA that can help to form a hairpin-like structure by base-pairing of complementary sequences.

Panmixis (Weismann, 1895) – pairing system with random partner selection, as opposed to assortative mating.

Parasexual (Pontecorvo, 1954) – refers to genetic recombination by nonsexual means, e.g., by Hybridization of cultured cells (see hybridization).

PCR (polymerase chain reaction) (Mullis, 1985) – technique for in vitro propagation (amplification) of a given DNA sequence. It is a repetitive thermal cyclic process consisting of denaturation of genomic DNA of the sequence of interest, annealing the DNA to appropriate oligonucleotide primers, and replication and extension of the DNA segments complementary to the primers.

Penetrance (Vogt, 1926) – the frequency or probability of expression of an allele (compare with Expressivity).

Peptide – a compound of two or more amino acids joined by peptide bonds.

Phage – see Bacteriophage.

Phenocopy (Goldschmidt, 1935) – a nonhereditary phenotype that resembles a genetically determined phenotype.

Phenotype (Johannsen, 1909) – the observable effect of one or more genes on an individual or a cell.

Phosphodiester bond – the chemical bond linking adjacent nucleotides of DNA or RNA.

Phytohemagglutinin (PHA) – a protein substance obtained from kidney beans

(*Phaseolus vulgaris*). It is used to separate red from white blood cells. Nowell (1960) discovered its ability to induce blastic transformation (see Transformation) and cell division in lymphocytes. It is the basis of phytohemagglutinin-stimulated lymphocyte cultures for chromosomal analysis.

Plasmid (Lederberg, 1952) – autonomously replicating circular DNA structures found in bacteria. Although usually separate from the actual genome, they may become integrated into the host chromosome.

Plastids – any of several types of organelles found in plant cells, e.g., chloroplasts.

Pleiotropy (Plate, 1910) – expression of a gene with multiple seemingly unrelated phenotypic features.

Point mutation – alteration of the genetic code within a single codon. The possible types are the exchange of a base: a pyrimidine for another pyrimidine (or a purine for another purine) as a *transition* (Frese, 1959), i.e., thymine for cytosine (or adenine for guanine); or the exchange of a pyrimidine by a purine or visa versa: *transversion*, i.e., thymine by adenine or visa versa (Frese, 1959). Besides the two types of exchange, a point mutation may be due to the insertion of a nucleotide base or the deletion of one or several base pairs.

Polar body (Robin, 1862) – an involutional cell arising during oogenesis that does not develop further as an oocyte.

Polyadenylation – the addition of multiple adenine residues at the 3' end of eukaryotic mRNA after transcription.

Polycistronic messenger – mRNA that includes coding regions from more than one gene (in prokaryotes).

Polygenic (Plate, 1913; Mather, 1941) – refers to traits that are based on several or numerous genes whose effects cannot be individually determined. The term multigenic is sometimes used instead.

Polymerases – enzymes that catalyze the combining of nucleotides to form RNA or DNA (genetic transcription and DNA replication).

Polymorphism, genetic (Ford, 1940) – existence of more than one normal allele at a gene locus, with the rarest allele exceeding a frequency of 1%. A polymorphism may exist at several levels, i.e., variants in DNA sequence, amino acid sequence, chromosomal structure, or phenotypic traits (p. 156ff).

Polypeptide – see Peptide.

Polyploid (Strasburger, 1910) – refers to cells, tissues, or individuals having more than two copies of the haploid genome, e.g., three (triploid) or four (tetraploid). In man, triploidy and tetraploidy in a conceptus are usually lethal and as a rule lead to spontaneous abortion.

Polytene (Koltzoff, 1934; Bauer, 1935) – refers to a special type of chromosome resulting from repeated endoreduplication of a single chromosome. Giant chromosomes arise in this manner (see Chromosome).

Population (Johannsen, 1903) – individuals of a species that interbreed and constitute a common gene pool (compare with Race).

Positive negative effect – a mutation exerting a negative functional effect in one allele (loss of function).

Premature chromosomal condensation (Johnson and Rao, 1970) – induction of chromosomal condensation in an interphase nucleus after fusion with a cell in mitosis. Condensed S-phase chromosomes appear pulverized (so-called chromosomal pulverization).

Prevalence – see Incidence.

Pribnow box – part of a promoter (TATAAT sequence 10 bp upstream of the gene) in prokaryotes.

Primary transcript – the original RNA transcribed from a eukaryotic gene before processing (splicing, addition of the cap, and polyadenylation).

Primer – a DNA or RNA oligonucleotide that after hybridization to an inversely complementary DNA has a 3'–OH end to which nucleotides can be added for synthesis of a new chain by DNA polymerase.

Prion – an infectious agent that consists of protein only, without nucleic acid.

Proband – see Propositus.

Prokaryote – one-celled organism without a cell nucleus or intracellular organelles.

Promoter – a defined DNA region at the 5' end of a gene that binds to transcription factors and RNA polymerase during the initiation of transcription. The –10 sequence is the consensus sequence TATAATG about 10 bp upstream of a prokaryotic gene (Pribnow box).

Prophage – a viral (phage) genome integrated into the bacterial (host) genome.

Propositus, proband – the individual in a pedigree that has brought a family to attention for genetic studies.

Proto-oncogene (cellular oncogene) – a eukaryotic gene. It may be present in truncated form in a retrovirus, where it may behave as an oncogene.

Prototrophic – refers to cells or cell lines that do not require a special nutrient added to the culture medium (compare with Auxotrophic).

Provirus – duplex DNA derived from an RNA retrovirus and incorporated into a eukaryotic genome.

Pseudogene – DNA sequences that closely resemble a gene but are without function due to an integral stop codon, deletion, or other structural change. A processed pseudogene consists of DNA sequences that resemble the mRNA copy of the parent gene, i.e., it does not contain introns.

Pseudohermaphroditism – a condition in which an individual has the gonads of one sex and phenotypic features of the opposite sex.

Quadriradial figure – the configuration assumed when homologous segments of chromosomes involved in a reciprocal translocation pair at meiosis. Rarely, such a figure may occur during mitosis.

Race – a population that differs from another population in the frequency of some of its gene alleles (L. C. Dunn: Heredity and Evolution in Human Populations, Harvard University Press, Cambridge, Mass., 1967). Accordingly, the concept of race is flexible and relative, defined in relation to the evolutionary process. The term race can be used to classify groups, whereas the classification of individuals is often uncertain and of biologically dubious value.

Reading frame – sequence of DNA nucleotides that can be read in triplets to code for a peptide (compare with Open reading frame).

Receptor – a transmembrane or intracellular protein involved in transmission of a cell signal.

Recessive (Mendel, 1865) – refers to the genetic effect of an allele at a gene locus that is manifest as phenotype in the homozygous state only.

Reciprocal translocation – mutual exchange of chromosome parts.

Recombinant DNA – A DNA molecule consisting of parts of different origin.

Recombination (Bridges and Morgan, 1913) – the formation of new combinations of genes as a result of crossing-over between homologous chromosomes during meiosis.

Recombination frequency – frequency of recombination between two or more gene loci. Expressed as the theta (θ) value. A θ of 0.01 (1% recombination frequency) corresponds to 1 centimorgan (cM).

Regulatory gene – a gene coding for a protein that regulates other genes.

Renaturation of DNA – combining of complementary single strands of DNA to form double-stranded DNA (compare with Denaturation).

Repair (Muller, 1954) – correction of structural and functional DNA damage.

Replication – identical duplication of DNA.

Replication fork – the unwound region of the DNA double helix in which replication takes place.

Replicon (Huberman and Riggs, 1968) – an individual unit of discontinuous DNA replication in eukaryotic DNA.

Reporter gene – a gene used to analyze another gene, especially the regulatory region of the latter.

Repulsion (Bateson, Saunders, and Punnett, 1905) – term to indicate that the mutant alleles of neighboring heterozygous gene loci lie on opposite chromosomes, i.e., in trans configuration (see cis/trans).

Resistance factor – a plasmid gene that causes antibiotic resistance.

Restriction enzyme, or restriction endonuclease (Meselson and Yuan, 1968) – endonuclease that cleaves DNA at a specific base sequence (restriction site or recognition sequence).

Restriction map – a segment of DNA characterized by a particular pattern of restriction sites.

Restriction site – a particular sequence of nucleotide bases in DNA that allow a particular restriction enzyme to cleave the DNA molecule at, or close to, that site (recognition site).

Retrovirus – a virus with a genome consisting of RNA that multiplies in a eukaryotic cell by conversion into duplex DNA.

Reverse transcriptase – an enzyme complex that occurs in RNA viruses and that can synthesize DNA from an RNA template.

RFLP – restriction fragment length polymorphism. The production of DNA fragments of different lengths by a given restriction enzyme, due to inherited differences in a restriction site.

Rho factor – a protein involved in termination of transcription in E. coli.

Ribosome (Roberts, 1958; Dintzis et al., 1958) – complex molecular structure in pro- and eukaryotic cells consisting of specific proteins and ribosomal RNA in different subunits. The translation of genetic information occurs in ribosomes.

Ring chromosome – a circular chromosome. In prokaryotes the normal chromosome is ring-shaped. In mammals it represents a structural anomaly and implies that chromosomal material has been lost.

RNA (ribonucleic acid) – a polynucleotide with a structure similar to that of DNA except that the sugar is ribose instead of deoxyribose.

Satellite (Navashin, 1912) – small mass of chromosomal material attached to the short arm of an acrocentric chromosome by a constricted appendage or stalk. It is involved in the organization of the nucleolus. The stalk region can be stained by specific silver stain (NOR stain, nucleolus-organizing region). The size of the satellite, the length of its stalk, and the intensity of its fluorescence after staining with acridine are polymorphic cytogenetic markers.

Satellite DNA (sDNA) (Sueoka, 1961; Kit, 1961; Britten and Kohne, 1968) – DNA that is either heavier (GC-rich) or lighter (AT-rich) than the main DNA (see DNA). Not to be confused with the satellite regions of acrocentric chromosomes.

SCE (sister chromatid exchange) (Taylor, 1958) – an exchange between the two chromatids of a metaphase chromosome. After two replication cycles in a cell culture in the presence of a halogenated base analog (e.g., 5-bromodeoxyuridine), both DNA strands of one chromatid will be substituted with the halogenated base analog, whereas only one DNA strand of the other chromatid will be substituted. As a result, the two chromatids differ in staining intensity, and it is possible to determine where an exchange between the two chromatids has occurred (see p. 339).

Segregation (Bateson and Saunders, 1902) – the separation of alleles at a gene locus at meiosis and their distribution to different gametes. Segregation accounts for the 1:1 distribution of allelic genes to different chromosomes.

Selection (Darwin, 1858) – preferential reproduction or survival of different genotypes under different environmental conditions.

Selection coefficient – quantitative expression (from 0 to 1) of the disadvantage that a genotype has (compared with a standard genotype) in transmitting genes to the next generation. The selection coefficient (s) is the numerical amount by which biological fitness (1–s) is decreased; i.e., a selection coefficient of 1 indicates a complete lack of biological fitness.

Selective medium – a medium that supports growth of cells in culture containing a particular gene.

Semiconservative (Delbrück and Stent, 1957) – the normal type of DNA replication. One DNA strand is completely retained; the other is synthesized completely anew.

SINE (short interspersed nuclear element) – short repetitive DNA sequences (LINE).

snRNPs (small nuclear ribonucleoprotein particles) – complexes of small nuclear RNA molecules and proteins.

Somatic – refers to cells and tissues of the body, as opposed to germinal (referring to germ cells).

Somatic cell hybridization – formation of cell hybrids in culture.

Southern blot (Southern, 1985) – method of transferring DNA fragments from an agarose gel to a membrane after the fragments have been separated according to size by electrophoresis.

Speciation (Simpson, 1944) – formation of species during evolution. One of the first steps toward speciation is the establishment of a reproductive barrier against genetic exchange. A frequent mechanism is chromosomal inversion.

Species (Ray, 1670) – a natural population in which there is interbreeding of the individuals, which share a common gene pool.

S phase (Howard and Pelc, 1953) – phase of DNA synthesis (DNA replication) between the G_1 and the G_2 phase of the eukaryotic cell cycle.

Spliceosome– an aggregation of different molecules that can splice RNA.

Splicing – a step in processing a primary RNA transcript in which introns are excised and exons are joined.

Splice junction – the sequences at the exon/intron boundaries.

Stop codon – see Termination codon

STS (sequence tagged site) – a short segment of DNA of known sequence.

Submetacentric – refers to a chromosome consisting of a short and a long arm because of the position of its centromere.

Synapsis (Moore, 1895) – the pairing of homologous chromosomes during meiotic prophase.

Synaptonemal complex (Moses, 1958) – parallel structures associated with chiasmata formation during meiosis, visible under the electron microscope.

Syndrome – within human genetics, a group of clinical and pathological characteristics that are etiologically related, regardless of whether the details of their relationship have yet to be identified.

Synteny (Renwick, 1971) – refers to gene loci that are located on the same chromosome, whether or not they are linked.

Tandem duplication – short identical DNA segments adjacent to each other.

TATA box – a conserved, noncoding DNA sequence about 25 bp in the 5' region of most eukaryotic genes. It consists mainly of sequences of the TATAAAA motif. Also known as Hogness box (compare with Pribnow box, of prokaryotes).

T cells – T lymphocytes.

Telocentric (Darlington, 1939) – refers to chromosomes or chromatids with a terminally located centromere, without a short arm or satellite. They do not occur in man.

Telomerase – a ribonucleoprotein enzyme that adds nucleotide bases at the telomere.

Telomere (Muller, 1940) – the terminal areas of both ends of a chromosome containing specific consensus sequences (see p. 189).

Teratogen (Ballantyne, 1894) – chemical or physical agent that leads to disturbances of embryological development and to malformations.

Termination codon – one of the three triplets signaling the end of translation (UAG, UAA, UGA).

Terminator – a DNA sequence that signals the end of transcription.

Tetraploid (Nemec, 1910) – having a double diploid chromosome complement, i.e., four of each kind of chromosome are present (4n instead of 2n).

Transcript – an RNA copy of a segment of the DNA of an active gene.

Transcription – the synthesis of messenger RNA (mRNA), the first step in relaying the information contained in DNA.

Transcription unit – all of the DNA sequences required to code for a given gene product (operationally corresponding to a gene). Includes promoter and coding and noncoding sequences.

Transduction (Zinder and Lederberg, 1952) – transfer of genes from one cell to another (bacteria) by special viruses, the bacteriophages.

Transfection – introduction of pure DNA into a living cell (compare with Transformation).

Transformation – this term has several different meanings in biology. In genetics, three main types of transformation are distinguished: 1) *malignant transformation*, the transition of a normal cell to a malignant state with loss of control of proliferation,
2) *genetic transformation* (Griffith, 1928; Avery et al. 1944), a change of genetic attributes of a cell by transfer of genetic information, and
3) *blastic transformation,* the reaction of lymphocytes to mitogenic substances (e.g., phytohemagglutinin or specific antigens) leading to cell division.

Translation – the second step in the relay of genetic information. Here the sequence of triplets in mRNA are translated into a corresponding sequence of amino acids to form a polypeptide as the gene product.

Translocation – transfer of all or part of a chromosome to another chromosome. A translocation is usually *reciprocal,* leading to an exchange of nonhomologous chromosomal segments. A translocation between two acrocentric chromosomes that lose their short arms and fuse at their centromeres is called a *fusion* type translocation (Robertsonian translocation).

Transposon – a DNA sequence with the ability to move and become inserted at a new location of the genome.

Triplet – a sequence of three nucleotides comprising a codon of a nucleic acid and representing the code for an amino acid (triplet code, see codon).

Trisomy (Blakeslee, 1921) – an extra chromosome, in addition to a homologous pair of chromosomes.

UPD (uniparental disomy) – see Disomy.

Upstream – 5' direction of a gene.

Vector – a molecule that can incorporate and transfer DNA.

Western blot – technique to identify protein antigens, in principle similar to the Southern blot method (q.v.). Also called immunoblot.

X chromatin (formerly called Barr body or sex chromatin) (Barr and Bertram, 1949) – darkly staining condensation in the interphase cell nucleus representing an inactivated X chromosome

Xenogenic – refers to transplantation between individuals of different species.

X-inactivation (Lyon, 1961) – inactivation of one of the two X chromosomes in somatic cells of female mammals during the early embryonic period by formation of X chromatin.

YAC (yeast artificial chromosome) – a yeast chromosome into which foreign DNA has been inserted for replication in dividing yeast cells. YACs can incorporate relatively large DNA fragments, up to about 1000 kb.

Y chromatin (F body; Pearson, Bobrow, Vosa, 1970) – the brightly fluorescing long arm of the Y chromosome visible in the interphase nucleus.

Z DNA – alternate conformation of DNA. Unlike normal B DNA, (Watson–Crick model), the helix is left-handed and angled (zigzag, thus Z DNA).

Zoo blot – a Southern blot containing conserved DNA sequences from related genes of different species. It is taken as evidence that the sequences are coding sequences from a gene (see p. 77).

Zygote (Bateson, 1902) – the new diploid cell formed by the fusion of the two haploid gametes, an ovum and a spermatozoon, at fertilization. The cell from which the embryo develops.

Zygotene (de Winiwarter, 1900) – a stage of prophase of meiosis I.

References for the Glossary

Bodmer W.F., Cavalli-Sforza, L. Genetics and the Evolution of Man. San Francisco: WH Freeman; 1976.

Brown T. Genetics. A Molecular Approach. 2nd ed. London: Chapman & Hall; 1992.

Dorland's Illustrated Medical Dictionary. 28th ed. Philadelphia: WB Saunders; 1994.

Emery AFH. Elements of Medical Genetics. 6th ed. Edinburgh: Churchill Livingstone; 1983.

Lewin B. Genes V. Oxford: Oxford University Press; 1994.

Passarge E. Definition genetischer Begriffe (Glossar), pp. 311–323. In: Elemente der Klinischen Genetik. Stuttgart: G. Fischer; 1979.

Rieger R, Michaelis A, Green MM. Glossary of Genetics and Cytogenetics. 5th ed. Berlin: Springer Verlag; 1979.

Rothwell NY. Understanding Genetics. A Molecular Approach. New York: Wiley–Liss; 1993.

Suzuki DT, Griffiths AGF, Miller JH, Lewontin RC. An Introduction to Genetic Analysis. New York: WH Freeman; 1986.

Tanaka Y, Macer D. Sense, nonsense and antisense. Trends Genet. 1994; 10: 417.

Watson JD. Molecular Biology of the Gene. 3rd ed. Menlo Park, California: W.A. Benjamin; 1976.

Whitehouse HLK. Towards an Understanding of the Mechanisms of Heredity. 3rd ed. London: Edward Arnold; 1973.

Index

Page numbers in bold type signify the main entry.
Human diseases referred to in the plates or text are shown in bold face.

E

Q

R

U